Cell Culture Bioprocess Engineering

Second Edition

Cell Culture Bioprocess Engineering

Second Edition

Wei-Shou Hu

CRC Press
Taylor & Francis Group
Boca Raton London New York

CRC Press is an imprint of the
Taylor & Francis Group, an **informa** business

CRC Press
Taylor & Francis Group
6000 Broken Sound Parkway NW, Suite 300
Boca Raton, FL 33487-2742

International Standard Book Number-13: 978-1-4987-6285-4 (Paperback)

Library of Congress Cataloging-in-Publication Data

Names: Hu, Wei-Shou, 1951- author.
Title: Cell culture bioprocess engineering / by Wei-Shou Hu.
Description: Second edition. | Boca Raton : CRC Press, [2020] | Includes bibliographical references and index.
Identifiers: LCCN 2019047451 (print) | LCCN 2019047452 (ebook) | ISBN 9781498762854 (paperback ; alk. paper) | ISBN 9780429162770 (ebook) | ISBN 9781498762861 (adobe pdf) | ISBN 9780429546594 (mobi)
Subjects: MESH: Cell Engineering--methods
Classification: LCC QH585.2 (print) | LCC QH585.2 (ebook) | NLM QU 300 | DDC 616.02/77--dc23
LC record available at https://lccn.loc.gov/2019047451
LC ebook record available at https://lccn.loc.gov/2019047452

Visit the Taylor & Francis Web site at
http://www.taylorandfrancis.com

and the CRC Press Web site at
http://www.crcpress.com

Contents

8 Cell Culture Bioreactors

9 Fed-Batch Culture Processes

10 Continuous Cell Culture Processes

11 Oxygen Transfer in Cell Culture Bioreactors

12 Scaling Up and Scaling Down for Cell Culture Bioreactors

Appendix A: Cell Culture Process Data Analysis

Preface

THIS BOOK draws its content from the Cellular Bioprocess Technology short course that has been offered at the University of Minnesota for over three decades. Each year, a number of prominent academics and industry leaders come to the course to share their knowledge and best practices. This book is distilled from the topics and materials developed for those lectures. This second edition reflects the developments in cell technology since the publication of the first edition in 2012.

This book, like the short course, aims to serve practitioners of mammalian cell-based biologics production. A major goal of this book is to present the fundamentals of process technology and biology to industrial practitioners of diverse backgrounds and to students from different academic disciplines in a succinct and easy-to-grasp way. Another goal is to bring new knowledge from cutting-edge research and practice into the very practical setting of today's industrial laboratories. Over the course of developing a new therapy—from the discovery in a laboratory to developing a production process, successfully commercializing the product, and manufacturing the product with consistently high quality and efficiency—scientists and engineers of different disciplines and in different functional roles must effectively communicate in order to develop, advance, and innovate.

The first chapter in this book gives a historical overview of the evolution of the science and technology of cell culture and the trends of process technology in recent years. The second part of the book covers fundamental knowledge of cell biology, metabolism, cells' nutritional needs, the determinants of product quality, and the means to create cell lines as the workhorse for product manufacturing. The third group of chapters delves into the process engineering principles of stoichiometry, reactor kinetics, and scale-up. Since this book is written for readers of diverse backgrounds, the aim is to help engineers understand the key biological features of cells without requiring an extensive prior background in biology, and also to help non-engineers grasp basic engineering principles without a prior two years of college math. This book was written to allow scientists to gain a process perspective while reading through biology-related topics, and to allow engineers to appreciate biological considerations while perusing engineering topics.

As a second edition, this book is based on and extended from its first edition, except for the addition of Chapter 4. Some chapters, especially those regarding engineering subjects (Chapters 5, 8, 11, and 12), bear a higher degree of resemblance to their counterparts in the first edition, reflecting that fundamental principles do not change over time. Some figures (including 1.3, 2.1, 2.3, 2.6, 2.7, 2.11–15, 3.2, 3.9, 3.11, 3.19, 3.21, 3.23, 5.3, 8.9–14, 11.3, 12.1, 12.2, 12.4, and 12.8) are also adopted from the first edition.

The content of the course and this book has been shaped by many who participated in teaching the course over the years since 1986. The delivery of the materials in this book has been polished through lecturing the more than three thousand biotechnologists who have attended the course in the past. I thank them sincerely for their participation. For this second edition, I would like to thank Susan Abu-Absi, Nitya M. Jacob, Rashmi Kshirsagar, Bhanu Chandra Mulukutla, Gargi Seth, and Weichang Zhou in particular. Susan prepared the draft of the second half of Chapter 4 and is acknowledged as a contributor to that chapter.

Many former and current members of my research laboratory at the University of Minnesota contributed to the preparation of course materials. For this edition I thank Tung Le, Sofie O'Brien, and Conor O'Brien in particular.

Many skillful hands have worked on this book. Kimberly Durand first took the notes to digital form two decades ago and continues to support the effort. For this edition, Victoria Roberts single-handedly took the book from a draft Word document and hand-sketched graphics to the laid-out and typeset product. She even learned some biology and engineering along the way. The dedication of both is remarkable.

Like its first edition, this book is dedicated to the students, fellows, and staff formerly and currently in my research group at the University of Minnesota. Without this group of individuals to help advance the content of the course, this book would not have been in the shape that it is today. Finally, I wish for my lovely family, Jenny, Kenny, Lauren, two granddaughters, Ya-Ling and Wen-Li, and my wife, Sheau-Ping, to share the joy of the book's completion.

Wei-Shou Hu
Department of Chemical Engineering and Materials Science
University of Minnesota

Cell Culture Bioprocess Engineering

Second Edition

Contents

1
Overview of Cell Culture Processes

An Introduction to Cell Culture: A Historical Perspective

Early Enabling Innovations in Cell Culture

Cultured mammalian cells are the workhorse for the manufacturing of the class of pharmaceuticals known as biologics, including viral vaccines and many protein medicines. What may not be known to many is the art-like nature of growing cells at the dawn of cell culture, a skill that required a great deal of creativity, inquisitiveness, passion, and perseverance, in addition to all the good traits of a scientist. Around the turn of the twentieth century, explants of animal tissues began to be cultured on glass surfaces that were submerged in an animal's tissue fluid. Cells grew from the tissue clumps and remained viable and observable for a few days.[1] From the outgrowth of tissue clumps, some cells could be isolated and eventually dissected out and expanded to new glass surfaces. The capability of passaging or expanding cell population was an important step towards genuine cell culture. Among the first cell lines established that could be continuously expanded in culture was the mouse L cell.[2] The early cell lines could be sustained in the lab only by continuous passaging; they could not be frozen, stored away, and later thawed to resume growth. In that era, the complex nutrient mixture could not be sterilized by heat. Rather, serum, ascetic fluids, or chicken embryo extract were carefully isolated from animals in order to maintain sterility for cell cultivation (Figure 1.1).[3] Imagine the amount of work involved in maintaining cultured cells!

The discovery of cryopreservation for freezing animal sperm and later other cells allowed cell growth to be paused and resumed in the lab.[4] Another important advance was the use of trypsin for cell passaging, instead of relying on dissection to dissociate cells from the surface. The first human cell line, HeLa, derived from human cervical cancer, fully took advantage of this.[3, 5]

Critical to cell culture advances was the arrival of membrane-filtration-based medium sterilization, first by ultrafiltration with a nitrocellulose membrane and later by microfiltration. While saline and media for microorganisms could be autoclaved for sterilization, the complex nutrients needed by animal cells are destroyed at high temperatures. Membrane sterilization accelerated the development of a chemical nutrient medium consisting of glucose, amino acids, vitamins, and balanced salts.[6] This not only advanced our knowledge of the nutritional needs of cells, but also greatly simplified the logistics of growing cells, leading to the establishment of many important cell lines and eventually to the industrialization of cell culture.

The early cell lines that could be continuously passaged, including mouse L and HeLa, were derived from cancerous tissues (Panel 1.1). Morphologically, they looked abnormal and were distinct from primary cells first grown from normal tissues. Later continuous cell lines were isolated from various animal tissues, including baby hamster kidney (BHK) from the Syrian hamster,[7] Vero from green monkey kidney,[8] and Chinese hamster ovary (CHO) from

Panel 1.1. Cell Substrate Example

Primary Cells
- Tissue explant, limited cell expansion

Cell Lines
- e.g., Mouse L, HeLa, BHK, CHO

Differentiated Cell Lines
- e.g., HepG2 (liver), PC12 (neuronal)

Stem Cells
- Multipotent, pluripotent
- Capable of directed differentiation *in vitro*

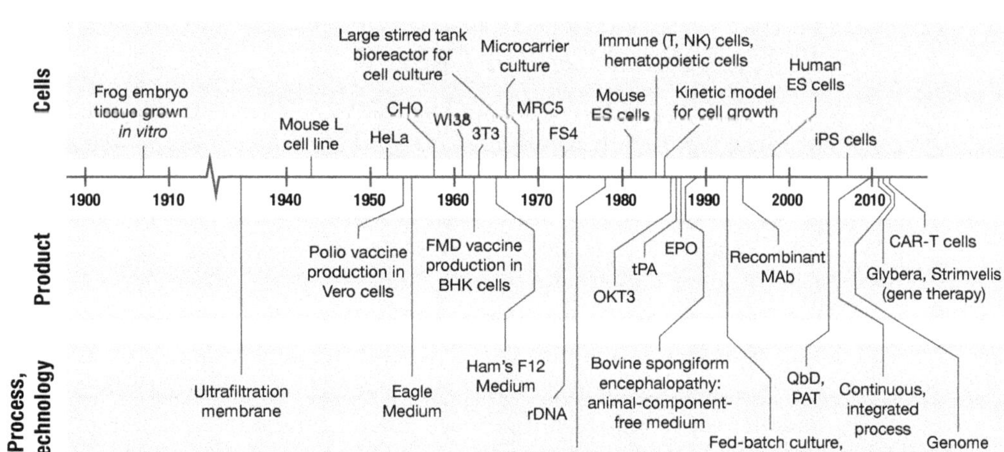

Figure 1.1. Milestones in cell culture technology.

Chinese hamster ovary.[9] These cell lines carried mutations that allowed them to bypass their cells' internal growth control mechanisms. These cells were not phenotypically normal. They often did not exhibit contact inhibition. With abundant nutrient supply, they grew into multiple layers of cells on a surface. Later, 3T3, a cell line which is adhesion dependent, exhibits contact inhibition, and does not undergo senescence, was established.[10] But the karyotype (or chromosome composition) of 3T3 cells, as well as other continuous cell lines, was aneuploid, not diploid. Around the same time, phenotypically normal human fibroblastic cell strains, such as WI-38 and later MRC-5 and FS-4, were isolated.[11, 12, 13] These cells were diploid and exhibited contact inhibition but were not continuous cell lines like 3T3. They senesced after repeated passaging in culture. For many decades, these cell lines and cell strains served important roles in biological science and medical research, and many were used in the industrial production of viral vaccines and other biologics.

Differentiated Cell Lines

The early cell lines were important tools for the *in vitro* study of cell physiology, biochemical mechanisms, and cancers. However, those cell lines didn't possess the differentiated functions that tissues do. Later, cells of various differentiated tissues would be isolated, including HepG2 from hepatocellular carcinoma,[14] Jurkat from human T cell leukemia,[15] and PC12 from rat pheochromocytoma.[16] Most of those differentiated cell lines were isolated from cancerous tissues instead of normal ones. Compared to their normal counterparts, they were easier to isolate due to their reduced dependence on various growth factors and cytokines needed for *in vitro* growth. Nevertheless, they carried the phenotypic characteristics of the tissue that they were derived from, and these differentiated cell lines became valuable tools for biomedical research.

Differentiated Cells in Culture

The arrival of rDNA technology in the 1970s made the production of various protein reagents a simpler task. It expanded the repertoire of biochemical reagents and greatly increased our ability to culture cells that require growth factors and cytokines. Primary cells, such as endothelial cells, keratinocytes, chondrocytes, and hepatocytes, began to be isolated from normal human tissues and cultured *in vitro*. Primary tissue cells isolated from tissues typically retain many tissue-specific activities, and thus may be exploited for tissue engineering applications such as repairing tissues or augmenting tissue functions. However, most differentiated cells have very limited proliferative potential and show phenotypic instability in culture.

Stem Cells

The quest for cells with differentiated properties soon extended from isolating terminally differentiated cells to stem cells that could be differentiated to specific lineages. Stem cells are classified by their differentiation potential: totipotent, pluripotent, and multipotent. Totipotent cells can become any type of cell, including extraembryonic tissues. Pluripotent cells (embryonic stem cells, induced pluripotent stem cells) can become any type of cell found in any of the three germ layers of ectoderm, mesoderm, and endoderm.[17] Multipotent cells can differentiate into different cell types of the same lineage. For example, hematopoietic stem cells (HSCs) can differentiate into any type of blood cell by exposure to different combinations of growth factors, such as stem cell factor (SCF), thrombopoietin (TPO), and granulocyte colony-stimulating factor (G-CSF). Multipotent HSCs have been isolated from bone marrow, peripheral blood, and umbilical cord blood. Mesenchymal stem cells (MSCs), also multipotent, were first isolated from adult bone marrow[18] but are now found in almost all tissues. They have the ability to differentiate mainly into osteogenic, chondrogenic, and adipogenic cells.[19] They also have the capacity to produce a large number of bioactive molecules and are being explored as signaling cells for possible immunomodulatory and trophic effects.

Pluripotent stem cells

Stable mouse embryonic stem cells (mESCs) were first derived from the inner cell mass (ICM) of early mouse embryos.[20] Nearly twenty years elapsed before two laboratories derived human embryonic stem cell lines (hESCs).[21, 22] These embryonic stem cells were pluripotent and capable of differentiating into cell types of all three germ layers. They were karyotypically normal when grown under the appropriate culture conditions. However, hESCs were derived from fertilized human eggs, making their use ethically controversial. In contrast, induced pluripotent stem cells (iPSCs) are derived by introducing four exogenous genes (OCT4, SOX2, KLF4, and c-Myc (OSKM)) into adult somatic cells. They can be readily derived from different individuals and are therefore free of ethical controversy. After the introduction of these genes, a small fraction of the transfected cells proliferated, bypassed apoptosis and cell senescence, lost somatic cell characteristics, and were reprogrammed into pluripotent cells.[23–25] Since then, the reprogramming of adult somatic cells to iPSCs has been accomplished with some modifications of gene combinations, or even with cocktails of small molecules that induce epigenetic changes.

Guiding Cell Differentiation in Vitro

Some cell lines can undergo differentiation *in vitro* when they are exposed to the appropriate signals. For example, a subclone of 3T3 cells, designated 3T3-L1, can be induced to differentiate into adipocytes.[26] PC12 cells, upon exposure to nerve growth factor or dexamethasone, can terminally differentiate into neuron-like cells. The potential of using cell lines capable of differentiation for *in vitro* drug testing—even for therapy—has been long recognized.

Most primary cells isolated from tissues have a very limited proliferation capacity. Primary hepatocytes, chondrocytes, and neuronal cells all proliferate to a very limited extent in cultures and cannot be cultured for a long period. In some cases, it is possible to enrich a subpopulation of cells from a cell mixture isolated from tissues using chemical cues that preferentially stimulate their proliferation. For example, T lymphocytes and natural killer (NK) cells isolated from peripheral blood can be cultured and expanded into larger populations. The isolation and culture of those cells from the immune system two decades ago paved the way for cell therapy today.

Adult stem cells isolated from tissues, like HSC and MSC, can also be guided to differentiate into cells of related lineages (MSC to bone and muscle lineages and HSC to lymphoid lineage). These adult stem cells are multipotent and can only be directed to differentiate into their respective related lineage. With pluripotent stem cells, directed differentiation to lineages of all three germ layers is now possible, although most differentiation protocols still generate only tissue-like cells with an insufficient extent of maturation.

It is also possible to directly reprogram adult somatic cells from one lineage to another.[27] Instead of ascending to the ESC-like state, the cells are transduced with a combination of gene factors and reprogrammed directly into the target differentiation state upon culturing in a differentiation medium.[28] Beta cells, cardiac cells, neurons, and hepatocytes have all been derived directly from somatic cells.[29–33]

Cell Culture as Process Technology

Early cell culture explorations aspired to establish a biological platform for scientific research, but even then the potential applications of cell culture were never far from the minds of the scientists involved. Soon after primary cells became culturable, viruses were produced in primary cell culture. Cell culture quickly began to take the place of animals and embryonated chicken eggs as the production vehicle of viral vaccines. Notably, foot and mouth disease (FMD) viruses were produced in primary calf kidney cells[34] and polio vaccines were produced in primary monkey kidney cells in the 1950s.[35] Subsequently, continuous cell lines became the production vehicle of viruses, including BHK cells for FMD virus[36] and Vero cells for polio virus.[37] Human fibroblasts MRC-5 were used in human vaccine production, while interferon was produced by FS-4 cells.[13] However, as we have witnessed in the past two decades, it was recombinant DNA (rDNA) technology and the use of mammalian cells for the production of therapeutic proteins that propelled cell culture to its place as a major manufacturing workhorse. Recent advances in T cell therapy offer hope that therapeutic cells will become a new class of product of cell culture processes.

Viral Vaccines

The most effective way to fight viral diseases has been widespread immunization by vaccines. The majority of viral vaccines are now produced in cell culture. While animal tissues are no longer used, embryonated eggs are still part of the production of many viruses, including influenza virus. Because of their prophylactic nature and impact on public health, vaccines do not command the same price as therapeutic proteins. The combined commercial value of vaccines is substantially lower than the total of pharmaceuticals, with total annual sales worldwide in the order of ~30 billion US dollars in 2015. Influenza virus vaccines, among the few viral vaccines with an annual sales value of more than 1 billion US dollars, have only about 10% of their doses produced in cell cultures. The rest are still produced in chicken eggs. Table 1.1 lists a number of viral vaccines for human use. Although viral vaccines have made major strides in advancing health care, many viral diseases, such as HIV, are still in need of vaccines.

The majority of viral vaccines employ whole viruses, either live attenuated viruses or virus particles that have been inactivated by formalin or another chemical treatment to render the virus incapable of infection but still capable of eliciting the immune response. Live attenuated viruses used in immunization have been adapted, often by a long serial passaging in a nonhuman host, to reduce their virulence to humans. The attenuated virus is still capable of replication, but often at a slower rate. They continue to replicate after being administered, thus requiring a smaller amount for immunization than if an inactivated virus had been used.

Viral vaccines may also be subunit vaccines consisting of protein subunits produced in microbial or insect cell systems that retain antigenicity for immunization. Another type of vaccine, VLP,

Table 1.1. Viruses Produced in Cell Culture

Virus	*Cell substrate*
Poliomyelitis	Vero
Rabies	Vero, MRC-5
Measles	Chicken embryo fibroblasts
Mumps	Chicken embryo fibroblasts
Rubella	WI-38, MRC-5
Adenovirus (vaccine)	MRC-5
Adenovirus (gene therapy vector)	HEK293
Rotavirus	Vero
Smallpox (vaccinia)	Chicken embryo fibroblasts, BHK
Varicella	MRC-5, 2BS
Influenza	MDCK, Vero

consists of proteins that are produced by recombinant DNA technology in microbial or cell systems but assembled into viral particles without the virus genome. A prominent example is the VLP vaccine against human papillomavirus virus, which is produced in yeast or insect cells.

The dose of viral vaccine for each immunization is low. Cell culture facilities for viral vaccine manufacturing are therefore relatively small in comparison to those for protein biologics. Nevertheless, the cell culture technologies established for vaccine production in the 1960s and 1970s laid the foundation for subsequent large-scale bioprocessing for therapeutic protein production. The basic cell culture processes of cell banking, medium design, process control, suspension culture, and adherent growth on microcarriers in stirred tank bioreactors were all first established for the manufacturing of vaccines and later adopted for recombinant protein production.

Many viral diseases in the industrialized world have all but vanished in the past few decades because of the availability of vaccines. However, many basic vaccines that are routinely administered to the general population in industrialized countries are out of reach for a large fraction of the world's population. Reducing the cost of those vaccines and improving their distribution to make this basic human necessity available to everyone is still a major challenge. Nowadays, an outbreak of an infectious disease in any part of the world has the potential to quickly become a worldwide pandemic. Establishing manufacturing technology capable of quickly producing vaccines for new strains of viruses and providing a rapid response to affected regions during a pandemic remains a challenge.

Protein Molecules as Therapeutics

Proteins derived from blood, tissues, and cell cultures were used for therapeutic purposes long before the arrival of recombinant DNA technology. Examples include insulin for treating diabetes, urokinase for stroke, factor VIII for blood coagulation disorder hemophilia, and interferons for viral infection. The first wave of recombinant DNA therapeutic proteins, including human growth hormone and insulin, were produced in *Escherichia coli* (Table 1.2). The subsequent products were also of human origin but required complex post-translational modifications—such as complex disulfide-bond formation and glycosylation—that could not be carried out by microbial cells (Table 1.3). Hence, mammalian cells were employed.

The early development of cell-culture-based therapeutic proteins focused on monoclonal antibodies produced by hybridoma cells. The hybridoma cells were derived from the fusion of non-antibody-secreting, but continuously proliferating, myeloma cells and the specific antibody-secreting, but non-dividing, B lymphocytes. Those fused cells that could both grow and produce the desired antibody were isolated for production.

Since the B cells were obtained from immunized mice, the method produced mouse antibodies rather than human or humanized antibodies.

Hybridoma technology thus soon gave way to recombinant DNA-based methods using Chinese hamster ovary (CHO), mouse myeloma, and a few other cell lines. By introducing the transgene coding for the

Table 1.2. Therapeutic Protein Biologics Produced in Non-Mammalian Hosts

Product	*Activity / Use*
Granulocyte colony-stimulating factor (Neupogen)	White blood cell growth for neutropenia
Insulin (Humulin)	Diabetes
α-Interferon (Intron-A)	Anticancer, viral infections
Somatropin [human growth hormone] (Humatrope/Protropin/Nutropin)	Growth deficiencies
Interleukin-2 (Proleukin)	Kidney cancer
Human papillomavirus VLP	Vaccine
Hepatitis B virus surface antigen	Vaccine

Table 1.3. Non-Antibody Products Produced in Mammalian Cells*

Trade name	*Type*	*Therapeutic use*
Aldurazyme	Laronidase	Mucopolysaccharidosis I
Cerezyme	β-glucocerebrosidase	Gaucher's disease
Myozyme	α-galactosidase	Pompe disease
Fabrazyme	α-galactosidase	Fabry disease
Naglazyme	N-acetylgalactosamine 4-sulfatase	Mucopolysaccharidosis VI
Orencia	Ig-CTLA4 fusion	Rheumatoid arthritis
Luveris	Luteinizing hormone	Infertility
Activase	Tissue plasminogen activator	Acute myocardial infarction
Epogen/Procrit	EPO	Anemia
Aranesp	EPO (engineered)	Anemia
Pulmozyme	Deoxyribonuclease I	Cystic fibrosis
Avonex	Interferon-β	Relapsing multiple sclerosis
Rebif	Interferon-β	Relapsing multiple sclerosis
Follistim/Gonal-F	Follicle stimulating hormone	Infertility
Benefix	Factor IX	Hemophilia A
Enbrel	TNF receptor fusion	Rheumatoid arthritis
Tenecteplase	Tissue plasminogen activator (engineered)	Myocardial infarction
ReFacto	Factor VIII	Hemophilia A
Advate	Factor VIII (engineered)	Hemophilia A

*all listed produced in CHO cells

Panel 1.2.
Evolution of Therapeutic Proteins

- Native proteins
  - Factor VIII, tPA, EPO
- Engineered native proteins
  - B-domain deleted factor VIII, glycosylation-modified EPO, recombinant IgG
- Fusion proteins
  - TNFR-Fc fusion protein, IL-1R-Fc fusion protein
- Bispecific antibodies
- Drug-conjugated antibodies

product and amplifying its copy number, the transduced CHO cell can be made into a high producer of the transgene product. Furthermore, the transgene can be engineered to increase its binding affinity to the antigen or to humanize its protein sequence. After the introduction of tissue plasminogen activation (tPA) by Genentech in 1987, erythropoietin (EPO) and factor VIII soon followed to reach the market (Figure 1.1).

The early recombinant cell culture products were native human proteins administered to patients who suffered from deficiency of the protein due to congenic disorders or some disease conditions (Panel 1.2). The second wave of therapeutic proteins was antibodies. Antibody products make up the bulk of protein drugs in clinical use today. Early antibody products, called chimeric antibodies, retained the sequence of the immunized species (mostly mouse, since mouse hybridoma was the primary source of antibody molecules) in the variable region of the molecule while using a human sequence in the constant region. Later generations of recombinant antibody molecules were humanized (retaining only the hypervariable region of the mouse sequence) or completely human. Table 1.4 lists some examples of antibody-based therapeutic proteins. Trastuzumab (trade name Herceptin) binds to an overexpressed protein, HER2, which is so named because of its structural similarity to the human epithelial growth factor receptor 1 (HER1), on the surface of breast cancer cells. About 20% of breast cancer cells overexpress the HER2 protein and can be treated with the antibody. Its success in clinics paved the way for subsequent growth in antibody drugs and antibody Fc-region-based drugs. More recently, antibodies against PD-L1 (programmed death ligand 1) that has become overexpressed in many cancer cells have become an important class of medicines. PD-L1 is present in many normal cells, where it binds to PD-1 on T cells to suppress the T cell mediated immune response. However, some cancer cells also express PD-L1 to evade the immune response by T cells. The binding of an antibody to PD-L1 or PD-1 prevents the suppression of T cell activation by cancer cells, thus allowing the cancer cells to be killed by T cells.

The success of antibodies as a class of therapeutics was facilitated by decades of vast investment in biomedical research by the federal government and a better understanding of disease mechanisms. Once a binding target in a disease pathway is identified and the antigen isolated, an antibody against the antigen can be obtained and its affinity toward the antigen optimized. The antibody-based drug discovery is thus mechanism-based and design-oriented. This is in contrast to the discovery of traditional biochemical drugs such as antibiotics, immunosuppressors, and anticancer drugs, which rely largely on screening using binding assays or bioassays. The success rate of a candidate antibody biologic is much higher than that of a biochemical drug. Nevertheless, it is sobering to recall that some had taken the failure of the 1992 clinical trial of Centoxin, an antibody against the *Staphylococcus aureus* toxin for treating septic shock, as the obituary of antibody therapeutics. The real success of antibody therapeutics and the rapid increase in the number of FDA-approved biologics did not truly occur until the second half of 1990s, nearly twenty years after the arrival of recombinant DNA technology and the first wave of venture capital-funded biotech companies.

Antibodies are among the most abundant proteins in blood circulation. They are highly soluble and can be secreted by B cells at high levels.

Table 1.4. Therapeutic Antibody Products

Trade name	*mAb type*	*Therapeutic use*	*Host cell*
Orthoclone OKT3	Muromomab CD3	Reversal of acute kidney transplant rejection	Hybridoma
ReoPro	Anti-Abciximab	Prevention of blood clots	SP2/0
Rituxan	Anti-CD20 mAb	Non-Hodgkin's lymphoma	CHO
Zenapax (Daclizumab)	Humanized, anti-α-subunit T cell IL-2 receptor	Prevention of acute kidney transplant rejection	NS0
Simulect (Basiliximab)	Chimeric, anti-α-chain T cell IL-2 receptor	Prophylaxis of acute organ rejection in allogeneic renal transplantation	SP2/0
Synagis (Palivizumab)	Humanized, anti-A antigen of RSV	Prophylaxis of lower-respiratory-tract disease	CHO
Remicade	Anti-TNF-α mAb	Active Crohn's disease	SP2/0
Herceptin	Anti-HER2 mAb	Metastatic breast cancer	CHO
Mylotarg	Anti-CD33	Acute myeloid leukemia	CHO
Campath	Anti-CD52 mAb	Chronic lymphocytic leukemia	CHO
Zevalin	Anti-CD20 murine mAb	Non-Hodgkin's lymphoma	CHO
Humira	Anti-TNF-α mAb	Rheumatoid arthritis	CHO
Xolair	Humanized, Anti-IgE mAb	Moderate/severe asthma	CHO
Bexxar	Anti-CD20 mAb	Follicular non-Hodgkin's lymphoma	CHO
Raptiva	Anti-CD11a mAb	Chronic psoriasis	CHO
Erbitux	Chimeric antibody raised against human EGF receptor	EGF receptor–expressing metastatic colorectal cancer	CHO
Avastin	Anti-VEGF	Metastatic colorectal cancer and lung cancer	CHO
Soliris	Antibody binding to C5	Paroxysmal nocturnal hemoglobinuria	NS0
Vectibix	Anti-EGFR mAb	Metastatic colorectal cancer	CHO

Portions of the antibody molecule, namely the Fc region, are major components of many therapeutic fusion proteins. In those fusion proteins, the functional domain (or fragment) of a protein is joined to the carrier domain through a linker segment (Figure 1.2). The Fc fragment provides many of the properties of an antibody. A prominent example is the fusion molecule of the Fc fragment of IgG and the tumor necrosis factor α (TNFα) binding fragment of the TNFα receptor (TNFRα). The molecule binds to TNFα and suppresses its inflammatory effect. Such non-natural proteins are increasingly being explored as medicine. Bispecific antibodies (BsAbs), as the name implies, use antigen-binding sites from two different antibody molecules to simultaneously target two components of the cellular pathways and thereby improve clinical efficacy. Bispecific T-cell engagers (BiTEs) are a special class of bispecific antibodies which engage the T cells of the host's immune system in order to treat cancers. Increasingly, antibodies are being derivatized so as to contain drugs. These antibody-drug conjugates (ADCs) deliver cytotoxic agents specifically to diseased cells through their recognition of a particular antigen on the cell surface. Many biologics in the pipeline include antibody-drug conjugates, bispecific antibodies, and bi- or tri-specific immune cell engagers.

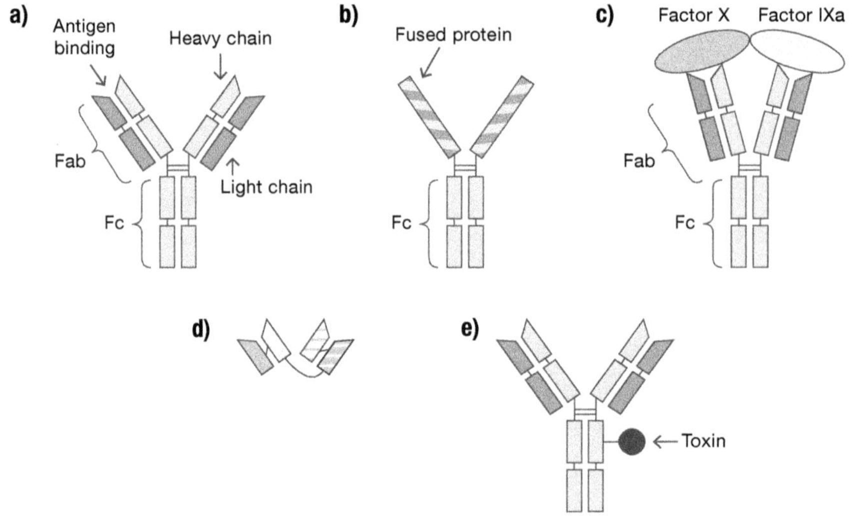

Figure 1.2. Conventional and new generation antibodies. (a) A traditional monoclonal antibody (mAb), a tetramer of two heavy and two light chains. (b) A fusion protein where the Fc region of the antibody is fused to different proteins/receptors. (c) A bispecific antibody which combines two antigen-recognizing elements into a single construct, enabling it to bind to two or more targets. In this example, one of the antigen-recognizing elements recognizes blood coagulation factor Factor X, and the other binds to Factor IXa, mimicking the function of Factor VIII to enable blood coagulation. (d) A bispecific antibody which exclusively recruits the T cell activity is called a Bispecific T-cell engager. BiTEs do not have a Fc region. (e) An antibody covalently linked to a chemical toxin, or an antibody-drug conjugate (ADC).

Biosimilars and the Expanded Reach of Protein Therapeutics

The patent that gives the market exclusivity of a mammalian cell-based therapeutic protein has a limited life span of 20 years after it is granted. After the expiration of a patent for a particular protein therapeutic, the generic version of the product, called the biosimilar, can be marketed by other manufacturers without the permission of the innovator of the original medicine. The availability of the generic version drives down the price of the drug and makes it more accessible to patients outside of industrialized countries. The regulation of biosimilars differs from that of generic drugs. The clinical effectiveness of chemical drugs (such as cholesterol-controlling statins and pathogen-killing antibiotics), whose molecular weight is relatively small compared to proteins, is usually assured once their chemical structure and purity are ascertained. The generic version of these small molecule drugs, upon verifying its chemical identity, purity, and drug product quality, can enter the market after the patent of the original drug expires (Panel 1.3). In contrast, both the structure and manufacturing process of protein biologics are complex. The biological activity and clinical efficacy of a biologic medicine is not readily predictable when a therapeutic protein is produced using a different cell line and a different process by a different manufacturer. The status of molecular folding, glycan composition, and even the composition of minute impurities may affect activity and immunogenicity. Biosimilars thus require regulatory approval and must undergo clinical trials, although on a smaller scale than the original (innovator) medicine.

A biosimilar must match the same product quality criteria as the original drug—it must have the same protein sequence, similar post-translational modification patterns (such as the profile of glycans, phosphate, or sulfate), and other quality indexes. However, such information is not available from the manufacturer of the original innovative medicine. Biosimilar makers thus are obliged to acquire the product from the market (often at different times and locations to cover a wide range of variabilities) and perform extensive product characterization to develop a product quality profile. They must also establish a producing cell line and

Panel 1.3. Biologics, Drugs, and Patents

Original		Off-Patent Version
• Drugs (small molecules) - Require clinical trials	→	• Generic drugs - No clinical trials needed prior to market entry
• Biologics (proteins) - Require clinical trials	→	• Biosimilars - Limited clinical trials

develop a process that generates a product that meets the same quality indexes as its counterpart from the innovator. It is possible that a biosimilar may have improved clinical performance compared to the innovator molecule; these are called "biobetters."

Producing a biosimilar with an acceptable post-translational glycosylation pattern is a challenging task, but one essential for receiving regulatory approval. The pathway for glycosylation is very complex. Many factors can affect the glycosylation pattern of a product, including the expression of glycosylation genes, the metabolic state of the cell, and the supply of nucleotide sugars. Controlling the flux, or even just modulating the flux in the glycosylation pathway, is not an easy task. Coupling the power of '–omics'-based analyses with a systems approach could possibly create a predictive tool to help steer the glycosylation pattern of the biosimilar to better match that of the innovator molecule.

Gene Therapy

Gene therapy aims to treat congenic diseases by delivering a correct copy of the gene to the cell of the patient via the introduction of a DNA molecule with the correct genetic sequence. The affected tissue can then be "corrected" to perform its proper function, or, in the case of the affected gene product being a secreted protein, the correct protein can be produced in the patient's body at a functional level. Due to the large number of genetic diseases that may benefit from gene therapy, there have been numerous ongoing studies and clinical trials. Most efforts have focused on diseases caused by a single gene mutation. For example, the cystic fibrosis transmembrane conductance regulator (CFTR) gene treats cystic fibrosis, and the factor VIII gene treats hemophilia. Some approaches seek to "permanently" incorporate a corrected copy of the gene into the cells of the patient, while others aim to transiently express the correct gene to rectify the disease conditions for a limited period of time. With the advances in genome engineering technology, the exploration of gene therapy has been extended from giving a correct copy to the repair or silencing of the incorrect copy in the host cell.

Panel 1.4. Gene Therapy

- Viral vectors
  - Adenovirus, adeno-associated virus, retrovirus, lentivirus
- Non-viral vectors
  - Nanoparticles
- Treatment
  - *Ex vivo* and *in vivo*

Gene therapy may be achieved by isolating a patient's target cells and performing gene transduction *ex vivo* before the "corrected" cells are returned to the patient (Panel 1.4). It may also be accomplished by suspending the replacement gene (or other DNA elements) in nanoparticles or a plasmid or viral vector, and then injecting them into the patient and allowing the DNA element to enter a fraction of the cells in the body and perform its function. Most clinical trials that have been conducted with viral vectors used adenovirus, lentivirus, or adeno-associated virus.

Producing viruses for gene delivery is similar to producing viruses for vaccines. However, there are also important differences. While the number of viruses per dose for vaccines is rather small, that for gene therapy in most applications is very high (100 to 1000 times higher). For one dose, the volume of culture fluid needed is thus much larger for gene therapy than for vaccine applications. This also means that the number of impurities carried in culture fluid that must be removed from the final product is much larger in a gene therapy product than a vaccine. In the manufacturing of live viruses, at the end of production the culture fluid cannot undergo the same type of extensive purification carried out in the production of therapeutic proteins. Thus, the purification and concentration of viruses for gene therapy face greater challenges. In a typical virus production process, a very large fraction of the virus particles produced do not contain complete viral genomes and are not infectious. While empty virus particles are still immunogenic, and can be effective for vaccination, these non-infectious particles do not deliver the DNA cargo needed for gene therapy. Increasing the infectious virus particle content is thus important in gene therapy virus production.

Recent approval of gene therapy products by regulatory agencies has been very encouraging for cell culture bioprocessing. However, the cell culture bioprocess profession still faces major challenges in the manufacturing technology required for virus production for gene therapy.

Cell Therapy

Bone marrow transplant, which has been practiced for more than thirty years, is one of the oldest forms of cell therapy (Panel 1.5). Following the discovery of HSC, its expansion and differentiation into separate lineages for transplantation was long envisioned. Some also predicted the enrichment and expansion of a particular lineage of cells, such as platelets, T cells, or NK cells. As the capability of scientists to isolate and culture various differentiated cells increased, many also attempted to culture and expand hepatocytes or beta cells for cell transplantation to functionally rectify liver failure or diabetes. Functional differentiated cells were also incorporated into scaffolds to construct tissue mimicries for implantation. The research field of employing functional cells for the construction of tissue analogues is more frequently referred to as tissue engineering.

The emergence of various multipotent and pluripotent stem cells gives hope to regenerative medicine. The use of stem cells as the source of differentiated cells for cell

Panel 1.5. Cell Therapy

- Cell transplant (autologous cryopreserved cell)
  - e.g., bone marrow transplant
- Donor cell transplant (heterologous)
  - e.g., liver cell transplant, bone marrow transplant
- *Ex vivo* expanded cell transplant
  - e.g., stem cell or stem-cell-derived differentiated cell transplant
- *Ex vivo* genetically altered cell transplant
  - e.g., Car-T cell therapy

therapy can potentially alleviate the problem of donor cell procurement in those applications. PSCs can provide an almost unlimited supply of liver and NK cells for regenerative applications. Many such applications would involve the infusion of differentiated cells into regenerated tissues or organs. Cell therapy thus entails overlapping applications with tissue engineering, regenerative medicine, etc.

An area of cell therapy that has shown much promise is the treatment of cancers using cells from immune systems, either the native cells or genetically engineered ones. Cells of immune systems may be cultured *ex vivo* and genetically modified to express cytokines, receptors, and chimeric tumor-antigen receptors for cancer immune therapy. T cells that have been engineered to express a chimeric T cell receptor with a single-chain variable fragment (scFv) and a high affinity for tumor-associated antigen (TAA), called chimeric antigen receptor T (CAR-T) cells, have been used successfully to recognize and kill tumor cells.[38]

Cell therapy, with its initial clinical success and great promise in cancer treatment, still faces major challenges. These are not only in producing a large number of cells and engrafting them at the target site, but also in cell isolation, expansion, and final preparation as a drug product. The challenge is especially acute in the autologous (from patient) applications that require a large number of cells. Without the economy of scale of production, manufacturing a large number of doses runs the cost of goods for those autologous treatments exceedingly high. Even for allogeneic (from external donor) applications, many challenges await cell culture bioprocess technologists in bringing the field forward. The manufacturing process must generate cells of high viability and therapeutic activity. This is in contrast to the production of protein therapeutics, in which the cell viability at the end of a fed-batch process is sometimes only somewhat higher than 50%. The recovery of those highly active cells for cell therapy poses another challenge. Unlike the production of protein biologics, in which the product can be subjected to various salts, pHs, and osmolality conditions that are considered abusive to cells, cells for cell therapy can only be subjected to a minimal number of steps while achieving the necessary degree of purification. Therefore, many challenges in process technology are yet to be overcome in the wide applications of cell therapy.

Industrial Cell Lines

Cell lines have been used for the industrial production of human and veterinary vaccines for more than half a century. For human viral vaccine production, human diploid cell strains (MRC-5 in particular) are the preferred cell substrate. Traditional viral vaccines consist of whole virus particles where the viral genome is packaged in the virus particle after the replication of its genetic materials in the production cell. The entire viral particle is then injected into the patient to elicit an immune response. There is a low-level risk that the virus genome might recombine with the genetic elements of the production cell and bring the genetic element of the host cell into the virus particle. This poses a potential risk of transmitting an activated oncogenic or other adventitious genetic element to the patient. To minimize such a risk, the vast majority of human virus vaccines are produced in normal diploid human cells. Vero and MDCK cells (along with chick embryos) are notable exceptions of non-human continuous cell lines used for human vaccine production (Table 1.5). In the case that a continuous cell line is used for human vaccine production, its passage number is carefully monitored and controlled to alleviate concerns raised by the report that prolonged passage of Vero cells poses an increased risk of tumor formation.[39] However, advances in molecular analytics in recent years are rapidly extending our capability to characterize cells as well as products. There is an increasing interest in exploring the use of continuous or even tumorigenic cell lines for the production of viral biologics. For veterinary vaccines, the repertoire of host cells is much larger. Both cell lines and tissue-derived cell strains with limited life spans are widely used (Table 1.6).

Table 1.5. Major Cell Strains and Lines for Human Biologics Production

Human Vaccines	
Primary cells	Green monkey kidney cells (no longer used), chicken embryo cells
Cell strains	MRC-5 (human lung fibroblast)
Cell lines	Vero (monkey kidney epithelial cells) MDCK (dog kidney epithelial cells)
Recombinant Proteins	
Cell line origin	
Human	HEK293
Mouse	C-127, NS0, hybridoma cells, SP2/0
Chinese hamster	CHO
Syrian hamster	BHK
Gene Therapy	
Human	HEK293, PER C6
Insect	Sf9

Most cell lines used for the production of recombinant therapeutic proteins are derived from rodents, including the mouse, Chinese hamster, and Syrian hamster. Human cells are only used for the production of a handful of products. The vast majority are produced using Chinese hamster ovary (CHO) cells lines (Panel 1.6). CHO cells were a favorite for cytogenetic research even before they were commonly used for therapeutic protein production. A large number of CHO cell mutants were isolated in cytogenetic research. This was in contrast to other mammalian cell lines; due to their diploidity, mutants are usually difficult to obtain since only dominant phenotypes are observable. However, the rate of obtaining mutants is relatively high for CHO cells. They were sometimes described as a functional haploid.

Using a non-human cell line to produce a therapeutic protein carries the risk of having contaminants from the host cell (nucleic acids, proteins)

Table 1.6. Cell Lines Used in the Production of Veterinarian Vaccines

Vaccines	*Cell line*
Bovine viral diarrhea virus	MDBK
Bovine parainfluenza virus type 3	MDBK
Bovine rhinotracheitis virus	MDBK
Bovine respiratory syncytial virus	MDBK
Feline leukemia virus	FL72
Feline panleukopenia virus	CRFK
Feline chlamydia	CRFK
Canine parvovirus	CRFK
Canine distemper	Vero
Canine adenovirus type 2	Vero
Ehrlichia canis	DH82
Rabies	BHK21
Eastern equine encephalitis virus	Vero
Western equine encephalitis virus	Vero
Equine rotavirus	MA104
Equine rhinopneumonitis virus types 1 and 4	Equine Dermal
Equine influenza virus	MDCK
Foot and mouth disease virus	BHK21
Swine parvovirus	ST, PK
Swine influenza virus	MDCK

Panel 1.6. CHO Cells

- Aneuploid, heterogeneous population in karyotype, chromosome number
- Many mutants were isolated, sometimes called functional haploid
- Easy for cell adaptation (to suspension growth, varying nutritional requirements, etc.)
- Glycans on proteins are similar to human
  - Does not express α2,6-sialyltransferase as in human glycoproteins
  - Has only α2,3-sialyltransferase

in the product even after extensive purification. This necessitates the reduction of contaminating host cell proteins and DNA in the product to an acceptable level to minimize the possibility of their causing any immunogenic response or other adverse effect. The required level is based on per dose, rather than per unit mass or volume of the therapeutic protein to be administered. Thus, the higher the protein dose is, the harder it is to reduce the host cell contaminants to an acceptable level. Many therapeutic antibodies are administered at gram level per dose, in contrast to tens of milligrams for growth factors and cytokines (Table 1.7). For such high dosage products, it may be advantageous to use a human cell line for production. However, the advantages of easy genetic manipulation and the high probability of obtaining a hyper-producing cell line thus far outweigh the requirement of removing residual host cell proteins.

Many viruses, including lentivirus, adenovirus, and adeno-associated virus, are used as carriers to deliver the corrective genetic material to the host tissue *in vivo* or host cell *in vitro* for gene therapy applications. The production of these viruses is often carried out in HEK293 cells. The HEK293 cell line was derived from human embryonic kidney cells immortalized by the insertion of a segment of adenovirus type 5 DNA into its genome.[40] In some cases, insect cells are also used for virus vector production.

No particular cell line of embryonic stem cells, iPSCs, or MSCs has yet become a commonly used or "standard" cell line in clinical trials involving allogeneic stem cells or other tissue cells. In the industrial production of biologics, cells for production are characterized by their cytogenetic characteristics, scrutinized for their history in terms of exposure to animal serum, viruses, or other adventitious agents, etc., and then expanded for banking as master and working cell stock. Such meticulous cell characterization and banking may not be possible for autologous cell therapy applications. For allogeneic cell therapy applications, even though no standard cell lines for cell therapy have emerged yet, it is expected that the cell line stock for such applications will follow a similarly rigorous procedure.

Table 1.7. Dose of Some Antibody Products

Product	*Disease indication*	*Approximate dose*
Amevive	Psoriasis	7.5mg / 0.5ml; 15mg / 0.5ml
Enbrel	Rheumatoid arthritis	25mg
Herceptin	Breast cancer	440mg / 30mL
Humira	Rheumatoid arthritis	40mg (1ml prefilled syringe)
Remicade	Crohn's disease, rheumatoid arthritis	100 mg / 20mL
Rituxan	Non-Hodgkin's lymphoma	100mg / 10mL; 500mg / 50mL
Synagis	Respiratory syncytial virus	100mg
Xolair	Allergic asthma	150mg / 5mL

Other Production Systems

Mammalian cells are the workhorse for the production of protein therapeutics that require post-translational modifications (e.g., glycosylation) or complex protein folding. Many proteins that do not require extensive post-translational modifications are produced in *E. coli* or in yeast *Saccharomyces cerevisiae*, including insulin, human growth hormone, and some cytokines. Even some proteins that requires glycosylation and other post-translational modifications are produced in other host cell systems that are capable of carrying out those functions. Some of those systems have been explored as the production vehicles of therapeutic proteins (Table 1.8, Panel 1.7).

Insect Cell Culture

The use of insect cells for protein expression dates back over three decades ago, to around the time that yeast and mammalian cells were being explored as a production vehicle for therapeutic proteins.[41] The expression usually involves the cloning of a transgene (or transgenes) into a viral vector such as Baculovirus. The vector is then packaged in insect viruses by transfecting the insect cells with the vector. The virus produced is then used to infect the host cell in the production. Virus replication in the host cell results in a large number of transgene copies. By using a strong promoter to drive the expression of the transgene, a high level of protein expression can be achieved.

A popular cell line for heterologous protein expression, Sf9, was derived from the insect *Spodoptera frugiperda*. Insect cells are capable of

Table 1.8. Recombinant Proteins Produced in Yeast

Product	*Use*
Medway (recombinant human serum albumin)	Blood expander
Hepatitis B vaccine	Hepatitis B
Interferon-alpha	Hepatitis C / cancer
DX-88	Hereditary angioedema (HAE), a debilitating condition characterized by acute attacks of inflammation
Recombinant human insulin	Diabetes
Recombinant collagen	Medical research reagents and dermal filler
Botulism vaccine	Botulism vaccine product

Panel 1.7. Other Therapeutic Protein Production Technologies

- *E. coli* for aglycosylated products
- Yeasts
  - *Saccharomyces cerevisiae*
  - *Pichia pastoris*
- Insect cells
- Transgenic animal
  - Transgenic goat
- Transgenic plant
  - Transgenic tobacco cell culture

glycosylation; however, the glycoforms of the proteins produced in insect systems are somewhat different than those produced in mammals. For example, the N-glycans synthesized in insect cells have high mannose content and fucose with different glycosidic bonds than those seen in mammals. Insect cell lines are thus not used in the production of therapeutic proteins, though they are used in the production of proteins for research use or toxicity studies. They are also frequently used in the production of recombinant proteins for use as antigen proteins or virus-like-particles in vaccine applications. Additionally, they are used in the production of human and veterinary viruses as vaccines or as transgene vector production vehicles in gene therapy. The insect cell culture remains attractive because the cultivation is relatively straightforward and the process development time can be comparatively short.

Yeasts

The yeast *Saccharomyces cerevisiae* has been used to produce serum albumin, cytokines, and some virus-like particles for use as vaccines (Table 1.8).[42] However, the glycans on the recombinant proteins synthesized in yeasts bear significant differences to those seen in mammals. Thus, *Saccharomyces cerevisiae* is not used for the production of therapeutic proteins that require extensive glycosylation. The yeast in the genus *Pichia* is capable of synthesizing N-glycans that are not the mannose-rich types produced in *Saccharomyces*. Advances have been made in 'humanizing' the glycosylation characteristics in the yeast *Pichia pastoris* for the production of therapeutic proteins. The development of a multistep genetic engineering process that first eliminated non-human glycosylation enzymes from *Pichia pastoris* and then introduced human glycosylation enzymes led to the synthesis of proteins with humanized glycans. The secretion rate of the synthesized recombinant proteins does not match that seen in recombinant mammalian cells; however, this might be compensated for by the high cell concentration achievable in the bioreactor. With further improved secretion capacities and glycosylation patterns, these engineered yeast strains may be capable of producing proteins with consistent glycosylation patterns, or even with uniform glycans.

Transgenic Animals

The use of transgenic animals, including goats, pigs, and rabbits, for the production of biotherapeutics has been in development for three decades. Most frequently, the product protein is tissue-specifically expressed in the mammary glands for secretion into milk.[43] These production systems need a low initial capital investment compared to the traditional manufacturing plant necessary for biologics. The downstream purification process is relatively simple, as the proteins are fully glycosylated. The product ATryn (antithrombin III), which was produced in transgenic goat's milk by GTC Biotherapeutics, has been approved by the regulatory agencies in the USA and Europe.

An advantage of transgenic animal production is its high titer in milk, on the order of 2–10 g/L. However, over the years the titer in cell culture processes has increased to a range of 5–10 g/L, thereby diminishing this particular advantage of transgenic animal production.

Manufacturing

Upstream Process

The basic process flowsheet for cell culture manufacturing is almost identical to that for microbial fermentation (Figure 1.3). Although laboratory-style culture devices, including roller bottles and multiple plates, are still used in some vaccine and cell therapy processes, the vast majority of vaccine, gene vector, and protein production processes use stirred bioreactors. Compared to microbial processes, the metabolic load of cell culture, in terms of nutrient and oxygen consumption rate per unit reactor volume, is at least one order of magnitude lower, as is the power input for mechanical agitation.

Viral vaccines, cytokines, and growth factors are administered to patients in relatively small quantities. The manufacturing scale for those low-dose products is therefore also relatively small. In contrast, antibodies are often administered in high doses, as shown in Table 1.7, to elicit longer-lasting effects. Thus, in general, the size of manufacturing plants

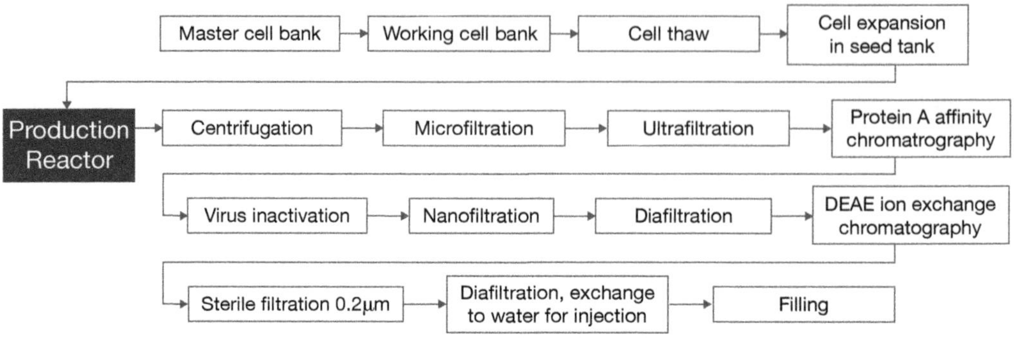

Figure 1.3. A typical cell culture manufacturing process for recombinant antibodies.

for antibody products tends to be larger than those for viral vaccine production, etc., except in the case of some orphan drugs that are used only for small patient populations.

A typical process uses a few small reactors for cell expansion before switching to the production reactor. The process cycle tends to be longer than in microbial fermentation. A fed-batch culture may last 10–14 days in the production scale and 30–40 days in total (including seed cultures). Perfusion cultures are operated as continuous processes and last from two to six months.

The fundamental principles of process engineering are virtually identical for cell culture and microbial fermentation. In both processes, the stoichiometric relationship is used as the guide for medium design, the kinetics of growth and production as the foundation for process optimization, and transport phenomena as the key for reactor design. Even the reactor technology for the production of cell culture biologics was mostly adopted from microbial fermentation. However, cell culture engineering, with its great contribution to the bioeconomy, has outshone microbial fermentation in process innovation. The technological advances in the past two decades in product separation and analysis and manufacturing facility design, the adoption of platform concepts in cell line and process development, the wide application of single-use reactors, and the recent exploration of continuous processes have been all driven by cell-culture-based production. Cell culture processes are still expanding to meet the increasing number of new, innovative drugs and biosimilars. With cell and gene therapy on the horizon, cell culture engineering will continue to be a key driver in bioprocess technology.

Single-Use Systems and Continuous Process

The increasing number of licensed products in the past decade has heightened the demand for production capacity. This increased demand has not seen a proportional increase in the volumetric capacity of the manufacturing plant, but has instead been met by increased productivity and improved efficiency in facility utilization. The manufacturing process for cell culture products has undergone some important changes in the past decade (Panel 1.8). To shorten the time spent in cell expansion before reaching the production bioreactor, some have employed a large quantity of cell stock in a working cell bank to initiate the seed train in larger bioreactors. In some cases, the production reactor is inoculated at a much higher cell concentration by employing some form of cell perfusion in the preceding seed culture (often called N-1) stage in order to shorten the time it takes to reach the peak of productivity.

The past decade has seen an increase in the use of disposable or single-use reactors. Used only in inoculum preparation and in small-quantity production a decade ago, single-use stirred reactors, at a volume of up to two thousand liters, are becoming common for manufacturing. This is feasible because of the relatively low power requirement for agitation in cell culture process, allowing the plastic vessel to sustain the mechanical stress of agitation. A plant based on disposable bioreactors offers the advantage of lower capital investment and faster plant construction in setting up the manufacturing facility, plus increased flexibility and modularity in process implementation.

Single-use bioreactors do have some shortcomings. Most of them cannot be easily pressurized to transfer the fluid between vessels using fixed pipes. Many have limited choices of impeller design. For some cells, especially those grown on microcarriers, the agitation mechanism may not be optimal. Nevertheless, the wide application of single-use bioreactors has changed the landscape of biomanufacturing. Due to the size limitation, to increase the production volume beyond the maximum size of the vessel one must resort to using multiple reactors. The alternative is to increase the concentration or the throughput of the product. This is achieved by increasing the cell density and prolonging the duration of fed-batch culture processes, or by running the process in a continuous fashion.

Panel 1.8. Disposable Single-Use Systems

- Reduced capital investment for manufacturing plants
- Reduced lead time for plant construction
- Transfered many manufacturing logistics to single-use equipment suppliers
- Reduced reactor size, with more operational constraints
- Compensates for smaller reactor volume with increased cell density and product throughput if a large product volume is needed, by:
  - High cell density perfusion
  - Hybrid process of perfused fed-batch

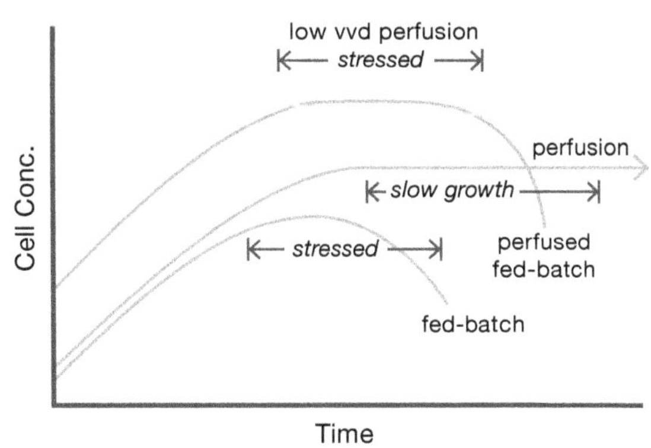

Figure 1.4. Cell culture manufacturing processes: fed-batch, perfusion, and hybrid (perfused fed-batch).

Most current cell culture biomanufacturing processes are operated in a fed-batch mode (Figure 1.4). After a few days of rapid cell growth, the culture is fed with medium containing high concentrations of nutrients a few times over the production period to increase the culture volume step-wisely to its maximum. The addition of medium increases the osmolality of the culture broth. This, along with increased metabolite accumulation, slows down cell growth. During the slow growth and stationary phase period, the product accumulates to a high level.

Cell culture processes are rarely operated as a simple continuous culture because the cell concentration at a steady state is too low to achieve a high productivity. A continuous cell culture process is performed as a perfusion culture, in which a cell separator is used to recycle a concentrated cell stream separated from the effluent back into the bioreactor to increase the cell concentration. However, only a small number of current processes employ perfusion culture. Most of the perfusion processes have been those involving labile proteins that need a short residence time in the reactor or proteins that are expressed only at very low levels. The lack of an effective and robust in-line device that can concentrate cells and return them to the bioreactor has contributed to the slow adoption of the perfusion process.

In the past decade, in-line cell separation devices have advanced markedly. This has made continuous perfusion culture easier to implement and allowed for a new hybrid mode of perfusion/fed-batch to be carried out. A hybrid perfusion/fed-batch culture often starts with a higher cell concentration than a typical fed-batch culture. This is accomplished by using a cell separation device to concentrate the seed culture stream and inoculate more cells into the reactor. The operation is similar to a fed-batch culture, though the high cell concentration results in a high accumulation of metabolites, requiring the continuous removal of culture medium to reduce their levels. Continuously flushing out salts, metabolites, etc., through medium withdrawal alleviates the growth inhibition and prolongs the production phase.

Product Recovery

The recovery process of cell culture products is simpler than that for bacterial-based recombinant proteins. The majority of cell culture processes now use media with a low concentration of proteins. At a high concentration in the range of 5–10 g/L, the product is the predominant protein in the spent medium at the end of the cell culture process. The product isolation and purification process is thus substantially simpler, since only relatively small quantities of contaminating cellular proteins and other medium components must be removed.

An example of the recovery of antibody IgG is shown in Figure 1.3. After the cell culture process, the cell suspension from the reactor is chilled in a holding tank to prevent cell lysis. Centrifugation or microfiltration is used to remove the cells. To prepare for subsequent chromatographic steps and membrane processes, the particulate matters in the supernatant are removed by microfiltration. The process stream is then concentrated using ultrafiltration. For some very high-productivity processes, this step may not be necessary, as the product concentration is already high. The retentate from ultrafiltration is then fed into a protein A affinity chromatography column. This chromatography step captures the product at a very high yield and at a high degree of purification. After washing, the adsorbed IgG is eluted with a low-pH buffer for several hours. This inactivates endogenous retroviruses from within the host cell that may form virus particles. Sometimes, another means of virus inactivation is used. After virus inactivation, a nanofiltration step is used to remove the virus particles. Because protein aggregation sometimes occurs, a gel-permeation chromatography step may also be employed to remove aggregated IgG (not shown). Non-human host cells invariably shed cellular proteins and DNA fragments into the medium and contaminate the product. To reduce the level of those contaminants, anion exchange chromatography is used. After purification, the product solution is filter sterilized for final preparation in a formulation solution.

With the renewed interest in continuous perfusion culture, there have been studies conducted concerning the possibility of making the product recovery process continuous. As can be seen in the typical recovery process, virtually all unit operations are continuous-flow open systems but operated in a periodic batch mode. The strategy is to connect those unit operations and adapt them to a continuous feed stream from the bioreactor by using multiple smaller pieces of equipment for each unit operation. They would cycle through the different stages of the operations such that there would always be a piece of equipment receiving the feed from the preceding operation. Some have envisioned using rather low-capacity equipment which cycles through at a short period so that the equipment completes its life cycle quickly. In such a way, the equipment can be disposable and be replaced after a short duration of continuous operation. Whether such a vision will be implemented, and how much operational and cost benefit it may bring, has yet to be seen.

Product Quality in Cell Culture Processing

Quality of the Product

THE QUALITY OF A CELL CULTURE PRODUCT is judged by its capability to deliver its target biological activities while minimizing the risk of causing any biological harm. In most cases, a direct measurement of the biological activity, in the form of a bioassay or even an animal assay, is not readily assessable. Instead, one examines some physical features and

Panel 1.9. Protein Product Quality Characteristics

- Identity
- Functional activities
- Purity/contaminants
- Chemical structural variations

chemical structural features that are readily measurable to assess the product quality. For example, peptide finger printing is used to establish the identity of a protein product after production (Panel 1.9).

The quality of the product is not static and is liable to change over time. Such quality variability may occur at different stages of the cell cultivation process through changes in product synthesis. It may also occur during product recovery or even after the protein has been purified and stored in the formulation solution. Depending on the nature of the chemical or physical change, product quality variation may occur at fast or slow kinetics. Therefore, quality is not only measured on the drug substance and the product after it is formulated. For variables that change in slow kinetics, quality is also assessed in inventory conditions. For example, after a long storage of a purified and formulated protein, color may develop or the solution may become cloudy due to aggregation. Such quality changes can only be assessed over time.

In general, the product must be void of some known contaminating species or be capable of keeping them below certain levels. For example, protein therapeutics produced in rodent cells may carry some host cell constituents even after product purification. The host's proteins, DNA contaminants, and endogenous viruses that are not removed in the product purification process and are present in the final product must be kept below an acceptable level in order to minimize the risk to the patient incurred by the presence of those contaminants.

Structural variants of a protein (e.g., an amino acid in the sequence of a protein molecule which differs from the one specified by the coding sequence) are frequently seen in proteins produced by cell culture. Even if no adverse effect of the structural variant is known, the variant must be kept below a certain level that is accepted as low-risk.

The glycan structures of the proteins invariably exhibit heterogeneity. In the case that the glycan structure affects the protein's biological activity, the content of the glycan with the desired or undesired structure must be kept within a bound. Even in the case that the glycan structure does not directly influence the biological activities of the protein, the extent of the heterogeneity of the glycan must be specified, and different lots of product produced in different times or at different manufacturing sites must all meet the same specifications.

Structural features and biological activities

The products of cell culture processes—proteins, viruses, and cells—are all structurally complex. The assessment of their quality is not easy. Ultimately, the quality of a virus intended for use as a vaccine is determined by its ability to elicit an immunogenic response. For gene therapy, the quality of the product virus is dictated by its capability to infect the target cells and express the cargo gene. The mechanism by which different therapeutic proteins deliver their biological effect is very diverse. Some bind and neutralize or remove the target molecules, while others bind to the target cell and elicit the killing by killer cells. Direct evaluation of such biological functions in a process setting is not feasible. Instead, one uses the mechanistic knowledge of the mode of action to develop metrics of product quality based on the structural characteristics that are critical to the biological functions. For example, for influenza virus, in addition to the virus particle count, the hemagglutinin level can be used as a quality index. In gene therapy, the ratio of transgene copy to the virus particle may be used as a measure of quality. In the production of therapeutic IgG for applications involving antibody-dependent cell-mediated cytotoxicity (ADCC, also referred to as antibody-dependent cellular cytotoxicity) activity, the fraction of fucosylated glycan is used as a quality measure because the presence of fucose on the glycan significantly reduces the ADCC activity of the IgG. Understanding the mode of action of the product involved is thus important in using the structural attributes of the product as quality metrics.

Glycosylation Profile

The glycans attached to a glycoprotein are structurally heterogeneous. This is seen in glycoproteins in blood circulation as well as in therapeutic proteins produced in cell culture. The glycan structure of some proteins plays a critical role in the protein's therapeutic efficacy (Panel 1.10). For example, the presence of mannose 6-phosphate on the glycan of a number of lysosomal enzymes is critical in mediating the protein binding to the mannose 6-phosphate receptor on cell plasma membrane for uptake, a step necessary for targeting the enzymes to lysosomes. The abundance level of mannose 6-phosphate on the glycan of those enzymes is thus critical in treating lysosomal storage diseases. In some cases, the glycan structure does not directly influence the protein's biological activities but affects its pharmacokinetic behavior. A higher content of sialic acid on erythropoietin and many other proteins, for example, increases the circulation half-life of those proteins. Without the sialic acid, the exposed galactose would bind to the asialoglycoprotein receptor on the surface of liver cells, leading to the internalization and degradation of the protein.

A protein may have multiple sites in its sequence for glycosylation. The glycan structure

Panel 1.10. Glycosylation Profile

- Glycans are not uniform, but heterogeneous *in vivo* and in recombinant proteins
- Glycan distribution must be within a prescribed range of the product
- Roles of glycans
  - Affect solubility
  - Affect pharmacokinetics, e.g., sialic acid content
  - Affect biological activities, e.g., fucose in Asn292 glycan
  - Critical for biological activities, e.g., mannose 6-phosphate for lysosomal enzymes

on a given site is not uniform. Rather, it consists of a number of somewhat different structures. This heterogeneity is the result of its biosynthesis reactions in the Golgi apparatus. The profile of glycans may also vary somewhat in different production runs, or under different production conditions or cell lines. The impact of such variation on its clinical efficacy is dependent on the mode of action of the protein. As discussed above, for some it is very important; for others, it may affect the blood circulation half-life, but not the biological activities. However, from a regulatory perspective, the glycan distribution of the product must be within a specified bound. Thus, confining glycan distribution to an acceptable range is important for the quality control of the product. A better understanding of the mode of action of the product can help define the acceptable range of heterogeneity for the product.

Protein Structural Variants

Prior to the selection of the production cell line of a given product, the DNA sequence of all the copies of the product gene integrated into the genome should be verified to ensure the transgene integrity. One thus expects that the primary sequence and the secondary, tertiary, and even quaternary structures of the protein will be uniform. However, structural variants of the protein do occur, both in natural and in recombinant proteins. They are caused by translational errors, incomplete post-translational enzyme processing, post-secretion extracellular enzymatic or chemical reactions, and even post-process events (Panel 1.11). At a higher order structural level, protein molecules may form dimers or oligomers, and may scramble disulfide bonds or even form large aggregates and precipitate. At a primary sequence level, sequence variants include incorrect amino acid incorporation and chemical modifications of amino acids in a protein.

Panel 1.11. Protein Variants

- Amino acid misincorporation
  - Mutations, transcript editing
  - Amino acid limitations
- Charge variants
  - Acidic or basic variants may be present in a detectable proportion
  - Asn deamidation, glycation of lysine, C-terminal lysine clipping, N-terminal pyroglutamic acid, C-terminal proline amidation
- Higher order structural changes
  - Cystein-linking, aggregation

Many production cell lines have multiple copies of the product gene. With high-throughput DNA sequencing technology, the DNA sequences of the many copies of transgenes in the genome of a candidate production cell line are nowadays verified to ensure their integrity. Once a cell line is selected, the probability is very low that a mutation will subsequently occur to a copy of the transgene. Nevertheless, if under a rare occasion a missense mutation (i.e., a mutation causing a change of an amino acid in the protein) occurs in one of those copies, it will inevitably result in the presence of a fraction of mutated protein molecules.

A sequence variant may also arise from amino acid depletion in the medium, as reported for the substitution of asparagine by serine in a small fraction of IgG. In rare cases, incomplete leader peptide cleavage from the protein may be seen.

A number of commonly seen structural variants of recombinant proteins are listed in Panel 1.11. Many proteins have a lysine residue at their C-terminus. These lysine residues are cleaved intracellularly by a carboxypeptidase. The cleavage, however, is often incomplete. Uncleaved lysine at the C-terminus is frequently seen in naturally occurring proteins in circulation as well as in the production of recombinant proteins.

Industrial fed-batch cultures often employ high glucose concentrations. The secreted product protein molecules have therefore had a long exposure to a high glucose concentration. Glycation, the attachment of a sugar to an amino group in the side chain of an amino acid, is commonly seen. Glycation is also seen in proteins in circulation in the human body.

Some of the amino acid modifications result in changes in the net charge of the protein. These charged variants may be acids that elute in an anion exchange chromatographic column faster than the "normal" or reference product protein, or be basics that elute faster in a cation exchange column. Acidic charged variants may also arise from changes in the glycosylation pattern, for example by increased sialic acid or sulfate content.

Process Control and Product Quality

The heterogeneity of the glycan profile in a protein is mostly the outcome of intracellular biosynthetic events. Degradation of glycans (especially removal of sialic acid by sialidases) by enzymes released from cells after the protein was secreted into culture broth may occur when the cell viability is low. Protein structural variants mostly occur after the protein has been translated, either due to intracellular events or in culture broth after the protein has been secreted into the medium. Protein-processing enzymes may be released by cells via secretion or due to cell lysis. Extracellular proteolytic cleavage has been shown to lead to degradation of factor VIII and cause alterations to the ratio of single chain/double chain molecules of tPA and protein C. Some higher order protein structural changes, such as aggregation, oxidation of methionine, and deamidation of asparagine, may even take place in the product recovery process.

The increasing emphasis on product quality calls for a better understanding of how intracellular events affect glycan profiles and how the extracellular culture environment influences the occurrence of protein variants. Ideally, this should advance our ability to control the process conditions and deliver consistent productivity and product quality. The emphasis on the development of process technology is increasingly focused on process analytical technology (PAT) for on-line monitoring and control. Going hand-in-hand with PAT is Quality by Design (QbD).

The aim of QbD is to develop process understanding and identify the process variables that contribute most to product quality. This may lead to the development of strategies to control those variables and confine product quality attributes to an acceptable range.

In order to implement QbD, one needs knowledge of the attributes that define product quality and an understanding of the relationships between the process variables and the quality attributes. Initially, one may rely on empirical data that correlates process variables and quality attributes. Over time, a mechanistic process model becomes critical to developing a predictive control strategy. The quality and the productivity are not separable in such an effort. The mechanistic understanding of the culture environment in regards to productivity and quality is incorporated into a process model that translates the input of process perturbations into the necessary control actions to sustain productivity and product quality. The advances in genomics-, proteomics-, and metabolomics-based global analytical methods, in conjunction with a systems approach, will facilitate the implementation of PAT and QbD in cell culture bioprocessing.

From Discovery to Clinical Products

THE COST OF DEVELOPING NEW MEDICINE is staggeringly high. A typical drug discovery scheme begins with the initial disease treatment conception and the production of a discovery quantity of product for biological and toxicity testing and pharmacokinetic evaluation. Then comes the decision to proceed with clinical trials, the construction of production cell lines and initial process development for upstream, downstream, and formulation, and the development of chemical and bioassays for the product. Simultaneously proceeding are phase I trials, phase II trials, commercial manufacturing planning and implementation, documentation preparation, and finally submission to a regulatory agency. The traditional flowsheet staggers different teams and different stages of drug development roughly in sequence (Figure 1.5a).

A key to reducing the cost of developing medicines is to condense the timeline, compressing the typical 6–8 years of development time into 2–3 years, as the regulatory agency now permits for drugs with breakthrough drug status. Such accelerated development has changed the way cell culture technologists, discovery chemists, pharmacists, and protein engineers work together. The time duration for most tasks in drug development cannot be shrunk proportionally to accommodate the shortened timeline; for instance, cells' doubling time cannot be made shorter to allow for a shorter cell line development time. As a result, the compression of the overall development timeline is not achieved by proportionally shortening the completion time of each task, but by overlapping them in time. Before the cell line construction is completed, the process development and the product assay development all have to start. The project flow thus changes from one task team handing over their completed object to the next team to different task teams simultaneously working together in a cross-communicative way (Figure 1.5b).

An important consequence of this change is that the scientists and engineers from different segments of development are increasingly working in concert. Each person has to gain the vocabulary and knowledge of the other segments of the drug development process. Two decades ago, at the dawn of biotechnology, the call was for engineers to learn biology and for scientists to understand some vocabulary of process engineering.

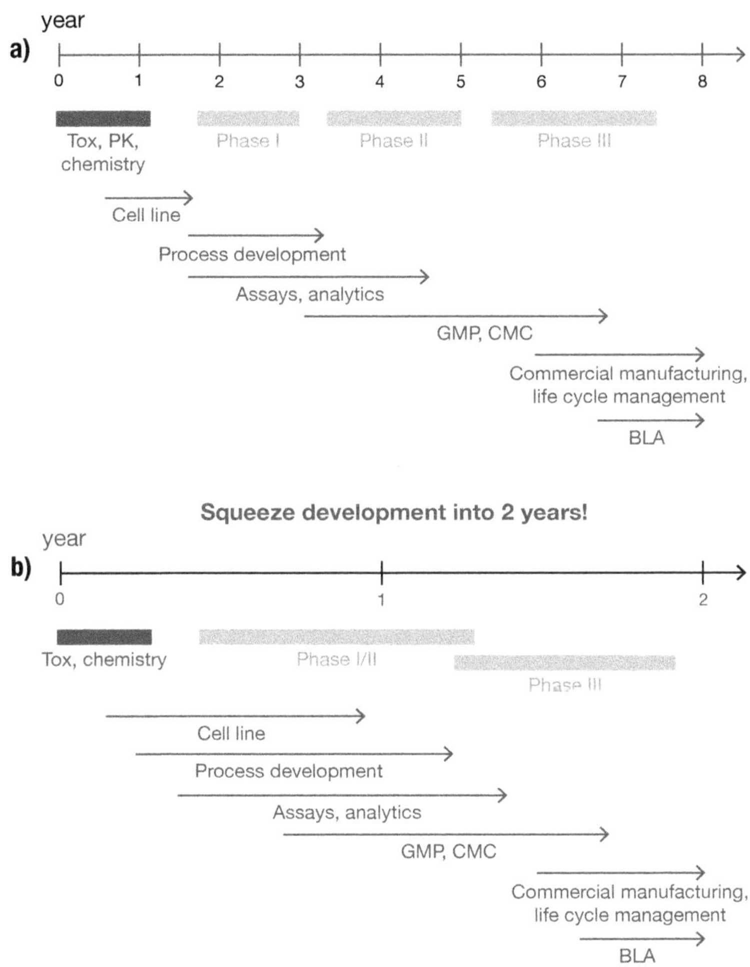

Figure 1.5. (a) Traditional biologics development. (b) Accelerated biologics development. Tox: toxicology; PK: pharmacokinetics; GMP: good manufacturing practice; CMC: chemistry, manufacturing, and controls; BLA: biologics license application.

Nowadays, engineers and scientists work hand-in-hand in the same segment of industry. This new cross-communication is far more than cross-disciplinary; it is cross-communication across segments of product development. Both upstream and downstream process technologists will need to understand the jargon of the regulatory specialists and that of the pharmacists developing bioassays, and vice versa. In the coming years, a new breed of cross-fertilized scientists and engineers will emerge to lead the advances in our field.

References

1. Harrison, R.G., *Observation on the living developing nerve fiber.* Proc. Soc. Exp. Biol. Med., 1907. **4**: p. 140–43.
2. Earle, W.R., et al., *Production of malignancy in vitro IV. the mouse fibroblast cultures and changes seen in the living cells.* J. Nat. Can. Inst., 1943. **4**: p. 165–212.
3. Scherer, W.F., J.T. Syverton, and G.O. Gey, *Studies on the propagation in vitro of poliomyelitis viruses: IV. Viral multiplication in a stable strain of human malignant epithelial cells (strain HeLa) derived from an epidermoid carcinoma of the cervix.* The Journal of Experimental Medicine, 1953. **97**: p. 695.
4. Smith, A.U. and C. Polge, *Survival of spermatozoa at low temperatures.* Nature, 1950. **166**: p. 668–69.
5. Gey, G.O., W.D. Coffman, and M.T. Kubicek, *Tissue culture studies of the proliferative capacity of cervical carcinoma and normal epithelium.* Cancer Research, 1952. **12**: p. 264–65.
6. Eagle, H., *Nutrition needs of mammalian cells in tissue culture.* Science, 1955. **122**: p. 501–04.
7. Macpherson, I. and M. Stoker, *Polyoma transformation of hamster cell clones—an investigation of genetic factors affecting cell competence.* Virology, 1962. **16**: p. 147–51.
8. Yasumura, Y. and M. Kawakita, *The research for the SV40 by means of tissue culture technique.* Nippon Rinsho, 1963. **21**(6): p. 1201–19.
9. Puck, T.T., *The genetics of somatic mammalian cells.* Adv. Biology. Med. Physics, 1957. **5**: p. 75–101.
10. Todaro, G.J. and H. Green, *Quantitative studies of the growth of mouse embryo cells in culture and their development into established lines.* J. Cell Biol., 1963. **17**: p. 299–313.
11. Hayflick, L. and P.S. Moorhead, *The serial cultivation of human diploid cell strains.* Experimental Cell Research, 1961. **25**: p. 585–621.
12. Jacobs, J.P., C.M. Jones, and J.P. Baille, *Characteristics of a human diploid cell designated MRC-5.* Nature, 1970. **227**: p. 168–170.
13. Vilcek, J. and E.A. Havell, *Stabilization of interferon messenger RNA activity by treatment of cells with metabolic inhibitors and lowering of the incubation temperature.* Proc. Natl. Acad. Sci. USA, 1973. **70**: p. 3909–13.
14. Aden, D.P., et al., *Controlled synthesis of HBsAg in a differentiated human liver carcinoma-derived cell line.* Nature, 1979. **282**: p. 615–16.
15. Schneider, U., H.-U. Schwenk, and G. Bornkamm, *Characterization of EBV-genome negative "null" and "T" cell lines derived from children with acute lymphoblastic leukemia and leukemic transformed non-Hodgkin lymphoma.* International Journal of Cancer, 1977. **19**: p. 621–26.
16. Greene, L.A. and A.S. Tischler, *Establishment of a noradrenergic clonal line of rat adrenal pheochromocytoma cells which respond to nerve growth factor.* Proceedings of the National Academy of Sciences, 1976. **73**: p. 2424–28.
17. Mitalipov, S. and D. Wolf, *Totipotency, pluripotency and nuclear reprogramming.* Adv. Biochem. Eng. Biotechnol., 2009. **114**: p. 185–99.

18. Caplan, A., *Mesenchymal stem cells.* Journal of Orthopaedic Research: Official Publication of the Orthopaedic Research Society, 1991. **9**: p. 641–50.

19. Hass, R., et al., *Different populations and sources of human mesenchymal stem cells (MSC): A comparison of adult and neonatal tissue-derived MSC.* Cell Communication and Signaling : CCS, 2011. **9**: p. 12.

20. Evans, M.J. and M.H. Kaufman, *Establishment in culture of pluripotential cells from mouse embryo.* Nature, 1981. **292**: p. 154–56.

21. Shamblott, M.J., et al., *Derivation of pluripotent stem cells from cultured human primordial germ cells.* Proceedings of the National Academy of Sciences of the United States of America, 1998. **95**: p. 13726–31.

22. Thomson, J.A., et al., *Embryonic stem cell lines derived from human blastocysts.* Advancement of Science, 1998. **282**: p. 1145–47.

23. Takahashi, K. and S. Yamanaka, *Induction of pluripotent stem cells from mouse embryonic and adult fibroblast cultures by defined factors.* Cell, 2006. **126**: p. 663–76.

24. Yu, J., et al., *Induced pluripotent stem cell lines derived from human somatic cells.* Science, 2007. **318**(5858): p. 1917–20.

25. Takahashi, K., et al., *Induction of pluripotent stem cells from adult human fibroblasts by defined factors.* Cell, 2007. **131**(5): p. 861–72.

26. Green, H. and M. Meuth, *An established pre-adipose cell line and its differentiation in culture.* Cell, 1974. **3**(2): p. 127–33.

27. Ieda, M., et al., *Direct reprogramming of fibroblasts into functional cardiomyocytes by defined factors.* Cell, 2010. **142**(3): p. 375–86.

28. Zhu, S., et al., *Mouse liver repopulation with hepatocytes generated from human fibroblasts.* Nature, 2014. **508**: p. 93–97.

29. Zhou, Q., et al., *In vivo reprogramming of adult pancreatic exocrine cells to beta-cells.* Nature, 2008. **455**: p. 627–32.

30. Masip, M., et al., *Reprogramming with defined factors: From induced pluripotency to induced transdifferentiation.* Molecular Human Reproduction, 2010. **16**: p. 856–68.

31. Song, K., et al., *Heart repair by reprogramming non-myocytes with cardiac transcription factors.* Nature, 2012. **485**: p. 599–604.

32. Margariti, A., et al., *Direct reprogramming of fibroblasts into endothelial cells capable of angiogenesis and reendothelialization in tissue-engineered vessels.* Proceedings of the National Academy of Sciences of the United States of America, 2012. **109**: p. 13793–98.

33. Kalani, M.Y.S. and N. Martirosyan, *Direct conversion of fibroblasts to functional neurons.* World Neurosurgery, 2012. **77**: p. 7–8.

34. Sutmoller, P. and Bartelling S.J., *The history of foot and mouth disease vaccine development: A personal perspective*, in *Foot and mouth disease: Control strategies.* Eds. B. Dodet and M. Vicari, Elsevier SAS, 2003.

35. Baicus, A., *History of polio vaccination.* World J. Virol., 2012. **1**(4): p. 108–14.

36. Mowat, G.N. and Chapman, G., *Growth of foot and mouth disease virus in a fibroblastic cell line derived from hamster kidneys.* Nature, 1962. **94**: p. 253–55.

37. Klein, R. and M. Teodorescu, *Propagation of poliovirus of Cercopithecus monkey kidney cells in rolling bottles.* Archives Roumaines de Pathologie Experimentales et de Microbiologie, 1969. **28**: p. 247–52.

38. Leen, A.M., C.M. Rooney, and A.E. Foster, *Improving T cell therapy for cancer.* Annu. Rev. Immunol., 2007. **25**: p. 243–65.

39. Montagnon, B.J. and J.C. Vincent-Falquet, *Experience with the Vero cell line,* in *Safety of biological products prepared from mammalian cell culture.* F. Brown, et al., Editors. 1998, Karger: Basel. p. 119–23.

40. Russell, W.C., et al., *Characteristics of a human cell line transformed by DNA from human adenovirus type 5.* Journal of General Virology, 1977. **36**: p. 59–72.

41. van Oers, M.M., G.P. Pijlman, and J.M. Vlak, *Thirty years of baculovirus-insect cell protein expression: From dark horse to mainstream technology.* Journal of General Virology, 2015. **96**(Pt 1): p. 6–23.

42. Wang, G., M. Huang, and J. Nielsen, *Exploring the potential of Saccharomyces cerevisiae for biopharmaceutical protein production.* Current Opinion in Biotechnology. **48**: p. 77–84.

43. Bertolini, L.R., et al., *The transgenic animal platform for biopharmaceutical production.* Transgenic Res., 2016. **25**(3): p. 329–43.

Contents

2
Cell Biology for Bioprocessing

Introduction

THERE ARE MANY DIFFERENT TYPES of cells that constitute the tissues and form the various organs in a mammal. They all arise from the same fertilized egg and then differentiate into cells with vastly different properties, using the same set of genetic information but arriving at very different destinies. Over a century ago, scientists began to isolate cells from explanted tissues. In Chapter 1 we briefly discussed the enabling advances that made cell culture a research tool and production host, including the development of media and its membrane sterilization and cell trypsinization and passaging. Most cells that outgrew from tissues could only be cultured for a very small number of generations, but eventually some cells were successfully established that could be passaged over many generations. After cryopreservation methods were established, cells could be "banked" for future use, allowing for the generation of the reproducible results that are essential to research and manufacturing.

Most cells established in the early decades of cell culture were derived from tumors. Those cells appeared morphologically different from primary cells that had grown out of normal tissues. Even those that were isolated originally from normal tissues and could be continuously passaged appeared more like cells that had originated from tumors. With increasing knowledge of growth factors and extracellular matrices, normal and differentiated cells began to be cultured in the 1970s. The isolation of adult stem cells and then mouse embryonic pluripotent stem cells, concurrently with the wide availability of many more growth factors and cytokines, began to change the landscape of cell culture for research. The drive to produce therapeutic proteins in large quantities in the 1990s changed our view of cells. They are now the primary workhorse for biomanufacturing. This chapter will discuss the key aspects of a cell's structural components that affect growth and the cell's capability to produce the protein product.

Tissue Cells and Their Isolation

Cell Lines, Cell Strains, and Senescence

THE VAST MAJORITY OF CELLS isolated from tissues of mammals are anchorage-dependent, meaning that they require surface adhesion in order to multiply. They are typically isolated from tissues by an enzymatic dissociation of the tissue (Figure 2.1). After dissociation and the removal of undissociated chunks, cells are plated on a compatible surface overlaid with media. The cell clumps in media suspension attach to the surface and gradually some cells grow out from the tissue clumps. Subsequently, cells extend their body length, spread, and begin to multiply. As they begin to cover the entire surface area the growth rate slows down, eventually forming a "monolayer" of cells over the surface. Upon reaching confluence, cell division stops. While the cell bodies of neighboring cells may cross each other, their nuclei never overlap. This is called contact inhibition of cell growth. The contact inhibited cell can be dissociated from the surface by treatment with trypsin. After being plated on a larger surface and provided with fresh medium, cell growth resumes until confluence is again reached. This process can be repeated to expand the population. Each round of detachment and expansion is called a "passage." The number of cell doublings taking place in each passage is determined by the split ratio, or the ratio of surface area expanded in each passage (see reference 1 for further reading).

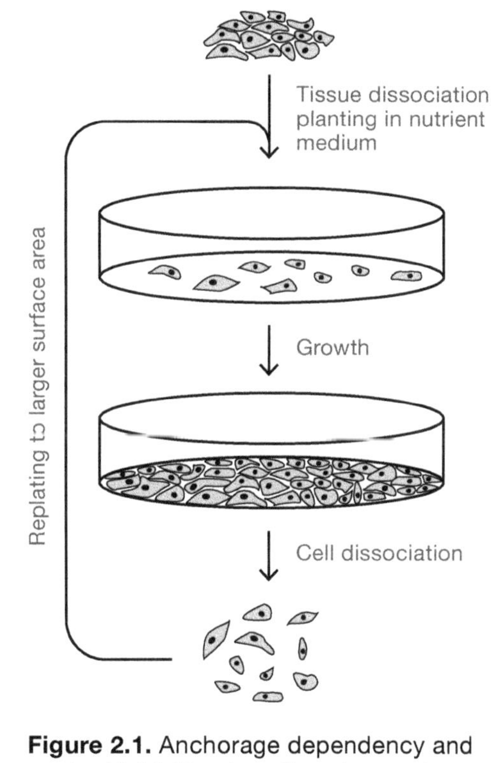

Figure 2.1. Anchorage dependency and contact inhibition in cultured normal diploid cells.

Cells isolated directly from tissues are called primary cells (Panel 2.1). Most primary cells stop growing after a very small number of cell doublings. This is especially true for cells isolated from specialized tissues, such as from the liver and ovaries. They often quickly lose their typical tissue-associated characteristics in culture.

Panel 2.1. Cell Types by Their Proliferative Potential *in Vitro*

Primary Cells

- Isolated directly from tissue with very limited capacity of cell expansion *in vitro*

Cell Strain

- Cells outgrown from isolated tissues that can be passaged and still maintain their normal behavior and diploid chromosome makeup
- Eventually they reach senescence

Continuous Cell Line

- Cells isolated from cancer or normal tissue that can be serially expanded without reaching senescence
- Not diploid and often lose their normal morphology

Multipotent Stem Cells

- Stem cells that can differentiate into different types of the same lineage

Pluripotent Stem Cells

- Stem cells that can differentiate into all different cell types as an adult
- Either embryonic stem cells or induced pluripotent stem cells (iPSCs)

Panel 2.2. Quiescent Cells and Growth Potential

- In mammals, most cells are quiescent or at a growth-arrested state. Only some cell types in some tissues, such as epithelial tissue in the intestines, divide actively.
- Cells that can be isolated and cultured *in vitro* have undergone changes to overcome growth arrest. They have somewhat different physiological states than their quiescent counterparts even if they appear normal.
- Cancer cells proliferate *in vivo* and are more amenable to culture growth

Most cells in tissues *in vivo* are quiescent, meaning that they are not dividing (Panel 2.2). The quiescent state is not merely the passive result of a lack of conditions necessary for growth, such as a missing essential nutrient. Rather, the quiescent state is imposed on the cell by the regulation of the organism. For example, stem cells in our body are in a quiescent state most of the time, as a result of cell cycle regulation through cyclin-dependent kinase inhibitors, transcription factors, and tumor suppressors. Cells that grow out from tissue explant have undergone changes in their cell growth control mechanism that enable them to proliferate in culture. In some cases, the cells outgrown from isolated tissue can be expanded serially over many passages and still retain normal morphology and cell behavior. They can even be cryopreserved in liquid nitrogen and be "banked" for future research, distribution, or manufacturing purposes. However, such repeated passaging and cell expansion in culture cannot be continued indefinitely. Eventually, senescence is reached. Before senescence, these cells are morphologically normal. Their karyotype is diploid, i.e., having two sets of normal chromosomes. Virtually all normal diploid cells derived from mammal tissues have a limited life span in culture, with the exception of embryonic stem cells and induced pluripotent stem cells.

In contrast to tissue cells, cancer cells proliferate *in vivo*. They typically carry a variety of mutations that allow them to bypass the growth control mechanism that would normally keep them in a quiescent state. Therefore, they are more readily isolated from tumor tissue for culture *in vitro*. Sometimes, lines of cells can also be established from normal tissues by "immortalization" through viral or oncogene transformation. Those cells that can be grown continuously without senescence typically do not have the same morphology as their normal counterparts. Their karyotype is not diploid and their growth is not contact inhibited. With adequate supplies of nutrients and growth factors, they can even overlay each other to form multiple layers of cells. Such cells include commonly used industrial cell lines like CHO cells from Chinese hamster ovary, Vero cells from green monkey kidney, and human kidney cell HEK293.

Without using special culture conditions to enrich differentiated cells, fibroblast cells are the predominant cell type isolated from normal mammalian tissues. Fibroblasts isolated from a mouse embryo can be cultured *in vitro* for about 60 doublings (Figure 2.2). As that limit approaches, the cells begin to fail to reach confluence and eventually cease to grow. This senescence, or a limit in the proliferating potential of these cells, is called "Hayflick's phenomenon." It is a common phenomenon for all normal diploid cells (again, with embryonic stem cells and induced pluripotent stem cells as exceptions) obtained from vertebrates.

In a historic experiment carried out half a century ago, the continued passaging of mouse fibroblast cells beyond crisis gave rise to a small fraction of survivors. These cells eventually grew, expanded, and could be cultured continuously *in vitro* without a limited life span. They were given the name 3T3 because the cells were passaged every three days by expanding at 3 x 10^5 cells per 20 cm^2 dish. These cells appear normal and are subjected to contact inhibition of growth under typical culture conditions.

However, unlike the mouse fibroblasts of the pre-crisis period, 3T3 cells have an abnormal, non-diploid set of chromosomes. Cells that succumb to Hayflick's constraint (e.g., those that are diploid and have a

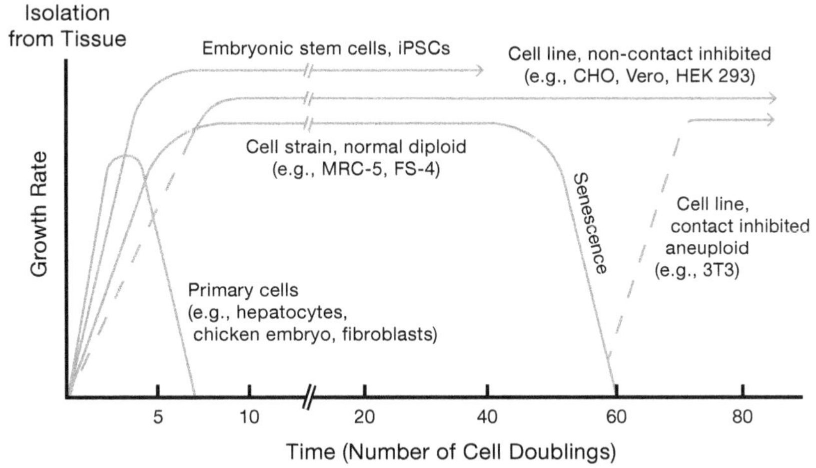

Figure 2.2. Life spans and growth rates of different cell types.

limited life span) are called "cell strains." The cells that can grow in culture indefinitely are referred to as "cell lines," including those reestablished after passing the crisis and derived directly from normal or tumor tissues. In general, cell lines are aneuploid (do not have a normal set of chromosomes).

Cells that are subject to senescence thus appear to "count" their number of doublings. Senescence is regulated by cellular events and is often thought to be related to aging. It is essentially a growth arrest, but one that differs from the quiescent state. Cells in a quiescent state can revert to their growth phase, but cells entering senescence undergo an irreversible growth arrest. Senescence may also occur in response to stress or overexpression of oncogene, or be induced by telomere shortening. Telomeres are special repetitive sequences at the end of chromosomes. They are not replicated by DNA polymerase during DNA replication but are synthesized by telomerase. The DNA polymerase reaction does not accurately reproduce the number of tandem repeats of the telomeres. There is thus much variation in telomere length among cells. As the number of passages increases, telomeres become shorter unless they are repaired by telomerase. For instance, in embryonic stem cells, telomerase activity is high in order to maintain the telomere length. Unlike cell strains, embryonic stem cells do not exhibit senescence.

Stem Cells and Differentiated Cells

Cells isolated from normal or cancerous differentiated tissues often retain their differentiated properties to different extents depending on the chemical and physical environment. Many differentiated cells isolated from cancerous tissues, including HepG2 cells (from hepatocellular carcinoma), Jurkat cells (human T lymphocyte from leukemia), and PC12 (neuronal cells isolated from pheochromocytoma of the rat adrenal medulla, can be induced to form dendrites in culture) have been valuable in biomedical research for decades. They are all continuous cell lines. Although they retain some differentiated phenotypes of the tissue that they were derived from, many important tissue characteristics are lost. The advances in the research tools for cell characterization and growth factors in the past three decades have allowed for the cultivation of various differentiated cells from normal tissues, including endothelial cells, keratinocytes, melanocytes, mononuclear cells, and smooth muscle cells. These primary tissue cells retain many tissue specific activities and are valuable not only in research but also potentially for tissue repair or regeneration. However, as will be discussed later in this chapter, most differentiated cells *in vivo* are at a G_0 phase of cell culture and are not proliferating. Cells isolated from those tissues have very limited proliferative potential and show phenotypic instability in culture. For possible regenerative medicine applications, differentiated cells isolated from tissues are not sustainable cell sources.

Stem cells must be capable of self-renewal (i.e., the ability to make more of themselves) and differentiation into their target cells. They exist in many adult tissues and are quiescent in their niche, expanding and differentiating only when needed. Some adult stem cells have been isolated and grown in culture for decades, including hematopoietic stem cells (HSCs) and mesenchymal stem cells (MSCs) isolated from bone marrow. They have limited proliferative potential, and after some generations in culture they lose their renewal and differentiation capacity. These adult stem cells are multipotent, meaning that they can differentiate into multiple, but a limited number of, cell lineages. Even with a less-than-unlimited proliferative potential, adult stem cells have the capacity to give rise to a large number of differentiated cells (such as bone and muscle cells from MSCs) for allogeneic applications.

In contrast to adult stem cells, embryonic stem cells isolated in a very early stage of embryo development have a virtually unlimited self-renewal capacity in culture and are pluripotent, meaning that they can differentiate into cell types of all adult tissues. They do not undergo senescence and have tremendous potential in regenerative applications. However, as they are derived from fertilized human eggs, their use is ethically controversial. To alleviate these ethical concerns, a process was devised to reprogram adult somatic cells into an undifferentiated embryonic state. These induced pluripotent stem cells (iPSCs) can be obtained by epigenetic reprogramming of adult tissue cells like fibroblasts through the introduction of four exogenous genes (OCT4, SOX2, KLF4, and c-Myc (OSKM)). The iPSCs have a similar potential to differentiate into cells of different lineages as ESCs. Both ESCs and iPSCs can be directed to differentiate into various types of cells, including hematopoietic stem cells, hepatocytes (liver cells), and pancreatic cells by using series of cocktails of growth factors to mimic *in vivo* differentiation conditions.

Cells for Biologics Manufacturing

The cells commonly used for the production of biologics are derived from different tissues of different species. Cells isolated from different species differ in their chromosome number and genome sequence. However, at a physiological and transcriptome level, cells from the same tissue of different species are strikingly similar. Their similarity is much greater than with different cell types from the same species. For example, chicken embryo fibroblasts look morphologically very similar to human fibroblasts from the lung or foreskin, while the epithelial MDCK cells look rather different from dog fibroblasts even though they are both derived from the same species.

Table 2.1. Cells Commonly Used in Bioprocessing

Cell name	*Species*	*Cell type*	*Tissue isolated*
WI-38	Human	Fibroblast	Lung
MRC-5	Human	Fibroblast	Lung
FS-4	Human	Fibroblast	Foreskin
HEK293	Human	Epithelial	Kidney
Vero	Monkey	Epithelial	Kidney
MDCK	Dog	Epithelial	Kidney
NS/SP2/0	Mouse	Lymphoid	Myeloma
CHO	Chinese hamster	Epithelial	Ovary
BHK	Syrian hamster	Fibroblast	Kidney

Panel 2.3. Cell Types of Different Tissues

- Among the ~200 different types of cells, fibroblasts, epithelial cells, and myeloma cells are the most frequently used cell types in biologics production
- Cells in culture bear closer physiological and morphological characteristics to the tissue they were derived from than to their species

Even though there are about two hundred types of cells in a vertebrate animal, most cells that are used for the production of biologics are either epithelial or fibroblastic in nature (Table 2.1, Panel 2.3). These two cell types are generally amenable to isolation from tissues and to *in vitro* culture, as demonstrated during the early explorations on tissue cell isolation more than half a century ago. NS0 and CHO are the two prominent host cell lines used for therapeutic recombinant protein production. They exhibit different behaviors and were derived from two different tissues and two different species. CHO cells were isolated from the ovary of a Chinese hamster, NS0 cells from a mouse myeloma. Cells used for recombinant protein production are primarily epithelial and lymphatic. Both fibroblasts and epithelial cells are also frequently used for viral vaccine production. Cell line selection for recombinant protein production versus vaccine production are based on fundamentally different needs. The former demands easy genetic manipulation and amenability to high protein secretion, while the latter requires cells with the matching tropism of the virus and a minimal antiviral response.

Epithelial cells line the "boundary" of tissues, while fibroblasts make up a larger part of the connective tissue. Epithelial cells form tightly connected sheets, which often become damaged, die, and are replenished by "new" ones. Fibroblasts are mostly quiescent. They migrate into wounds and begin to grow only when they are stimulated by various cues. The terminally differentiated plasma cells from which myeloma cells are derived secrete antibodies against a particular antigen. The antibodies are needed only for a limited period of time after the host's exposure to the antigen. These cells undergo apoptosis two to three weeks after their differentiation into active antibody-secreting cells so that the host does not continue to have a large quantity of unneeded antibody molecules in circulation. Many native characteristics of the original-tissue cell are often still evident in tissue-derived cell lines in culture.

Cell Membrane

AN ANTIQUATED BUT PREVIOUSLY common-held belief regarding the growing of mammalian cells in a bioreactor was that cells were extreme fragile and vulnerable to mechanical stress because the only thing preventing cellular content from dissolving into the aqueous environment was the lipid bilayer membrane. Yet, in a modern manufacturing plant, cells thrive in bioreactors of tens of cubic meters in volume under highly turbulent conditions. The cytoplasmic membrane surrounding a cell is much more than a mere double layer of lipids and the integrity of a cell is not only dictated by its membrane wrapping.

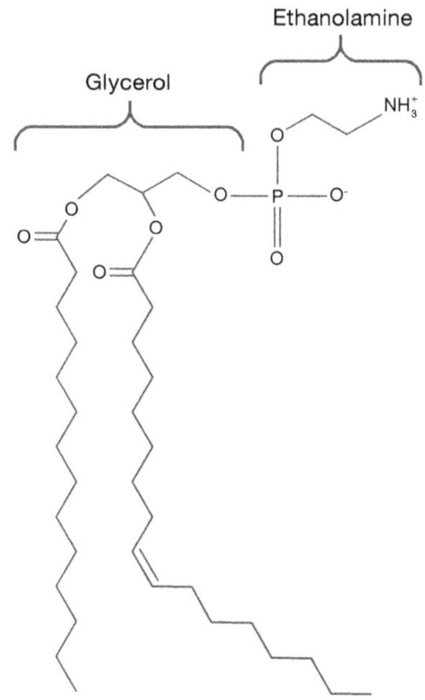

Figure 2.3. A phospholipid molecule with glycerol conjugated with an ethanolamine, a saturated fatty acid, and an unsaturated fatty acid.

Lipid Bilayer

The lipids that make up the lipid bilayer are amphipathic phospholipids. Each phospholipid molecule consists of a glycerol backbone linking a hydrophilic head group that includes a charged phosphate group and a hydrophobic tail group consisting of two fatty acids (Figure 2.3). The fatty acids at the hydrophobic tail provide the hydrophobic interactions necessary to form the ordered structure of a membrane at a mild temperature. When suspended in an aqueous solution, amphipathic molecules can form micelles. In such micelles, the hydrophilic heads face outward and the hydrophobic tails project inward. The organization of a micelle makes it easy to enclose hydrophobic molecules inside while not having an aqueous environment both inside and outside. In contrast, a sphere formed by a lipid bilayer membrane has the hydrophobic tails projecting toward the middle of the bilayer and the hydrophilic heads on the external as well as the internal surfaces. It can readily have an aqueous environment both inside and outside (Figure 2.4).

A lipid bilayer membrane behaves like a fluid (Panel 2.4). If the lipid molecules in a specific location were to be labeled with a fluorescent

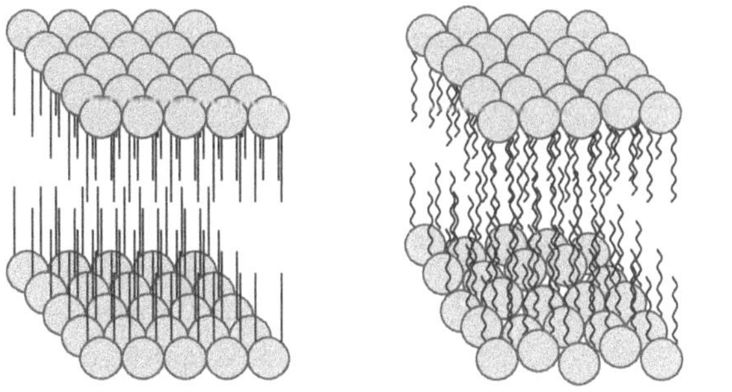

Figure 2.4. Lipid bilayer membrane at a gel state and a liquid crystal state.

Panel 2.4. Properties of a Lipid Bilayer

- Fluid
- As temperature decreases, the bilayer transitions from a fluid state to a gel state
- The extent of unsaturation and length of fatty acids affects the transition temperature of the membrane
- The magnitude of diffusion of various solutes in the cell membrane resembles that of a liquid

Panel 2.5. Lipid Bilayer Composition

Phospholipids
- Constitute the majority (35–70%)

Glycolipids
- Neutral glycolipids (e.g., galactocerebroside)

Gangliosides
- Have sialic acids

Types of Phospholipids
- Glycerol as backbone
  - Phosphatidyl ethanol amine
  - Phosphatidyl serine
  - Phosphatidyl choline
- Serine as backbone
  - Ceramide
  - Sphingomyelin

dye, the fluorescence would disperse quickly due to molecular diffusion. The lateral diffusion coefficient of a phospholipid molecule in a bilayer membrane is about 10^{-8} cm^2/s. Gas species diffuse about equally fast in a lipid bilayer as in water. Even large protein molecules diffuse in a lipid bilayer membrane.

Although a lipid bilayer membrane is a fluid, by forming a tightly packed, ordered structure it becomes a very good barrier that keeps most molecules from freely passing into or out of the cell. A lipid molecule does not flip-flop (i.e., change its side of the lipid bilayer) without the aid of a membrane-bound phospholipid translocator. The permeability of most biological molecules across a lipid bilayer membrane is rather low. Even the smallest nutrients, like glucose and simple amino acids, cannot pass through frequently enough to support cell growth.

All major biological macromolecules (e.g., DNA, proteins, polysaccharides) are biopolymers made of covalently-bonded monomers. A lipid bilayer membrane is not a polymer, but rather an assembly of lipid monomers. The noncovalent nature of the cell membrane allows it to expand, shrink, break, change shape, and fuse rapidly. As cells grow, the cell membrane can expand rapidly, dynamically changing its composition and shape to meet cellular needs without breaking down many covalent bonds as in polymeric biomolecules.

The lipid bilayer membrane makes up not only the cytoplasmic membrane enwrapping the cell. It also envelops various organelles to compartmentalize regions in the cell for specialized functions. Major organelles in a cell include the nucleus, mitochondrion, endoplasmic reticulum (ER), Golgi apparatus, and lysosome. Many of these organelles are in a constant dynamic process of membrane budding and fusion. For example, in trafficking between organelles and in protein secretion, the "cargo" is carried inside membrane vesicles while transiting from one organelle to another.

Three types of lipids make up the lipid bilayer membrane in cells and organelles: phospholipids, glycolipids, and gangliosides (phospholipids being the most common) (Panel 2.5). There are several different types of phospholipids. The most abundant type is glycerophospholipid, which uses glycerol as the backbone. Its 3-hydroxyal group of glycerol is linked to a phosphate group, to which an ethanolamine or serine is attached. The other two hydroxyl groups of glycerol are linked to two fatty acids through an ester bond. Typically, one of those two fatty acids is saturated and the other is unsaturated, with a cis double bond in between C14 and C24 (Panel 2.6). The degree of unsaturation affects the packing of the lipid bilayer. Saturated fatty acids allow more dense packing, while the double bonds in the unsaturated fatty acids create kinks, which reduce packing and increase the membrane fluidity.

Like other lipids, the lipid bilayer membrane undergoes a phase transition from a liquid (or liquid crystal) phase to a gel phase at a relatively moderate temperature (Figure 2.4). A lipid bilayer can be in a gel state or in a liquid crystalline state depending on the temperature and degree of hydration. A lipid bilayer's phase transition temperature is affected by its composition of fatty acid and cholesterol. As temperature decreases, the lipid bilayer changes from a liquid-crystalline state to crystalline (or gel) state. A higher content of shorter, unsaturated fatty

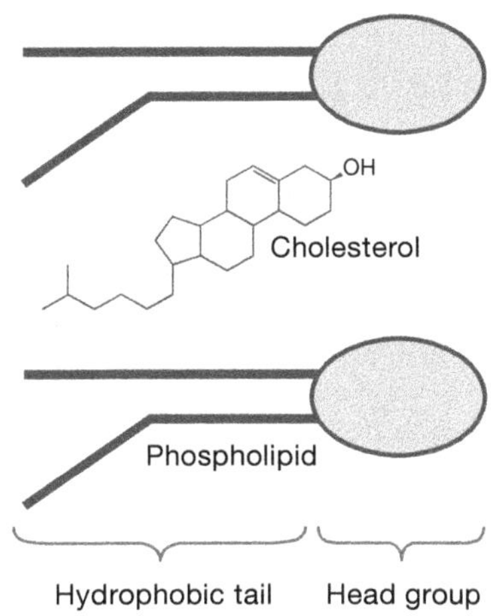

Figure 2.5. Schematic diagram of a cholesterol molecule interacting with two phospholipid molecules in one leaflet of a lipid bilayer membrane.

Panel 2.6. Lipid Composition and Membrane Fluidity

- One saturated, one cis-unsaturated (C14-C24) typically constitute the tail of the phospholipid
- Fatty acids (the tail group) on the lipid affect the packing of lipids in the bilayer membrane. Saturated fatty acids allow more dense packing; double bonds in unsaturated fatty acids create kinks, reduce packing, and increase fluidity.
- Cholesterol has a small head polar group linked to a rigid planar region of steroid rings followed by a more flexible non-polar tail. It interacts with phospholipids to stabilize the region closer to the head group as well as to make the lipid bilayer less inclined to become crystalline. Overall, it increases the membrane permeability to small compounds and makes the membrane less fluid.

Table 2.2. Biochemical Composition of a Hepatocyte Plasma Membrane

Total lipids	*Total proteins*	*Protein / lipid mass ratio*	*Cholesterol / phospholipid molar ratio*	*Cholesterol in total lipids*	*Phospholipids in total lipids*
30–40% (by mass)	50–60% (by mass)	1:2	0.4–0.8	12–20%	50–70%

Adapted from *The Liver: Biology & Pathology*, 4th Ed., p. 78 (2001).

acids increases the fluidity of the lipid bilayer and decreases its phase transition temperature.

Another molecule that plays a key role in the membrane properties of animal cells is cholesterol (Figure 2.5). Cholesterol has a small polar hydroxyl head group linked to a rigid planar region of steroid rings that are further linked to a more flexible non-polar tail. Cholesterol interacts with phospholipids to stabilize the region closer to the head group and to make the lipid bilayer less inclined to become crystalline. Overall, cholesterol increases the membrane permeability to small compounds and makes the membrane less fluid. Cholesterol content varies in different lipid bilayer membranes. Its level in the cytoplasmic membrane is high, but it is very low or absent in the organelle membranes.

Membrane Proteins

A typical biological membrane has ~50% lipids and ~50% proteins, by mass (Table 2.2, Panel 2.7). In terms of the number of molecules, the lipid-protein ratio is about 50:1, since proteins have much higher molecular weight than lipids. The protein content of a membrane is greatly affected by the tissue and organelle of origin. The mitochondrial membrane, through which many molecules (e.g., amino acids, pyruvate, various ions, and many other proteins) pass at a high flux, has a high protein content of about 75% by mass. On the other hand, the myelin membrane, which serves as a protective sheath between the nerve cell and its surroundings, has a low protein content of about 25%.

Panel 2.7. Membrane Proteins

- A typical biological membrane has ~50% proteins by mass; in terms of molecules, lipid : protein = 50 : 1
- A metabolically active mitochondrion's membrane is 75% protein
- Na^+/K^+ ATPase acts as a pump, using ATP to pump 3 Na^+ out of and 2 K^+ into the cell
- The electric potential across the plasma membrane is about –80mV.

Lipid bilayer membranes separate cellular content from their surroundings and divide the organelles from the cytosol. Not only do they act as a barrier for the physical retention of a cell's contents, but they can also create different chemical environments on either side of the membrane. For example, cells maintain about an 80 mV electric potential across the plasma membrane and about 140 mV across the mitochondrial membrane. The ER membrane separates an oxidative environment (inside the ER) from a reduced one (in the cytosol).

The maintenance of various chemical, electrical, and redox potentials across a membrane is accomplished by various membrane proteins. Rat small intestinal enterocyte has about 150,000 Na^+ pumps per cell, which collectively allow each cell to transport about 4.5 billion Na^+ ions out of the cell each minute. The sodium and potassium membrane gradients generated by those pumps, as well as the electric potential across cytoplasmic and mitochondrial membranes, are fundamental to cellular bioenergetics.

Membrane Dynamics

The cellular membrane is in a dynamic state; membrane constituents are continuously being added and removed. This is not only for membrane expansion and cell growth, but also for turnover and vesicle trafficking. Like other cellular components, the turnover of the cellular membrane is necessary to replace lipid molecules that have been oxidized or damaged, or to allow cells to change their membrane composition to adapt to new environments. Cellular membrane proteins are also turned over (Panel 2.8).

Inter-organelle trafficking and the secretion of proteins into the extracellular environment also contribute to a membrane's dynamic state. Protein molecules that are destined for secretion or for the cell surface travel between organelles (ER and Golgi apparatus) and to the cytoplasmic membrane by residing inside membrane vesicles. Upon reaching the inner surface of the cytoplasmic membrane, those vesicles fuse with the cytoplasmic membrane and release their contents to the outside of the cell.

In the liver, each hepatocyte synthesizes ~120 x 10^3 albumin molecules per minute (translating to about 15 pg/cell/day). All of those molecules are wrapped in membrane vesicles of 280–400 nm and delivered to the basal plasma membrane of the cell. The infusion of those membrane vesicles would cause the membrane surface to expand at a rate of 0.5%/min (Panel 2.9). However, since hepatocytes are typically in a G_0 state (i.e., not dividing), the size of their cytoplasmic membrane does not need to increase to accommodate cell growth. Even if they are proliferating, the rate of membrane expansion caused by membrane vesicle fusion would be far too high for cell growth. Therefore, the lipid molecules that are added to the cytoplasmic membrane must be recycled back into the intracellular organelle (Golgi bodies) to maintain the cytoplasmic membrane in a homeostatic state.

Similarly, cells active in endocytosis can internalize up to 0.8%/min of a plasma membrane. The loss of lipids from the membrane caused by endocytosis must be replenished to maintain the size of the cell's outer envelope. Thus, cellular membranes, both cytoplasmic and organelle-related, are dynamic and maintained in a homeostatic state.

Panel 2.8. Factors Contributing to the Dynamic State of the Cellular Membrane

- Cell growth and membrane expansion
- Lipid turnover
- Inter-organelle trafficking of membrane vesicles
- Secretion, endocytosis

Panel 2.9. Homeostasis of Cellular Membranes

- Professional secretory cells in the body can add 0.5% of their plasma membrane per minute due to the fusion of secretory vesicles with the plasma membrane; they must be recycled to maintain a balance
- Phospholipids in the membrane are subject to turnover

Cytoplasm and Organelles

The cytoplasm and nucleus are both enclosed by the cytoplasmic membrane in the cell. The cytoplasm can be largely divided into two groups: organelles and the highly viscous cytosol. Cytosol has a very high concentration of proteins (100–300 mg/mL). For comparison, the protein content in blood plasma is only 90 mg/mL. Cytosol also contains the inorganic solutes, building blocks, and intermediates and metabolites of metabolic reactions.

The cytosol is not only full of soluble components (Panel 2.10). It also contains large assemblies (or aggregates) of particles. The ribosome is the main machinery for making proteins; it is a complex particle consisting of many ribosomal proteins and ribosomal RNAs (rRNA). Each cell contains thousands of ribosomes ~30 nm in size. Many ribosomes are located on the cytosolic surface of the endoplasmic membrane and appear as black spots when viewed under an electron microscope. Some enzymes also form large complexes that can be seen under an electron microscope, such as pyruvate dehydrogenase complexes.

Panel 2.10. Cytosol

- Not a simple solution
- Some protein complexes (e.g., ribosomes, pyruvate dehydrogenase) are aggregates that are visible as particles under a transmission electron microscope
- The cytoskeletal network is interspersed with cytosol

Also rich in the cytosol are the fiber-like structures of the cytoskeleton. These large protein particles, enzyme complexes, cytoskeletal proteins, and organelles make the cytoplasm of a cell very crowded and render its solution phase very dense in mass. Under light microscopy, an animal cell appears to be primarily cytoplasm wrapped in a membrane, with a nucleus sitting near the center spanning over half of the cell's diameter. Other than the nucleus, various organelles include mitochondria, the endoplasmic reticulum, the Golgi apparatus, peroxisomes, endosomes, etc., which are visible only by electron microscopy (Figure 2.6).

Nucleus

The nucleus is the largest organelle in the cell, and the only one besides the mitochondrion that has a double membrane, with both an outer and inner nuclear membrane instead of only a lipid bilayer membrane. The nucleus contains the genome, which is separated into a number of DNA molecules that each form a chromosome. Being diploid, mammals have two sets of chromosomes in each somatic cell, one of maternal and one of paternal origin. The DNA molecules are segregated into nuclear compartments. The size of a haploid genome of a rodent species of interest to bioprocessing is about 2.8 Gbp (giga (10^9) base pair). That of a human is about 3.3 Gbp. That equates to about 6–7 pg of DNA for a diploid cell. If stretched, the total DNA (from both sets of

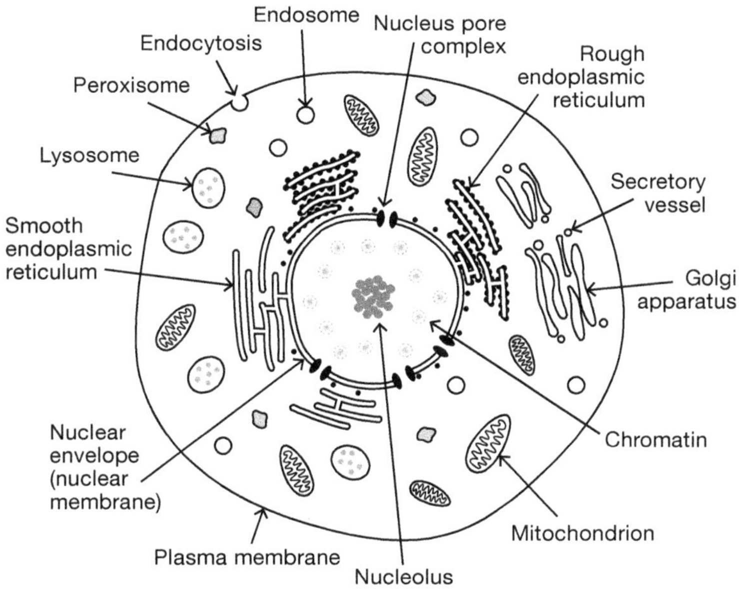

Figure 2.6. Organelles in a mammalian cell.

chromosomes) would extend to over 2 m in length. This large amount of DNA is packed into the small space of the nucleus by forming DNA-protein (histone) complexes.

The cytosol side of the outer membrane of the nucleus is associated with many ribosomes. As will be discussed later, the endoplasmic reticulum is the site of active translation. These ribosomes on the outer membrane continue to present in the cytosol side of the endoplasmic membrane surface. Much trafficking occurs between the nucleoplasm and cytosol through nuclear pore complexes that transverse both the outer and inner nuclear membrane (Panel 2.11). DNA replication, the synthesis of various RNAs (mRNA, tRNA, rRNA, non-coding (nc)RNA, snRNA, etc.), and ribosome assembly occur in the nuclear compartment. They are segregated from the metabolic processes and protein synthesis occurring in the cytosol. The nucleotides and deoxynucleotides that are synthesized in the cytosol are imported into the nucleus for RNA and DNA synthesis. Many of the RNA products, including mRNAs and tRNAs, are exported to the cytosol where translation takes place. Ribosomes are assembled from ribosomal proteins and rRNAs in nucleoli and are

Panel 2.11. Characteristics and Biochemical Roles of the Nucleus of a Human Cell

- The largest organelle in the cell
- Each cell contains ~6–7 pg DNA organized into 46 chromosomes
- When connected and totally stretched, it is ~2 m long
- Histone proteins help pack it into the nucleus
- Major reactions occurring in the nucleus:
  - DNA replication
  - RNA synthesis
  - Ribosome assembly
- Raw materials and products of these reactions are trafficked through nuclear pores

subsequently exported to the cytosol to participate in protein synthesis. Since translation occurs in cytosol, the ribosomal proteins are also synthesized in cytosol before being imported into the nucleus for ribosome assembly. The task of sorting out which segments of DNA, or which genes, are to be transcribed into RNA at a given moment occurs in the nucleus. A large array of transcription factors and other transcription regulators are also synthesized in the cytoplasm and then imported into the nucleus where they bind to specific genetic loci to perform their role in transcription.

Thus, the volume of material exchange between the nucleus and the cytoplasm is large. This large volume of traffic, from small molecules to proteins to complex protein-RNA ribosome assemblies, must be sorted so that different species move in the right direction.

Mitochondria

The mitochondrion is the organelle present in the highest numbers in a cell. With about 1,700 per cell, they make up 20% of the cell's volume (Panel 2.12). Mitochondria are about the size of bacteria and are thought to have originated from bacteria-like structures that were acquired by primitive eukaryotes.

Mitochondria serve as the cell's power plants. The most reactive reactions in the cell (e.g., oxidizing nutrients and generating energy through electron transfer and oxidative phosphorylation) take place in the mitochondria. Cells with different energy needs have different numbers of mitochondria. In a high-energy demanding cell, there can be as many as 3,000 mitochondria.

Like the nucleus, mitochondria have a double membrane (Figure 2.7). The outer membrane maintains their macroscopic morphology, while the inner membrane is highly invaginated, often called cristae, to generate more surface area. The inner membrane is the barrier for solute transport. The outer membrane is rich in pores made of porin proteins that allow molecules smaller than 5000 daltons to pass through freely. The chemical environment of the space enclosed by the inner and outer membranes is thus largely the same as that of cytosol. The space inside the inner membrane, called the matrix, is rich in enzymes and is the place

Panel 2.12. Mitochondria

- The most abundant organelle in a cell (about 1,700 per cell)
- Use up to 20% of the cell's volume
- In the catabolism of glucose to carbon dioxide, the oxygen atom in CO_2 is contributed from water molecules. The oxygen reacts through an electron transport chain to form water with NADH and $FADH_2$ in the mitochondria.
- An active mitochondrion has a -140 mV electric potential across its inner membrane and 0.5–1.0 units of pH gradient (the pH inside the mitochondria is higher [H+ concentration is lower] and pH is pumped against the concentration gradient)
- The membrane potential cannot be increased too much. The homeostasis of mitochondria is therefore critical.
- Cells meet long-term increased energetic needs by the biogenesis of mitochondria

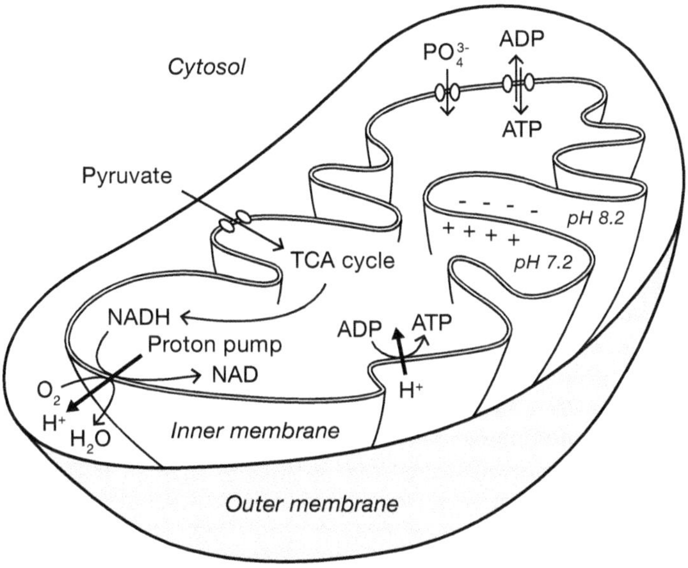

Figure 2.7. The proton and electric potential (charge) gradients, along with the material flow, cross the mitochondrial inner membrane. The direction of the fluxes of major species are indicated by arrows. Note that NADH oxidation coupled with electron transfer pumps protons against the proton and charge gradient, while the movement of protons to drive ATP synthesis is in the direction of the proton and charge gradient.

where the final breakdown of metabolic intermediates derived from glucose metabolism to CO_2 occurs. It is also where the final degradation of fatty acids and amino acids occurs when they are used to derive energy under some physiological conditions.

The chemical potential energy of pyruvate oxidation through the TCA cycle is stored in NADH. At the mitochondrial inner membrane, the electrons at a high energetic state in NADH pass down the electron transfer chain consisting of a number of protein complexes (NADH dehydrogenase, cytochrome C reductase complex, and cytochrome C oxidase complex) and pump H^+ out of the inner membrane. At the end of the electron transport, the electron reacts with oxygen to form H_2O. The ATP-generating process occurs via ATP synthase residing in the inner mitochondrial membrane. The total surface area of all mitochondrial inner membranes in a cell is greater than that of the cytoplasmic membrane. Mitochondria are thus rich in potentially damaging free radical species. The cell confines these potentially damaging reactions to the mitochondria.

The mitochondrion resembles a bacterium not only in its size but also in its possession of a genome in the form of a circular DNA molecule. Each mammalian mitochondrion contains one or more mitochondrial genomes of about 15–17 kbp that encode a number of mitochondrial proteins and RNA molecules. The control of mitochondrial DNA replication is separate from the regulation of genomic DNA replication. The mitochondrion has its own ribosomes and tRNAs for translation, and RNA polymerase for transcription. However, many mitochondrial proteins are encoded in the cell's genome and synthesized in the cytoplasm before being imported into the mitochondria. Like other organelles, mitochondria cannot be generated from their components alone or from the genetic content in the nucleus; rather, a cell must already possess a mitochondrion in order to generate more mitochondria for its proliferation. The biogenesis (i.e., replication) of mitochondria is independent of cell division.

An active respiring mitochondrion has an about -140 mV electric potential (negative inside) and a pH difference of almost 1 unit across its inner membrane. The pH of the mitochondrial matrix is higher (i.e., the H^+ ion concentration is lower inside) than in the cytoplasm, as H^+ are pumped against the concentration gradient through the electron transport chain. The pH gradient and the electric potential are critical for the transfer of the chemical potential energy from NADH to ATP.

When the energetic need of a cell becomes high over a long period of time, it responds by increasing the number of mitochondria. Under normal (non-starvation) culture conditions, the flux of the carbon energy source (primarily pyruvate) and oxygen into the mitochondria appears to be within a narrow range. While the glucose flux into glycolysis can vary over a wide range, the pyruvate flux into mitochondria is more restricted.

Lysosomes, Peroxisomes, Autophagosomes, Endosomes

Lysosomes are small organelles where the degradation of many cellular materials takes place. Its lumen is a low pH environment (pH ~4.5) and it contains many enzymes that hydrolyze proteins, nucleic acids, and lipids (Panel 2.13). It has proton pumps in the membrane to maintain a low interior pH. It is the site of degradation of both ingested materials and cellular materials that are no longer needed by the cell. Most cellular materials have a finite life span, regardless of whether they are catalyzing chemical reactions or playing structural or mechanical roles. Over time, any cellular material can be oxidized or chemically modified in other ways in some part of its structure. The accumulation of such "damages" may render a protein non-functional. Thus, most proteins and other cellular materials, such as RNAs and lipids, are turned over after a finite period of time. Many of such processes occur in lysosomes. For some proteins, this occurs in proteasomes. Proteasomes are complexes of proteolytic enzymes that are capable of degrading proteins. Proteins that need to be turned

Panel 2.13. Lysosomes

- Low pH
- The site of degradation of endocytosed material and other cellular materials destined for degradation
- Participates in protein secretion, apoptosis

Panel 2.14. Peroxisomes

- Site of fatty acid oxidation
- Rich in oxidative enzymes

over are tagged by ubiquitin and sent to the proteasome for degradation. Lysosomes are not trash cans, but rather more like recycling centers. They play an important role in lipid homeostasis. Lipids taken up from the extracellular environment and other organelles are processed and then redistributed to maintain the proper lipid composition in the membranes of different organelles.

Panel 2.15. Endosomes

- In endocytosis, the invaginated plasma membrane forms small organelles
- Moves inward along the microtubule network
- Site of extensive cargo distribution and sorting
- Some material is sent to lysosomes
- Some is recycled to the plasma membrane

Like lysosomes and proteasomes, autophagosomes are involved in a cell's "house cleaning." However, they have double layer membranes which they wrap around the damaged materials. They can be large, as they may need to engulf damaged organelles, ingested microorganisms, or damaged proteins (a process called autophagy). They then deliver the content to a lysosome.

Through endosomes, eukaryotic cells can take up external particles by wrapping the particles within cellular membranes and taking them up as vesicles enclosed in lipid bilayers (Panel 2.14). The endocytosis pathway, discussed later in this chapter, also participates in the recycle of cellular materials involved in protein secretion. Some cells in higher organisms are specialized "scavengers" that engulf foreign particles or dead cells.

In fat cells, the catabolism of lipids occurs in peroxisomes (Panel 2.15). The reactions involved in such metabolisms generate large amounts of reactive oxygen and therefore need to be contained within these specialized organelles.

Endoplasmic Reticulum

The endoplasmic reticulum (ER) extends out of the nuclear membrane and is the organelle where the folding of proteins for secretion or delivery to many organelles and surfaces takes place (Panel 2.16). It is also a major site of lipid synthesis. It is largely classified into the smooth ER and the rough ER, based on its morphology under transmission electron microscopy. The smooth ER is rich in enzymes involved in chemical transformation reactions. In liver cells, the smooth ER takes on the role of detoxification; in the ovaries and testes, it makes hormones. The rough ER has ribosomes binding to its cytosolic surface and therefore appears rough in electron microscopy. It is the site of the synthesis, folding, and processing of proteins destined for some organelles, the plasma membrane, and secretion. It is estimated that roughly 30% of all cellular proteins are synthesized in the rough ER. Professional secretors in the body, such as pancreatic beta cells that secrete insulin and antibody-secreting plasma cells, have abundant ER. As B cells (non-antibody secreting)

Panel 2.16. The Endoplasmic Reticulum

Smooth ER

- Function varies with tissue. In liver cells it detoxifies; in ovary and testes it produces hormones.

Rough ER

- Site of folding of integral membrane proteins, proteins for some organelles, and secreted proteins
- Professional secretors in the body, such as pancreatic beta cells, hepatocytes, and antibody secreting plasma cells, all have abundant ER. As B cells differentiate to become plasma cells, the ER and Golgi apparatus expand drastically, at least by 15-fold.

Panel 2.17. Some Characteristics of the ER

- In a hepatocyte, the surface of the ER is about 63,000 μm^2 per cell, or about 40 times that of the plasma membrane
- The ER lumen is very viscous and gel-like. The diffusion coefficient of a fluorescent probe is 9–18 times lower than that for water.
- The ER has a much higher oxidative environment than the cytoplasm, appropriate for disulfide bond formation
- High free Ca^{2+} environment
- Many proteins are present at very high concentrations (PDI, GRP94, GRP74)
- A major site of protein folding, other post-translational processing, Ca^{2+} homeostasis, and cholesterol synthesis

differentiate to become plasma cells, the ER and Golgi apparatus expand drastically, more than 15-fold (Panel 2.17).

Hepatocytes secrete many proteins, including albumin, and many of the blood's protein components. In the liver, some hepatocytes specialize in protein secretion, while others play major roles in oxidative detoxification. These hepatocytes have distinctive ERs. Those involved in protein secretion have an abundance of rough ER, while those that specialize in xenobiotic metabolism have an abundance of smooth ER.

The ER is also a major site of protein post-translational modifications (Panel 2.18). The ER lumen is rich in proteins that facilitate protein folding. Disulfide bonds are formed in the ER, and the initial step of glycosylation on the asparagine residue (N-glycosylation) occurs there. The ER is involved in Ca^{2+} homeostasis, as its high Ca^{2+} serves as the reservoir to keep the cytosolic Ca^{2+} concentration low unless it is needed. The ER is also the site of lipoprotein and cholesterol synthesis.

The Golgi Apparatus and Protein Post-Translational Modification

The Golgi apparatus appears as dozens of stacks of three to seven flattened sacks in different locations throughout the cytoplasm (Panel 2.19). They are often near the ER, and the stacks are linked into a ribbon-like structure. The Golgi apparatus is loosely divided into four compartments: cis, medial, trans, and the trans-Golgi network (TGN) (Figure 2.8). The secretory and membrane proteins are transferred to the Golgi after they are folded in the ER, first to the cis Golgi and then to

Panel 2.18.
Protein Processing in the ER

- Cleavage of signal peptides
- Addition of high mannose core oligosaccharide to a Asn-x-Ser/Thr N-linked glycosylation site
- Trimming of terminal glucose and mannose residues from initial glycan
- Fatty acid addition
- Disulfide bond formation

Panel 2.19. Some Characteristics of the Golgi

- Compartmentalized into functionally distinct regions: the Golgi stack (consisting of cis, medial, and trans cisternae) and the trans-Golgi network (TGN)
- Proteins and lipids are sorted in the Golgi for delivery to different cellular locations
- From the Golgi, proteins go to secretion (exocytosis) or other organelles
- The Golgi apparatus breaks down during mitosis and reassembles afterwards

the medial and trans compartments through membrane vesicles. Then, in TGN, the proteins are sorted and transported to different organelles or to the plasma membrane for secretion. As the protein passes through the different Golgi compartments, it is chemically modified through glycosylation, sulfation, etc. (Panel 2.20). The enzymes needed for those chemical modifications are not distributed evenly in the four Golgi compartments. Thus, different reactions take place in different compartments. An estimated 100–200 glycosyltransferases, transporters of various nucleotide sugars, are the membrane proteins that constitute the majority of Golgi enzymes.

Protein Secretion through the ER and Golgi Apparatus

IT HAS BEEN ESTIMATED THAT 20–30% of all proteins encoded in a eukaryotic genome are membrane proteins that are processed through the ER and Golgi. For an industrial recombinant protein producing cell line, the secreted recombinant product constitutes a very large mass fraction of proteins passing through the ER and Golgi. These cells devote a large portion of their protein processing capacity to the secreted protein product.

Although some proteins are translocated into the ER after they are completely synthesized in the cytoplasm, most (including typical recombinant protein products) are translocated as nascent protein molecules (Figure 2.8, Panel 2.21). Proteins that are destined for secretion, the cytoplasmic membrane, and organelles have a leader sequence of five to thirty amino acids at the amino terminus. These signal sequences serve as markers of each protein's destination. After the mature mRNA with the polyadenylated tail (polyA tail) is exported to the cytosol from the nucleus,

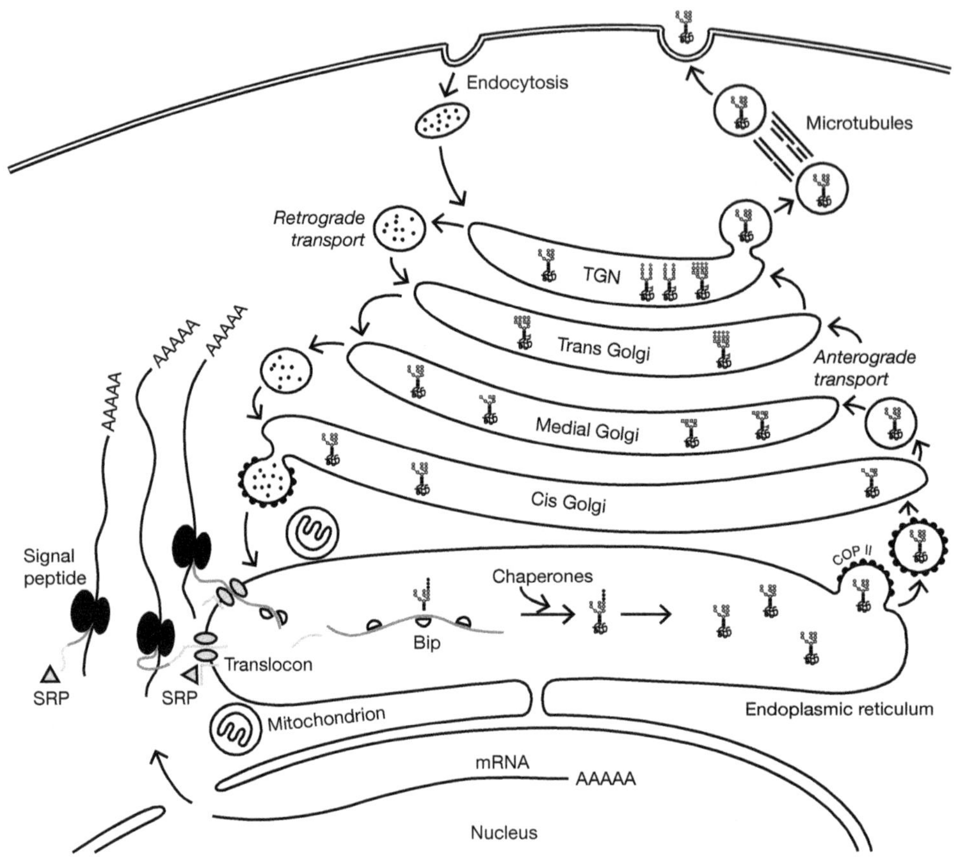

Figure 2.8. Synthesis and secretion of proteins to the extracellular environment. See Panel 2.21 and text for details.

Panel 2.20. Protein Processing in the Golgi

- Addition glycoform modification
- Sulfation of tyrosines or carbohydrates
- Glycosylation
- Peptide proteolytic cleavage
- γ-carboxylation of glutamic acid
- β-hydroxylation of aspartic acid

a ribosome binds to the translation start site of the mRNA and starts translation. The signal peptide of the nascent protein is synthesized and then recognized by signal recognition particles (SRP) near the ER. The binding of SRP to the signal sequence causes a pause in translation and docks the nascent protein (which has only the beginning segment of the entire sequence) along with the ribosome and the mRNA to a translocator (also called a translocon or translocation channel) on the ER membrane. The nascent polypeptide is then transferred through the translocator into the ER lumen, but the ribosome and mRNA are left on the cytosolic surface of the ER. Subsequently, translation elongation resumes and the elongating polypeptide passes through the channel of the translocator into the ER lumen. As the translation proceeds downstream of the mRNA, new ribosomes continue to bind to the translation initiation site of the mRNA, initiating the synthesis of the signal peptide and repeating the process of translocating the nascent protein into the ER. Thus, on the cytoplasmic side of the ER each mRNA molecule has multiple ribosomes actively translating the protein next to the translocator.

Folding of the polypeptide starts immediately upon translocation into the ER lumen. The signal peptide on the elongating polypeptide in the ER lumen is cleaved upon entry into the ER. A class of ER chaperones and other proteins that facilitate protein folding act on the nascent protein molecules to prevent them from aggregating and to facilitate folding. Their actions require cellular energy (ATP). An important member of the ER luminal chaperones, BiP (also known as GRP78), is also a component of the translocator complex. BiP binds to hydrophobic amino acids in incompletely folded proteins to prevent them from aggregating. In addition to BiP, major ER luminal chaperones include protein disulfide isomerase (PDI), which facilitates the formation of the correct disulfide bond.

Throughout the process of protein secretion, N-glycosylation takes place on the protein, starting in the ER and continuing into the Golgi (Figure 2.8, also Figure 3.21). It is called N-glycosylation because the oligoglycan is linked to the asparagine residue of specific sequences on the protein. As the protein is being synthesized and folded, a preassembled oligoglycan containing three terminal glucose residues is transferred and covalently linked to the asparagine (Figure 2.9). The addition of glycan to the protein increases its stability in a soluble form. It is also an important marker of protein folding. While the protein undergoes folding, the glucose residues are cleaved. A glucosyl transferase adds a glucose to the aglycosylated glycan on any protein that

Panel 2.21. Protein Secretion

- Nascent protein molecules destined for the ER have a special ER signal sequence being synthesized in an organized polysome. They are recognized by SRP (signal recognition particle), a ribonucleoprotein.
- SRP binding transiently arrests elongation, directing the ribosome/nascent polypeptide complex (RNC) to a receptor on the ER membrane and transferring the growing polypeptide to a translocon
- SRP is released from the RNC
- The nascent polypeptide begins to pass through the translocon and elongate into the ER lumen
- Signal peptide on the elongating polypeptide is cleaved
- Protein folding and post-translation modification begins as the polypeptide continues to elongate
- Major ER luminal chaperons: BiP, calnexin, calreticulin, and PDI
- The ribosome is released once the translation is complete
- Folded proteins (with an inner core of glycan if it is a glycoprotein) concentrate at the exit site of the ER and are thought to bud into vesicles and translocate to the Golgi as pre-Golgi intermediates
- The Golgi apparatus is in a dynamic state. Transport is retrograde (recycling its own proteins) and anterograde.
- After reaching the trans Golgi network (TGN), secretory proteins are packaged into post-Golgi vesicles and moved along the cytoskeletal network through the cytoplasm to fuse with plasma membrane and be secreted

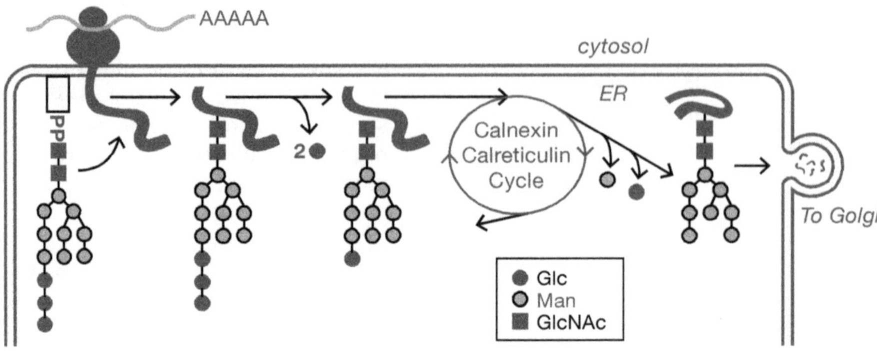

Figure 2.9. Glucose trimming from an N-glycan as a signal of correct protein folding.

is not correctly folded, leaving the completely folded protein with an N-glycan that does not have any terminal glucose. Calnexin and calreticulin are lectins (glycan-binding proteins). They bind to the glycan on the protein that has only one glucose residue and retain them in the ER for further folding. Once the protein is completely folded and has no terminal glucose, it is ready for transfer to the Golgi apparatus. The protein molecules that are improperly folded are then exported to the cytosol and degraded in a proteasome.

Protein molecules that have completed the folding process are localized in some regions of the ER where the ER membrane is coated with a protein COPII. Membrane vesicles that contain the folded proteins and are coated with COPII are then budded out from the ER and exported to the cis-Golgi. In the Golgi, further post-translational modifications take place. The cis-, medial-, trans-Golgi, and TGN comprise an array of tubules which are tethered as microbioreactors in a series, and in which different enzymes carry out different modifications on the protein. The medial-Golgi, typically containing three to seven stacks of cisternae, is the main site of glycan elongation for glycoproteins. Vesicle budding from a compartment and fission with another are the main forms of protein trafficking from the ER to the Golgi and among Golgi compartments. The surface of the Golgi compartment is decorated with proteins (golgin) that form coiled-coils that act almost like tentacles to help capture the vesicle.

In addition to secretion, proteins are also bound for different organelles and different regions of the plasma membrane. In TGN, proteins are sorted into membrane vesicles according to their destination. The vesicles are coated with different proteins to mark their destination. The transport of the membrane vesicle is facilitated by microtubules. As will be discussed later, microtubules can polymerize and depolymerize quickly to become track-like, and as such can be used to guide the vesicle cargos to the right location.

As the trafficking happens, the membrane and the contents of an early compartment are translocated to a later compartment of the ER, the Golgi, different organelles, and the plasma membrane. Such a material flow needs to be balanced by a similar material flow in the opposite direction. Thus, the protein secretion process involves not only anterograde transfer of membrane vesicles to the later compartments and plasma membrane, but also endocytosis and retrograde transport to return the membrane and other constituents to their originating compartment in order to maintain each compartment at a homeostatic state. The ER, Golgi, lysosomes, and endosomes are thus all part of the secretory network. They communicate through the dynamic trafficking of membrane vesicles. After the translation is completed, a finite amount of time is required for a protein molecule to be folded, glycosylated, and then excreted. The complete translation of an average protein about 350 amino acids in length takes only about a minute. In comparison, the secretion of a synthesized protein molecule takes from tens of minutes to a couple of hours depending on the nature of the protein. The α1-protease inhibitor is among the fastest secreted proteins, with a half-life of about 28 min (Table 2.3). The secretion of transferrin and IgG takes about two hours. In general, a secreted protein molecule spends more time in the ER than in the Golgi apparatus.

One can view the process of protein cargos traveling through different compartments towards the TGN as a production stream passing through a series of bioreactors. There are two idealized scenarios on how the protein molecules pass through, one like a plug flow reactor (PFR), the other a continuous stirred tank reactor (CSTR). These two reactor types will be discussed in Chapter 8. In a PFR model, all molecules move along like a marching band in a parade, with those entering the reactor at the same time exiting it together. In a CSTR model, the molecules enter the reactor and then run around in a circle inside, with each molecule having an equal chance of being randomly taken out regardless of how long it has been in the reactor. If the protein secretion process in the Golgi apparatus behaves like a PRF, all molecules will be subjected to the same degree of enzymatic modifications; if it behaves like a CSTR or a number of CSTRs in a series, some will spend more time in the reactor and

Table 2.3.
Secretion Times of Liver Proteins

Protein	*Half-life in ER (min)*	*Half-life in Golgi (min)*
Transferrin	110	45
Ceruloplasmin	80	30
Anti-trypsin	30–40	10
IgG	Total ~120	

Panel 2.22. Processing Time of Secreted Proteins

- Translation of a protein molecule takes only seconds, but the secretion process takes tens of minutes to hours
- Different proteins spend different amounts of time in the ER for folding, and in the Golgi apparatus for other post-translational modifications
- Different molecules of the same protein have different holding times in the ER and Golgi. The half-life of a protein in the ER and Golgi is the average of all molecules.

possibly be modified by enzymes more extensively than others. The nature of protein secretion is likely to be neither a PFR nor a CSTR. Nevertheless, the holding time of a secreted protein is likely to distribute over a range.

Experimental observation has indeed shown that the secretion time of different molecules of the same protein is not uniform, but instead distributes over a range (Panel 2.22). This was illustrated by arresting protein secretion using a drug and then removing that drug; upon resumption of protein secretion, the fluorescently labeled protein molecules travelled through different cellular compartments at different rates. The glycosylation reactions in the Golgi, from a core oligoglycan to the completely extended product, involve more than ten enzymes distributed over different compartments. Since different molecules spend different amounts of time in each compartment, they are subject to varying degrees of different reactions. This distribution of residence time in the Golgi contributes to the heterogeneity of glycosylation.

Transport across Cellular Membranes

The lipid bilayer membrane that separates the cell from its environment, and an organelle from cytosol, presents a barrier to material exchange across it. It has a very low permeability for large and small molecules alike (Figure 2.10). Among the nutrients and metabolites of cells, only oxygen, fatty acids, ethanol, and some lipids can freely pass through the membrane fast enough to meet the needs of cell growth. Specialized transport mechanisms are needed to mediate the movement of the vast majority of nutrients and the excretion of metabolites.

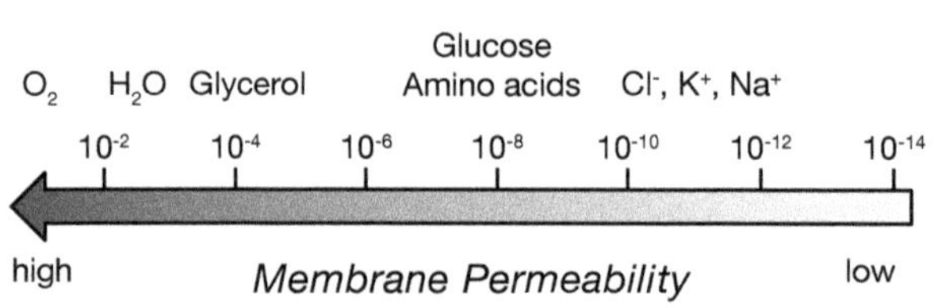

Figure 2.10. Orders of magnitude of the permeability of some solutes across a lipid membrane.

Cells have a large number of transporters (sometimes called permeases) that allow small molecules (up to about 1 kDa, e.g., sugar, oligosaccharides, amino acids, oligopeptides, nucleotides, cholesterol, ions, organic acids, etc.) to cross the cytoplasmic membrane and the membranes of various organelles. Macromolecules are transported across membranes by membrane vesicles (as in the secretion process through the ER and Golgi apparatus), pinocytosis, or exocytosis. Specific receptors are involved in taking up large proteins and complexes such as low-density lipoprotein (LDL) and transferrin.

Types of Solute Transporters

Active and passive transport

The cellular transport of solutes across a membrane can be classified into two types: passive transport, which transports a solute along its concentration gradient, from the high concentration side of the membrane to the low concentration side, and active transport, which passes the solute from the low concentration side to the high concentration side (Figure 2.11, Panel 2.23). The former is thermodynamically favorable, whereas the latter must be coupled to an exergonic process to make the transportation possible. One such coupled process that provides the chemical potential energy necessary for active transport is the hydrolysis of ATP. Another is the co-transfer of a second solute along a large concentration gradient across the membrane. The chemical potential energy change of the co-transport of the second solute provides a sufficient net energy change to drive the transport of the first solute against its concentration gradient. For example, the intestinal sodium-dependent glucose transporter relies on the more than 10-fold higher concentration of sodium in the exterior to drive glucose to move from a low-level area in the gut lumen to the interior of intestinal cells with a higher glucose level. Thus, the driving force of sodium transfer "pushes" glucose to move against its concentration gradient.

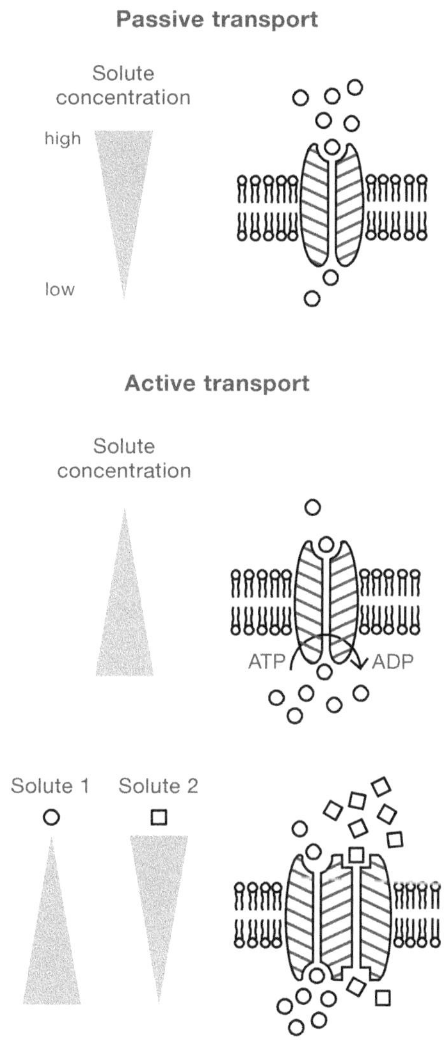

Figure 2.11. Active transport and passive transport. Active transport requires the supply of chemical potential energy derived from ATP hydrolysis or co-transport of a solute such as H^+ or Na^+.

Transporter and channel proteins

From a structural perspective, the transport of a solute along the concentration gradient can be mediated by a carrier protein (transporter) or by a channel protein (Figure 2.12, Panel 2.24).

Channel proteins open into a duct-like structure across the membrane that is specific to a particular solute, such as water, Na^+, or K^+. Channel proteins exist in either an open or closed state. Once the channel is open, the transfer of solutes is very fast in the direction of decreasing concentration. The flux is very

Panel 2.23. Types of Solute Transport

Active Transport

- Moves molecules against their concentration gradient
- Requires ATP or ion gradients (H^+, Na^+) as an energy source to drive transportation

Passive Transport

- Moves chemical species from the high concentration side to the low concentration side of a membrane
- May involve channel proteins or facilitated diffusion transport

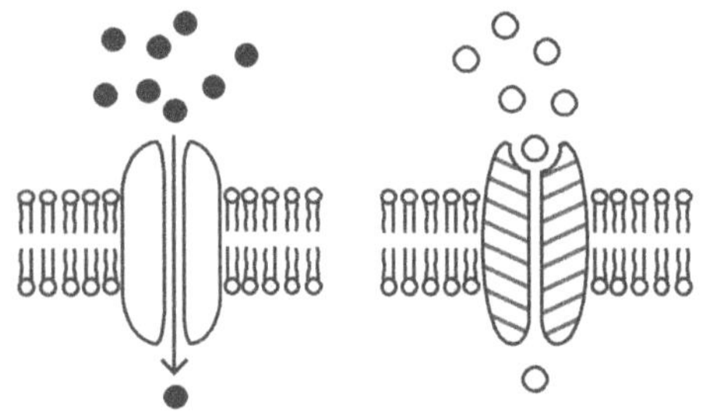

Figure 2.12. Channel proteins and transporter proteins.

Panel 2.24. Three Classes of Transport Processes

Channel-Protein-Mediated Diffusion

- Molecules or ion-specific
- Once a channel is open, very fast flux

Transporter-Facilitated Diffusion

- Provides specific bonding to a solute
- Solute binding causes protein conformation change to allow for transfer to the other side of the transporter
- Opening the transporter allows the release of the solute to the other side of the membrane

fast and is affected by the number of channel protein molecules on the membrane and by the time period that the channel is open.

Carrier-mediated transporters mediate facilitated diffusion. First, the solute from the high concentration side of the membrane is diffused into the transporter and docks to the binding domain of the transporter. Next, the solute is translocated to the low concentration side of the membrane. Once on the low concentration side, the solute is exposed and is free to diffuse away. Under normal culture conditions, glucose and some amino acids are transported by facilitated diffusion.

Transporters for facilitated diffusion mediate reversible transport. The rate of transport by a transporter is dependent on the concentration difference of the solute across the membrane. Typically, the transporter has a high affinity for the solute on the donating side, which helps facilitate the binding of the solute, and a lower affinity on the downstream side to facilitate the release of the solute. If the concentration of the solute on one side is much greater than the other, the rate generally depends only on the solute level in the donating side. The mechanism of such transport is similar to a typical enzyme-catalyzed conversion of a substrate to a product. The dependence of the transport rate to the solute concentration can be described by a Michaelis–Menten kinetic equation. At low concentrations, the transport rate is in proportion to the solute concentration. At high concentrations, the rate is constant as the transporter becomes saturated. A half-saturation constant (K_M) is used in the kinetic expression of the transport rate as a function of solute concentration. K_M is the solute concentration when the transport rate is half of its maximum.

A ubiquitous transporter for glucose is the glucose transporter 1 (GLUT1). The half-saturation constant (K_M) for GLUT1 is about 0.1–1 mM. In the 0.1–1 mM range, the glucose import rate of the cell increases with increasing glucose concentration. In the concentration range often used in cell culture media (1–10 g/L, or 5.5 mM to 55 mM), the glucose transport rate is near its maximum and is not affected by glucose concentration.

Uniporters and co-transporters

Transporters for facilitated diffusion and active transport can also be categorized according to the number of solutes each carries and the direction of solute flow (Figure 2.13, Panel 2.25).

Uniporters transfer a single solute from a high concentration side to a low concentration side, e.g., the GLUT1 transporter for glucose and the GLUT5 transporter for fructose. Symporters and antiporters transfer two solutes simultaneously. A transporter that delivers the two solutes in the same direction is called symporter. Conversely, an antiporter transfers two solutes in opposite directions.

Collectively, symporters and antiporters are called co-transporters. Co-transporters are often used to transport charged organic molecules. Dissociable solutes move with a counter ion to maintain electric charge neutrality. When a charged solute moves from one side of the membrane to the other, the charge neutrality must be maintained. Otherwise, a net charge will accumulate across the membrane and create an electric potential that will retard further transfer of the solute. For example, lactic acid exists as lactate in an aqueous solution at neutral pH. If lactate moves from one side of a membrane to the other without being accompanied by a positively charged species such as H^+, the cell membrane will become negatively charged on the lactate-receiving side and positively charged

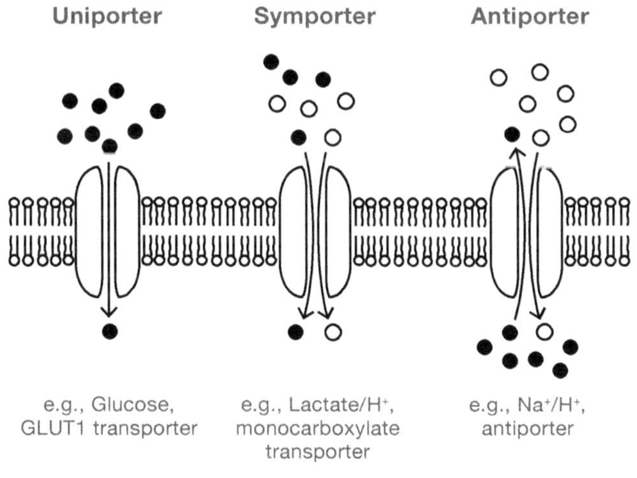

Figure 2.13. A uniporter and the two types of co-transporter: symporter and antiporter.

Panel 2.25. Transporters

Uniporter
- Transfers a single molecule (e.g., glucose, fructose)
- The solute is mostly uncharged

Bispecies Transporter (Co-Transporter)
- Requires a stoichiometric exchange of two species simultaneously
- Important in charge balance
  - Symporter
    - Two species transported in the same direction
  - Antiporter
    - Two species transported in opposite directions

on the lactate-donating side, eventually creating a voltage across the membrane that prevents further transport of the negatively charged lactate. Thus, many dissociated organic acids are transported by a co-transporter so that a charge does not build up across the membrane.

Two mechanisms are commonly employed to maintain a charge neutrality across a membrane: a) co-transport with a counter ion (such as H^+ for lactate) by a symporter, and b) co-transport with an ion of the same positive or negative charge, but in the opposite direction (such as Cl^- for HCO_3^-).

In co-transporter mediated transfer, the transport rate is not only affected by the concentration difference of the solute across the membrane, but also by the concentration gradient of the co-transported ion. For example, the transport rate of lactate by the monocarboxylate transporter (MCT) is not only affected by the concentration of lactate, but also by the pH difference between the cytosol and the medium.

"Pump"-like transporters

We use pumps to transport water. Sometimes a pump is used to overcome gravity and move water higher, and other times the water is going downhill anyway, but the pump is used to move it faster. In solute transport across a membrane, energy is sometimes used to overcome the concentration gradient (active transport) such as in the case of pumping water uphill; in other cases, energy is used to make the transport faster (Figure 2.14).

The Na^+/glucose transporter (SGLT) in the epithelium of the intestine can take up glucose from the digestive track even when glucose levels are

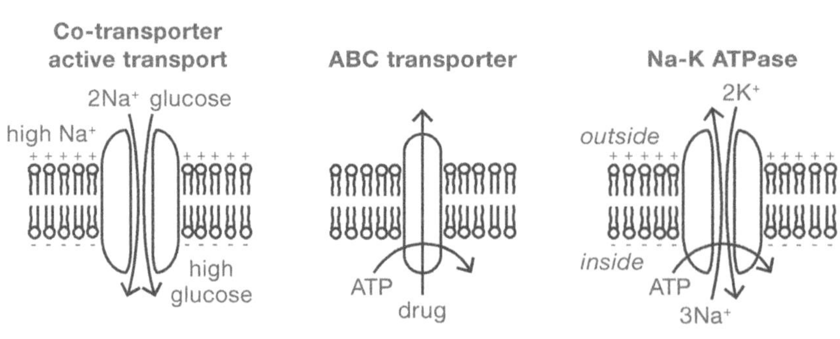

Figure 2.14. "Pump"-like transporters.

lower than in the cell. This is accomplished by using the Na^+/glucose transporter to transport two Na^+ from the lumen of the intestine into the cell, where the Na^+ level is very low (Figure 2.14). As the sodium ion is transported, a glucose molecule also binds to the transporter and is transported simultaneously. The propensity of Na^+ to move across the membrane is so high that it can drive glucose to move against a large concentration gradient. The driving force for Na^+ to cross the membrane is not only from the concentration difference of Na^+, but also from the attracting power of the negative membrane potential, the –80 mV across the membrane.

Another type of pump-like transporter for active transport is the ATP-binding cassette (ABC) transporter, which transports some hydrophobic compounds by utilizing ATP as the energy source (Figure 2.14). Its mutation is involved in many human diseases. It is also involved in the development of resistance to many drugs by pumping the drug out of the cell, causing a low intracellular concentration of the drug. Even if its efflux is along its concentration gradient, the transport rate will be low. With an ATP-driven transporter, the rate is enhanced. After prolonged exposure to methotrexate, some cancer cells can develop drug resistance by pumping the drug out of the cells using ABC transporters, thereby becoming drug-resistant. A third type of pump-like transporter, Na-K ATPase, will be described below.

Ion transport and membrane potential

The intracellular fluid contains a variety of inorganic ions. Some trace elements (Fe^{3+}, Zn^{2+}, Co^{3+}, Se^{2+}) are present only at very low concentrations. Many ion species (Na^+, K^+, PO_4^{3-}, Mg^{2+}, HCO_3^-, Cl^-) are present in the millimolar range (Panel 2.26). All rely on transporters to cross the cell membrane. Some are also transported by channel proteins in a regulated fashion for controlling membrane potential and maintaining its homeostatic level in the cell. Those ions are present in the cytoplasm in very different concentrations. They also face varying concentration differences across the cell membrane. For instance, Na^+ and K^+ have opposite directions of concentration gradient across the plasma membrane. The intracellular concentration of K^+ is 20–50 times higher inside than outside of the cell, while the Na^+ concentration is 10–15 times higher outside the cell versus inside.

Along with the concentration gradients of major ions, cells also maintain an electric potential gradient of –80 mV across their plasma membrane (negative inside the cell). This electric potential is fundamental to the transport of many compounds across the membrane. Because of the large capacitance in the lipid bilayer membrane, it takes the transfer of only an extremely small number of positively charged ions (e.g., K^+, Na^+) from the cytoplasm to the extracellular space to create the 80 mV charge difference. The number of ions transferred to generate

Panel 2.26. Transporters for Ions

- The concentrations of many major ion species (H^+, Na^+, K^+, Ca^{2+}, PO_4^{3-}, Cl^-) across membranes are polarized. Na^+ and K^+ have opposite directions of concentration gradients across the plasma membrane, about a 10-fold difference. For Ca^{2+}, the intracellular concentration is so low that the gradient is extremely steep.
- Na^+-K^+ ATPases play a major role in maintaining Na^+ and K^+ gradients. ATPase utilizes the hydrolysis of ATP as its energy source.
- Iron is extremely reactive and participates in many redox reactions. In biological systems, it exists in its "bound" form to binding proteins or chelators. In its free form, it catalyses the formation of peroxide and peroxidizes unsaturated fatty acids.

the membrane potential is many orders of magnitude lower than the total number of ions in the cell, and their transfer does not cause any change in their intracellular concentration.

Sodium-potassium ATPase plays a key role in maintaining the intracellular levels of Na^+ and K^+ (Figure 2.14). It transfers 2 K^+ into the cytoplasm and 3 Na^+ out of the cytoplasm using ATP as the energy source. It is present in the cytoplasmic membranes of all mammalian cells. The cytoplasmic membrane is not absolutely impermeable to ions. It allows Na^+ and K^+ to diffuse across at a very slow rate. Due to their small difference in the membrane permeability, K^+ diffuses faster than Na^+. Intracellular K^+ also leaks out through potassium channels. The combined actions of Na^+/K^+ ATPase and leakage through channels and the membrane facilitate the balance of the Na^+ and K^+ gradient and the membrane potential. Cells also have chloride ion pumps to pump negatively charged Cl^- out of the cell. In some cells, Na^+ and Cl^- channel proteins also facilitate the maintenance of the membrane electric potential in the correct range. The electric potential across the membrane plays an important role in solute transport, as illustrated by the Na^+/glucose transporter.

Cell Shape, Mechanics, and Movement

Cells present themselves in many different shapes. Those circulating in blood are mostly spherical. Many epithelial cells are cubical, while fibroblast cells grown on Petri dishes spread out like a sunny-side-up fried egg, complete with elongated spindles. The shape of a cultured cell tells a lot about the cell type, the status of its differentiation, and its health. Before modern biochemical and molecular biological analytical tools were available, the observation of cell shape was, and still is, a primary tool for judging the "state" of a cell. The shape of a cell is the result of the force balance among the cell's internal structural organization and the interactions between the cell and adjacent cells as well as between the cell and the surface that it attaches to. Cell shape is thus a reflection of a cell's mechanical properties.

Cell shape is dynamic. Cells move from one place to another for various reasons. In embryo development, they may move to a new site

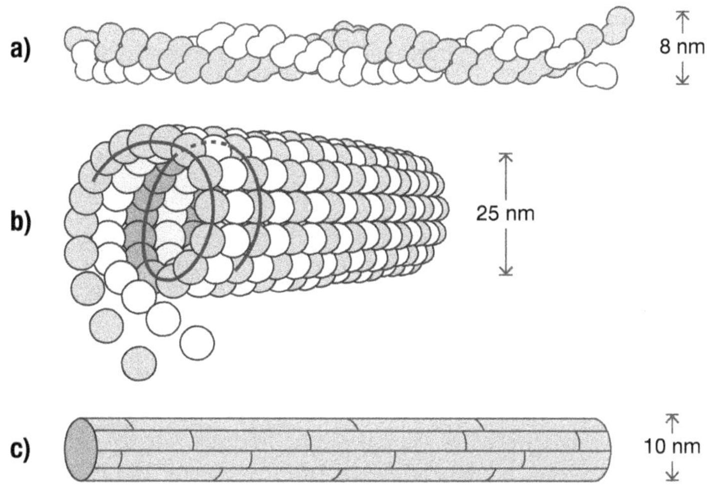

Figure 2.15. (a) An actin filament, (b) a microtubule, and (c) an intermediate filament.

to differentiate into a new tissue. In an adult, they may move to repair a damaged tissue. Two daughter cells move away from each other after cell division. As a cell moves or divides, its mechanical properties, shape, and interactions with other cells and the surface also change. In short, a cell's internal mechanical structure, interactions with adjacent cells, and interactions with the surface it attaches to together influence its shape, movement, and intracellular as well as extracellular mechanical actions.

A cell's internal mechanical structure is governed by the proteins of the cytoskeleton. Cell–cell interactions are influenced by junction proteins and a number of cell–cell adhesion molecules. Cells secrete extracellular matrices that overlay a surface with characteristics that facilitate the differentiation of itself or other cells. These extracellular matrices are also subject to dynamic modifications during their development.

These aspects of cell culture used to be of lesser interest to process engineers when the focus was solely on highly adapted recombinant cells growing in suspension. With increasing interest in cells for regenerative medicine and cell therapy applications, maintaining a cell's differentiation status is critical and a better understanding of cell–cell and cell–surface interactions is important.

Cytoskeleton

Three major components make up the cytoskeleton of a cell: actin filaments, microtubules, and intermediate filaments. These give cells their shapes and transmit forces within them (Figure 2.15). Both actin filaments and microtubules are assemblies of their respective monomers of globular proteins (actin and tubulin, respectively). The assembly of monomers requires chemical potential energy in the form of ATP (for actin) or GTP (for tubulin). The stacking of monomers into the assembly is done without forming covalent bonds among monomers. Intermediate filaments are also polymers but are made of monomeric filament proteins. Actin filaments, microtubules, and intermediate filaments are all helical-type fibers with multiple strands that can reach substantial lengths. Actin filaments, at about 8 nm in diameter, are the smallest of the three. Microtubules are hollow, with a diameter of about 25 nm. Intermediate filaments have a diameter of about 10 nm.

Actin filaments and microtubules can quickly grow or shrink in length by assembling or disassembling monomers. Because monomers form the assembly in a head-to-tail fashion, their filaments are polarized. The filaments and the monomer interact with a number of auxiliary proteins that modulate the filament into a dynamic or static state and determine the network type that the filaments form. The monomer proteins also contain a nucleic acid (ATP in actin, GTP in tubulin) that is converted to ADP or GDP in the assembling process.

Panel 2.27. Actin Filaments

- Two-stranded filaments (F-actin) of helical polymer of actin (G actin) (50 kD)
- 5–9 nm flexible structures organized into linear bundles, 2-D networks, and 3-D gels
- In non-motile cells, actin filaments form bundles called stress fibers, creating a loose mesh of filaments that underlie the cell membrane
- Distributed all over the cell, but concentrated in the cortex beneath the plasma membrane
- Projections from cells such as microvilli, lamellipodia, microspikes, and filopodia are maintained by rigid bundles of actin filaments
- In actively moving cells, stress fibers disappear and actin filaments concentrate at the leading edge.
- There are many actin-related proteins which affect the polymerization and motor functions of actin filaments
- Actin is involved in motor functions such as muscle contraction

Actin filaments

Actin filaments are comprised of two or three strands that, like rope, are twisted to form a single-string filament (Figure 2.15, Panel 2.27). This geometry enhances their structural integrity while maintaining a high degree of flexibility. A large number of auxiliary proteins provide control of the nucleation of actin monomers and direct their assembly into different types of networks with varying degrees of actin filament length, cross linking, bundling, and tethering to the cell membrane. These auxiliary proteins help regulate the spatial distribution of different types of actin filament networks in the cell.

A high concentration of actin is often seen immediately underneath the cell's plasma membrane, forming a gel-like structure called the cortex. In immunofluorescent microscopy, following the staining of cells with anti-actin antibodies, actin appears in web-like bundles. The cortex acts as the first absorber of external mechanical perturbations and gives the lipid bilayer its local shape.

Many cell types, such as fibroblasts, extend their body and spread flat on a surface, both in tissues and in culture. The edge of an adherent cell has an irregular shape, much like an egg laid on a pan. In the protruding regions, actin fibers

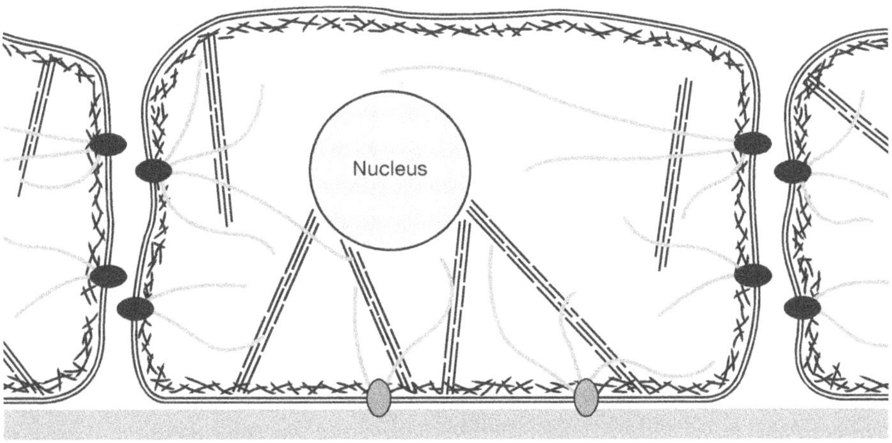

Figure 2.16. Cytoskeleton of a cultured fibroblast cell in culture in an epithelial cell layer. Actin filaments are shown as intersecting hash marks, intermediate filaments as fluid gray lines, and microtubules as dashed compound lines.

localize in the lamellipodia and filopodia. In the protruding filopodia, actin filaments form intensely labeled parallel bundles of fibers. While in the lamellipodia, they form cross-linked non-parallel networks. The two types of network can reorganize, the filopodium region protruding out from a lamellipodium during cell movement. In stationary cells, the actin fibers form long stress fibers that may span a large fraction of the entire cell length. The stress fibers connect to the focal adhesion (the location where the cytoplasmic membrane is in contact with the substrate surface) and establish a tension force between the cell and the extracellular matrix. Stress fibers also connect to adjacent cells through adhesion junctions.

Microtubules

Microtubules are hollow tubes formed by assembling 13 threads of α- and β-tubulins (called protofilaments) together (Figure 2.15 and Figure 2.16, Panel 2.28). Like actin filaments, the head-to-tail assembly of the monomers renders a microtubule directional, giving rise to a "+" and a "−" end. The two ends of the microtubules can extend or shrink quickly by assembling or disassembling reactions. The negative ends of many microtubules converge in an area called a microtubule organizing center (MTOC). The largest MTOC is the centrosome, from which a large number of microtubules extend their "+" ends outward. Many auxiliary proteins bind to or interact with microtubules and tubulins. A number of microtubule-associate proteins (MAPs) stabilize microtubules or facilitate the interactions of microtubules to cell components, such as binding to the cell membrane. Other auxiliary proteins promote disassembly, nucleation, bundling, or cross-linking. Microtubules are thus major

Panel 2.28. Microtubules

- Long, hollow tubes of polymerized subunit tubulin (MW 50 kD), about 25 nm in diameter, more rigid than actin filaments
- Typically long and straight, many have one end (–) attached to a single microtubule organizing center (e.g., centrosome)
- α- and β-tubuline (GTP) forms heterodimers; the dimer assembles in a head-to-tail fashion into chains called protofilaments, 13 of which make up the microtubule wall
- Can polymerize to extend and depolymerize to retract its length rapidly
- Play a key role in intracellular organelles and small vesicle transport
- For secretory proteins, the movement of post-Golgi vesicles to the plasma membrane is mediated by microtubules

Panel 2.29. Intermediate Filaments

- Play a structural role by stabilizing and transmitting mechanical force
- The subunit is not a globular protein, but fibril, different from tubulin and actin
- 10 nm diameter, has a head, a tail, and an α-helical rod; can be made of a wide variety of proteins in various tissues

structural components that affect cell shape, both statically and dynamically.

Microtubules are also the major player in a cell's motor functions. They accomplish this by serving as the track that motor proteins, like kinesin and dyneins, can slide along with expenditure of ATP. The motor protein, by forming complexes with other proteins that bind to various cargos, such as excretory membrane vesicles or organelles, can then transport the cargo along the microtubule.

Intermediate filaments

Unlike actin filaments and microtubules, which are ubiquitous to all eukaryotic cells, intermediate filaments are only present in some animals, including vertebrates, nematodes, and mollusks (Figure 2.15 and Figure 2.16, Panel 2.29). Although all intermediate filaments share common structural features, they are comprised of many different molecules and their expression is tissue specific. For example, laminins are present in the nucleus, vimentins in many mesenchymal cells, and keratins in epithelial cells. The major role of intermediate filaments is to transmit mechanical force and provide cells with their mechanical characteristics. For example, the intermediate filament keratin gives the outer layer of skin its toughness. While the diverse intermediate filaments are not conserved in their amino acid sequence, they are conserved in major protein domains and share common characteristics in their molecular organization. A common feature of intermediate filaments is that they form a head-to-tail coiled heterodimer, and then a pair of dimers form an antiparallel and symmetrical tetramer. Thus, intermediate filaments fundamentally differ from actin fibers and microtubules in that the latter two are polar with a "+" and a "–" end. Tetramer subunits are stacked together to form a filament. Each intermediate filament fiber is made of multiple fibrils, which are in turn made of a series of subunit proteins.

These intermediate filament fibers are flexible. Unlike microtubules and actin fibers, which can have tens of micrometers of persistence length, intermediate filaments can bend within a short distance and are capable of absorbing energy exerted by external forces and transmitting it to other regions of the cell. In a tissue or in interconnected cells in culture, intermediate filaments also help to transmit forces between cells.

Extracellular Matrices and Integrins

The vast majority of cells in the body are embedded in an acellular tissue structure composed of proteins and polysaccharides (Figure 2.16, Panel 2.30). In tissue differentiation, these structures may become calcified bone, cartilage, the connective tissue underlying the epithelium layer, or other specialized tissues, or form the basement membrane (also known as basal lamina) that cells sit on. A variety of glycoproteins (proteins linked with oligosaccharides) and proteoglycans (a core protein with one or more glycosaminoglycan (GAG)) secreted by various cells provide the core components of ECM (Panel 2.31). For example, osteoblasts secrete collagens, other bone-forming proteins, and hydroxyapatite; chondroblasts secrete collagen, hyaluronic acid, and proteoglycans; and fibroblasts secrete collagens and fibronectin. The tissue cells not only secrete different ECM components, but also organize them into different structures to give the special characteristics of different tissues. For example, collagen fibers may be aligned in one direction rather than being randomly coiled.

Panel 2.30. Extracellular Matrix

- The substrate for adhesion of many cell types
- Provides cues for growth, differentiation, and development
- Rich in electrocharges, allowing growth factors, cytokines, etc., to be "stored" inside them

The chemical composition as well as the physical organization of ECM as a cell's local environment affects the differentiation and maintenance of differentiation properties. When cultured *in vitro* on plastic or glass surfaces, cells secrete those materials onto the surface after adhering to form the ECM on which they reside. For *in vitro* culture of embryonic stem cells, laminin-coated culture dishes are used to maintain their pluripotency; while in culturing primary hepatocytes isolated from liver, type I collagen is used. Many ECM components are highly negatively charged. This allows many protein growth factors to be adsorbed to the ECM and released to surrounding cells, perhaps even serving as a chemoattractant. Therefore, they also play a role in providing cues for cell migration and differentiation.

Panel 2.31.
Components of the ECM

ECM Proteins
- Collagen
- Laminin
- Fibronectin

Proteoglycan
- Chondroitin sulfate

Glycosaminoglycan
- Heparin
- Hyaluronic acid

Integrins and cell–ECM interactions

The biological role of ECM is not merely to provide a surface appropriate for cell adhesion, but also to provide cell-surface interactions that are linked to the expression of special cell characteristics and growth control (Panel 2.32, Figure 2.17). Cells of the same type may exhibit very different shape and other cell behavior when cultured on different ECMs. The different responses to different ECMs is mediated by

Panel 2.32. Integrin

- Provides mechanical and signaling interactions between the intracellular and extracellular environments
- Can sense extracellular ECM components and transmit signals to the cytoskeleton
- Consists of different α and β chains for heterodimers of different integrins

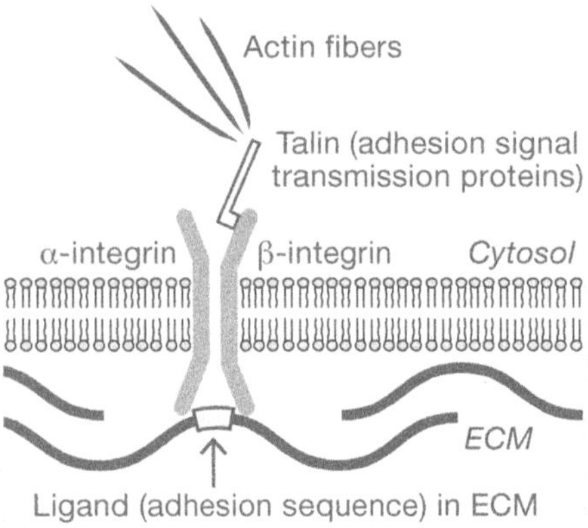

Figure 2.17. Integrin and interactions between ECM and the cytoskeleton.

transmembrane cell surface receptors, called integrins, that connect the extracellular ECM to the intracellular cytoskeleton and relay the chemical and physical features of the extracellular microenvironment to intracellular signaling events. The integrins are heterodimers of different α and β chains. Different combinations of α and β chains gives the specificity of binding to different ECM molecules. For example, $\alpha_5\beta_1$ integrin binds to fibronectin, while $\alpha_6\beta_1$ binds to laminin. On the intracellular side, integrin binds mostly to actin filaments with the help of some auxiliary binding proteins. Different cell types express different α and β chains and thus form different integrin complexes with different binding affinities for various ECM components. As a result, different cell types often have different ECM requirements for adhesion, growth, or maintaining their differentiated properties. To establish strong and stable interactions between ECM and the cytoskeleton, multiple cell–matrix junction complexes cluster together to form focal adhesion.

The integrin heterodimer can exist in either an active state that engages ECM, actin filaments, and other associated proteins, or in an inactive state. This switch of state allows cells to exert or disengage internal and external mechanical interaction. The switch of state is facilitated by coupling to a cell's signaling proteins. The cell–matrix junction is associated with signaling kinases or other signaling system proteins. The signal transmission occurs from both sides of the cellular membrane, from intracellular to extracellular environment and vice versa. The binding of integrin to ECM proteins can activate integrin to allow for binding by actin filaments. Conversely, internal signaling activation of integrin can also initiate the connection of actin filaments and ECM through integrin.

The inward pulling of actin filaments through a cell–ECM junction results in a tension force that is transmitted through cytoskeletal fibers. In the case of anchorage-dependent cells, the tension force provides the necessary signal to allow the cell cycle to proceed.

Cell Movement

The vast majority of cells are capable of movement on surfaces. In general, cell movement can be the result of an attraction to diffusible chemicals or favorable ECM on the surface, or random movement. Cell movement involves a restructuring of the cytoskeleton, a protrusion of the membrane, the establishment of surface adhesions on one side of the cell, and finally the detachment of the cell membrane from the adhesion complexes in the rear end of the cell. The actin fibers and plasma membranes of moving cells extend the cell into a more elongated shape and form lamellipodia that may also contain microspikes called filopodia (Figure 2.18, Panel 2.33). The reorganization of the actin fibers happens within a few minutes in the leading edge of a moving cell.

On an "open" surface (i.e., uncrowded), cells move randomly. They exhibit locomotion and contact avoidance, meaning that, when two moving cells encounter each other, both will move away in opposite directions. Furthermore, the two daughter cells of a dividing mother cell move away from each other when cell division is complete. Cell migration is regulated by many factors, including growth factors. For example, epithelial cells respond to hepatocyte growth factor by moving away from each other and becoming more scattered, instead of forming the cell clusters typical of epithelial cells.

Cell migration is not intentionally controlled or manipulated in cell bioprocessing. In the case that adherent cells are plated on a surface for cultivation, they may redistribute themselves after cell plating. However, since their movement is largely random, their ability to distribute themselves through contact avoidance is limited to a local region of tens of cell length. It is still important to distribute them uniformly during cell inoculation.

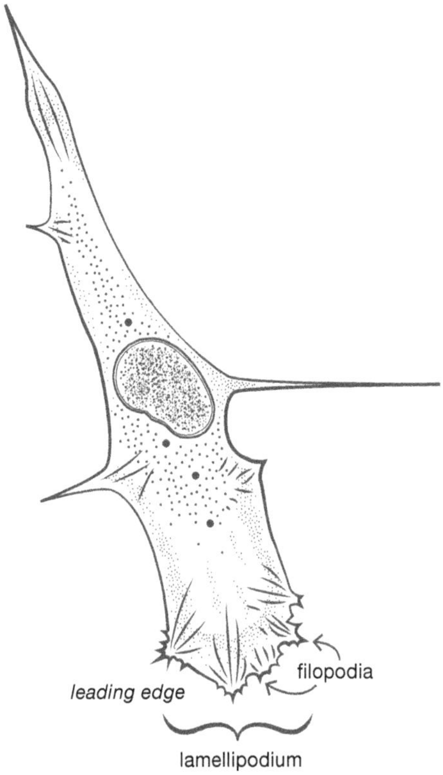

Figure 2.18. A fibroblast in locomotion. Lamellipodia and a few filopodia extend from the leading edge.

Panel 2.33. Cell Movement

- Filopodia extend as a result of actin fibers growing at the cell front, and establish a "grip" on the surface
- The subsequent dissociation of cell–substrate contact at the rear of the cell allows the cell's center of mass to move forward

Growth Control

THE GROWTH OF AN ANIMAL from embryo to adult is regulated at cellular levels in different tissues through the control of the increase in both cell mass and cell number. The increase in cell mass and cell number may not always be synchronous, but over the course of a few cell generations they occur simultaneously. Although we see only the growth aspect of the development, at a cellular level the growth and sustenance of an

Panel 2.34. Growth Control

Growth is the Balance of

- Positive factors: promote increasing cell mass/cell division, suppress cell death
  - Availability of nutrients, growth factors, mechanical tension through the cytoskeleton and surface adhesion
- Negative factors: promote programmed cell death, suppress growth
  - Sensing presence of death signal, apoptosis-triggering stress

organism involves not only cell growth, but also cell death. Both the growth and death of a cell are tightly regulated, or programmed. The most well-studied programmed cell death is apoptosis. Cell growth control and apoptosis will be the focal points of the following discussion.

The growth of a cell is the manifestation of a delicate balance between positive and negative regulations that respond to signals both from outside and inside the cell (Panel 2.34). Positive regulations stimulate cell growth and proliferation and suppress cell death mechanisms, while negative regulations suppress growth and promote cell death. External signals from the environment tell cells the availability or absence of nutrients and other factors necessary for DNA replication and biomass synthesis. Meanwhile, internal signals assess the cell's structural integrity and physiological readiness for cell division and modulate cellular programs to increase cellular component content, divide, or die. Apoptosis is regulated in a similar way to growth control, by pro- and anti-apoptotic factors that respond to both internal signals of cell damage and external signals of developmental cues.

Cell Cycle and Growth Control

Eukaryotic cells progress cyclically through four stages in their undertaking to grow in number: G_1, S, G_2, and M phases (Figure 2.19, Panel 2.35). G_1 and G_2 refer to the gap phase. S and M phases derive their designation from DNA synthesis and mitosis, respectively. The four stages constitute a cell cycle which is repeated every time a single cell becomes two daughter cells. In G_1 phase, cells increase in mass and cell size. Its duration is variable depending on the chemical and physical environment of the culture. The cell will also check whether the internal and external conditions are ready for proceeding to the next phase of the cell cycle, S phase. A restriction point is considered to exist before the entry into S phase. Cells pass the restriction point and enter the S phase only if cellular conditions are right, the necessary external positive mitogenic factor is available, and the negative inhibitory factor is absent. For example, PDGF (platelet derived growth factor) is required for the proliferation of fibroblasts, EGF (epidermal growth factor) for epithelial and many non-epithelial

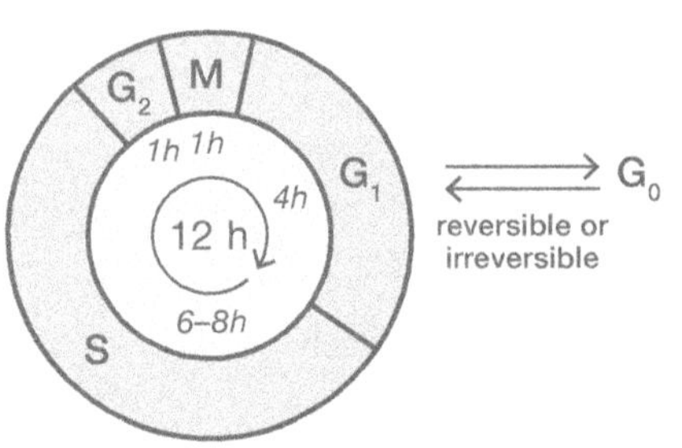

Figure 2.19. The cell cycle for a mammalian cell.

cells, and EPO for red blood cell precursors. The entry into S phase for anchorage-dependent cells additionally requires established contact between integrin and ECM and the appropriate tension in the cytoskeletal network. Countering the actions of the positive factors are those factors that provide signals to cause growth arrest. For example, TGFβ (transforming growth factor-β) inhibits the proliferation of many cell types. Cell-cell contact after anchorage-dependent cells have grown to a confluent state causes growth to cease. For cells that require the establishment of cell–cell contact for growth, the dissociation of adherent junctions between cells disrupts the cells' internal signaling networks and causes growth arrest. The process of mitosis and cytokinesis takes about an hour for a mammalian cell, while the S phase takes around 6 hours. While the doubling time (i.e., a period of cell cycle) changes under different culture conditions, G_1 phase is the stage that is extended most.

Panel 2.35. The Cell Cycle

- Proliferative cells go through the G_1, S, G_2, and M phases of the cell cycle
- Quiescent cells exit G_1 and enter G_0
  - G_0 can be reversible and the differentiated cells returned to the G_1 phase (e.g., hepatocytes), or irreversible for terminally differentiated cells (e.g., nerve cells)
  - Major professional secretory cells in the human body (hepatocytes, plasma and pancreatic cells) are at G_0 phase
- The progression of the cell cycle is regulated by different cyclins, Cdks, and CDI inhibitors, each of which is dynamically expressed
- Cell cycle control is linked to programmed cell death control through factors that affect both

Cells that are in a long period of quiescence, such as terminally differentiated cells, divert from G_1 to exit the cell cycle and enter G_0 stage. G_0 stage may exist in two different types: a resting state that allows cells to eventually reenter the cell cycle and proliferate, or a state that has permanently inactivated the cell cycle reentry mechanism. For example, terminally differentiated cells such as neuronal cells in the body are at a permanent G_0 stage. Fibroblasts in the body and in culture that are under amino acid starvation conditions enter a resting G_0 stage. They can reenter G_1 phase and resume growth under appropriate conditions.

During S phase, DNA replication takes place. DNA replication in the somatic cells of an organism has a very high fidelity, at the sequence level as well as the macroscopic structural level of chromosome. All $>10^{12}$ cells in our body have basically the same genome sequence and chromosomal organization. At the end of S phase, a cell has duplicated its genome or DNA content in the correct sequence and with an extremely low error rate. This may not be the case for some cancer cells or cell lines in culture. For example, at DNA sequence levels, segments of genomic DNA in CHO cells can be seen to have gained or lost copies compared to the rest of the genome. When CHO cells, which are known for having an aneuploid karyotype, are derived from a clone, they originally have the same chromosomal organization. After a few generations, however, their offspring have very different numbers of chromosomes and composition.

Following G_2 phase, the replicated genome, in the form of two sets of chromosomes, segregates in the M phase (called cytokinesis). As the nucleus divides into two, each of them receives one set of the chromosomes that constitutes a diploid genome. The cytoplasmic materials and nuclei are then distributed to two daughter cells. The process of cytokinesis has a very high fidelity. However, this is not the case in continuous cell lines and cancer cells. Errors in DNA synthesis and repair in those abnormal cells give rise to abnormal chromosomes, and mistakes in cytokinesis result in a heterogeneous population where different cells have different makeups of chromosomes (i.e., abnormal karyotypes). Such lack of fidelity in genome replication control is part of the nature of those cells. After isolating a single cell to start a new cell population, the chromosome makeup of the population becomes heterogenous again, even though the entire population was from a single ancestor karyotype.

Cyclins and CDKs

The progression through the cell cycle (G_1, S, G_2, and M) is positively regulated by cyclins and cyclin-dependent kinases (CDKs) and negatively controlled by CDK inhibitors (CDIs). Stage-specific cyclins for the G_1, G_1/S, S, and M phases rise and fall during the cell cycle and interact with different CDKs to form cell cycle stage-specific Cyclin-Cdk complexes. The different Cyclin-Cdk complexes are further regulated at activity levels by activating and inactivating phosphorylation and by complexing with CDI. Additionally, Cyclin-Cdk complexes are subject to proteolytic regulation (Figure 2.20).

Each of these regulatory proteins displays a characteristic dynamic profile throughout the cell cycle (Figure 2.21). The activity profile of the Cyclin-Cdk complex is the result of the expression and interactions of its components, which are further subject to activation, inactivation, and degradation, all choreographed to a rhythm. Importantly, the choreography is dynamic, able to respond to environmental cues such as nutrient abundance or starvation, DNA damage, or chromosomal segregation disorder.

An important cell cycle checkpoint occurs during the transition from G_1 to S phase. The pivotal players in the G_1/S phase transition are the regulatory retinoblastoma protein (Rb) and the Cdk4/6-Cyclin D complex (i.e., G_1-Cdk complex) (Figure 2.20). In growth-arrested or quiescent cells, Rb is unphosphorylated and binds to and inhibits its E2F. E2F is a transcription factor that activates the transcription of Cyclin E and other proteins for S phase. When in a Rb bound state, it suppresses their transcription. Upon stimulation of mitogen and other cues, ERK (a pivotal player of the growth regulatory signaling system) activates the G_1-Cdk complex (via a transcription factor like c-Myc) to phosphorylate Rb. The phosphorylated Rb dissociates from E2F, leading to the activation

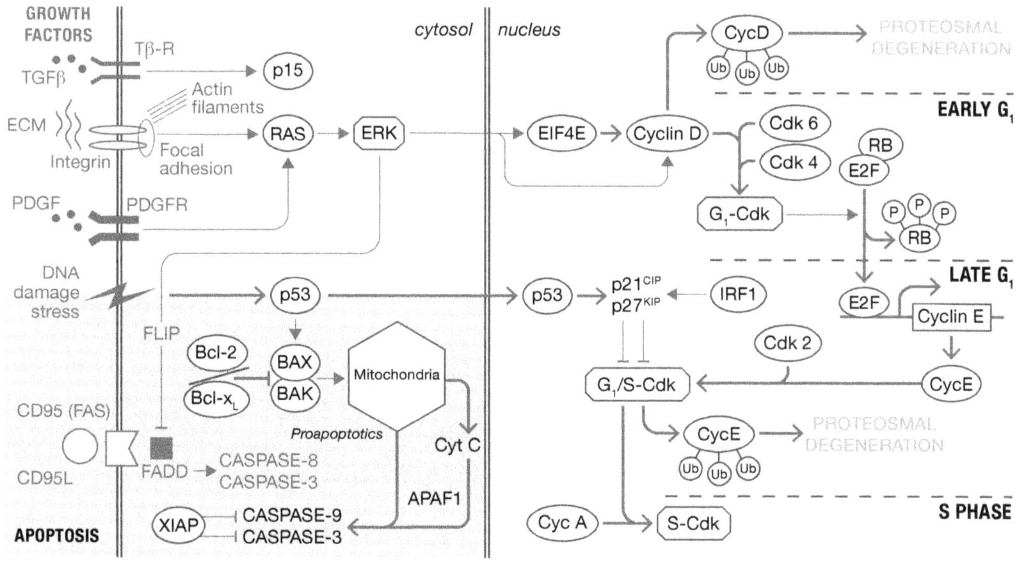

Figure 2.20. Schematic representation of the interactions between cell growth regulation and apoptosis regulation. Cdk: cell cycle-dependent kinase, IRF1: interferon regulatory factor 1, RB: retinoblastoma, ERK: extracellular signal-regulated kinase, FADD: FAS-associated death domain protein, FLIP: FLICE-inhibitory protein, EIF4E: eukaryotic translation initiation factor 4E, Cyc: cyclin, XIAP: cross-linked inhibitor of apoptosis proteins, APAF1: apoptotic peptidase activating factor 1, BAX: Bcl-2 associated X protein, BAK: Bcl-2 homologous antagonist/killer, Ub: ubiquitination, Cyt C: cytochrome C.

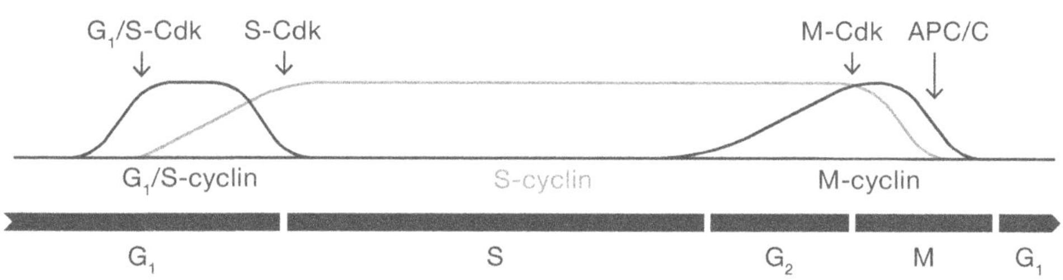

Figure 2.21. Dynamic expression of cyclins in different phases of the cell cycle.

of expression of Cyclin E that in turn drives Cdk2 activation and the formation of cyclin E-Cdk2 (G_1/S Cdk complex). E2F positive feedback on the phosphorylation of Rb causes hyperphosphorylation of Rb and activation of Cdk2. The S phase cyclin, Cyclin A, begins to accumulate in late G_1 phase, but is bound by an inhibitor. The G_1-Cdk complex facilitates the removal of the inhibition to enable the accumulation of S-Cdk complexes, marking the entry into S phase.

The requirements of internal and external conditions for passing the restriction point are strictly followed in normal cells, but not in cancer cells and continuous cell lines. Virtually all cell lines used for the production of therapeutic proteins, including CHO, BHK, HEK293, and mouse myeloma cells, have lost their normal growth control. Their cell cycle checkpoint controls have been compromised and the requirements of mitogens and cell-adhesion complex formations have been relaxed. However, the molecular mechanisms responsible for growth control relaxation are not completely characterized in every cell line. CHO cells have an amplified genomic region that encodes for c-Myc gene, an oncogene, and a mutated TP53, a tumor suppressor gene. The product of gene TP53, p53, mediates external stress signals to cell cycle regulators to constrain the cell cycle. Its genome is also highly structurally abnormal in terms of DNA duplication and deletion and chromosomal reorganization. What additional genomic alterations exist in the cell cycle control in CHO cells that give them their highly erratic growth behavior (e.g., growth factor- and anchorage-independence) is still not clear.

Programmed Cell Death and Apoptosis

Most cell death in an organism is part of a regulated or programmed process, rather than simply the result of extensive damages. Apoptosis is the process of regulated cell death in response to developmental cues or to accumulating non-lethal stresses (Panel 2.36). There are many reasons a cell might be programmed to die for the benefit of the organism. In some developmental events, more cells are generated than will eventually constitute a differentiated tissue, and the excess ones will undergo apoptosis and die. The dead cells are subsequently ingested by neighboring cells and disappear. This is seen in neuronal tissue development, where nerve cells are produced in large numbers. The cells require positive survival signals that are limited in number. Those which receive a sufficient quantity of positive signals survive to constitute the neuronal tissue, while those that do not undergo apoptosis and die out. Some cells are generated to serve their function only for a set period of time. After they outlive their intended active time span, they are programmed to die. For example, immune cells, like antibody-producing plasma cells and natural killer cells, serve their functions for a period of time and then enter apoptosis and die. Cells that have been

Panel 2.36.
Programmed Cell Death and Apoptosis

- Most cell death is programmed
- Purposeful cell death
  - Developmental need, provides space for other cells to grow
  - No longer needed in the organism
  - Infected by a pathogen or otherwise damaging to the organism (e.g., cancerous)
- Programmed cell death includes apoptosis and necroptosis
- Two types of apoptosis:
  - Extrinsic pathway: mediated by death receptor
  - Intrinsic pathway: stress signal causes cytochromes and proapoptotic components in the mitochondria to release, activating proapoptotic proteins and triggering caspases to degrade cellular components
- Caspases-mediated cell destruction exhibits characteristic DNA fragmentation, nuclei condensation, and blebbing. *In vivo*, the dead cell is phagocytosed by other cells.

infected by pathogens and damaged also enter programmed cell death, thus reducing the spread of the pathogen to other uninfected cells.

The types of programmed cell death are traditionally grouped by the morphological changes of dying cells, although the molecular mechanisms have become better known in recent years. Apoptosis is marked by specific cell morphological changes: mitochondrial membrane rupture, DNA condensation and fragmentation, chromatin shrinkage, and membrane bulging. The final intracellular event involves a series of cascades leading to cellular destruction. *In vivo*, the dead cell and debris are often engulfed by neighboring phagocytotic cells before their complete rupture. The second type of cell death, necroptosis, is similar to necrosis, i.e., cell death caused by accumulation of damages. It is marked by the rupture of the cytoplasmic membrane and the release of the cell's contents, thus possibly contributing to inflammatory responses such as seen in the programmed cell death of immune cells.

Apoptosis associated with developmental events is mediated by the binding of extracellular ligands to the death receptor on the cell membrane, called the extrinsic pathway. The other pathway, the intrinsic pathway, is activated under cell stresses, such as nutrient depletion, growth factor deprivation, virus infection, or metabolite accumulation. The intrinsic pathway is also referred to as the mitochondrial pathway. The stress induces an intracellular signal that activates the intrinsic pathway and thus causes cell death.

The final acts of self-destruction through both types of pathways are similar in all apoptosis mechanisms. However, the initiating "signal" of the death receptor pathway and the mitochondrial pathway are different (Figure 2.20).

Extrinsic pathway

In many developmental events, individual cells serve their function for only a finite period of time. These cells are programmed to enter apoptosis after their functional life span is reached. Their survival is dependent on the presence of positive factors and the absence of negative effectors. Developmentally related apoptosis is largely regulated by death receptors on the cell surface. The death receptor pathway is mediated by the binding of the ligand to death receptor.

Cells subject to developmentally regulated apoptosis express death receptors, such as the Fas death receptor, on their surface. The binding of an external Fas ligand to the death receptor recruits an adaptor molecule, Fas-associated death domain (FADD), to the cytoplasmic end of the receptor. The presence of FADD causes pro-caspase 8 or 9 to associate with the death receptor, forming a death-inducing signaling complex (DISC). The caspase is then proteolytically activated, triggering the activation of a series of downstream effector caspases (3, 6, and 7). The activation of

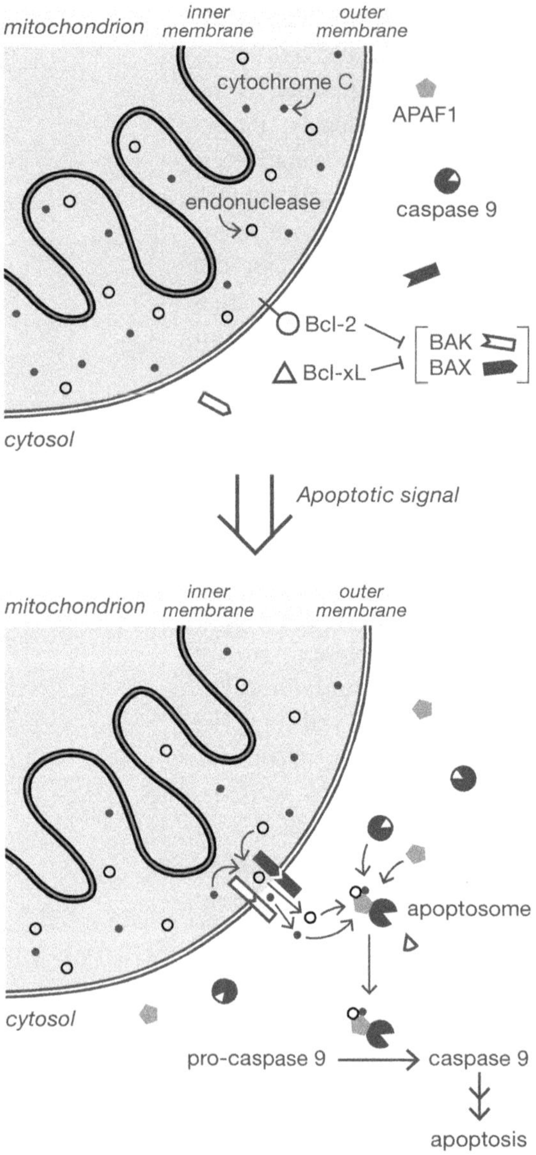

Figure 2.22. An intrinsic (mitochondrial) pathway of apoptosis.

these effector caspases leads to the final stages of cell destruction.

Intrinsic (mitochondrial) pathway

In addition to their role in energy metabolism, mitochondria also play a key role in the regulation of apoptosis. Some pro-apoptotic proteins are sequestered in the space between the outer and inner membranes of mitochondria. Cytochrome C, a hemeprotein that is an important component of the cytochrome C complex in the electron transport chain, is associated with the inner membrane of mitochondria. The release of cytochrome C and those pro-apoptotic proteins in stressed cells initiates the intrinsic pathway of apoptosis (Figure 2.22). The cytochrome C released into the cytoplasm proceeds to form a complex with APF1, pro-caspase 9, and dATP, known collectively as the apoptosome. In the apoptosome, the inactive pro-caspase 9 becomes activated and subsequently activates downstream caspases.

The mitochondrial apoptosis pathway involves positive pro-apoptotic and negative anti-apoptotic factors. The Bcl-2 family that consists of over 20 pro- or anti-apoptotic proteins is a major player in the mitochondrial apoptosis pathway. The pro-apoptotic subfamily includes BAX, BAK, and BOK, which all contain BH1, 2, and 3 homology domains.

Upon exposure to death signals, BAX undergoes conformational changes and translocates to the mitochondria, where it inserts itself into the outer mitochondrial membrane and forms channels. These channels allow the leakage of cytochrome C and other pro-apoptotic molecules.

Two anti-apoptotic proteins, Bcl-2 and Bcl-xL, counter the actions of the pro-apoptotic components. Bcl-2 is localized on the mitochondrial membrane and inhibits the release of pro-apoptotic molecules from the mitochondria by maintaining membrane integrity. Bcl-xL is localized in the cytoplasm and binds to pro-apoptosis members of the Bcl-2 family. The involvement of multiple protagonistic and antagonistic factors ensures tight control of the apoptotic event.

For cells in culture, apoptosis also occurs under some stress conditions. It is tempting to delay apoptosis and prolong the duration of cell cultivation by using chemical means or genetic modification. For example, a large number of caspase inhibitors have been explored for use *in vitro* as well as in animal models. Overexpression of anti-apoptotic genes in cultured cells has been shown to delay the decline of cell viability in the late stages of fed-batch culture.

Concluding Remarks

THIS CHAPTER HAS PROVIDED an abbreviated overview of the key biological features of cells that are essential for biotechnologists to practice cell culture process. The structure and makeup of cells gives them their functional versatility but also constrains their capability. While exploiting their biological versatility, we must also understand the cells' structural and functional constraints and biological limits. We must keep in mind that the objective of cell bioprocessing is to fully harness cells' biological potential for technological applications. For that, the limit may not be bounded by a cell's nature; rather, engineering can be used to facilitate the fulfillment of our goal. By equipping ourselves with a better knowledge of cells' capabilities and limits, we will be able to push the technological boundary further.

References

1. Freshney, I., *Biology of cultured cells*, in *Culture of animal cells: a manual of basic technique and specialized applications*. 2010, John Wiley & Sons, Inc. p. 11.
2. Alberts, B., Johnson, A., Lewis, J., Morgan D., Raff M., Roberts, K., and Walter, P. *Molecular biology of the cell (6th Edition)*. 2015, Garland Science.

Contents

3
Cell Physiology

Introduction

Cells in culture take up sugar, amino acids, lipids, and nucleosides from their growth medium. They metabolize these compounds to derive energy and use them to generate more cell mass, divide into more cells, and produce products. Proteins, which are polymers of amino acids connected by peptide bonds, constitute over 50% of the dry mass in a typical cell. Protein synthesis is an energetically costly process. Each peptide bond costs at least 3 ATP, which is nearly 1/10 of the energy generated by the oxidization of one glucose. A high-producing recombinant cell may produce over 40 pg per day of IgG protein. Each cell has about 400 pg of cell mass (or about 200 pg of cellular proteins) and can double its biomass in a day. The production of protein products is therefore a major energetic load for cells.

A classic cell culture medium contains 2–5 g/L of glucose, and somewhat lower levels of amino acids (about 0.8 mM, or 1 g/L). The consumption of these nutrients typically generates only about 2–5 x 10^9 cells/L, or approximately 0.2–0.5 g/L of cell dry mass. The efficiency of producing cell mass from glucose and other nutrients is rather low.

Glucose is the most important source of energy for most cells. Cultured cells also consume a significant quantity of glutamine, second only to glucose. The pathway for glucose metabolism (called glycolysis) is a key metabolic pathway and is virtually identical in all mammals. The same pathway, with a couple of additional entry reactions, is used when another sugar, such as galactose or fructose, is used as the carbohydrate source.

The complete oxidation of one glucose molecule consumes 6 O_2 and generates 6 H_2O and 6 CO_2 (Panel 3.1). The complete oxidation of glutamine uses 4.5 O_2 and produces 2 H_2O and 5 CO_2. For cells in culture, however, the majority of consumed glucose is not completely oxidized; rather, it is converted to lactate and excreted. By converting to lactate instead of completely oxidizing to CO_2, much less energy is derived from each mole of glucose. This is the root cause of the low efficiency in the production of cell mass from glucose.

This type of "wasteful" metabolism is common to almost all vertebrate cells in culture. For bioprocessing, the accumulation of metabolic byproducts inhibits cell growth and impedes productivity.

Panel 3.1. Reaction Stoichiometry of Breaking Up of Glucose and Glutamine

Glucose Oxidation

$$\underset{\text{(glucose)}}{C_6H_{12}O_6} + 6\,O_2 \rightarrow 6\,CO_2 + 6\,H_2O \qquad \textbf{(Eq. 3-1)}$$

Glucose Anaerobic Metabolism

$$C_6H_{12}O_6 \rightarrow 2\,\underset{\text{(lactate)}}{CH_3 \cdot CHOH \cdot COOH} \qquad \textbf{(Eq. 3-2)}$$

Glutamine Oxidation

$$\underset{\text{(glutamine)}}{CO(NH_2) \cdot CH_2 \cdot CH_2 \cdot CH\,(NH_2) \cdot COOH}$$

$$+\,4.5\,O_2 \rightarrow 2\,NH_3 + 5\,CO_2 + 2\,H_2O \qquad \textbf{(Eq. 3-3)}$$

Cells invariably produce lactate from glucose when growing rapidly. However, under some conditions (e.g., in the stationary phase of fed-batch culture), lactate may also be consumed. It is not unusual that under seemingly identical culture conditions using the same cell line and the same standard operating protocols, cells may have different metabolic behaviors. In some runs, lactate is produced throughout the fed-batch culture, from the rapid growth phase until the stationary phase. But in other runs, lactate production in the exponential growth phase switches to lactate consumption in the stationary phase.

When the metabolic behavior of a cell culture process fluctuates, it is commonly seen that the high productivity runs switch from lactate production to lactate consumption, while low productivity runs remain in lactate production mode throughout the culture (Figure 3.1). Such observations attest to the importance of cell metabolism in influencing productivity. This chapter discusses the biochemical pathways that are of the utmost importance to cells in culture.

Glucose and Energy Metabolism

GLUCOSE IS MAINLY CATABOLIZED through three pathways: glycolysis, the pentose phosphate pathway (PPP), and the tricarboxylic acid (TCA) cycle (Panel 3.2). In glycolysis, one mole of glucose is converted into two moles of pyruvate. In this segment of catabolism, only a small fraction of the chemical potential energy of glucose is converted into the "usable" form of chemical potential energy in the cell (i.e., ATP). Two moles each of ATP and NADH are generated from each mole of glucose.

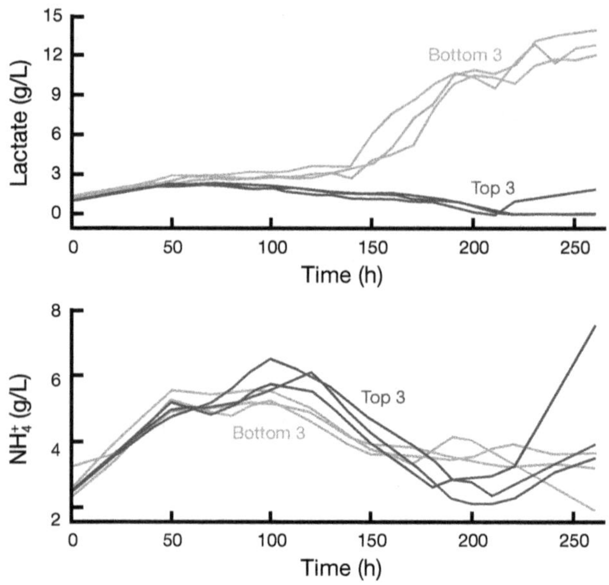

Figure 3.1. Lactate and ammonium profiles in manufacturing runs. The three runs with high product titers (dark gray) and low product titers (light gray) are shown. Lactate was produced and accumulated, and then either continued to increase or was consumed. The ammonium concentration profile was similar among all six runs. The lactate profile correlates to productivity, but not ammonium. Data from: Charaniya, S., et al., *Mining manufacturing data for discovery of high productivity process characteristics.* J Biotechnol, 2010. **147**(3-4): p. 186–97.

Panel 3.2. Metabolic Pathways of Glucose Oxidation

Main Metabolic Pathways of Energy Metabolism

- Glycolysis
  - May operate without oxygen
  - Low energy yield
- TCA cycle (tricarboxylic acid cycle, Krebs cycle)
  - Oxygen-dependent
  - Transfers chemical potential energy into NADH
- Pentose phosphate pathway
  - Generates five-carbon sugars and NADPH for reductive biosynthesis

Pyruvate may enter the TCA cycle for further oxidation, or it may become a shunted product as lactic acid (at a neutral pH it exists as lactate). During the TCA cycle, the carbon skeleton of glucose is finally broken down into CO_2. PPP is a shunt from glycolysis. It generates five-carbon sugars for nucleoside synthesis and supplies NADPH for many biosynthesis reactions and to maintain a balanced redox state in the cell.

In eukaryotic cells, glycolysis and PPP take place in the cytosol, while the further oxidation of pyruvate to CO_2 occurs in the mitochondria. It is in the mitochondria that the majority of the chemical potential energy of glucose is converted to ATP for use in cellular synthesis and other energy-dependent cellular processes.

Oxidation of Glucose

Glycolysis

Glucose from the medium is imported into cells via the glucose transporter. During glycolysis, high-energy compounds (ATP and NADH) are generated. However, the first segment of glycolysis actually consumes two ATP for each glucose (Figure 3.2, Panel 3.3). The two ATP are used to add a phosphate group to each end of the glucose molecule. The first phosphorylation converts glucose to glucose 6-phosphate (G6P). After isomerization to fructose 6-phosphate (F6P), the second

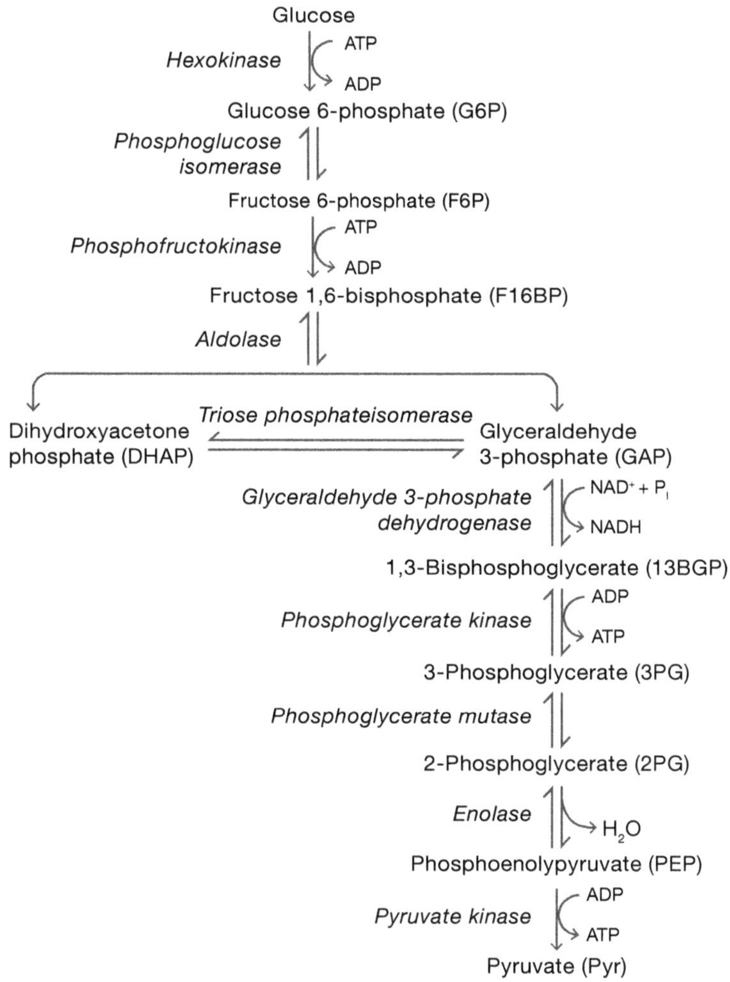

Figure 3.2. Glycolysis pathway. Only the chemical species are shown, not their stoichiometric relationships. The abbreviation of each compound is also shown. One mole of glucose yields two moles of pyruvate.

Panel 3.3. Glycolysis

- Each mole of glucose
  - Consumes 2 moles of ATP (to activate glucose to fructose 1,6-bisphosphate)
  - Produces 2 moles of NADH, 4 moles of ATP (by converting $F16P_2$ to 2 moles of pyruvate)
- Net:
  - Produces 2 pyruvate
  - Produces 2 ATP, 2 NADH
  - Requires 2 ADP, 2 NAD, plus glucose as reactants

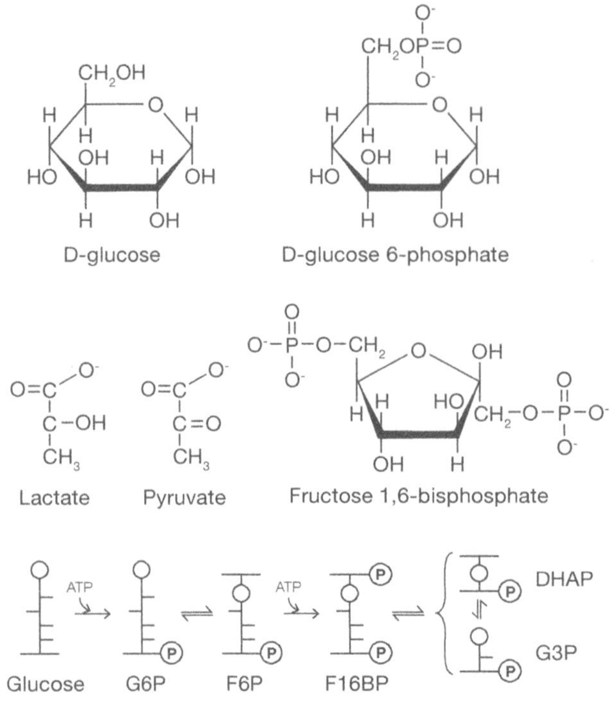

Figure 3.3. The structures of key compounds in glycolysis and a schematic representation of the hexose-splitting reactions in glycolysis. The carbonyl group in sugar is represented by a circle.

phosphate is added to give fructose 1,6-bisphosphate (F16BP). The two phosphate groups, being nucleophilic centers, help pull their surrounding electron clouds toward the two ends of the molecule, thereby making the carbon-carbon bond in the middle of the glucose molecule susceptible to enzymatic cleavage (Figure 3.3). The six-carbon F16BP becomes two three-carbon compounds: glyceraldehyde 3-phosphate (GAP) and dihydroxyacetone-phosphate (DHAP). These two compounds are interconvertible through a reversible reaction. The continued utilization of GAP toward downstream reactions effectively draws DHAP toward GAP. The original 1 mole of glucose now becomes 2 moles of GAP and moves further downstream in glycolysis. The conversion of 2 GAP to the end product of 2 pyruvates also converts 2 NAD^+ and 4 ADP to 2 NADH and 4 ATP. The net energetic consequence of the conversion of glucose to 2 pyruvate in glycolysis is the generation of 2 ATP (recall 2 ATP are consumed to activate glucose) and 2 NADH. Note that NAD^+ is often denoted as NAD in the text.

TCA cycle

The further oxidation of pyruvate takes place in the mitochondria. After passing through a transporter and entering the mitochondrial matrix, pyruvate is first decarboxylated to become acetyl-CoA while releasing 1 CO_2 (Figure 3.4a). The decarboxylation reaction breaks a carbon-carbon (C–C) bond. The energy of the C–C bond is preserved in the 1 NADH generated and in the high energy bond of acetyl-CoA. Acetyl-CoA is then fed into the TCA cycle where it is broken down into 2 CO_2 after going through one round of the TCA cycle. The pathway is cyclic, with four- to six-carbon skeletons cycling in a loop (Figure 3.4a). At the beginning of the cycle, acetyl-CoA uses its high energy bond to form a new C–C bond on the four-carbon oxaloacetate (OAA) to become six-carbon citrate. Citrate has three carboxylic acid groups, hence

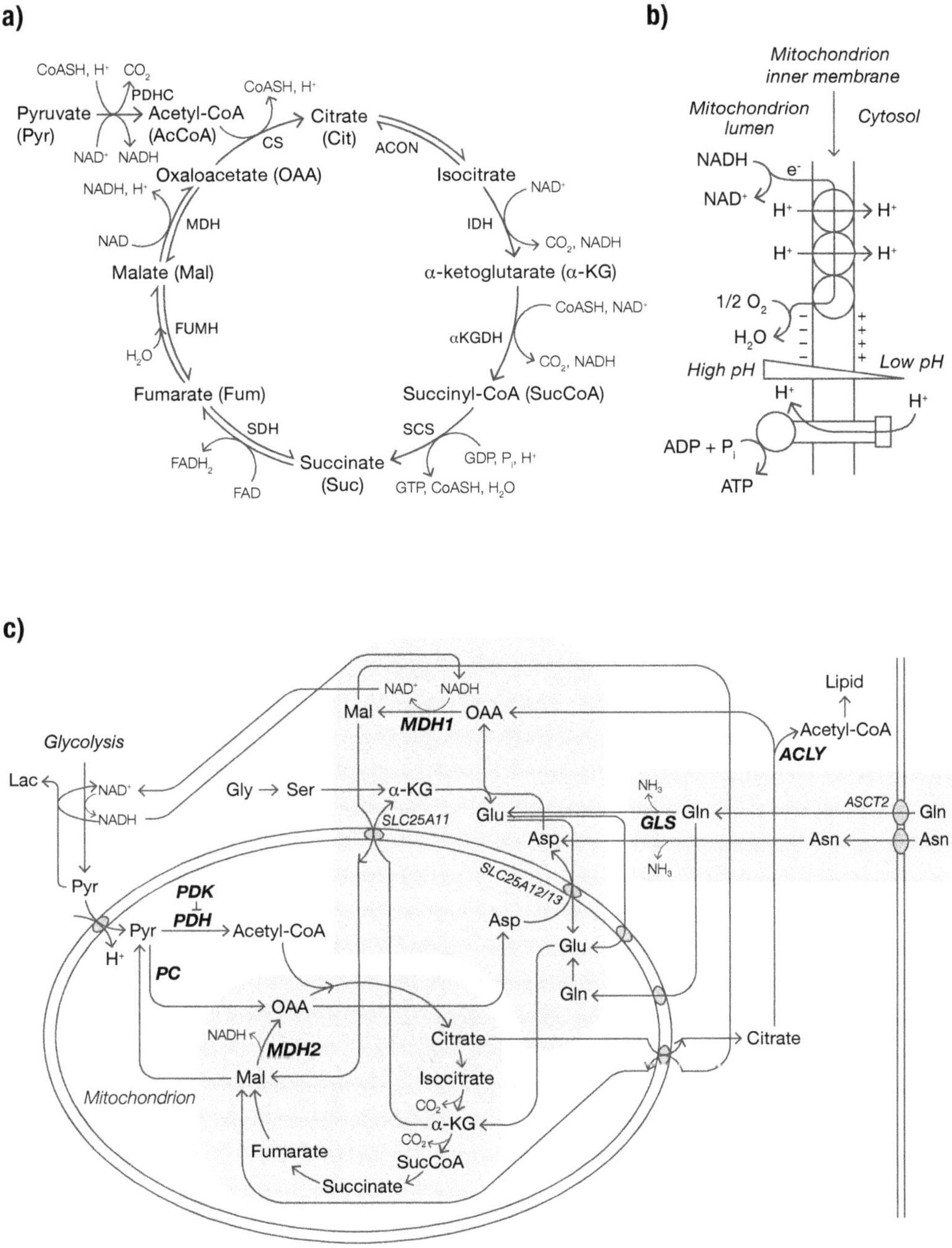

Figure 3.4. (a) The tricarboxylic acid cycle. (b) Electron transfer chain and oxidative phosphorylation. (c) The metabolite reaction network centering around a mitochondrion. Many metabolites, including α-KG, Glu, Mal, Asp, OAA, and citrate, are shared in multiple reactions/co-transports.

the name "tricarboxylic acid cycle" (Panel 3.4). The TCA cycle is also known as the citric acid cycle and the Krebs cycle. The structure of citrate and some other TCA cycle intermediates and their derived amino acids are shown in Figure 3.5.

The 6 carbon atoms in glucose which enter the mitochondria as 2 pyruvate are converted into 6 CO_2 after one round of the TCA cycle. This is as seen in the oxidation reaction of glucose, where 1 mole of glucose generates 6 moles of CO_2 (Panel 3.1). However, no molecular oxygen participates in any reaction in glycolysis or the TCA cycle. The 6 CO_2 are released through decarboxylation reactions; they are derived from the carbon skeleton of pyruvate, citrate, and α-ketoglutarate without the participation of molecular oxygen. In two of these reactions, the carboxylic acid group next to a 2-keto group is released (catalyzed by pyruvate dehydrogenase and α-ketoglutarate dehydrogenase) and the energy from the breakup of the CO–COO^- bond is preserved in the high-energy compounds acyl-CoA (acetyl-CoA and succinyl-CoA, respectively) and NADH. In the other case, the isocitrate dehydrogenase reaction, one of the three carboxylic acid groups in isocitrate is released and 1 NADH is generated. In the second half of the TCA cycle, succinyl-CoA is further oxidized to regenerate OAA via fumarate and malate and to preserve the chemical potential energy in NADH, $FADH_2$, and GTP.

Panel 3.4. The TCA Cycle

- Pyruvate enters a mitochondrion
- 1 CO_2 is released in pyruvate dehydrogenase reaction, becoming acetyl-CoA (2 C compound)
- Acetyl-CoA enters the TCA cycle by combining with OAA (4 C compound) to become citric acid (6 C compound, has 3 carboxylic acid groups)
- In each cycle, 2 CO_2 are released to regenerate OAA and produce NADH and $FADH_2$

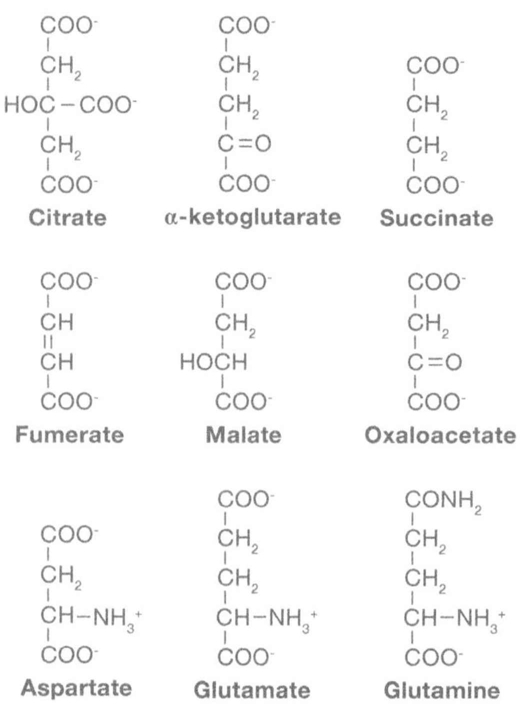

Figure 3.5. Compounds in the TCA cycle and the amino acids that can be derived from their carbon skeletons.

Electron transfer and oxidative phosphorylation

In the oxidation of glucose, if the C–C bond is broken by directly reacting with oxygen, as in the case of the combustion of coal or wood, a very high temperature is necessary to provide the activation energy needed to trigger the reaction. Furthermore, the chemical potential energy in the C–C bond would be released as heat. Cells utilize decarboxylation reactions to break C–C bonds and release CO_2 while preserving the chemical potential energy in NADH.

Oxygen is thereby used to extract the chemical potential energy from NADH and $FADH_2$ in order to generate ATP that can be used in cellular work. A total of 6 oxygens participate in the oxidation of the NADH/$FADH_2$ generated in glucose oxidation, thus accounting for the 6 O_2 required to oxidize 1 glucose as shown in the stoichiometric equation of glucose oxidation (Panel 3.1). This process of transferring chemical potential energy from NADH to ATP and reacting with oxygen is carried out using electron transport and oxidative phosphorylation (oxphos).

Extraction of the chemical potential energy of NADH and $FADH_2$ takes place through an electron transfer chain residing in the mitochondrial inner membrane (Panel 3.5). The high-energy electrons of NADH and $FADH_2$ enter the electron transport chain to move down the energy ladder, mediated by electron carriers including flavin, the iron-sulfur complex, heme, and copper ions which are embedded in a number of large enzyme complexes such as NADH dehydrogenase. At the end of the electron transfer chain, the electron is received by oxygen which then reacts with a proton to form H_2O (Figure 3.4b).

The energy released is then used to trigger a proton pump to drive H^+ out of the mitochondrial inner membrane. The export of H^+ from the mitochondrion creates a ~1.0 unit pH difference across the membrane, as well as about −120 mV of electric potential. Because of the higher pH (lower proton concentration) and excessive negative charge inside the mitochondrial membrane, there is a propensity for the proton ions outside the mitochondrion to cross the mitochondrial membrane and enter the mitochondrial matrix. The proton ion enters the mitochondrion through an ATP synthase embedded in the mitochondrion's inner membrane. While protons pass through ATP synthase, a series of protein conformation changes bring an ADP and a phosphate together to synthesize ATP. The electron transfer and the generation of ATP are often referred to as "oxidative phosphorylation" (Panel 3.6).

Panel 3.5. Electron Transport

- NADH and $FADH_2$ enter the electron transport chain, pass their high energy electron to electron carriers
- As an electron passes the energetic ladder, it pumps protons out of the mitochondrion, increasing the pH inside the mitochondrion by 1.0 and generating a negative charge of ~120 mV across the mitochondrion's inner membrane
- At the bottom of the energetic ladder, the electron and proton react with oxygen to form water

Panel 3.6. Oxidative Phosphorylation Pathway

- The higher concentration of protons in the cytosol and the negative charge inside the mitochondrion drives the proton to move into the mitochondrion
- The proton passes through ATP synthase and converts ADP and P_i to ATP

A general estimation of the energy yield in oxidative phosphorylation is 3 ATP per NADH and 2 ATP per $FADH_2$. However, the amount of ATP generated per mole of NADH or $FADH_2$ (hence also per mole of glucose) varies somewhat among different species due to their variable expressions of ATP synthase. For mammals, about 30–32 moles of ATP

Panel 3.7. Energetic Yield of Glucose Oxidation

Cytosol

$$\text{Glucose} + 2\ \text{NAD} + 2\ \text{ADP} + 2\ \text{P}_i \rightarrow 2\ \text{pyruvate} + 2\ \text{NADH} + 2\ \text{ATP} \qquad \textbf{(Eq. 3-4)}$$

Mitochondria

$$2\ \text{pyruvate} + 2\ \text{CoASH} + 2\ \text{NAD} + 2\ \text{ADP} + 2\ \text{P}_i \rightarrow 2\ \text{acetyl-COA} + 2\ \text{NADH} + 2\ \text{ATP} + 2\ \text{CO}_2 \qquad \textbf{(Eq. 3-5)}$$

$$2\ \text{acetyl-CoA} + 6\ \text{NAD} + 2\ \text{FAD} + 2\ \text{GDP} \rightarrow 6\ \text{NADH} + 2\ \text{FADH}_2 + 2\ \text{GTP} + 4\ \text{CO}_2 \qquad \textbf{(Eq. 3-6)}$$

- In glycolysis and the TCA cycle, glucose carbons never react with O_2
- The energy is preserved in 12 NADH/$FADH_2$ that is converted to ATP in electron transport and oxidative phosphorylation
- The overall energetic yield is ~30 ATP

are generated per mole of completely oxidized glucose (Panel 3.7). Older literature tends to list the number as 36 moles of ATP per mole of glucose. Under some physiological conditions, the electron transfer chain and oxidative phosphorylation are uncoupled. Instead of generating ATP, the energy from NADH is released as heat to maintain body temperature. This is seen in hibernating animals, for example.

The amount of energy, 2 ATP and 2 NADH (or the equivalent of 6 ATP, since 1 NADH in the cytosol can be roughly considered to be 2 ATP), produced from splitting glucose into two moles of pyruvate is only about 1/6 of what can be generated from the complete oxidation of glucose to CO_2 and H_2O. The vast majority of energy conversion in glucose oxidation therefore occurs in the mitochondria.

Pentose Phosphate Pathway (PPP)

The PPP is an important shunt from glycolysis that supplies five-carbon sugars (in the form of sugar phosphate) and NADPH. Five-carbon sugar is converted into ribose and deoxyribose to make nucleotides and deoxynucleotides for RNA and DNA synthesis. NADPH supplies the reduction potential used in many biosynthetic reactions (Figure 3.6, Panel 3.8).

The PPP consists of an oxidative segment and a monosaccharide transformation segment. In the first segment, glucose 6-phosphate from glycolysis is oxidized to generate 2 NADPH and then decarboxylated to form the five-carbon ribulose 5-phosphate. The molecular transformation segment converts the five-carbon sugar phosphates to 1 three-carbon (GAP) and 2 six-carbon sugar phosphates (F6P) to rejoin glycolysis (Panel 3.8).

Cells use two different nicotinamide-adenine dinucleotides as reductive chemical potential energy carriers: NADH and NADPH. NADH is used to store chemical potential energy in glycolysis, the TCA

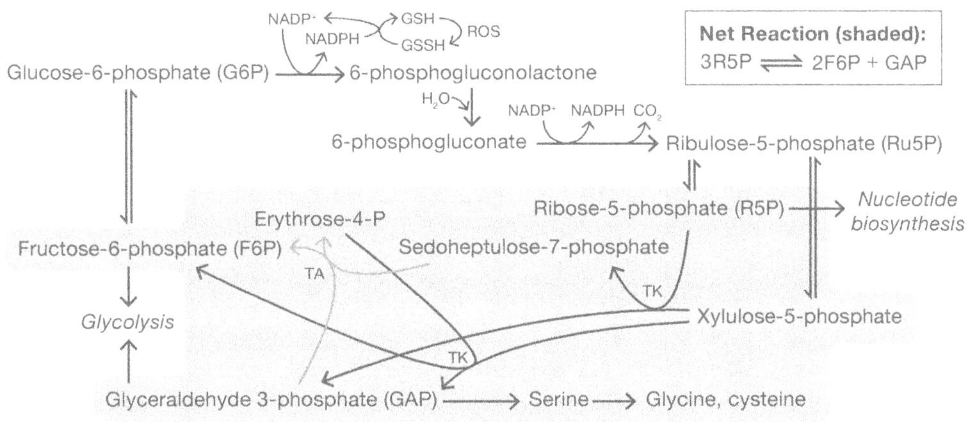

Figure 3.6. The pentose phosphate pathway with its oxidative branch and molecular transformation branch. Isomerization, transaldolase, and transketolase reactions are all reversible. Only one direction of the transformation (or non-oxidative) reactions (in the shaded area) is shown to illustrate the pairing of the reactants and the product of each reaction.

Panel 3.8. Pentose Phosphate Pathway (PPP)

Oxidative Segment

- Releases 1 CO_2, generates 5-carbon sugar phosphate for nucleotide synthesis
- Produces 2 NADPH

Overall Reaction:

$$G6P + 2\,NADP \rightarrow Ru5P + CO_2 + 2\,NADPH \quad \textbf{(Eq. 3-7)}$$

Molecular Transformation

- Interconverts 5-carbon sugar phosphate to 3-carbon and 6-carbon
- Enables NADPH and 5-carbon sugar to be produced at different ratios

Overall Reaction:

$$3\,Ru5P \rightleftharpoons 2\,F6P + GAP \quad \textbf{(Eq. 3-8)}$$

cycle, and lipid catabolism. Eventually, NADH is used to derive ATP in the mitochondria.

NADPH, on the other hand, carries a chemical potential that is used in biosynthetic reactions (e.g., the synthesis of lipids or nucleotides). NADPH is also used to reduce oxidized glutathione and to regenerate it. The reduced form of glutathione is important in maintaining the cell's reductive environment and in the suppression of reactive oxygen species (ROS).

The second segment of the PPP is a molecular conversion pool that allows a 2-C aldehyde (CHO–CH_2–) unit or 3-C keto units (CH_2OH–CO–CHOH–) to be translocated among a number of 3-C to 5-C aldoses and ketoses. These reversible reactions together allow the interconversion of carbohydrate molecules that are three to seven carbons in length. This "mixing pool" enables 5-C sugars from the first segment of the PPP to be connected to glycolysis through 6-C fructose 6-phosphate or 3-C glyceraldehyde 3-phosphate.

The first segment of the PPP generates 5-C ribulose and NADPH at a molecular ratio of 1:2. However, cells do not always need those two compounds at a 1:2 proportion. The molecular conversion in the second segment allows the pentose-phosphate:NADPH ratio to be increased or decreased according to cellular needs. If ribose is needed in a higher proportion, GAP and F6P can be used to generate more pentose phosphate. Conversely, if the pentose phosphate is needed in a smaller proportion, the extra pentose can be converted to F6P and GAP to re-enter glycolysis.

Lactate Formation

In the complete oxidation of glucose to 6 CO_2, 10 NADH and 2 $FADH_2$ are generated. Among them, 2 NADH are produced from glycolysis and the rest from the TCA cycle. They are oxidized in the electron transport chains in the mitochondria, consuming 6 O_2. This regenerates the NAD and FAD that are needed in order to continue the oxidation of glucose. Without the regeneration of NAD from electron transfer, the continued oxidation of glucose would run out of NAD as a reactant and stop. In the absence of oxygen, the reactions in the TCA cycle that are dependent on a continuous supply of NAD also cease to operate. Microorganisms that are capable of anaerobic metabolism can divert the pyruvate to a more reduced compound, such as ethanol or lactate. The fermentative metabolism, in the course of reducing pyruvate, also oxidizes NADH and regenerates NAD to keep glycolysis running to supply cellular energy (Panel 3.9).

Under some physiological conditions, some tissue cells can metabolize glucose just like fermentation in microorganisms. For example,

Panel 3.9. Yeast Fermentation Converting Glucose to Ethanol

$$\underset{\text{(glucose)}}{C_6H_{12}O_6} + 2\,ADP + 2\,P_i + 2\,NAD \rightarrow \underset{\text{(pyruvate)}}{2\,CH_3COCOOH} + 2\,ATP + 2\,NADH$$

$$2\,ATP \rightarrow \text{work} + 2\,ADP + 2\,P_i$$

$$2\,CH_3COCOOH + 2\,NADH \rightarrow \underset{\text{(ethanol)}}{2\,C_2H_5OH} + 2\,CO_2 + 2\,NAD$$

- Regenerate NAD to allow glucose utilization to continue

during exercise, glucose consumption in muscle cells increases to boost energy output, often exceeding the capacity of the TCA cycle in the mitochondria. To keep the higher level of energy generation going, pyruvate is converted to lactate. The reaction converts NADH back to NAD that can be recycled to keep glycolysis running. The lactate produced is then excreted out of muscle cells (Panel 3.10).

Mammalian cells in culture and cancer cells have a high glucose consumption rate and generate pyruvate at 1:2 molar ratio. But only a small portion of the pyruvate generated is transported into mitochondria to further oxidize to CO_2. Cells appear to have a limited capacity to translocate pyruvate into the mitochondria. The rest of the pyruvate generated in glycolysis is converted to lactate. This occurs in spite of the presence of sufficient oxygen. The phenomenon is therefore different from anaerobic fermentation in bacteria or yeast, and is referred to as "aerobic glycolysis."

Lactate synthesis is catalyzed by lactate dehydrogenase. This reversible reaction converts 1 pyruvate and 1 NADH into 1 lactate and 1 NAD. In the conversion of glucose to 2 pyruvate in glycolysis, 2 ATP and 2 NADH are generated while also consuming 2 ADP and 2 NAD. Continued glucose metabolism through glycolysis requires continued supplies of both ADP and NAD as reactants. ATP formed in glycolysis is used by cells to perform many tasks, such as synthesis, maintaining osmotic balance, etc. It is continually being consumed in various cellular reactions and is converted back to ADP to resupply the reactant for glycolysis.

NADH is normally converted back to NAD through the electron transport chain in the mitochondria. To be regenerated in the electron transfer chain, the reducing equivalent of the cytosolic NADH generated in glycolysis must first enter the mitochondria and the regenerated NAD must be exported out of the mitochondria, as will be discussed later. Like the transport of pyruvate into the mitochondria, cells' capacity for transferring the reducing equivalent of NADH into the mitochondria is limited. The lactate dehydrogenase reaction in cytosol allows for

Panel 3.10. Aerobic Glycolysis

Lactate Dehydrogenase Reaction

$$\text{Pyruvate} + \text{NADH} + \text{H}^+ \rightleftharpoons \text{Lactate} + \text{NAD}^+ \qquad \textbf{(Eq. 3-9)}$$

- Cultured cells and cancer cells undergo glycolysis and produce lactate even at high oxygen concentration
- This propensity toward lactate production is not for lack of oxygen (anaerobic glycolysis)
- At a high glycolysis flux, not all NADH can be oxidized by electron transfer in the mitochondria
- Lactate production serves to regenerate NAD to keep glycolysis going

Panel 3.11. Energetic Yield of Aerobic Glycolysis

Oxidation of Glucose to Pyruvate

Glucose + 2 ADP + 2 P_i + 2 NAD^+ →
2 pyruvate + 2 ATP + 2 NADH **(Eq. 3-4)**

Reduction of Pyruvate (Regeneration of NAD)

2 pyruvate + 2 NADH → 2 lactate + 2 NAD^+ **(Eq. 3-10)**

Net:

Glucose + 2 ADP + 2 P_i → 2 lactate + 2 ATP **(Eq. 3-11)**

NAD regeneration from NADH to take place, thereby enabling glycolysis to continue at a high flux (Panel 3.11). But, as a consequence, lactate must be excreted and it eventually accumulates in the medium.

Under typical culture conditions, proliferating cells in culture convert 85–90% of their glucose intake to lactate. A significant portion of the other 10–15% of glucose is converted to CO_2 while the rest is incorporated into cell mass. At the completion of glycolysis, 2 ATP and 2 NADH are generated. This is in contrast to about ~30 ATP generated upon the complete oxidation of glucose. The 90% of glucose converted to lactate generates 1.8 ATP (2 ATP x 0.9), while the other 10% of glucose generates about 3 ATP (30 ATP x 0.1). Aerobic glycolysis generates a significant portion of the total energy of glucose catabolism in proliferating cells.

Not all cells in our body convert a large portion of the glucose they take up into lactate. The vast majority of cells in our body are in a quiescent (non-proliferating) state. They consume less glucose than proliferating cells. Excessive glucose consumption and lactate production, known as the Warburg effect, was first observed between normal tissues and cancer cells. While normal cells have a lower glucose flux, cancer and other fast-proliferating cells consume a larger amount of glucose and convert much of that glucose to lactate.

Carbon Flow and the Supply of Biosynthetic Precursors

Among all of the pathways in the cellular metabolic network, glycolysis has the highest flux in terms of moles of substrate and reaction intermediates passing through. For proliferating cells in culture, the molar flux (based on the number of moles of each compound) or carbon flux (based on the number of moles of carbon atoms, i.e., the number of carbons in the compound multiplied by the number of moles of the compound) of glycolysis is normally several times higher than that of the TCA cycle. PPP flux usually constitutes only about 5% of glucose intake.

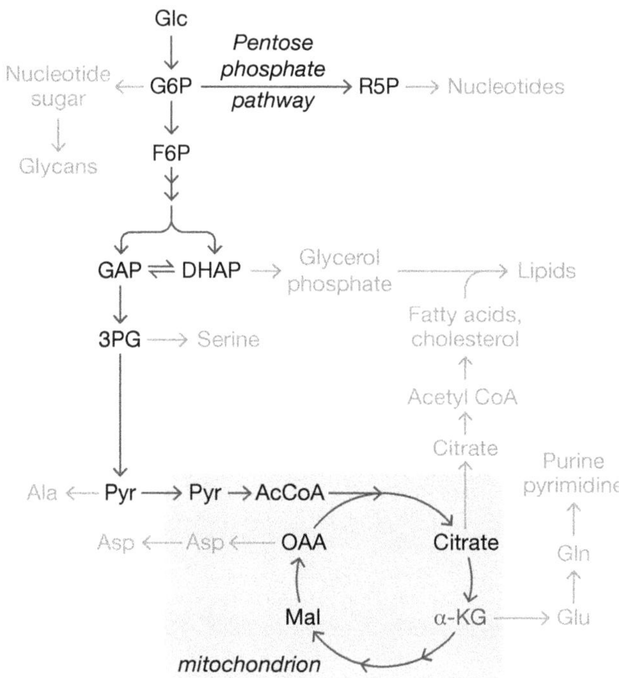

Figure 3.7. Energy metabolism pathways supply precursors for biosynthesis. The degradation of nutrients and cellular materials also feeds into these pathways.

Glycolysis and the TCA cycle also supply precursors to build cellular components. The culture medium does not necessarily supply cells with the right balance of all of the components that they need to synthesize cell mass and maintain cellular functions. The three main pathways for energy metabolism (glycolysis, the TCA cycle, PPP) also provide the precursors for the synthesis of some cellular materials. For example, DHAP in glycolysis supplies glycerol phosphate, which is used in the synthesis of phospholipids (Figure 3.7). Glucose 6-phosphate and fructose 6-phosphate are both sources of nucleotide sugars for glycan synthesis, such as UDP-galactose, UDP-glucose, and GDP-mannose. Except for liver cells (hepatocytes), cells in culture do not express enzymes involved in gluconeogenesis; that is, they cannot make hexose from lactate or amino acids. So, even if cells can derive energy from lactate and amino acids, they will still need hexose to synthesize ribose and glycans. α-ketoglutarate provides a carbon skeleton for glutamate and glutamine while oxaloacetate is used to make aspartate and asparagine.

Cells in culture take up a large quantity of amino acids from medium usually in excess of what is needed to make cell mass and product. Furthermore, the amino acids taken up are not necessarily of the stoichiometric amount that is needed for synthesis. The surplus of nitrogen is either excreted as ammonia or transferred as an amino group to

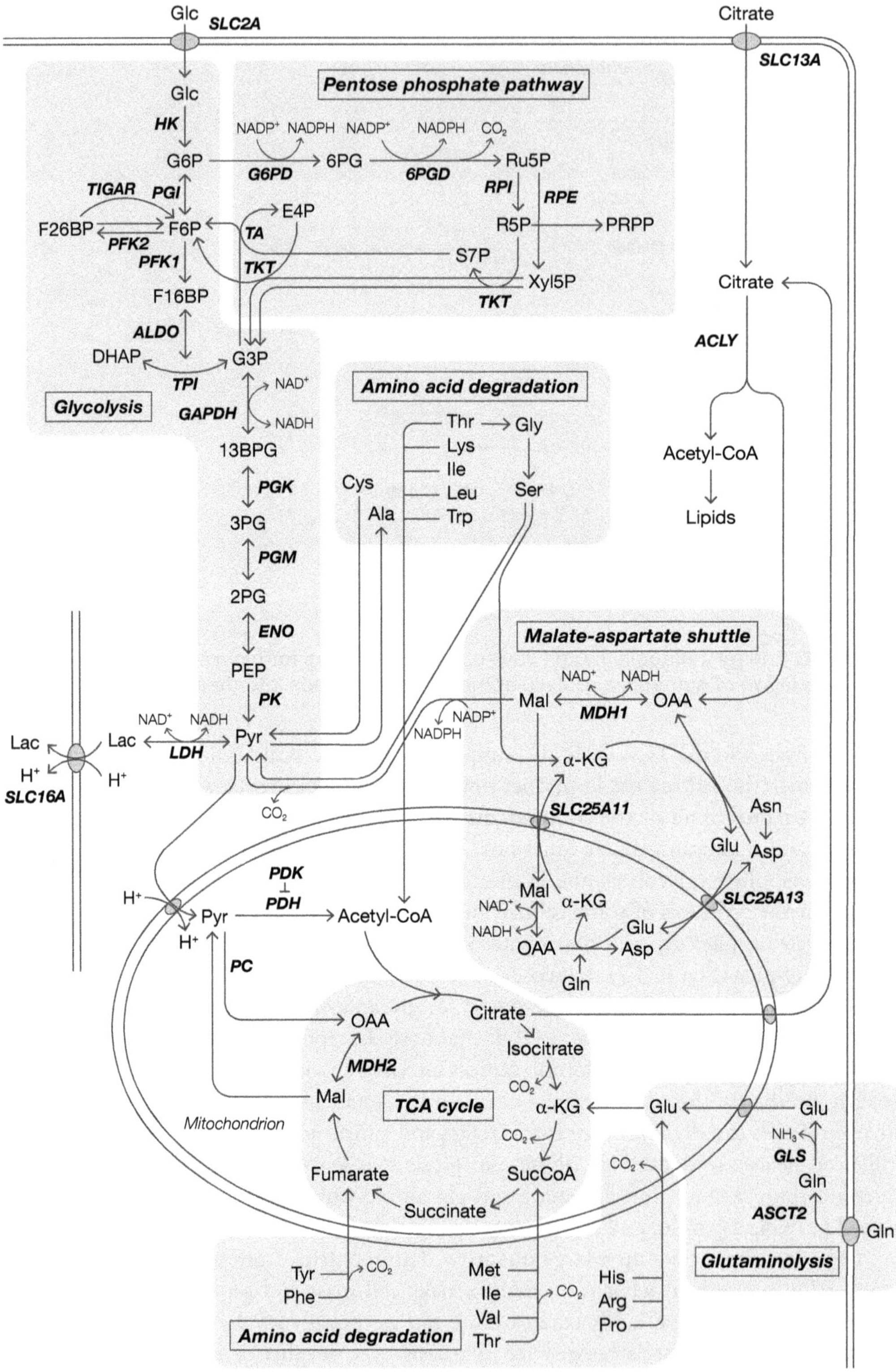

Figure 3.8. Key pathways in energy metabolism and their interconnections.

pyruvate to form alanine, and then excreted. Since alanine is much less growth-inhibitory than ammonium, pyruvate has some moderating effect on ammonium toxicity. The carbon skeletons of the amino acids enter the carbon metabolism pathway through pyruvate, acetyl-CoA, and some TCA cycle intermediates (α-ketoglutarate, succinyl-CoA, oxaloacetate) where they are degraded in the TCA cycle or converted to other amino acids.

At the start of the TCA cycle, one acetyl-CoA combines with one OAA to become one citrate (Figure 3.4a). At the end of the cycle, one OAA must be regenerated to take on another acetyl-CoA for a second round of the cycle. If a portion of the carbon flow initiated from citrate is diverted from the cycle, or if extra carbon enters into the cycle, then the amount of OAA returning at the end will not be what is needed for starting the second round of the TCA cycle.

Examples of carbon infusion include the entry of excess aspartic acid into OAA or glutamic acid through α-KG. A prominent diversion of carbons from the TCA cycle occurs in the export of citrate to the cytosol, where it is converted to acetyl-CoA for fatty acid synthesis. α-KG may also be converted to glutamate and glutamine. Glutamine is then taken to nucleotide synthesis.

To balance the withdrawal and the infusion of TCA cycle intermediates, a number of reactions are in place, including the conversion of pyruvate to oxaloacetate and the conversion of glutamine to glutamic acid and then to α-ketoglutarate (Figure 3.8). Under typical culture conditions, a significant flux of citrate exports from the mitochondria and glutamate to nucleotide synthesis are needed to support cell growth. The deficit in the OAA supply is made up for by a high consumption rate of glutamine. The replenishment of the carbon deficit in the TCA cycle, called anaplerosis, will be discussed later.

Regulation of Glucose Metabolism

Different tissues in multicellular organisms play different physiological roles that often require them to have their particular metabolic characteristic. For example, muscle cells can ramp up glucose consumption quickly to generate energy during exercise; liver can take up the breakdown product of fat and convert it to glucose (gluconeogenesis); and adipose tissue can convert excess glucose to lipids for storage. All of these tissues have the same glycolysis pathway to metabolize glucose, but the pathway of each tissue also bears special characteristics that allow it to fulfill each tissue's specific function. The diversity of the metabolic capability in different tissues is endowed by the expression of different isozymes in different tissues.

Isozymes and Allosteric Regulation

Different isoforms of the same enzyme or transport protein, although performing the same reaction, may have different catalytic rates (k_{cat}) or affinities for a substrate (K_M) (Panel 3.12). Note that K_M is the substrate concentration at which the reaction rate is at half of its maximum. An isozyme that has a low K_M for its substrate can catalyze the reaction at a low substrate concentration, while one that has a high K_M will become active only at higher substrate concentrations. For example, all lactate dehydrogenase isoforms catalyze the reversible reaction of the conversion of pyruvate and NADH to lactate and NAD, but they have different K_M's to pyruvate and lactate. LDHC favors the conversion of lactate to pyruvate, while LDHA favors the reverse reaction.

In the case that an enzyme can accept a number of different substrates, different isozymes may have different substrate preferences. For example, mammals have many different isoforms of the glucose transporter. They have different affinities to glucose. Many of them can also transport some other monosaccharides such as galactose or glucosamine.

Panel 3.12. Isozymes in Glycolysis

- Different tissue cells express different glycolytic isozymes with different kinetics and allosteric and gene expression regulations
- F26BP and F16BP play key regulatory roles
- Key regulatory enzymes of glycolysis with multiple isoforms: HK, PFK, PFKPB, PK
- Cancerous, fast-growing, and cultured cells express isozymes that favor aerobic glycolysis

An enzyme may also be subject to allosteric regulation, meaning that its activity is modulated by the binding of an effector molecule to the enzyme. For example, the binding of the product of the reaction can decrease the enzyme activity (i.e., feedback inhibition). In many biosynthetic pathways in microbial systems, the first enzyme in the pathway is feedback inhibited by the product of the pathway. Upon the accumulation of the product, its activity is decreased to reduce the flux through the pathway and to prevent wasteful accumulation of the product. Different isozymes may be subject to different allosteric regulations. Isoforms may also respond differently to hormonal or signaling stimuli.

Isozymes are thus important for giving different cells different metabolic characteristics. While in some cases isozymes of the same enzyme are products of alternative splicing in transcription, in other cases isozymes are coded by entirely distinct sequences in the genome.

Combination of Isozymes Gives Pathway Specific Characteristics

A pathway serving a relatively simple function, such as the aspartate amino acid biosynthesis pathway in *E. coli*, may be controlled by feedback regulation in one or a small number of "rate-limiting" enzymes. However, glycolysis in mammals is a complex pathway serving many functions under a wide range of physiological conditions, from during a feast to under severe starvation. In a growing cell, glycolysis not only supplies pyruvate for energy generation in the TCA cycle, but also supplies PPP, sustains the biosynthesis of glycerol-phosphate (for lipids) and serine (through 3-phosphoglycerate), and maintains the supply of pyruvate for the biosynthesis of many cellular constituents. The needs of each branched pathway differ under different physiological conditions and in different tissues, so the supply rate to each branched pathway also differs. That supply rate is adjusted by controlling the concentration of the glycolysis intermediate at various branching points. Mammals have evolved to cope with such a wide range of needs in glycolysis by expressing different combinations of isoforms. The interactions of feed-forward and feedback inhibition and activation exerted by those isozymes can give rise to very diverse and complex metabolic behaviors that depend heavily on the environment.

Cells in culture are derived from various tissues. They often bear the metabolic characteristics of their tissues of origin. However, cells *in vitro* are also different from their counterpart in the tissue. For one thing, the cultured cells are proliferating, unlike their quiescent counterparts *in vivo*. Many cells in culture are continuous cell lines or are even transformed. Their growth control differs from tissue cells. All these factors contribute to enzyme and isoform composition and the metabolic properties of cells in culture.

Isozymes are often named after the tissue in which they are the dominant isoform. However, the expression of isozymes is not tissue-exclusive. The liver isoform of PFK is widely expressed in many tissues. The expression of isozymes in a tissue cell is not limited to one form. Different isoforms of the same enzyme are often co-expressed in the same tissue or the same cell. Different combinations of isozymes give rise to different kinetics and regulatory behaviors that may meet different physiological needs.

With the available genomic tools, we can easily determine the relative expression of different isoforms of the key enzymes of glycolysis, and further evaluate how to influence cellular metabolism.

Key Isozymes in Regulating Glucose Metabolism

Four reactions in glycolysis play key roles in regulating its flux: hexokinase (HK), phosphofructokinase (PFK), pyruvate kinase (PK), and 6-phosphofructo 2-kinase/fructose 2,6-bisphosphate (PFKFB) (Panel 3.12). These four enzymes along with pyruvate dehydrogenase kinase (PDK) regulate the flux of glucose carbon and its distribution at the pyruvate node. We will take a simplified view to largely divide glycolysis into two types of metabolism: one high flux in proliferating cells and the other low flux in quiescent cells (Figure 3.9). These two types of metabolism are influenced by the isoforms involved, the composition of the medium, and the growth rate, among other factors. With some isoform

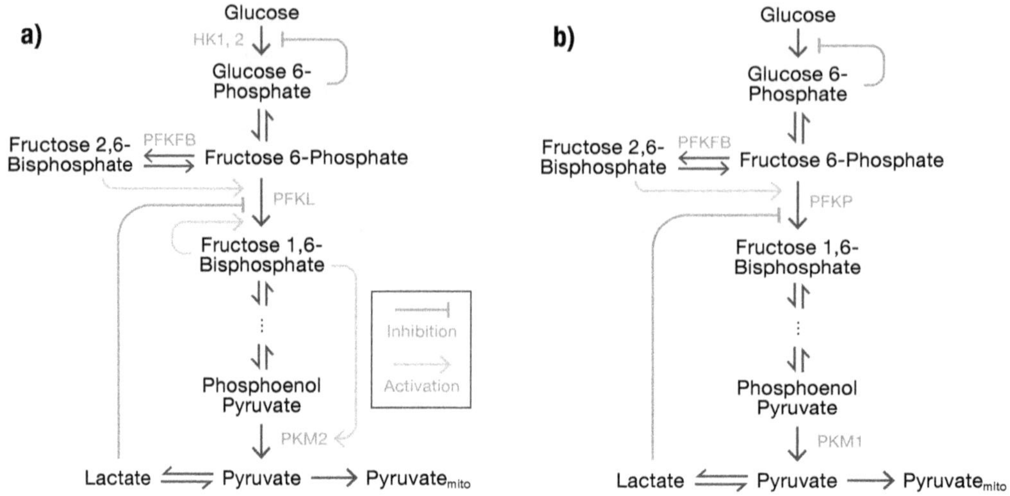

Figure 3.9. Allosteric regulation of glycolysis in (a) proliferating cells and (b) quiescent cells. The allosteric regulation present in the pathway is dependent on the isozymes expressed in the cell.

Table 3.1. Composition of Different Enzyme Isoforms Based on Transcript Expression (%)

	Heart Muscle	*Liver*	*Cerebral Cortex*	*CHO K1*	*CHO S*
PKM1	48	3	48	1	4
PKM2	52	37	52	99	96
PKL	0	60	0	0	0
PKR	0	0	0	0	0
PFKL	6	79	17	65	72
PFKM	64	16	36	23	20
PFKP	30	6	47	12	8
PFKFB1	0	52	0	12	27
PFKFB2	80	4	17	30	22
PFKFB3	18	43	79	55	46
PFKFB4	2	1	3	4	6
HK1	94	54	94	75	91
HK2	5	19	4	25	9
HK3	1	27	2	0	0

combinations, a number of reaction steps are activated by the accumulation of F2,6P and F1,6P (Figure 3.9a). Upon full activation, one may see a 5-fold or higher increase in glucose consumption, whereas with the isoform combination depicted in Figure 3.9b, the degree of activation is much lower. Below, we will describe the allosteric regulation of a few major enzymes that play key roles in determining the flux. Table 3.1 lists the compositions of the isozymes of glycolysis in a few human tissues.

Hexokinase

Hexokinase-1 (HK1) is present in virtually all cells. It has a low K_M for glucose and phosphorylates glucose for glycolysis even at a very low intracellular concentration. Liver cells additionally express glucokinase (GCK or HK4), which has a high K_M and is active only when glucose concentration is high. HK4 phosphorylates glucose for conversion to glycogen storage when glucose is in excess.

PFK

PFK converts F6P to become F16BP. It is pivotal in modulating the overall rate of glycolysis and is a key node in energy metabolism. Its activity is subjected to allosteric inhibition by ATP and citrate, and is activated by AMP. Citrate is a TCA cycle intermediate. The accumulation of citrate and ATP in the cytosol, along with a low concentration of AMP, is indicative of an abundance of cellular energy. Its activity is thus also suppressed by an abundance of cellular energy. PFK has three isozymes: liver (PFKL), muscle (PFKM), and platelet (PFKP). Among the three isoforms, PFKL and PFKM are activated allosterically by their own reaction product, fructose 1,6-bisphosphate (F16BP). Upon a flux increase and initial accumulation of F16BP, the feedback activation quickly escalates the activity of PFKL and PFKM to increase the glycolysis rate. PFKM is inhibited by lactate, a characteristic which may facilitate the reduction of glycolysis flux at high lactate levels. This enzyme likely plays an important role in triggering the lactate consumption seen in the late culture stage (when lactate accumulates to high levels) in some bioprocess fed-batch cultures.

6-Phosphofructo-2-kinase / fructo-2,6-phosphate phosphatase (PFKFB)

Fructose 2,6-bisphosphate (F26BP) is a shunted reaction product of glycolysis that plays a key regulatory role in rapidly adjusting the activity of PFK and the glycolysis flux (Figure 3.9). It is not metabolized further or used in any biosynthesis, but is converted back to fructose 6-phosphate (F6P). Its concentration is a sensitive indicator of the "state" of glycolysis and is used for regulating the glycolysis flux. Both the synthesis of F26BP and its reversion to F6P are catalyzed by the same enzyme, a bi-functional PFKFB (also known as PFK-2). The kinase activity of PFKFB catalyzes the synthesis of F26BP and the phosphatase activity catalyzes the hydrolysis of F26BP to F6P. All three PFK isozymes are activated by F26BP, which activates PFK-1 by allosterically increasing its affinity for F6P, even in the presence of inhibitors such as ATP or lactate.

PFKFB has four isozymes, each with different kinase and phosphatase activities, allowing each to respond differently to regulators. A high kinase to phosphatase ratio allows for a higher accumulation of F26BP to exert a more robust activation of PFK activity. The brain isoform, PFKFB3, has the highest kinase to phosphatase activity and is expressed in several tumor cells. This suggests that PFKFB3 may be accountable for the glycolytic phenotype of reported cancerous cell lines by allowing them to have high cellular F26BP levels.

Pyruvate kinase

The enzyme catalyzing the penultimate step of glycolysis, pyruvate kinase, has three isozymes in mammalian systems. The muscle isozyme is expressed as either of two splice variants, M1 or M2. Both muscle isoforms of PK are activated by phosphoenolpyruvate, but only PKM2 is also activated by F16BP. The M1 isoform is mostly expressed in adult tissues, whereas the M2 isoform is expressed in rapidly growing tissues, such as fetal and tumor tissues, and is also thought to be a critical player in the transformation leading to cancer.

Pyruvate dehydrogenase kinase (PDK)

Pyruvate dehydrogenase complex (PDHC), which catalyzes the conversion of pyruvate to acetyl-CoA (and generates a CO_2 and NADH), is a protein complex made of 30 copies of E1 (consisting of two subunits), 60 E2 and 12 E3 proteins. It has three serine residues that can be phosphorylated by PDK to render it inactive, and dephosphorylated by pyruvate dehydrogenase phosphatase to return it to active state. There are four different isoforms for PDK that have varying phosphorylation activity towards the three serine residues. They thus tune the activity of PDHC to regulate the carbon flux into the TCA cycle.

Growth Control and Metabolic Regulation

The regulation of cellular metabolism is tightly linked to the control of cell growth. The signaling pathways and transcription factors that regulate a cell's growth rate play regulatory roles in regulating glucose metabolism. The transformation that causes cells to switch from a quiescent state to a proliferating state also triggers metabolic changes to increase their glucose uptake and glycolysis flux (Panel 3.13).

p53 is a major tumor suppressor that plays key roles in cell cycle arrest, senescence, apoptosis, and the regulation of glucose metabolism and oxidative phosphorylation. It induces the overexpression of TIGAR (TP53-induced glycolysis and apoptosis regulator) under mild oxidative stress conditions (Figure 3.10). TIGAR contains a fructose 2,6-bisphosphatase catalytic activity domain, which mediates the degradation of fructose 2,6-phosphate and leads to a decrease in PFK-1 activity and the attenuation of the glycolytic flux.

p53 can also modulate the glycolytic rate by regulating the activity of PGM, GLUT1, and GLUT4 transporters. Furthermore, it up-regulates mitochondrial oxidative phosphorylation by upregulating the expression of SCO2 (synthesis of cytochrome c oxidase 2), which mediates the assembly and activity of the cytochrome c oxidase complex.

AKT is a serine/threonine kinase that plays a key role in the regulation of cell proliferation and glucose metabolism. AKT exists in an active/phosphorylated form and an inactive/unphosphorylated form. The

Panel 3.13. Signaling Pathway / Growth Rate Control of Glycolysis

- Insulin signaling
  - Positively regulates growth rate
  - Regulates glucose and amino acid metabolism through AKT
- p53 (tumor suppressor) suppresses glucose uptake and glycolysis
- MYC (proto-oncogene) stimulates glycolysis
- Fast-growing (tumor) cells have fast glycolysis (and lactate production)

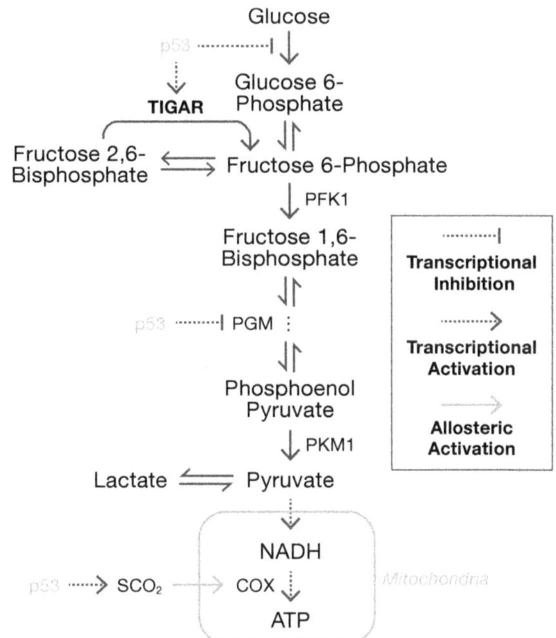

Figure 3.10. Tumor suppressor p53 negatively regulates glycolysis flux. TIGAR has phosphatase activity that converts F26BP to F16BP.

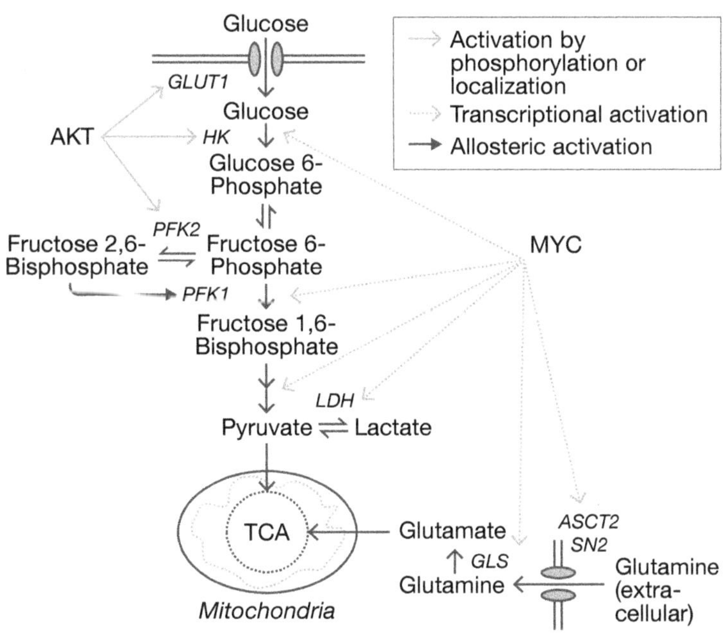

Figure 3.11. Signaling kinase AKT and transcription factor MYC positively regulate energy metabolism.

AKT signaling cascade has been shown to activate the transcription of GLUT1 and mediate the association of HK with the outer mitochondrial membrane (Figure 3.11). The phosphorylated form of AKT (pAKT) can increase the phosphorylation of PFKFB to shift its kinase/phosphatase ratio to increase the formation of fructose 2,6-bisphosphate levels, which in turn increases PFK activity and glycolysis flux. A decrease in the growth rate of the cells observed during the course of the culture is accompanied by a decrease in the pAKT activity.

MYC is a proto-oncogene whose pleiotropic regulatory roles include energy metabolism. It directly modulates the transcript level of glycolysis enzymes. Glycolytic enzymes have MYC canonical E-boxes in their promoter and are deregulated when MYC is overexpressed.

Regulation through the AKT signaling pathway provides a link between energy metabolism and cell proliferation in the tissue. Oncogenic transformation involves not only alterations in growth control but also invokes the expression of different isoforms or elevation of expression levels of key glycolysis enzymes that are hallmarks of fast growth. Virtually all proliferating cells in culture consume glucose at a high rate and convert the bulk of it into lactate.

Most cell lines of bioprocess importance have altered growth control that allows them to proliferate quickly in culture. During rapid growth, they exhibit a high glycolytic flux similar to other transformed cells. The decrease in glycolysis flux and the consumption of lactate as seen in some cultures occurs in the late culture stages when cell growth is already slow. Even for cells in culture, metabolism is linked to growth control.

Transport and Transporters

Energy metabolism takes place in multiple cellular compartments that are separated by lipid bilayer membranes. First, glucose must cross the cytoplasmic membrane to enter glycolysis in the cytosol. Next, the products from glycolysis (pyruvate and NADH, via the malate-aspartate shuttle) are transported into the mitochondria. Finally, metabolites such as lactate are excreted through the cytoplasmic membrane. This crossing of molecules across the membrane is mediated by a number of transporters.

Glucose Transporters

Glucose transporters mediate the influx of glucose across the cytosolic membrane. There are two types of glucose transporters: GLUT and SGLT (Panel 3.14). The GLUT transporters are uniporters for facilitated transport, allowing glucose to move along its concentration gradient. The second type of glucose transporter, SGLT, is a co-transporter with Na^+ that is expressed in intestinal epithelial cells.

Panel 3.14. Two Main Types of Glucose Transporters

- GLUT transporters mediate facilitative diffusion across the plasma membrane
- SGLT, the sodium-dependent glucose co-transporters, are expressed primarily in small intestinal absorptive cells or renal proximal tubular cells. They use Na^+-K^+ ATPase pumps for active transport of glucose.

Panel 3.15. Glucose Transporter

- GLUT1 is highly expressed in all cells
- K_M is small for GLUT1. At the glucose concentration of the medium, it operates at its maximum rate.
  - K_M is the substrate concentration that gives a reaction rate of 1/2 the maximum. It is a measure of substrate affinity.

Table 3.2. Glucose Transporters

	Class	*Tissue expression*	*Affinity (sugar, K_M)*
GLUT1	I	Ubiquitous	Glucose, 1–2 mM
GLUT2	I	Liver, pancreas, intestine, kidney	Glucose, 16–20 mM; glucosamine, 0.8 mM
GLUT3	I	Brain, neurons	Glucose, 0.8 mM
GLUT4	I	Heart, muscle, adipose	Glucose, 5 mM
GLUT5	II	Intestine, testis	Fructose, 10–13 mM
GLUT7	II	Intestine, testis	Glucose, 0.3 mM; fructose, 0.1 mM
GLUT9	II	Kidney, liver	Fructose, 0.6 mM
GLUT6	III	Brain, spleen, leukocytes	Glucose, 5 mM
GLUT8	III	Testis, brain, liver	Glucose, 6 mM
GLUT10	III	Liver, pancreas	Glucose, 0.3 mM
GLUT12	III	Heart, muscle, prostate	Not well known

The GLUT transporters have twelve transmembrane regions and intracellular carboxyl and amino termini. According to common sequence motifs, they are divided into three subclasses (Table 3.2). GLUT1 is ubiquitous, appearing in almost all cells (Panel 3.15). It can transport glucose and galactose in a concentration-dependent manner that is described by Michaelis-Menten kinetics (Figure 3.12). Its K_M for glucose is low (~1 mM). At the glucose concentration used in culture medium, the flux of GLUT1 is at its maximum. In some cells, GLUT1 is under the regulation of the transcription factor HIF-1 (hypoxia inducible factor). Under hypoxic conditions, the expression of GLUT1 is upregulated to increase the uptake rate of glucose. The K_M of GLUT1 for galactose is rather high. When galactose is used as the only sugar, even at a high galactose concentration, the uptake rate and lactate production rate is low.

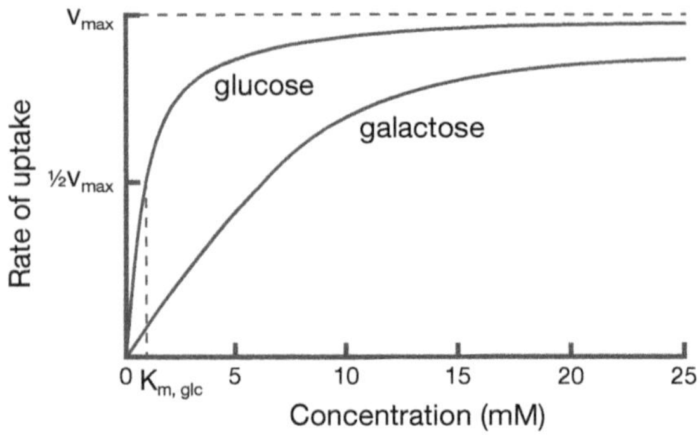

Figure 3.12. Michaelis-Menten kinetics plot for a GLUT1 transporter.

A few other notable GLUT transporters are insulin-responsive GLUT4 and fructose-transporting GLUT5. In addition to GLUT1, cells in culture and in different tissues may express other GLUT transporters at different proportions. The expression of different transporters will give them different responses to the concentrations of glucose or other sugars.

SGLT, another class of glucose transporter, transports two sodium ions and one glucose molecule into the cell. The Na^+ concentration is low intracellularly but is high in the medium and in body fluid. The large sodium concentration difference and negative electric potential across the cytoplasmic membrane gives rise to a high propensity of Na^+ to enter the cell. Thus, the chemical potential energy of the sodium concentration difference and the electric potential is used to drive the uptake of glucose against a concentration gradient.

SGLT transport is abundant in intestinal epithelial cells and is responsible for moving glucose from the gut into the intestinal epithelial cells. The glucose is then exported into the bloodstream on the other side of the cellular barrier.

Lactate Transporters

Lactate produced in glycolysis is excreted across the cytoplasmic membrane through monocarboxylate transporters (MCT). Lactate can also be consumed in some tissues and by cultured cells under some conditions. The uptake of lactate by cells is also mediated by MCT. Monocarboxylate transporters are largely divided into two categories: facilitated diffusion type monocarboxylate transporters and sodium-coupled monocarboxylate transporters (sMCT). The latter is expressed only in some tissue cells (kidney, colon epithelial cells) that can utilize the concentration gradient of sodium ions across its apical surface membrane to take up the solute. The solute transported by MCT and sMCT is not limited to lactate, and can also transport pyruvate, propionate, and other short-chain fatty acids.

Lactate is a negatively charged molecule. Its movement across the cellular membrane causes a charge imbalance and creates an electric potential across the membrane as a barrier for further transfer across the membrane, unless measures are taken to counteract the charge imbalance. The monocarboxylate transporters are a family of co-transporters that couple the transport of lactate to the transport of a hydrogen ion in the same direction to maintain the charge balance (Figure 3.13). MCT is thus a symporter; its mechanism of transport is facilitated diffusion.

A number of MCTs are expressed in different tissue cells. Four of them, with 12 transmembrane helices, have been shown to mediate reversible transport of lactate, pyruvate, and ketone bodies. By far, lactate is the most important species transferred. MCT1 favors the uptake of lactate, while MCT4 favors the excretion of lactate produced in glycolysis.

Lactate transport is enhanced by a large difference in lactate concentration between intracellular and extracellular environments. However, pH also affects the flux of lactate through MCT, with the enhancing or retarding effect being dependent on the direction of the proton gradient. MCT allows for lactate transport in both directions, for excretion as well as uptake. Keeping medium pH at a lower level reduces lactate production during the rapid growth period, but enhances lactate consumption in the stationary phase.

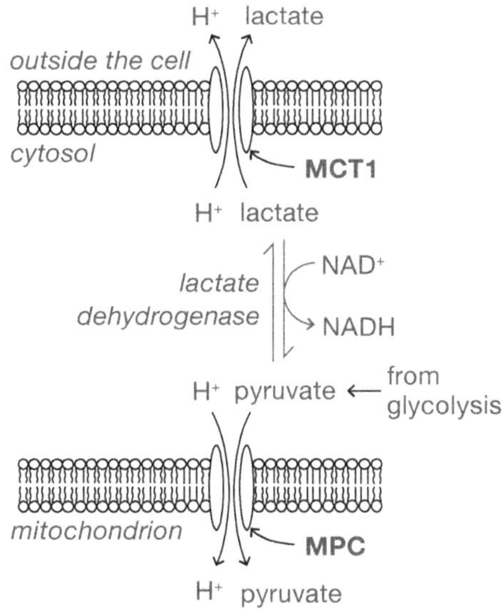

Figure 3.13. A monocarboxylate transporter for lactate and a mitochondria pyruvate carrier for pyruvate.

Mitochondrial Pyruvate Carrier (MPC)

Proliferating cells in culture typically channel about 1/10–1/20 of the carbons from their glucose intake to the mitochondria as pyruvate, where it is broken down to CO_2. The molar flux of pyruvate into the mitochondria is thus about 1/5–1/10 of that of the glucose consumption rate or 1/10–1/20 of the lactate excretion flux. Assuming an average cell diameter of 15 μm, a mitochondrial diameter of 1 μm, and 1000 mitochondria per cell, the total surface area of mitochondria in a cell is in the same order of magnitude as the cytoplasmic membrane. Thus, the pyruvate flux across the mitochondrial membrane (μmole/μm^2-h) is perhaps not as high as that of lactate passing through the cytoplasmic membrane.

The precise nature of the mitochondrial transport of pyruvate has been elusive until recently. Two proteins, mitochondrial pyruvate carriers MPC1 and MPC2, are involved in pyruvate transport. MPC1 and MPC2 have only 3 transmembrane helices. They are likely to form a protein complex (currently thought to be a heterodimer) to provide the transport activity. The transporter is also a symporter, co-transporting a proton to maintain charge neutrality. The 1.0 unit pH difference and –140 mV electric potential between the cytosol and the mitochondrial matrix provides a strong driving force for pyruvate transport into the mitochondria.

Not all pyruvate is generated from glycolysis. Pyruvate may also arise from catabolism of alanine and a few other amino acids. It may also be generated from acetyl-CoA shuttling, as will be discussed later in this chapter. Under some conditions, pyruvate may also be diverted to synthesize alanine. During rapid cell growth, these fluxes are relatively small compared to pyruvate synthesis from glycolysis. By and large, the flux of pyruvate into the mitochondria is estimated to be the difference between glucose flux in glycolysis and lactate production flux (Figure 3.9).

Metabolic Homeostasis and Lactate Consumption

Cells in culture have been selected for their capability to proliferate. The makeup of their glycolytic enzymes directs their metabolism to consume much glucose and convert most of it to lactate. Lactate accumulation in culture inhibits cell growth and hastens the decline of cell viability in the stationary phase. The phenomenon of high glucose consumption and high lactate production is seen in virtually all proliferating cells in culture. The problem is particularly acute in fed-batch cultures, in which continued glucose feeding raises both glucose and lactate to many times higher than their physiological levels. To achieve a high cell concentration and high productivity, it is beneficial to direct cell metabolism to a metabolic state that minimizes the accumulation of lactate.

States of Glucose Metabolism

High flux state

The "default" glucose metabolism is at a high flux state, in which cells consume glucose at a high level and convert most of it to lactate. Figure 3.14a illustrates a typical metabolic flux of a cultured mammalian cell line. If 100 moles of glucose are consumed in glycolysis, 200 moles each of pyruvate and NADH are generated. Here, for simplicity we neglect the diversion of carbons to the PPP, lipid biosynthesis, etc. Consider a scenario of a high flux state in which 90% (180 moles) of the pyruvate is channeled to lactate and excreted, and the rest (20 moles) of the pyruvate enters the mitochondria for the TCA cycle. Those 180 moles of pyruvate converted to lactate will also regenerate 180 moles of NADH to NAD. The remaining 20 moles of NADH will enter the mitochondria, where it is converted to NAD, consuming oxygen and generating H_2O and ATP. In total, 200 moles of NAD are regenerated to sustain glycolysis at a steady state flux of 100 moles of glucose. NADH generated in glycolysis in the cytosol cannot enter the mitochondria, as there is no appropriate transporter. Instead, it is transported via the malate-aspartate shuttle, as will be discussed next.

Neglecting the other cytosolic reactions that involve pyruvate, the molar fluxes of pyruvate and NADH into the mitochondria are equal (Panel 3.16). Overall, 10% of glucose is oxidized to CO_2 while 90%

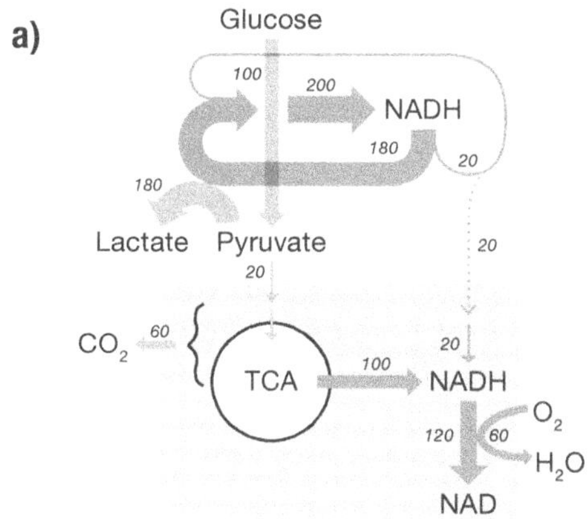

Net

$100\ C_6H_{12}O_6 + 60\ O_2 \longrightarrow$

$180\ C_3H_6O_3 + 60\ CO_2 + 60\ H_2O$

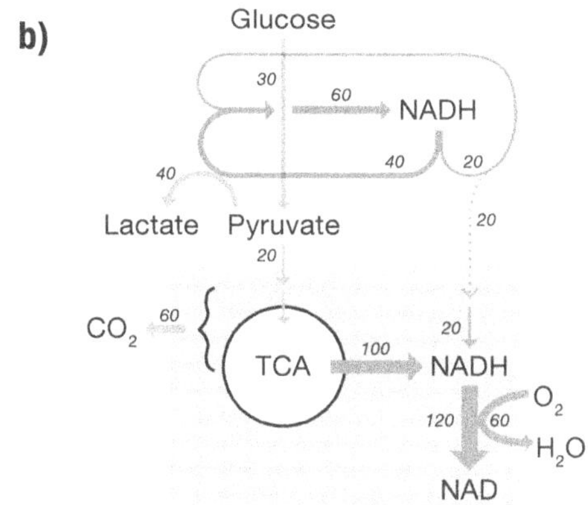

Net

$30\ C_6H_{12}O_6 + 60\ O_2 \longrightarrow$

$40\ C_3H_6O_3 + 60\ CO_2 + 60\ H_2O$

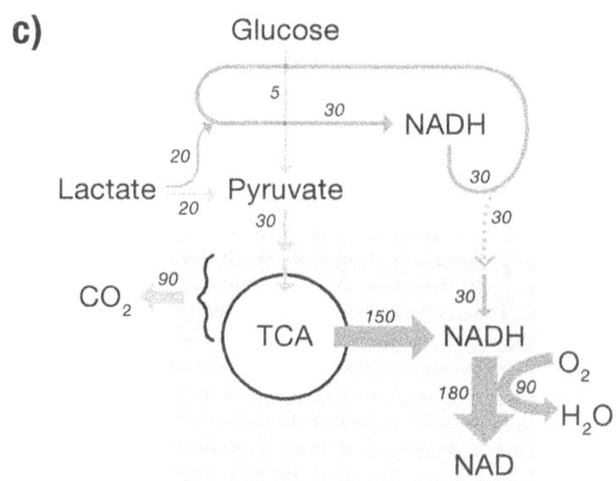

Net

$5\ C_6H_{12}O_6 + 20\ C_3H_6O_3 + 90\ O_2$

$\longrightarrow 90\ CO_2 + 90\ H_2O$

Figure 3.14. Three metabolic states of glucose metabolism: (a) high flux state, high glucose consumption, and high lactate production; (b) low flux state with low lactate production; (c) low flux state with low glucose consumption and low lactate consumption.

Panel 3.16. Pyruvate and NADH Balance

- Pyruvate and NADH are generated in glycolysis at a 1:1 ratio
- Both are transferred into the mitochondria and converted by lactate dehydrogenase (LDH)
- Altogether, 12 reducing equivalents (NADH / $FADH_2$) are generated to react with 6 O_2, generating 6 H_2O
- 2 of the 12 reducing equivalents are generated in the cytosol (in glycolysis), 10 in the mitochondria
- The 2 NADH generated in the cytosol need to be transferred into the mitochondria to feed into the electron transport chain, or oxidized to NAD in the LDH reaction

is converted to lactate. The 10% of glucose that is completely oxidized consumes 60 moles of O_2, as seen in the stoichiometric equation (Panel 3.1). This is carried out by the carbon and reducing equivalent entering the mitochondria to react with 60 moles of O_2 and produce 60 moles of CO_2 and H_2O (Eq. 3-1, Panel 3.1). The fluxes are thus all balanced.

Above all, there appears to be a limit in the capacity of MPC or in the malate-aspartate shuttle to allow a much greater flux of carbon from glucose to enter the mitochondria. Virtually all cells in culture resort to high flux metabolism to generate energy. There have been many hypotheses to explain the benefit of the Warburg effect to fast-growing cells, and why no organism has evolved a way to support fast proliferation without producing a large amount of lactate.

Low flux, lactate-producing state

Cells may enter a low glycolytic flux state under some conditions. Under most laboratory culture conditions, such as in a Petri dish, where cell density, lactate, and other metabolites are at low levels, the metabolism is mostly at a high flux state. As the metabolites accumulate and the growth rate decreases, as in a fed-batch culture, the combined effects of reduced AKT activation and possible lactate inhibition of PFKL lead the metabolism to a low flux state.

Even fast-growing cells can be guided to a low flux state by controlling glucose at low levels. Such a glucose flux control in yeast has been known for decades as the Crabtree effect. Even with plenty of oxygen, *Saccharomyces cerevisiae* ferments glucose to ethanol if glucose is kept at high levels. By controlling glucose at low levels to reduce its consumption, cells return to an oxidative metabolism with a reduced glucose flux and without ethanol production. This approach of restricting glucose supply and controlling glucose at low levels to reduce glycolysis flux and thereby reduce lactate flux has been demonstrated in hybridoma and CHO cell culture. By controlling glucose at very low levels or by replacing glucose with another sugar such as galactose or fructose that is only taken up slowly, even fast-growing cells can be manipulated at a low flux state.[1–4]

As the cell growth rate decreases in culture, the AKT activation of the glycolytic enzymes also reduces,[5] and it becomes more common to observe a switch to a low flux state. This may be facilitated by employing a slowly consumed sugar in the culture medium.

A typical flux distribution of cells in a low flux state, regardless of whether it is at a fast-growing or slow-growing state, is shown in Figure 3.14b. The prerequisite of entering a low lactate-producing state is low glucose uptake. In Figure 3.14, we assume that the flux into the mitochondria is fixed by the capacity of MPC. In practice, the value may increase somewhat by a fraction but not by a large margin. Glycolysis generates 60 moles of NADH, of which 20 are regenerated to NAD through oxidation in the mitochondria (via the malate-aspartate shuttle) and 40 through lactate production. This allows a steady state flux of 30 moles of glucose through glycolysis.

Low flux, lactate-consumption state

Lactate transportation across the cytoplasmic membrane by MCT and lactate/pyruvate conversion by LDH are both reversible; one requires the co-transport of H^+ and the other the co-substrate NAD/NADH. They operate in unison in the same direction depending on the intracellular levels of NAD, NADH, pyruvate, and lactate, and the intracellular and extracellular concentration difference of lactate and H^+. Their concentrations, in turn, are affected by the glycolysis generation rate and the mitochondrial intake rate of pyruvate. At a high glucose flux state, the pyruvate production rate is so high, exceeding its mitochondrial intake, that lactate is always excreted as described above. Conversely, at a low glycolysis flux state, the pyruvate production rate is low and it is possible to observe lactate consumption.

In vivo, lactate consumption occurs under some condition in some tissues, such as in the muscle, liver, and brain. *In vitro*, lactate consumption by cultured cells is seen when the glycolysis flux is low, typically after the rapid growth stage is over and when lactate has accumulated to a significant level. The propensity and rate of lactate consumption is thus affected by pH, where a low pH (high H^+) will facilitate lactate import and consumption.

A scenario of flux at a lactate consumption rate is shown in Figure 3.14c. The reverse reaction of LDH generates pyruvate and NADH. Both pyruvate and NADH are further oxidized in the mitochondria. Since the overall capacity of MPC and the intake rate of pyruvate into the mitochondria is limited, the lactate consumption flux is also limited. It never reaches the magnitude of the flux of glucose at its high flux state. On a per cell basis, the specific lactate consumption rate is relatively small.

In a fed-batch culture, one may observe the three metabolic states happening in order (Figure 3.15). In the fast-growing state, the metabolism is at a high flux state. As the growth rate slows down, it enters a low flux state with little lactate production, and then it may enter the lactate consumption state.

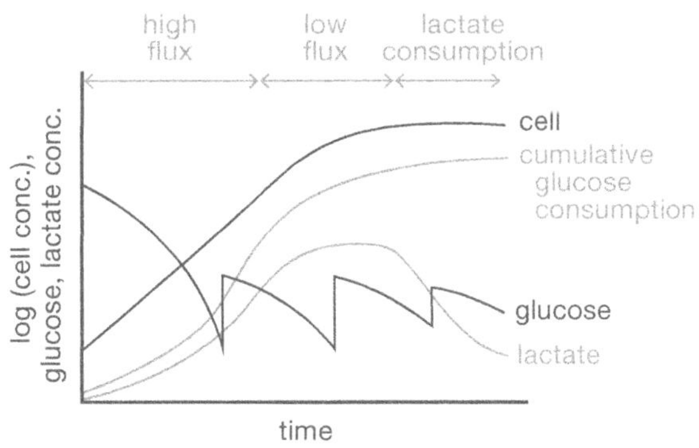

Figure 3.15. The occurrence of three metabolic states in fed-batch culture.

Glucose consumption in the lactate-consumption state

The lactate consumed by cells cannot be converted to glucose in most cultured cells. A number of reactions in glycolysis are irreversible. The conversion of pyruvate to glucose in the reverse direction of glycolysis, called gluconeogenesis, requires the expression of a few additional enzymes to counter these irreversible reactions. In mammals, gluconeogenesis primarily occurs in the liver. During the period that cells are consuming lactate, many intermediates derived from glycolysis are still needed for maintaining cellular functions. For example, dihydroxyacetone phosphate (DHAP) is needed for supplying glycerol 3-phosphate for lipid synthesis and NADPH, derived in the PPP, is needed for reductive biosynthesis and for maintaining the cell's redox balance. Furthermore, glucose 6-phosphate is required to synthesize the glucosamine and galactose that are used in glycan synthesis for the production of recombinant proteins. The glycolysis pathway thus remains active during the lactate-consumption stage. The glucose consumption rate is small, but not zero (Panel 3.17).

Panel 3.17.
Lactate Consumption in Cultured Cells

- Lactate consumption is associated with a low glucose flux and occurs in a slow growth stage
- It is transported into the cell by MCT, converted to pyruvate by LDH, and generates 1 NADH
- Both pyruvate and NADH enter the mitochondria for energy generation
- While consuming lactate, glucose is also consumed at a low rate to supply NADPH and other intermediates

NADH Balance Malate-Aspartate Shuttle

A total of 12 moles of reducing equivalent (10 NADH and 2 $FADH_2$) are produced when 1 mole of glucose is completely oxidized to CO_2 through glycolysis and the TCA cycle. The 12 mole reducing equivalents consume 6 moles of O_2 in oxidative phosphorylation, consistent with the stoichiometry of glucose oxidation (1 glucose/6 O_2). Among the 12 NADH/$FADH_2$, 10 are produced in the TCA cycle in the mitochondria and the other 2 NADH are produced in cytosolic glycolysis (Panel 3.16). The two reducing equivalents produced in the cytosol must then be transported into the mitochondria where their oxidation consumes the sixth molecule of O_2.

NADH does not pass through the inner membrane of the mitochondria. Rather, it passes its reducing potential through a carrier system called the malate-aspartate shuttle. This system takes the reducing equivalent into the mitochondria through an exchange of molecules between the mitochondria and the cytosol. On the cytosolic side, NADH is oxidized to NAD and transfers its reducing equivalent to malate by reducing oxaloacetate. Malate is then transported across the mitochondrial membrane via an antiporter with α-KG going in the opposite direction. Once inside the mitochondria, the reducing equivalent is transferred back to NADH by oxidizing malate to oxaloacetate (Figure 3.16).

The shuttle employs two antiporter systems involving malate/α-KG and aspartate/glutamate (hence the name "malate-aspartate shuttle") to ensure the carbon flow and reducing equivalent flow are balanced by one other. The net result of a cycle of the shuttle is the transfer of a reducing equivalent from the cytosol to the mitochondria, or the conversion of a NADH to NAD in the cytosol and a simultaneous conversion of a NAD to NADH in the mitochondria. All other components involved in the shuttle then return to their original state. This is shown in Figure 3.16, with the fluxes in opposite directions shown in gray with different arrowheads. On each side of the mitochondrial membrane, the same aminotransferase reaction (converting a glutamate/OAA pair to α-KG/aspartate pair, or vice versa) occurs, but operates in the opposite direction.

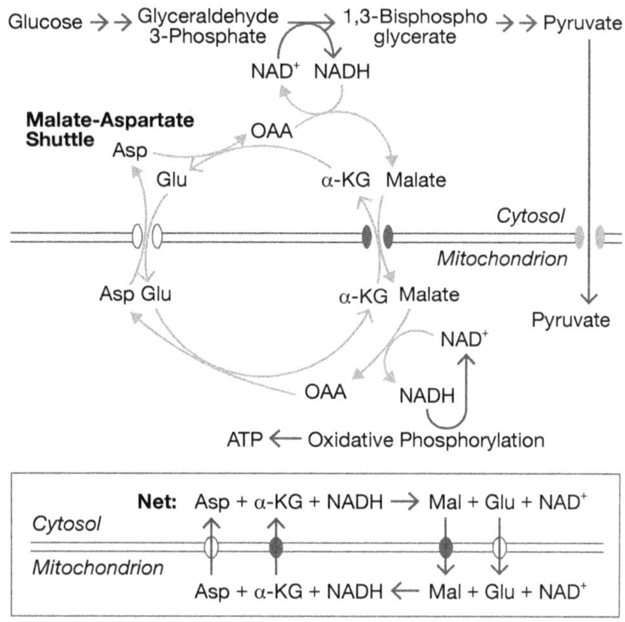

Figure 3.16. The malate-aspartate shuttle, for the transfer of the reducing equivalent of NADH from the cytoplasm to the mitochondria.

The transfer of the reducing equivalent of NADH from the cytosol into the mitochondria is therefore dependent not only on the NADH concentration but also on the concentrations of α-KG, malate, aspartate, and glutamate in the two compartments. Furthermore, diversions of fluxes of those shuttle components may also perturb the transport of the reducing equivalent.

Other Carbon Fluxes across the Mitochondrial Membrane

Acetyl-CoA shuttle

Acetyl-CoA is the building block of fatty acids and cholesterol. It is generated primarily through the oxidative decarboxylation of pyruvate in the mitochondria. However, fatty acid biosynthesis takes place in the cytosol. Acetyl-CoA does not pass through the bilayer membrane. Instead, it is exported to the cytosol via an indirect process called the acetyl-CoA shuttle (Figure 3.7). Citrate, formed by condensation of OAA and acetyl-CoA in the TCA cycle, is diverted from the TCA cycle and transported into the cytosol by an antiporter with one malate being transported in the opposite direction (Panel 3.18). Once in the cytosol, citrate is

Panel 3.18. Acetyl-CoA Shuttle

In the Mitochondrial Matrix

- Pyruvate carboxylase
  Pyruvate + CO_2 + ATP → OAA + ADP
- Malate dehydrogenase
  Malate + NAD → OAA + NADH
- Citrate synthase
  OAA + acetyl-CoA → citrate + CoA

On the Mitochondrial Inner Membrane

- Citrate-malate transporter
  - An antiporter for citrate export from and malate import to the mitochondria
- Malate, α-ketoglutarate transporter
  - Transports malate into the mitochondria
- Mitochondrial pyruvate carrier (MPC)
  - Transports pyruvate into the mitochondria

In the Cytosol

- Citrate lyase
  Citrate + CoA + ATP → OAA + acetyl-CoA
- Malate dehydrogenase
  OAA + NADH → malate + NAD
- Malic enzyme
  Malate + NADP → pyruvate + NADPH + CO_2

Panel 3.19. Metabolic Roles of Glutamine

- Essential for most cultured cells, consumed at a high rate
- Synthesized through glutamine synthetase *in vivo*
- Not an essential nutrient for mammals
- Used in nucleotide and protein synthesis
- Maintains carbon balance in the TCA cycle (anaplerotic metabolism)
- Releases NH_3 when entering the TCA cycle

split into OAA and acetyl-CoA, consuming one ATP. Acetyl-CoA is then used for lipid biosynthesis. OAA gets reduced to malate at the expense of one NADH. Malate is then transported back to the mitochondria, or is converted to pyruvate after releasing a CO_2 and consuming one NADPH. Pyruvate then recycles into the mitochondria. The process of making fatty acids using acetyl-CoA is thus energetically expensive.

Therefore, citrate and malate cross the mitochondrial membrane at a significant rate to sustain lipid biosynthesis. Citrate is not only present in the mitochondria, but also in the cytosol. Its accumulation can lead to allosteric inhibition of the conversion of F6P to F16BP in glycolysis.

Anaplerotic metabolism

As discussed earlier, the TCA cycle does not run perfect cycles all the time. When citrate, OAA, and other compounds are withdrawn from the cycle to serve other biosynthetic roles, they must be replenished in order to sustain the cycling at a steady state. This replenishment is mostly carried out by glutamine through anaplerotic reactions (Panel 3.19). Cells in culture consume glutamine at a high rate. The amount of glutamine consumed is in excess of that needed for biomass synthesis. Nearly half of the glutamine is converted to glutamate, which then enters the TCA cycle via α-KG to replenish OAA. α-KG can also be converted to isocitrate through an isocitrate dehydrogenase I catalyzed reaction at the expense of 1 NADPH. Isocitrate then becomes citrate for export out of the mitochondria. Another enzyme, pyruvate carboxylase, converts pyruvate to OAA and can also serve to replenish OAA to allow the TCA cycle to continue.

Other fluxes across the mitochondria

Each pyruvate catabolized through the TCA cycle generates about 15 ATP and 3 CO_2. These products are exported to the cytosol. The exportation of ATP necessitates the importation of an equal amount of ADP and PO_4^{3-} for its synthesis. Besides these major species, many other molecules (including amino acids and nucleotides) are transported into the mitochondria for DNA, RNA, and protein synthesis.

Transportation across the mitochondrial inner membrane is dynamic and complex. Many compounds crossing the membrane are charged, yet their transport should not disrupt the proton and electric potential gradient that is fundamental to the energy generation function of the mitochondria. The transport across the mitochondrial membrane must be tightly regulated. Our understanding of that regulation is still rather limited.

Amino Acid Metabolism

Glutamine and Its Role in Energy Metabolism

Most cultured cells consume glutamine at a very high level, second only to glucose. Its molar consumption rate is about 1/5 to 1/10 that of glucose for many cell lines. Glutamine is not an essential amino acid for mammals; it becomes essential when cells are isolated from tissue and begin to grow in culture. Many tissues express glutamine synthetase that converts glutamic acid to glutamine by incorporating an ammonium at the expense of an ATP. The transcript level of this enzyme varies in cultured cells. Some cell lines can be readily adapted to grow in the absence of glutamine, apparently by increased glutamine synthase activity, while most cells are cultured in glutamine-supplemented medium.

Glutamine is a major amino acid constituent of cellular proteins. Additionally, its amide group supplies the nitrogen in the synthesis of purine and pyrimidine bases, which are the backbone of nucleic acids. However, the amount of glutamine consumed by cells far exceeds what is needed for synthesizing cellular components.

A large portion of glutamine is converted to glutamate by glutaminase in the cytosol or mitochondria. Glutamate in the mitochondria is converted to α-KG via glutamate dehydrogenase (GDH), which releases an ammonium and NADH. Glutamate is also converted to α-KG via an aminotransferase reaction that transfers its amino group to the receiving OAA or pyruvate, forming aspartate or alanine, respectively. α-KG then enters the TCA cycle. The aminotransferase reaction retains the amino group in an amino acid, while the dehydrogenase reaction loses the amino group to ammonium. The former is likely to be favored in proliferating cells. Through α-ketoglutarate, glutamine is a major contributor to central metabolic flux by fulfilling its anaplerotic role as discussed above. The conversion of glutamine to α-KG releases one or two ammonium, depending on the path taken (aminotransferase or glutamate dehydrogenase). The ammonium generated is excreted to the extracellular environment. The ammonium that is released from glutamine contributes to the waste metabolite accumulation.

Amino Acid Metabolism

In addition to providing the building blocks for protein synthesis, amino acids are also used in the biosynthesis of many other cellular components. They contribute the nitrogen atom and amino group in the synthesis of nucleoside bases and nitrogen-containing sugars. They also provide part of their carbon skeleton to form groups like the methyl group (Panel 3.20). Mammals can synthesize only some of the twenty amino acids used in the translational synthesis of proteins. The nine (depending on the species) essential amino acids that mammals cannot synthesize must be acquired through diet. The non-essential amino acids alanine and aspartic acid (and asparagine) are synthesized from pyruvate and OAA, respectively by acquiring the amino group from donor amino acid glutamate through transaminase reactions. Glutamate is derived from glutamine supplied in the medium or formed by incorporating ammonium into α-KG at the expense of an NADH via a glutamate dehydrogenase reaction. The synthetic pathways of serine, cysteine, and arginine are expressed in many cells. Tyrosine is derived from phenylalanine. However, in cultured cells, the supply rates of these amino acids may be too low for proliferation.

Panel 3.20. Amino Acids in Cellular Biosynthesis

- Glutamine (amide group) is used as an amino group donor in adenosine (AMP), guanosine (GMP), and cytosine (CTP) biosynthesis
- Aspartic acid and glycine are also used in nucleic acid synthesis
- Methionine is a methyl group donor. Tryptophan is used in NAD synthesis.
- Glutamate participates in a large number of reactions. The flux of its synthesis or supply is expected to be high.

It is important to remember that an industrial high-producing cell line diverts a large portion of its amino acid pool to making the recombinant product. Even though non-essential amino acids can be synthesized by cells, their provision rate (i.e., the sum of uptake and synthesis) must be sufficiently high to prevent the misincorporation of amino acids into proteins. It has been shown that the depletion of the non-essential amino acid asparagine in culture medium can lead to amino acid misincorporation into the product protein.

Amino Acid Transport

Amino acids are taken up by cells through a large number of amino acid transporters (Figure 3.17). Most amino acid transporters transfer a group of amino acids with similar chemical characteristics, such large neutral (uncharged side chain) amino acids, cationic, or anionic amino acids. The uptake rate of a particular amino acid is thus not only dependent on its own concentration, but also on the concentrations of other amino acids that compete for the same transporter. One amino acid may be taken up through more than one transporter, albeit with different affinities. Transport of amino acids across the cytoplasmic membrane is thus rather complex, and changing the concentration of one or several amino acids may affect the uptake rates of many.

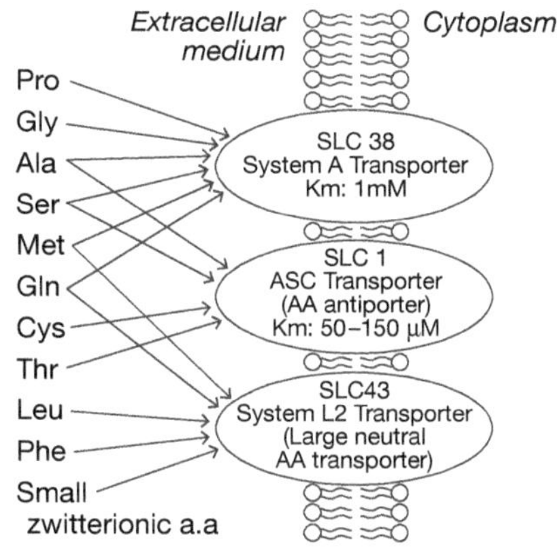

Figure 3.17. Major amino acid transporters.

Amino Acid Degradation and Growth Inhibition

In many industrial processes, especially in fed-batch cultures, concentrated amino acid solution is fed to the culture intermittently (see Chapter 9), causing some amino acids to accumulate to high levels

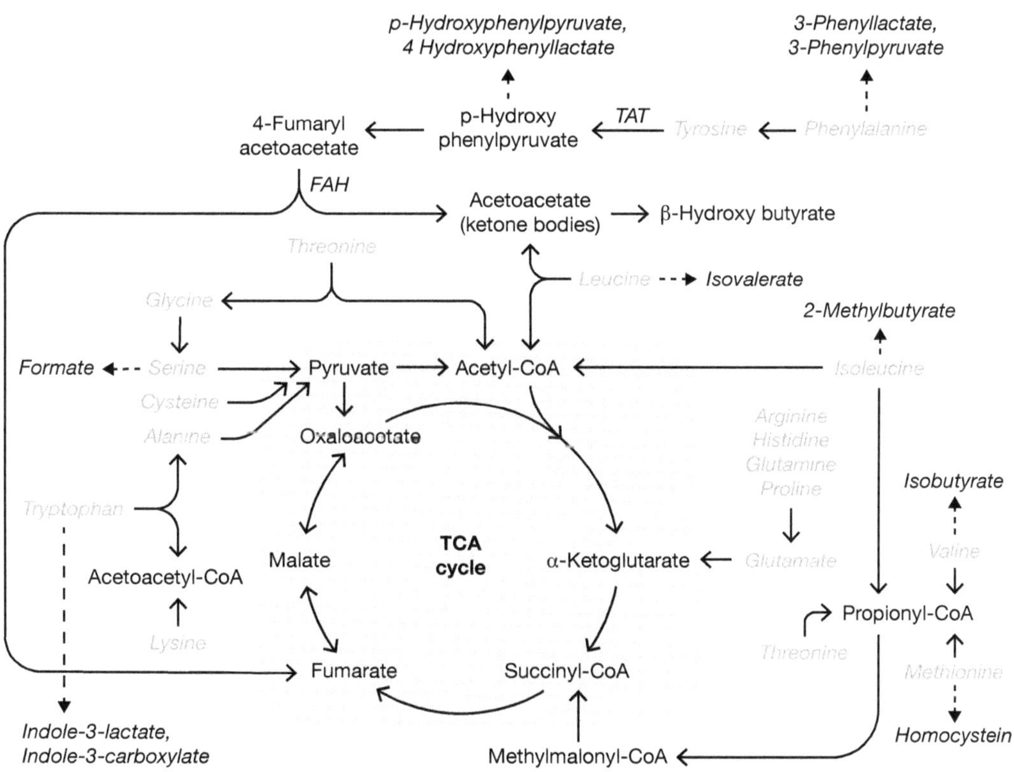

Figure 3.18. Entry of excess amino acids into catabolic pathways. The dashed line shows the amino acid degradation product that may be excreted into the culture medium.

Panel 3.21. Amino Acid Degradation

- Excess amino acids taken up by cells must be excreted
- The nitrogen (amino group) is removed from the carbon skeleton by transamination or oxidative/non-oxidative deamination. The excess nitrogen is excreted as ammonium ions or amino acids (e.g., alanine, proline, asparagine)
- The fate of the carbon skeleton is to enter the TCA cycle through pyruvate, aceto-acetyl-coA, oxaloacetate, or α-ketoglutarate
- The degradation of aromatic and aliphatic amino acids generates growth-inhibitory metabolites

transiently. This high level of amino acids may lead to excessive consumption. The amino acids taken up by cells in excess of cellular need are stripped of their nitrogen atoms and metabolized to pyruvate, acetyl-CoA, propionyl-CoA, or TCA cycle intermediates to enter energy metabolism (Figure 3.18, Panel 3.21). Some catabolic products are secreted into the medium. The catabolism of amino acids also releases ammonium. Ammonium and some catabolic products are growth inhibitory when accumulated to high levels. The catabolism of methionine, valine, and isoleucine gives rise to propionyl-CoA that is then carboxylated to methylmalonyl-CoA and further to succinyl-CoA, which enters the TCA cycle. In animals and humans, insufficient reactivity in the carboxylation reaction can lead to propionic acidemia. In culture, the accumulation of aliphatic amino acid degradation products may also lead to growth inhibition.[6] Similarly, the accumulation of the catabolic intermediates of aromatic amino acids (phenylalanine and tyrosine), especially fumarylacetoacetate, can be fatal in animals deficient in its degradation enzyme if left untreated. A number of inborn diseases of amino acid catabolism are caused by the accumulation of catabolic metabolites. The catabolism of amino acids typically takes place in the liver. Although cultured cells are not derived from individuals with inborn errors, the expression level of those catabolic enzymes in those cells may not be programmed to catabolize amino acids efficiently. At a very high cell concentration in the reactor, and with excessive feeding of amino acids, the catabolic products, both ammonium and reaction intermediates, may become growth inhibitory.

Lipid Metabolism

Lipid Transport

Lipids serve many functions in the body. Phospholipids are critical components of the cell membrane, glycerides store energy, and sterols are not only constituents of the cell membrane but also hormones. Although their role in cultured cells is primarily as the constituent of cell membranes, they are also important in signaling, protein trafficking, and protein-membrane interactions (Panel 3.22). However, as cells grow and the membrane expands, lipids must be supplied at a sufficient rate through biosynthesis and media provision. A balanced lipid composition in the cellular membrane is important in maintaining cellular vitality. However, our understanding of the balanced state of cellular lipids is rather limited.

Panel 3.22. Functions of Lipids

- Contribute to membrane fluidity
- Involved in protein trafficking and membrane fusion events (e.g., polyphosphoinositide)
- Involved in the attachment of cytoskeletal proteins to membranes (anionic lipids, e.g., phosphatidylserine)
- Form microdomains or 'rafts' enriched in specific subsets of membrane proteins (cholesterol and sphingolipids)

The content of lipids in cell culture media varies widely, from high-serum-containing media that is rich in various lipids and lipoproteins to very lean media with only minimal amounts of lipid precursors, such as ethanolamine and a small number of fatty acids. Fatty acids and lipid precursors like ethanolamine may be directly dissolved in media. Most lipids are supplied as serum lipoproteins, conjugated to serum albumin, as liposomes, or as solubilized conjugates, such as sorbitol-fatty acid esters. Although some cells can be cultured in lipid-free media, most cell culture media contains some fatty acids and lipids.

Cellular uptake of lipids is mediated by receptors of the lipoprotein lipid carriers, by endocytosis of membrane vesicles, or by diffusion. Cells readily take up fatty acids, phospholipids, and cholesterol from the medium and incorporate them into cellular lipids. The cellular uptake of fatty acids is a passive, non-energy-dependent process. After being taken up by cells, fatty acids quickly become esters; the intracellular levels of free fatty acids are quite low. Cholesterol is complexed to low density lipoprotein (LDL) in the body and is taken up by cells through the LDL receptor. For cells in culture, cholesterol is often supplied as a conjugate with serum albumin, or as complexes with cyclodextrin.

The lipid composition in the bilayer membrane affects membrane fluidity and permeability. Not all lipid bilayer membranes are the same. The lipid composition of the outer leaflet and the inner leaflet of the cytoplasmic membrane, for example, differ from each other. The plasma membrane is enriched in cholesterol, while the amount of cholesterol in the ER and other organelles is much less. There is very little cholesterol in the inner mitochondrial membrane. After being taken up by cells, lipids have to be distributed to different organelles to give them distinctive membrane lipid compositions. Their transport is mediated by a number of lipid transport proteins or by membrane vesicles for those organelles.

After processing in the endoplasmic reticulum (ER) and the Golgi apparatus, recombinant proteins are transported via membrane vesicles to the cytoplasmic membrane for secretion. Membrane homeostasis and organelle biogenesis, therefore, also affect the productivity of recombinant proteins.

Panel 3.23. Subcellular Localization of Lipid Metabolism

Cytosol
- NADPH synthesis (pentose phosphate pathway)
- Isoprenoid and early cholesterol synthesis
- Fatty acid synthesis

Mitochondria
- Fatty acid oxidation
- Acetyl-CoA synthesis
- Ketone body synthesis
- Fatty acid elongation

Endoplasmic Reticulum
- Phospholipid synthesis
- Cholesterol synthesis (late stage)
- Fatty acid elongation
- Fatty acid desaturation

Peroxisome
- Cholesterol precursors synthesis
- The final steps of cholesterol synthesis

Fatty Acid Metabolism

Most cells have the capability of synthesizing various fatty acids. Under starvation conditions, cells also perform β-oxidation to degrade fatty acids into acetyl-CoA in the mitochondria or peroxisomes (Panel 3.23). Acetyl-CoA then enters the TCA cycle and generates energy.

Fatty acids are synthesized from acetyl-CoA in the cytosol. The first step of fatty acid synthesis involves adding a CO_2 to acetyl-CoA to form malonyl-CoA, which then reacts with acetyl-CoA to become a four-carbon fatty acyl-CoA. This is a case where CO_2 is an essential substrate for biosynthesis, not merely a catabolic product. Fatty acid synthesis, therefore, involves the step-wise elongation processes of using three-carbon malonyl-CoA to add a two-carbon unit to fatty acyl-CoA in each cycle. NADPH is also used to provide energy in this reductive synthesis reaction. There are a number of fatty acid synthetases that can synthesize fatty acids to different lengths.

The fatty acid products from elongation reactions are all saturated fatty acids. Double bonds are then synthesized by unsaturation reactions after saturated fatty acids have been made.

Cholesterol and Its Biosynthesis

Cholesterol is a 27-carbon molecule that has a hydrocarbon chain linked to aliphatic rings (Figure 3.19). A hydroxyl group attached to the aliphatic ring allows it to interact with the head group of phospholipids while the hydrocarbon chain is well embedded in the membrane. Mammals require cholesterol as a constituent of cellular membranes and as a precursor for the synthesis of steroid hormones, bile acids, and lipoproteins (Panel 3.24). Cholesterol is relatively insoluble and resides exclusively in various cell membranes. Its regulation is particularly important since excess cholesterol forms solid crystals, leading to cell death.

Cholesterol constitutes ~10% of the dry weight of plasma membranes, and plasma membrane cholesterol accounts for 65% to 80% of total cellular cholesterol. Cells in culture obtain cholesterol either by *de novo* synthesis or through receptor-mediated uptake of exogenous low-density lipoproteins.

Cholesterol is synthesized from acetyl-CoA, which is condensed by 3-hydroxy-3-methylglutaryl (HMG)-CoA synthase (HMGCS) to form HMG-CoA. HMG-CoA is converted to mevalonate by HMG-CoA

a)

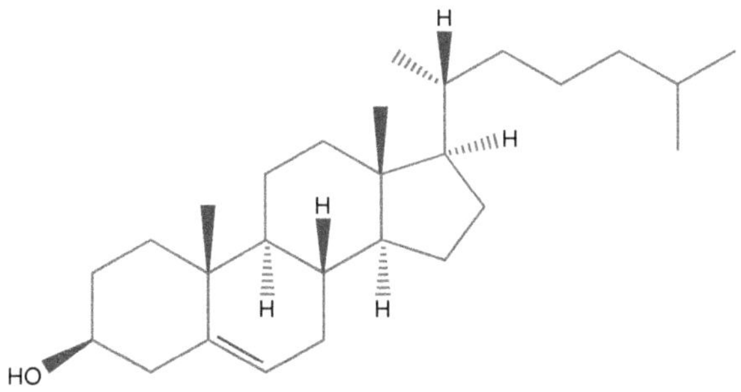

b)

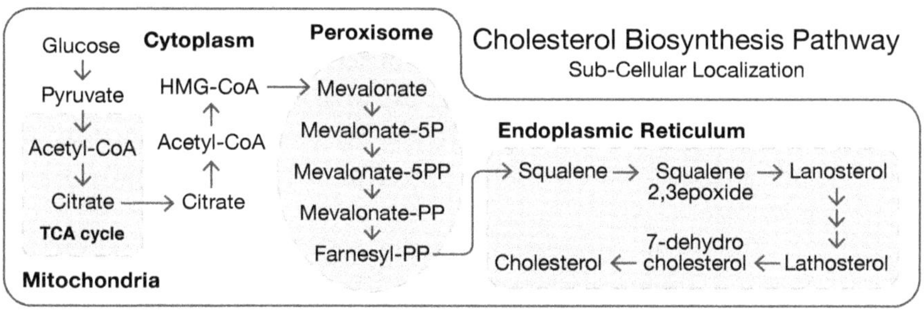

Figure 3.19. (a) The structure of cholesterol. (b) The segregation of the cholesterol biosynthesis pathway in the mitochondria, peroxisome, and endoplasmic reticulum.

reductase (HMGCR) (Figure 3.19b). This enzyme is the target of statins, the class of drugs that suppresses cholesterol synthesis in patients.

Further synthesis of mevalonate to farnesyl-diphosphate takes place in peroxisomes. Subsequent condensation of two molecules of farnesyl diphosphate to form squalene, lanosterol, lathosterol, and finally cholesterol occurs in the ER.

Out of the 18 key enzymes taking part in cholesterol biosynthesis, 5 enzymes reside in the peroxisome and 13 reside in the ER. HMGCS is upstream of HMGCR and is found in the cytosol. Thus, there are at least three different sub-cellular compartments involved in cholesterol biosynthesis.

Panel 3.24. Cholesterol

- A component of membranes and a precursor of steroid hormones, bile acids, and lipoproteins
- Resides exclusively in cell membranes
- Excess cholesterol forms solid crystals, leading to cell death
- Constitutes ~10% of dry weight of plasma membranes
- Cells obtain cholesterol by *de novo* synthesis and by receptor-mediated uptake of plasma lipoproteins

Although cholesterol in mammals is synthesized primarily in the liver, most cells have the capability of synthesizing cholesterol for their own growth requirements. NS0 cells lack an enzyme, 17-HSD, which converts lanosterol to lathosterol. In NS0 cells, 17-HSD is silenced through methylation of a CpG island upstream of its promoter, leading to the cell line's dependency on cholesterol for growth.

Glycan Biosynthesis and Protein Glycosylation

Glycan Heterogeneity

A VAST MAJORITY OF RECOMBINANT therapeutic proteins are glycoproteins. These proteins have carbohydrates, in the form of branched oligosaccharides, attached to them. The glycosylation of proteins, along with disulfide bond formation, is the most common post-translational modification in recombinant protein products. Glycans are classified as O-linked or N-linked glycans (Panel 3.25). O-glycans attach to the polypeptide through the -OH group of serine or threonine. N-glycans link to protein through the amide group of asparagine. For N-linked glycans, the asparagine is in an Asn-X-Thr/Ser recognition sequence, where X indicates no specificity. For O-glycans, no specific recognition sequence is known.

The glycans attached to proteins are oligosaccharides of varying sizes. They are structurally heterogeneous (Panel 3.26). Glycans that attach to the same attachment site of different glycoprotein molecules often have different monosaccharide compositions and sequences. Even glycans of the same sequence may have different linkage bonds (glycosidic bonds) between monosaccharides. Such heterogeneity is called microheterogeneity.

Multiple glycosylation sites are often present on a protein molecule. Not all glycan attachment sites on a protein molecule may be occupied. Different protein molecules may have different combinations of occupied

Panel 3.25. Protein Glycosylation

- Many secreted proteins and cell surface proteins are glycosylated
- All IgG antibodies produced by mammalian cells are glycoproteins, with an N-linked oligosaccharide attached to each heavy chain in the hinge region at Asn_{297}

N-linked Glycosylation

- Attachment of an oligosaccharide to the protein through the amine group of an asparagine

```
      glycan
        |
   —Asn—X—Thr/Ser
```

O-linked glycosylation

- Attachment of an oligosaccharide to the protein through the hydroxyl group of a serine or threonine

Panel 3.26. Heterogeneity In Glycoforms

Macroheterogeneity

- When multiple sites of glycosylation are present in a protein, the occupancy on different sites differs on different molecules
- Possible occupancy:

Glycosylation Site 1	Glycosylation Site 2
+	+
-	-
+	-
-	+

Microheterogeneity

- The structure of the glycan occupying the same site differs among different molecules and free sites; such differences in the occupancy of different attachment sites among protein molecules is called macroheterogeneity.

The microheterogeneity and macroheterogeneity of glycoproteins give them complex glycosylation patterns. The diversity of glycans is seen on secreted glycoproteins in the human body's circulation, as well as on cell surface proteins. The glycan pattern (i.e., the distribution of abundance levels of different glycan structures) of a protein may vary at different developmental or disease states. In biomanufacturing, glycoproteins produced by the same cell line are also heterogeneous in their structure. However, for the manufacturing of therapeutic biologics the glycan profile is specified in the filing of the product for regulatory approval. Upon drug approval, the glycosylation pattern of the manufactured product must be within the range of specification. Those who wish to produce a biosimilar biologic after the patent of an innovative product expires must reproduce the protein with comparable or better (biobetter) quality with the acceptable glycan profile. The heterogeneity of glycans thus poses a special challenge for biosimilar production.

Major Types of N-Glycans

A few commonly seen N-glycans are shown in Figure 3.20. A glycan is first formed in the ER, with 9 mannose extended from 2 N-acetylglucosamine. The 9 mannose is normally trimmed down to 3 and extended further. Glycans that are not trimmed and still have 5 or more mannose residues are called high mannose glycans. High mannose glycans are not abundant in humans, but appear prominently or even dominantly in

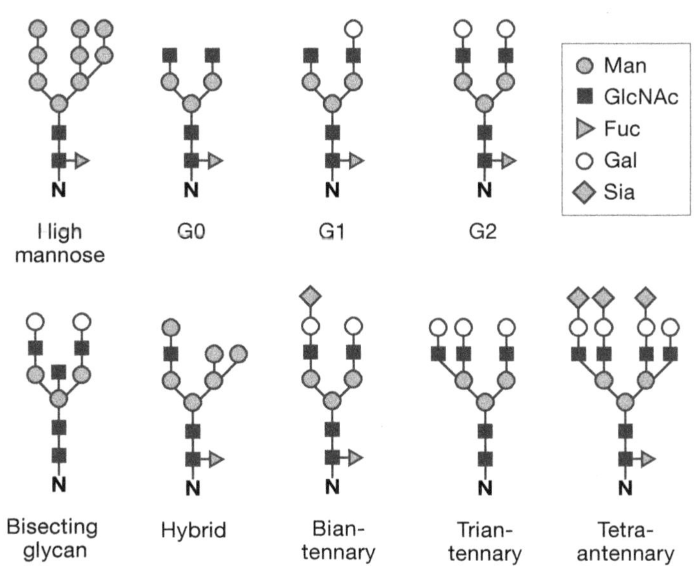

Figure 3.20. Different types of N-glycans in glycoproteins.

yeast, plants, and insects. They may cause an immunogenic response in humans. The trimmed mannose can then be extended to become a biantennary, triantennary, or tetraantennary glycan. These glycans, which are sometimes referred to as complex glycans, may be extended fully to have sialic acid at the terminal, or can be terminated sooner. Glycans with both untrimmed mannose and an antennary extending are called hybrid glycans. A glycan may be fucosylated or unfucosylated. The mannose at the base is extended to two antennaries through its C3 and C6 to two more mannoses. Sometimes the C4 of the mannose at the base is linked to a bisecting N-acetylglucosamine. Bisecting sugar will affect fucosylation and further glycan extension.

IgG molecules typically have simpler glycan structure with only two antennaries. After extending by an N-acetylglucosamine, it may be further galactosylated in 0 to 2 branches. These glycans with 0,1, or 2 galactose are referred to as G0, G1, and G2 structures.

Role of Glycans

The glycan structure on a glycoprotein affects its half-life in blood circulation and its immunogenicity. For IgG molecules, N-glycosylation in the Fc region affects their biological activities. The presence of glycans on interferon produced in mammalian cells prolongs its clearance from blood, as compared to its non-glycosylated counterpart produced in *E. coli*. Higher sialic acid content on erythropoietin (EPO) increases its circulation half-life. Under-sialylated glycoproteins are cleared by liver uptake via the hepatic asialoglycoprotein binding protein faster. It has been postulated that glycosylated recombinant proteins are better retained by the extracellular matrix, thus giving them a longer bioavailability *in vivo* than their unglycosylated variants.

Glycans on glycoproteins may also affect their biological activities. The effect of glycan structure on biological activities is relatively well understood in IgG. Many therapeutic antibody IgG molecules facilitate the killing of target cells through antibody-dependent cellular cytotoxicity (ADCC). ADCC activities of those antibodies are affected by the glycan structure at position Asn_{297} on the CH_2 domain in the Fc region of the IgG heavy chain. Unlike many other glycans on glycoproteins, which are often thought of as being "coated" on the exterior of proteins, the glycan at Asn_{297} is in a "pocket" inside the folded IgG molecule. However, the glycan structure at Asn_{297} affects the binding of IgG to the Fc receptor on the effector cells and its interactions with the complement component C1q. IgG molecules that do not have a fucose on their mannose core have a greater than 10-fold higher ADCC activity compared with those with a fucose. Bisection of the glycan on Asn_{297} also appears to affect ADCC activity. However, bisection and afucosylation occur together, making their effects difficult to distinguish. The effects of galactosylation on IgG activity is less clear. However, a decrease in galactosylation has been related to some autoimmune diseases.

Protein Folding and Glycosylation in the ER

N-glycosylation starts while the protein molecule is still being translated and folded in the ER (Figure 3.21). The translation and translocation into the endoplasmic reticulum of a protein molecule takes a finite amount of time. The translation rate is in the order of 10 amino acids per second. It takes about half a minute to synthesize an average protein and many minutes for a large protein like the heavy chain of IgG. During that period, protein molecules must be protected from misfolding or aggregation. Many chaperone proteins play important roles in the folding process. The addition of glycans to the protein facilitates the folding process (Panel 3.27). Glycans also serve as a quality control signal to ensure the molecules passed on to the Golgi apparatus for further processing are all properly folded.

N-glycosylation is initiated by the transfer of a preassembled oligosaccharide ($Glc_3Man_9GlcNAc_2$, an oligosaccharide of three glucose, nine mannose, and two N-acetylglucosamine) to the asparagine in a recognition sequence of a nascent protein in the ER lumen (Figure 3.21).[7]

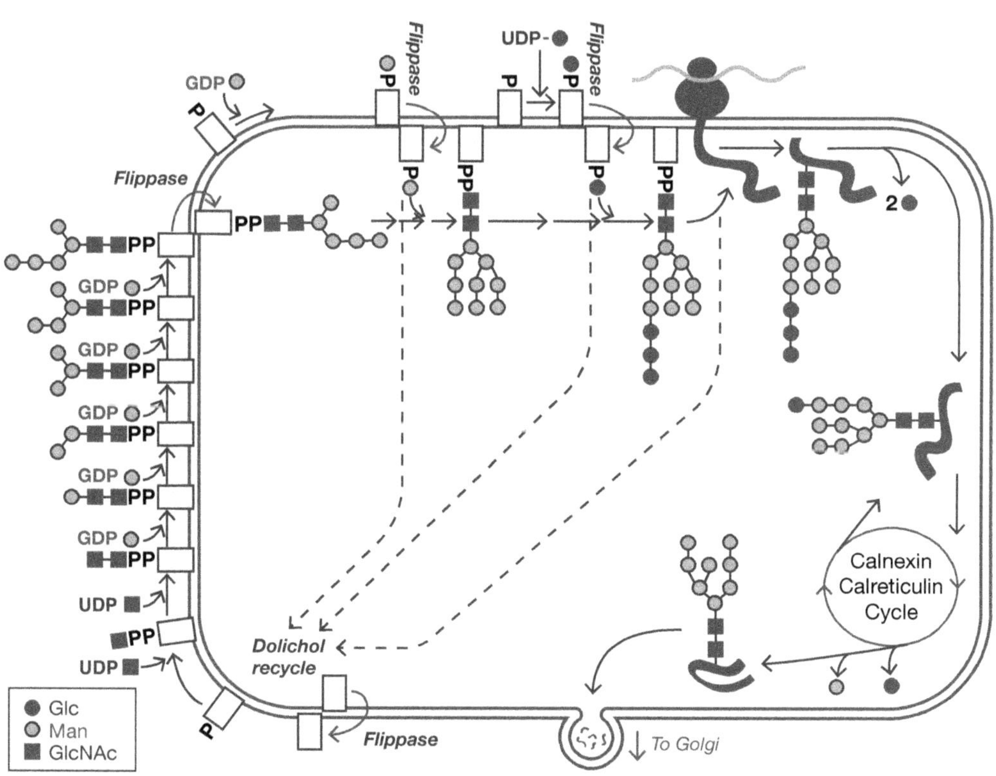

Figure 3.21. N-glycan processing of glycoproteins in the endoplasmic reticulum.

Panel 3.27. Effects of Glycans

- Facilitates protein folding in the ER
- Increases protein solubility
- Affects biological activities
  - Fucose for ADCC activity
- Affects half-life in circulation and pharmacokinetics

The assembly of the high mannose backbone starts on the outside surface of the ER. The glycan is linked to a membrane-anchored dolichol carrier through a pyrophosphate group. After the seven-sugar backbone is formed (with five mannose and two N-acetyl glucosamine), it flips over to the interior of the ER. No transporters are needed for the transport of the backbone glycan; rather, a flippase catalyzes their translocation into the ER lumen. Once inside the ER, the backbone acquires an additional four mannose and three glucose to become a mature core.

The mature core is then transferred to a binding site (Asn-X-Thr/Ser) on a nascent protein molecule. However, not all glycan binding sites may receive the core glycan. This might be due to competition between local protein folding, which could make the glycosylation site inaccessible and the transfer of the glycan core to the site impossible. Hence, on the same glycosylation site of a protein, some molecules may be occupied by a glycan while others may not.

After translation, protein molecules are assisted by many chaperone molecules to undergo the folding process. The three glucose on the glycan core serve as a quality control signal for the proper folding of these glycoprotein molecules. The three glucose are quickly removed from the glycan to generate a monoglucosylated intermediate. But a glucosyltransferase quickly adds a glucose to any protein molecule that is not yet correctly folded. The presence of the monoglucosylated glycan signals the protein molecule liable to bind to the ER lectins calnexin and calreticulin, and the oxidoreductase ERp57. ERp57 facilitates the formation of disulfide, a critical step in protein folding. The folding process continues until the absence of the glucose residue allows the folded molecule to dissociate from the lectin. The "correctly" folded protein molecule is then transported to the Golgi apparatus by a membrane vesicle.

Misfolded proteins, with the de-glucosylated mannose exposed, are subjected to the removal of their terminal α1,2-bonded mannose. Trimming of the mannose then serves as the signal to divert the unfolded molecules to the ER disposal system, through which they are exported to a proteasome for degradation.

Well-folded glycoprotein molecules are enclosed in membrane vesicles of the ER, where they bud and then translocate to the Golgi apparatus. Once there, they fuse with the Golgi body membrane and the glycoprotein cargos are released into the lumen of the Golgi apparatus.

Glycan Extension in the Golgi Apparatus

Inside the Golgi, mannose is trimmed further from the N-glycan core, reducing the number of mannoses from nine to three to form the tri-mannosyl ($Man_3GlcNAc_2$) core structure (Figure 3.22). However, incomplete trimming does occur, leading to the synthesis of high-mannose-type glycans ($Man_{5\text{-}9}GlcNAc_2$) (Figure 3.20). After mannose trimming, more monosaccharide units are added to the glycan one at a time. The extension of the glycan from the 3-mannose core creates complex-type glycans, while the extension from those with 4-5 mannose creates hybrid-type glycans.

Three monosaccharide units constitute most of the extended N-glycans: N-acetyl glucosamine, galactose, and sialic acid. Different glycosyltransferases are involved in these reactions. Each enzyme adds a different monosaccharaide and forms a specific glycosidic linkage to the growing core glycan. Each glycosidic bond is defined by the two sugars being linked and the position of the carbon on each sugar. The incoming monosaccharide provides the activated carbonyl group (by linking to a nucleotide), and the receiving carbohydrate moiety on the growing glycan on the protein provides a hydroxyl group for forming the glycosidic bond. For example, the two-terminal mannose residues each use its C1 (carbonyl carbon) to link to C3 and C6 of the base mannose, respectively (Figure 3.22).

Each terminal mannose can be further extended by the addition of a N-acetyl glucosamine (GlcNAc). The terminal receiving mannose has three hydroxyl groups, a pair of which (carbon 2 and carbon 4, or carbon 2 and carbon 6) can be extended by the addition of a GlcNAc. The extension reaction does not take place on all of the available reaction (receiving) sites. If one of each of the two-terminal mannoses of the Man_3-$GlcNAc_2$ core are extended by the addition of a GlcNAc, it becomes a biantennary glycan (Figure 3.20). If more than one hydroxyl group is

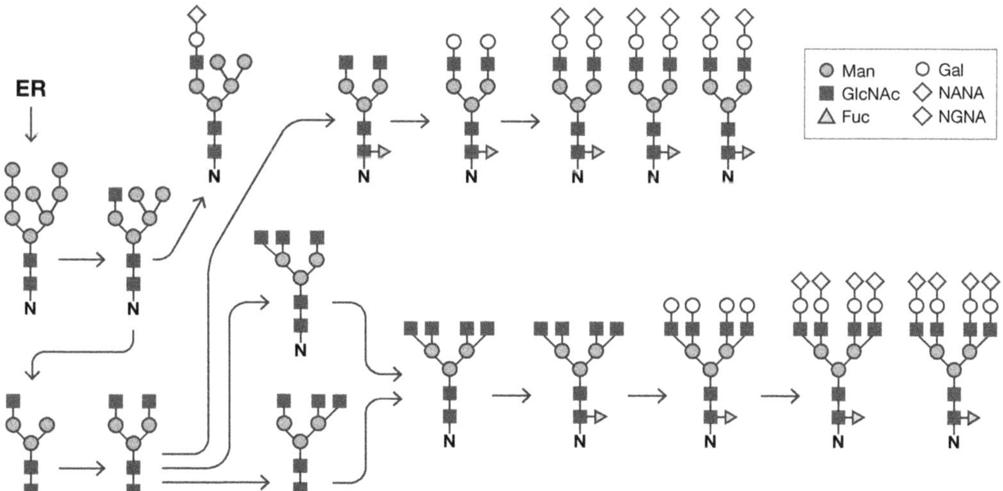

Figure 3.22. N-glycan extension in the Golgi apparatus.

extended on either one or both terminal mannoses, then the glycan becomes tri- or tetra-antennary.

There are a number of different glycosyltransferases, each of which recognizes different pairs of incoming nucleotide sugars and substrate glycans, and which catalyzes the formation of different glycosidic bonds. A number of glycosyltransferases do allow for some flexibility in glycosidic bond formation. In mammals, some glycosyltransferases have a large number of isozymes. Their expression may be tissue-specific. Some isozymes are co-expressed in some cells. Roughly a couple dozen glycosyltransferases are prominent in contributing to the glycan heterogeneity in recombinant proteins.

Each growing glycan, thus, has multiple available reaction paths for extension. Many of those different reaction paths will lead to different products (Figure 3.22). However, in other cases, the reactions of adding sugars to different branches of the glycan may occur in different orders but lead to the same product. In the extension of the first layer GlcNAc to the 3-mannose core, the addition of GlcNAc to the base mannose (catalyzed by N-acetylglucosaminyltransferase III (GnT-III)) suppresses further extension of the glycan. This bisecting GlcNAc thus has a strong effect on glycan structure. The enzyme fucosyltransferase has the opposite effect of increasing structural diversity; it adds a fucose to the base sugar (GlcNAc), thus making two versions of most glycans, fucosylated or unfucosylated. Both the bisecting GlcNAc and the core fucose affect the biological activities of many protein molecules (e.g., the ADCC activity of IgG, and the binding of EGF to EGFR).

The web of glycan extension reactions forms a complex network which, when drawn out graphically, indeed resembles a network of diverging and converging paths leading to a number of different fully-extended N-glycan structures (Figure 3.23). Adding to the complexity is the compartmentalization within the Golgi apparatus. The Golgi apparatus consists of stacks of membranous compartments commonly grouped into cis, medial, trans, and trans-Golgi network (TGN) cisternae. These cisternae are not biochemically homogeneous. The glycosyltransferase composition varies among different compartments, with the earlier enzymes more enriched in the early compartment and vice versa. As the secretory glycoproteins traverse through these Golgi compartments, the extension reactions will favor the addition of different glycans. An extending glycan that fails to acquire a particular sugar in an early compartment may thus not get extended further after moving on to a later compartment.

The reaction path for glycan extension in the Golgi apparatus is a rather long one. Protein molecules spend tens of minutes passing through the Golgi, but the residence time that each protein molecule spends in the Golgi is not uniform. Not all protein molecules will have fully extended glycans. In fact, often only a portion will be fully extended and

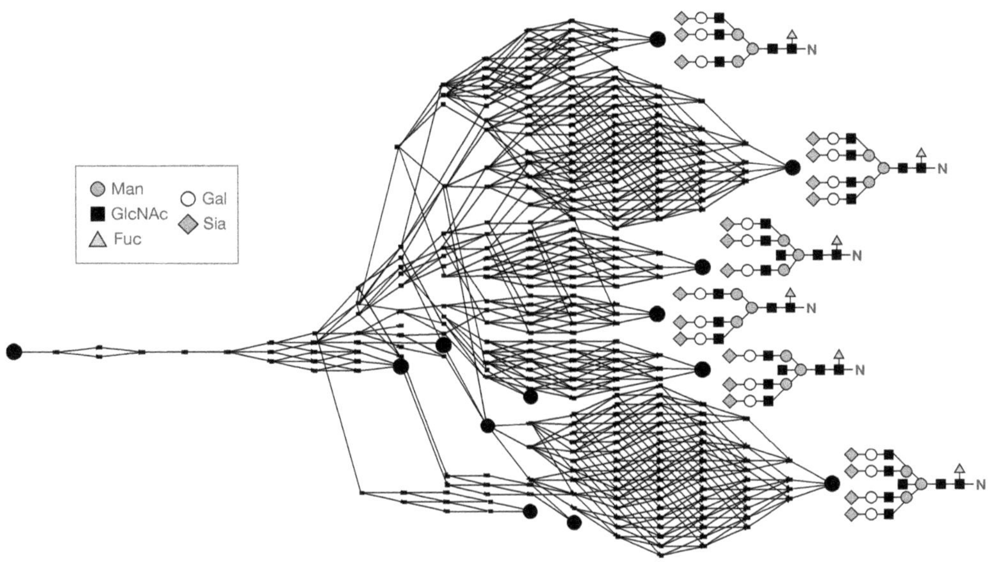

Figure 3.23. N-glycan extension reactions in the Golgi form a complex network. Only a small number of enzymatic reactions generate a very large number of different glycans. Each enzyme catalyzes the same reaction on many different extending glycans.

contain sialic acid. IgG molecules have glycans on Asn_{297} of the heavy chain constant region and in some cases also in the hypervariable region. Sialylation is rarely seen in the glycan on Asn_{297}, but is more often seen in the one in the hypervariable region.

A very large number of glycan structures can be formed in the N-glycosylation pathway; typically only a fraction of all these possible configurations constitute the dominant types. The patterns of glycans on recombinant DNA proteins produced in different production cell lines but derived from the same host cell line are likely to be different. The pattern is also affected by different culture conditions.

O-Glycosylation

O-glycosylation is initiated in the endoplasmic reticulum or in the Golgi apparatus. Unlike N-glycosylation, which is initiated by the translocation of a preassembled core-oligoglycan (Man_9Glc_3) to the nascent protein, O-glycosylation starts by adding monosaccharides directly to the protein and extends mostly through one-by-one addition of more sugars. There is no known consensus sequence for the addition of O-glycans to serine and threonine, although clusters of O-linked glycosylation are often seen in regions with a high density of serine/proline/threonine. The OH− groups of serine and threonine can be linked to glucose, N-acetylglucosamine, galactose, N-acetyl galactosamine, fucose, xylose, or mannose. O-glycans can thus be more complex and heterogeneous than N-glycans, because they lack the consensus attachment amino

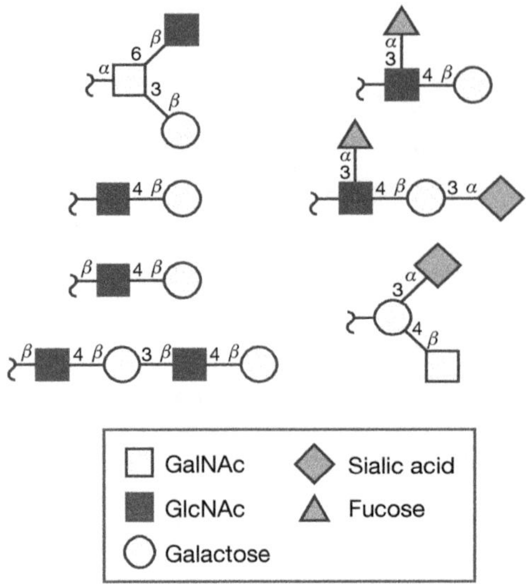

Figure 3.24. Some O-glycans synthesized in CHO cells. CHO cells have a relatively small number of enzymes involved in O-glycosylation.

acid sequence and a single common core glycan structure. O-glycans are abundantly represented in some glycoproteins. The O-glycans seen in recombinant proteins are mostly of the mucin-type structure with an N-acetylgalactosamine (GalNAc) as the first sugar. Etanercept (Embrel, a fusion protein of TNF α-Fc for rheumatoid arthritis treatment) has 3 N-glycosylation sites and 11 O-glycosylation sites. Most of those O-glycans are linked to serine/threonine through O-GalNAc, whereas in the O-glycan of factor VII, O-fucose and O-glucose are also seen (Figure 3.24).

The attachment of N-acetylgalactosamine to the −OH group of serine/threonine is initiated by GalNAc-transferases. The presence of GalNAc recruits other enzymes to add more GalNAc to nearby serines/threonines. Subsequently, a number of enzymes catalyze the addition of different sugars (glucose, galactose, GalNAc) to form different core O-glycans. This is followed by core extension (e.g., by adding more galactose) and O-glycan capping (e.g., by the addition of sialic acid, fucose, or sulfate). Like N-glycosylation, various glycosyltransferases are locally enriched in different compartments of the Golgi apparatus. Some glycosyltransferases are probably shared with N-glycan biosynthesis. Like N-glycan synthesis, the formation of the different cores of O-glycans is affected by the relative abundance of different enzymes in the tissue or the cell. The addition and extension of different glycans is also affected by the protein domain structure, although our knowledge with respect to the effect of protein domain on glycosylation is still rather limited.

Synthesis and Transport of Nucleotide Sugar Precursors

The substrate for glycosidic bond formation by glycosyltransferase is a nucleotide sugar. The carbonyl carbon of the sugar is the carbon that will form the glycosidic bond. Before it can react with the glycan substrate, the carbonyl carbon of the sugar substrate (glucose 1-phosphate, galactose 1-phosphate, N-acetyl-glucosamine-phosphate, or mannose 1-phosphate) needs to be "activated" by reacting with a nucleotide (NTPs: UTP, CTP, GTP) to form NDP-sugar or CMP-sialic acid at the expense of an equivalent of 1 ATP. Different sugars are linked to

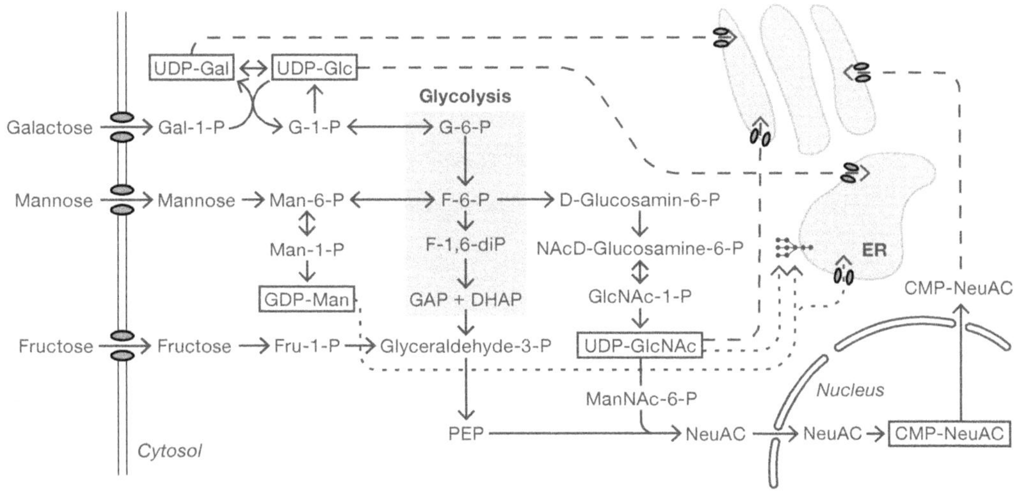

Figure 3.25. Biosynthesis of the precursors of glycans.

different NDPs. Uracil is used for glucose- and galactose-based sugars (e.g., UDP-glucose and UDP-galactose), guanyl for mannose and fucose, and cytidyl for sialic acid.

Mannose, galactose, N-acetylglucosamine, and fucose are synthesized in branches of the glycolysis pathway (Figure 3.25). All four sugars are activated at their first (carbonyl) carbon. Therefore, they link to glycans through the formation of (1→n) glycosidic bonds. For example, UDP-GlcNAc is added to a growing core by the formation of an N-acetylglucosamine β (1→n) mannose bond. The linkage can be through a number of possible carbons on mannose (e.g., 2, 3, 4, or 6) depending on the position of the mannose in the glycan. Upon giving away its sugar, an NMP is formed as the reaction product, and is transported back to the cytosol (Figure 3.25).

Sialic acid is a generic name for the derivative of 9-carbon neuraminic acid and is also the name of the most commonly seen N-acetylneuraminic acids. The second carbon of neuraminic acid is activated; thus CMP-2-sialic acid will form a sialyl (2→n) bond with galactose. The synthesis of all the precursor sugars occurs in the cytosol, including a nine-carbon neuraminic acid and N-acetyl neuraminic acid. Similarly, all nucleotide sugars are formed in the cytosol, except for CMP-sialic acid. The activation of sialic acid to CMP-sialic acid occurs in the nucleus.

The backbone of N-linked glycan is synthesized on the cytosolic side of the ER membrane through the membrane-anchored dolichol. The nucleotide sugars used in the formation of the backbone, GDP-mannose and UDP-N-acetyl glucosamine, are synthesized in the cytosol and directly react with dolichol or with the growing glycan backbone. The

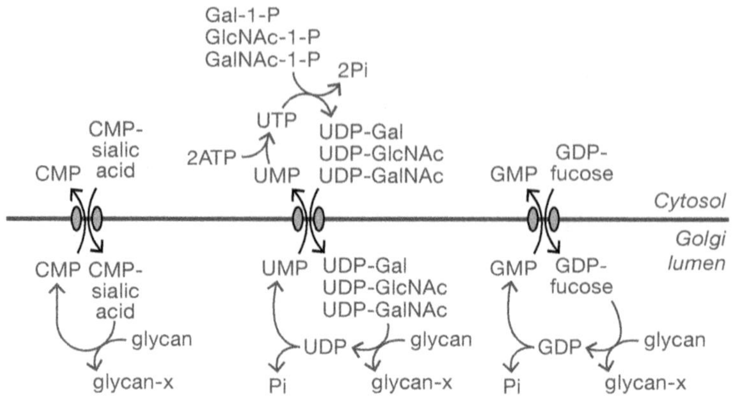

Figure 3.26. Transport of nucleotide-sugar precursors into organelles.

assembled backbone is then "flipped" into the ER and the subsequent reactions occur inside the ER. Transporters are used to supply the nucleotide sugars involved in the subsequent reactions, including GDP-mannose and GDP-glucose in the ER.

The nucleotide sugars for the extension reactions in the Golgi apparatus are also transported through transporters (Figure 3.26). These include CMP-sialic acid, GDP-fucose, UDP-N-acetylglucosamine, UDP-galactose, and the activated sulfate donor (3′-phosphoadenosine, 5′-phosphosulfate). All of these transporters are antiporters, requiring an equal molar exchange of the import of nucleotide sugar and the export of the corresponding NMP (as a product of glycosyltransferase reactions) during the process.

Glycan Diversity among Species

A concern about recombinant protein therapy is the possible elicitation of immunogenicity. An antibody elicited by and against the protein therapeutic can result in neutralization of the therapeutic protein and may result in an unintended drop in efficacy, thus causing serious adverse clinical effects.

The potential immunogenicity of recombinant therapeutics may arise from an aglycosylated protein core or from the glycans associated with it. There are at least two mechanisms by which glycans on a protein may affect the immunogenicity of a human therapeutic: 1) by being a foreign glycan structure, or 2) by shielding a segment of the protein that is otherwise antibody inductive.

Recombinant human therapeutic proteins produced in different organisms or cells are often differently glycosylated (such as those from CHO versus yeast) or aglycosylated (such as from CHO versus *E. coli*) (Panel 3.28).[8] Comparison of those proteins indicates that the "shielding" effect of minimizing immunogenicity is affected by the nature of the protein as well as by the source of the protein. The concerns about the immunogenicity of different glycoforms of the rDNA proteins produced in insect cells and in transgenic plants has hindered those technologies' application for rDNA therapeutic protein production. Glycosylated proteins produced in CHO and mouse myeloma cells are minimally immunogenic.

Panel 3.28. Diversity of Glycosylation in Species

- The makeup of glycosylation enzymes is somewhat different among species, even among different tissues of the same species
- Possible immunogenicity of non-native glycans (e.g., high mannose glycan from most yeast)
- Glycans on CHO-produced proteins
  - Have NGNA, humans have only NANA
  - Have only α(2,3) sialic acid, whereas human proteins have both α(2,3) and α(2,6)

The glycosylation pathway is highly conserved in mammals. Nevertheless, divergence among different species is seen. Furthermore, the expression pattern of glycosylation genes in different tissues or in cells derived from different tissues is often different. Host cells derived from different species or from different tissues may produce recombinant proteins with different glycosylation patterns. For example, the sialic acid in glycans produced in animal cells can be different from that produced in human cells. Human glycans have terminal N-acetylneuraminic acid (NANA). NANA can be further hydrolyzed to N-glycolylneuraminic acid (NGNA) in most other mammals. However, humans have a truncated CMP-sialic acid hydroxylase that renders it unfunctional.[9] Thus, glycoproteins produced in non-human cells have some NGNA that is not present in human glycoproteins. Similarly, glycoproteins expressed in CHO cells have only terminal α(2,3)-linked sialic acids, in contrast to α(2,6) and α(2,3) seen in humans, due to the near absence of expression of 2,6-sialyltransferase in CHO cells. Such differences in glycan composition have posed a concern; however, immunogenicity of recombinant proteins directly caused by variant glycans is still rare.

In the production of biosimilar biologics, a critical issue is the production of glycoproteins with "the same" glycoform distribution as the original innovative product. In almost all cases, the cell lines as well as the processes used for the production of the biosimilar are different from those employed in the production of the original innovative product. It is therefore a challenge to devise a robust process that can produce protein molecules with the desired or target glycosylation pattern. Many factors affect the glycosylation pattern of the product, including the expression pattern of glycosylation enzymes in the cell line, the nucleotide sugar supply, the cultivation conditions, and the metabolic state of the cells. The glycosylation pattern is also affected by signaling pathways and their modulators. Further complicating the issue is the fact that the glycosylation pathways in the endoplasmic reticulum and the Golgi apparatus are not only present to process product proteins, but also a very large fraction of all cellular proteins. Thus, the effect of modulating the pathway is not limited to the product protein but potentially includes other aspects of cell physiology. Due to the importance of glycoforms in the assessment of the quality of therapeutic proteins from a regulatory perspective, we will likely see growing efforts to better understand and control glycoforms in the near future.

Concluding Remarks

In this chapter, we presented a brief overview of the broad areas of cellular metabolic processes. We covered the core of energy metabolism, the process of glucose utilization through glycolysis, the PPP, and the TCA cycle, and examined how all of these affect cell growth behavior and productivity. Through interconnected pathways, the central corridor of energy metabolism also influences the synthesis and glycosylation of the product proteins. The excessive consumption of glucose and glutamine and the corresponding accumulation of lactate and ammonium in culture contribute to growth inhibition and low productivity. Lactate consumption in the late stage of culture has been positively associated with a high productivity. There are, therefore, ample incentives to better understand cell metabolism and search for new ways of manipulating it that can better redirect the process. In recent years, we have developed a better understanding of the link between glycolytic regulation and growth control. We have also established better tools to probe the relationship between metabolic flux distribution and other aspects of physiology that influence both productivity and product quality. With the benefit of global physiological perspectives, we continue to gain a deeper understanding of metabolism. Global views at a systemic level will significantly enhance our capacity to manipulate cell metabolism and thus increase productivity and product quality.

References

1. Zhou, W., et al., *Alteration of mammalian cell metabolism by dynamic nutrient feeding.* Cytotechnology, 1997. **24**(2): p. 99–108.
2. Gambhir, A., et al., *Analysis of cellular metabolism of hybridoma cells at distinct physiological states.* J. Biosci. Bioeng., 2003. **95**(4): p. 317–27.
3. Fleischaker, R.J.J. and A. Sinskey, *Oxygen demand and supply in cell culture.* Eur. J. Appl. Microbiol. Biotechnol., 1981. **12**: p. 193–97.
4. Wlaschin, K.F. and W.-S. Hu, *Engineering cell metabolism for high-density cell culture via manipulation of sugar transport.* J. Biotechnol., 2007. **131**: p. 168–76.
5. Mulukutla, B.C., M. CGramer, and W.-S. Hu, *On metabolic shift to lactate consumption in fed-batch culture of mammalian cells.* Metabolic Engineering, 2212. **14**: p. 138–49.
6. Mulukutla, B.C., J. Kale, T. Kalomeris, M. Jacobs, and G.W. Hiller, *Identification and control of novel growth inhibitors in fed-batch cultures of Chinese hamster ovary cells.* Biotechnology & Bioengineering, 2017. **114**: p. 1779–90.
7. Aebi, M., *N-linked protein glycosylation in the ER.* Biochim. Biophys. Acta, 2013. **1833**(11): p. 2430–37.
8. Chung, C.Y., et al., *SnapShot: N-glycosylation processing pathways across kingdoms.* Cell, 2017. **171**(1): p. 258 e1.
9. Irie, A., et al., *The molecular basis for the absence of N-glycolylneuraminic acid in human.* J. Biol. Chem., 1998. **273**: p. 15866–71.

Contents

4

Product Quality and Process

Susan Abu-Absi and Wei-Shou Hu

Introduction

THE QUALITY OF THE PRODUCT is a most important aspect of manufacturing for products of therapeutic applications, even more so for biologics. The complex structures of those products make the quality determinants harder to quantify. Ensuring the manufacturing process produces these biologics of consistent quality is critical for the manufacturer as well as for the public. Manufacturing processes should be designed to robustly deliver consistent yields and product quality. Increasingly cells and viruses with even higher degrees of complexity than proteins are employed as therapeutic agents. Although gene and cell therapy products are still few, maintaining and monitoring the quality of these products will only become more crucial. The vast knowledge and experience of therapeutic protein manufacturing will help set the path forward for the development of manufacturing processes for these new products.

Panel 4.1. Quality of a Therapeutic Product

- Structural identity
- Functional/biological activities
- Impurities
- Product homogeneity
  - Natural variability
  - Chemical variants

What constitutes product quality? Generally speaking, a product is established by its structural identity (Panel 4.1). Sometimes, structural variants of the product arise in the manufacturing process.[1] These must be controlled in the drug substance so as to restrict their levels to within acceptable limits. Most therapeutic proteins are glycoproteins whose glycans vary somewhat between individual protein molecules, but whose glycan heterogeneity must be within a certain bound. Inevitably, the product will have some contaminants, which must also fall within a set limit. Finally, the product must be tested and found biologically functional and safe before product release. The general quality requirements and characterization methods are detailed in Table 4.1.

The productivity of cell culture bioprocesses is currently very high. For many proteins, such as IgG, the titer in the reactor has exceeded the highest level in its native environment *in vivo*. At such high concentrations, physical interactions with other components may differ from those in older generation processes or those *in vivo*. These molecular interactions may pose challenges in protein aggregation or host cell protein contamination. The high productivity achieved nowadays is accomplished with a high cell concentration and often high-stress culture conditions such as high sugar content, high salt and waste metabolite concentration, and low temperature in the production bioreactor. Under more extreme culture conditions, the secreted product may be subject to chemical modifications as well as enzymatic modifications caused by enzymes released from dead cells. Some higher-order protein structural changes, such as aggregation or fragmentation, may even take place during the product recovery process. For some products, the productivity of the producing cell line has reached or even exceeded the secretion rate of professional secretors in the human body (e.g., liver cells and antibody- and insulin-secreting cells). It is unclear whether these hyperproducing cells, having been pushed to the limit of some of their functions related to protein synthesis and secretion, have exposed their product to a higher probability of deviation (e.g., a higher degree of heterogeneity in a glycan profile or a higher rate of amino acid misincorporation).

The first task in ensuring consistent product quality in manufacturing is the identification of the properties of a given product that play a key role in its safety and clinical efficacy. This is undertaken while the product is being developed, and the knowledge is applied while developing the process. In the manufacturing process, the product is made in bioreactors, purified through downstream steps, stored as a drug substance, and then converted to a drug product for use in clinical application. The operating conditions in production and downstream recovery that may lead to unsatisfactory product quality should then be identified and the controlled range of critical process parameters defined. After the product is launched, the product quality data is continually collected, compiled, and analyzed in order to better control product quality during the manufacturing process. The process of quality enhancement and process improvement thus continues throughout the product's life cycle.

This chapter will first discuss those chemical and physical properties of protein biologics which are important quality attributes. This will be followed by a discussion on Quality by Design (QbD), an approach to process development whereby the manufacturing process and controls are proactively designed to deliver the intended product. The last section

Table 4.1. Characterization of a Protein Biologic

Class	*Attributes*	*Assays*
Primary sequence and identity	Intact mass, peptide map, N-terminal sequencing, MS-based sequence identification (mutations, misincorporations)	SDS-PAGE and WB, HPLC, mass spectrometry
Higher-order structure	Secondary (α-helix structure), quaternary structure (e.g., disulfide bonds)	Circular dichroism (CD), differential scanning calorimetry (DSC), Fourier transform infrared spectroscopy (FTIR), analytical ultracentrifugation (AUC), disulfide bond analysis, free-thiol analysis
Purity/impurity	Residual host cell proteins (HCPs), residual DNA, other process impurities such as insulin, IGF-1, Protein A, excipients, methotrexate, and beta-glucan	Hydrophobic interaction chromatography (HIC), ELISA, HCP by LC-MS/MS
Structural variants	Aggregates / high and low molecular weight species	Size-exclusion chromatography (SEC), SDS-PAGE, capillary electrophoresis (CE)
	Charge variants; acidic and basic	Cation exchange chromatography (CEX), capillary zone electrophoresis (CZE), HIC, capillary isoelectric focusing (cIEF)
	Degradation and chemical modification, hot spots (deamidation, oxidation, isomerization, glycation, succinic acid modification, leader sequence, etc.)	Peptide map-MS/MS
	Glycoforms (N-linked glycan, sialylated species, high mannose, afucosylation, galactosylation, o-sialylation)	HILIC, CZE, UPLC, HPLC-MS/MS
Functional	Ligand binding kinetics	ELISA, surface plasmon resonance (SPR)
	MoA-based, cell-based function assay	Bioassay
	Fc effector function assay (specific for Fc-containing products) e.g., FcRN binding	Bioassay
	Complement dependent cytotoxicity (CDC), antibody dependent cell-mediated cytotoxicity (ADCC)	Bioassay
Safety	Endotoxin, bioburden, sterility, mycoplasma	
General tests	Affinity (Protein A) HPLC, protein content by UV280, visible and sub-visible particles, excipient, pH, appearance, osmolarity, residual moisture, clarity, color	

emphasizes the importance of maintaining consistent product quality throughout the entire life cycle of the product.

Quality of Protein Product

Identity of Protein Product

IN PROTEIN PRODUCTION through recombinant cells, the identity of the product gene is dictated by the gene sequence that encodes the product and by the downstream purification process. In the early stages of product development, the chemical structural identity of the product is affirmed at both the gene sequence level and the protein level. During manufacturing, the product is routinely tested to ensure its structural integrity as well as to detect the presence of structural variants.

Molecular and immunochemical identity and quantification

The identity of a protein product is typically established by its molecular weight and chemical structure. The purified product protein is typically run on polyacrylamide gel electrophoresis (PAGE) to assess its molecular weight and purity in terms of contamination by other proteins. The purified protein solution is run on a polyacrylamide gel in an electrical field, which separates individual proteins based on their mobility as determined by their size or molecular weight (Figure 4.1). The post-electrophoresis gel is stained with Coomassie Blue or Silver Stain to visualize the migration of the protein from its initial position. Typically, the protein sample loaded into PAGE is treated with sodium dodecyl sulfate that denatures the protein, dissociates non-chemically cross-linked protein subunits, and extends the protein to better reflect its size. Since larger proteins migrate at a slower rate, the molecular weight of a protein molecule can be determined by calibrating to a mixture of protein markers. PAGE also serves as a tool for checking protein purity. A highly purified product protein should show only the gel band(s) of the product without any contaminating band(s) of impurity. Some proteins consist of subunit proteins that are bound together by disulfide bonds (e.g., IgG). On a native (non-denatured) protein gel and a reduced gel (in which a reducing chemical breaks the disulfide bond), the protein thus displays different bands as the reduced monomers migrate at different rates down the gel. This type of analysis is often employed to detect low molecular weight species or fragments generated by enzymatic degradation.

Western Blot is another method used to identify the product protein. After the purified product protein is run on PAGE, the protein is blotted to a membrane. The membrane is then treated with an antibody against the product protein. The antibody, which had been previously labeled with an enzyme, isotope, or fluorogenic tag, binds to the protein band.

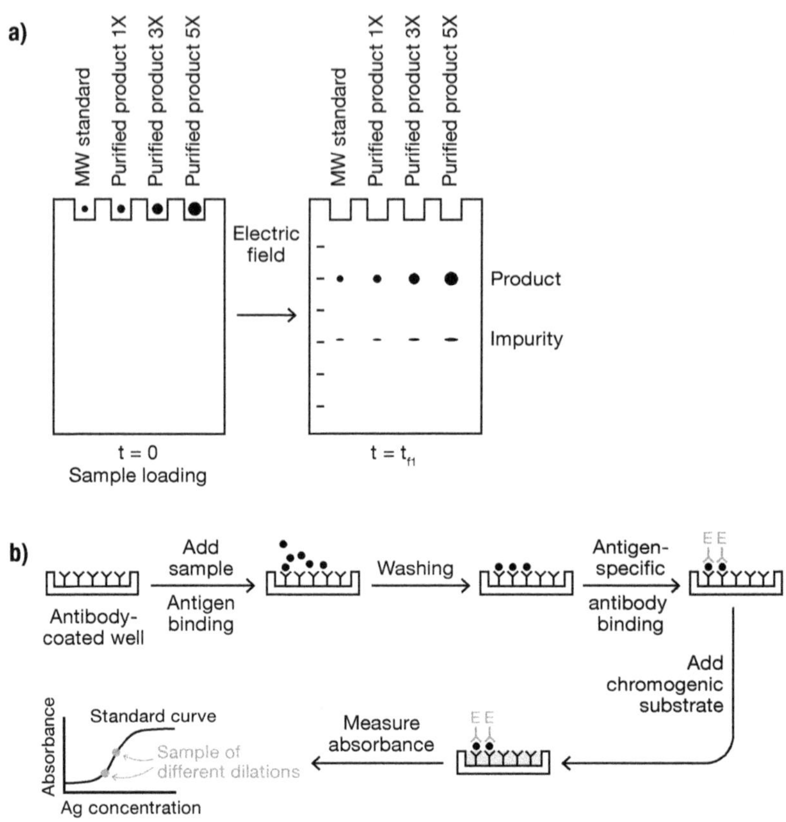

Figure 4.1. (a) Polyacrylamide gel electrophoresis for determining product protein molecular weight and purity. (b) Enzyme-linked immunosorbent assay (ELISA) for determining antigen concentration.

When the unbound antibody in the membrane is washed off and a chromogenic substrate of the antibody-conjugated enzyme is applied, the product protein bands appear in the correct molecular weight location.

Immunological method ELISA, which utilizes antibodies against the product protein, is widely used for measuring protein concentration. There are a number of different ways to implement ELISA. In one scenario, the antibody against the product protein is coated on a 96-well or 384-well assay plate (Figure 4.1b). The sample solution containing the product at different dilutions is then added to allow for binding to the antibody. Then a second antibody that recognizes a different epitope of the product and has been conjugated to an enzyme is added to bind to the product that was captured by the antibody coated on the plate. After a washing step, the amount of the enzyme of the second antibody remaining on the coated surface is quantified by measuring the rate of the enzymatic reaction. The enzymes commonly used typically convert a chromogenic substrate to a product with a visible signal. For example, horseradish peroxidase (HRP) converts a substrate to a colored molecule that can be detected using a spectrophotometer. A standard curve constructed with a serially diluted, purified antigen is used to determine the concentration of the antigen in the sample. Such "sandwich ELISA" is used for quantification of antibody products and many other proteins. Another method specific to antibody quantification utilizes affinity chromatography. Protein A, which preferentially binds to antibodies, is immobilized on a column support. At neutral pH, the antibody molecule binds to Protein A through the Fc region while host-cell proteins, cell culture media components, and buffer flow through the column. Captured antibodies elute at acidic pH and are detected by UV absorbance at 280 nm. A calibration curve is derived from standards and the corresponding peak areas using linear regression analysis. Sample and control product concentrations are calculated from the calibration curve.

Peptide sequence identity

The most accurate check of protein identity is based on the primary structure, or the protein's amino acid sequence. This is generally done by the classical Edman degradation using an automated protein sequencer or by liquid chromatography followed by mass spectrometry. In Edman degradation, complex proteins with multiple chains are separated into single chain molecules. Large proteins are then enzymatically digested into fragments of less than 50 amino acids long. Next, the target protein or peptide is immobilized onto a solid and subjected to chemical hydrolysis from the protein's N-terminus. The released amino acid is made to fluoresce and identified in HPLC. By cycling through the terminal amino acid degradation process, the entire sequence of the protein can be established.

Amino acid sequence identity can also be established by mass spectrometry. To establish the sequence identity of the protein product, the purified protein is subjected to proteolysis to fragment the protein into smaller size peptides (typically, trypsin is used to cleave the C-side of a peptide bond adjacent to lysine or arginine) (Panel 4.2). This process cleaves the protein to peptides of mostly 15–30 amino acids long (Figure 4.2). The resulting mixture of peptides is then subjected to reverse-phase HPLC or capillary electrophoresis for separation. Next, the peptides are separated into peaks and fed into the mass spectrometer (MS) for detection. Each peptide peak will have a characteristic mass over charge ratio (*m*/*z*) of the

Panel 4.2. Protein Identification by Peptide Mapping

- Peptide mapping entails exposure of the protein product to chemical hydrolysis at specific points along the protein backbone
- This generates a series of peptide fragments, which can be separated by various techniques (e.g., RP-HPLC, MS/MS)
- A standardized sample of the product generates a characteristic peptide map
- Peptide map fingerprinting can be used to detect point mutations in a product's gene that could lead to an altered primary structure (i.e., amino acid sequence)

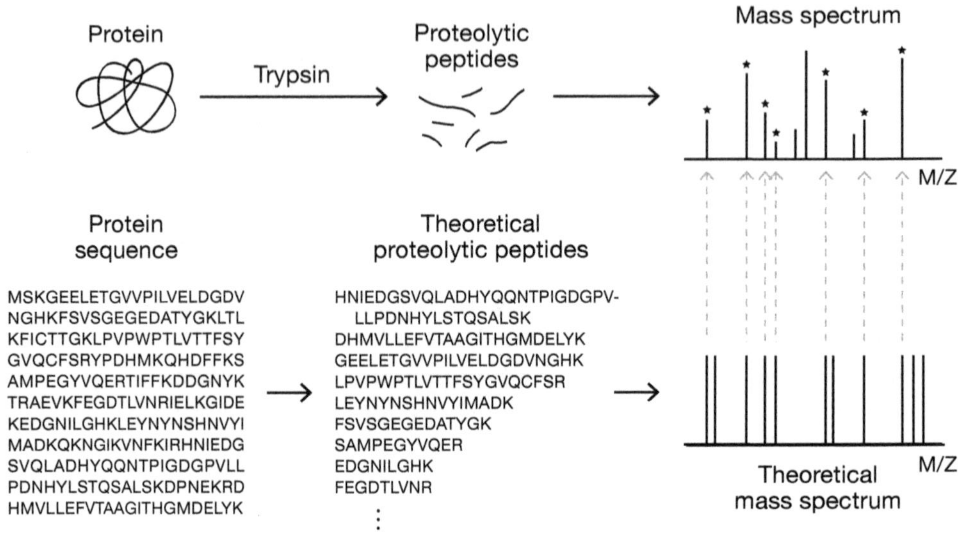

Figure 4.2. Determining the identity of a protein with a known sequence using mass spectrometry after tryptic digestion.

MQAFVGTLTK

B Series	**Y Series**
MQAFVGTLT	K
MQAFVGTL	TK
MQAFVGT	LTK
MQAFVG	TLTK
MQAFV	GTLTK
MQAF	VGTLTK
MQA	FVGTLTK
MQ	AFVGTLTK
M	QAFVGTLTK

Figure 4.3. Fragmentation of tryptic peptide in MS/MS to generate a contiguous series of b- and y-ions for *de novo* sequence determination.

corresponding tryptic peptide as compared to a reference standard. Further fragmentation using tandem mass spectrometry (MS/MS) further fragments each peptide into a series of peptide fragment ions. Since each fragmentation event breaks a peptide bond and generates two fragments, one on the N-side (y-ion) and one on the C-side (b-ion), these y- and b-ions bear *m*/*z* values that are characteristic of the fragmentation pattern of the corresponding peptide. From this pattern, the amino acid sequence can be deduced (Figure 4.3). Mass spectrometry can therefore be used for *de novo* sequencing in which the identity of the protein is unknown.

Mass spectrometry is also used in sequence verification. By searching protein databases to match fragmentation patterns, mass spectrometry-based methods can be readily used to find the identity of a protein or even to identity proteins in a mixture. It is also a sensitive method for protein identity verification. If an amino acid in a peptide is altered or chemically modified in a small fraction of protein molecules, the fragmentation pattern in the mass spectrum will be different from the original. Mass spectrometry can also be used to detect contaminating protein in a purified protein sample. It is thus an important tool in protein identification as well as in chemical structural variants detection.

Higher-order structure identity

The biological function and activity of a protein is dictated by its higher-order (secondary, tertiary, and quaternary) structure. Even if its amino acid sequence is unchanged, a protein molecule may have an altered secondary, tertiary, or quaternary structure. Conversely, the effects of an amino acid change on the functional activity of the protein is dependent on the site of the amino acid change and its effect on the higher order of protein structure. Changes in higher-order structure may result in different biological activity or even immunogenicity (Panel 4.3). In the comparative study of biosimilar protein therapeutics, it is important to demonstrate that the higher-order structure of the biosimilar is the same as the innovator drug. Traditional analytical tools for protein higher-order structure determination include X-ray diffraction and NMR. These methods are of limited use for routine characterization because of the amount of materials needed and the lengthy time required for analysis.

Circular dichroism in the far-UV region is used to reveal changes in protein secondary structure. Due to the chiral nature of proteins, they have different absorption of left-handed and right-handed circularly polarized light. The absorption spectra of a protein in circular dichroism spectroscopy is affected not only by its α-helix and β-sheet content, but also its three-dimensional structure. This type of investigation is thus a powerful tool in revealing secondary structural changes of a protein upon exposure to an agent of change.

Panel 4.3. Higher-Order Structures

- Influence functional activities
- Biosimilars are required to have same higher-order structure as innovator drug
- Typical methods of characterization
  - X-ray diffraction
  - NMR
  - Circular dichroism spectroscopy

Impurities

Therapeutic proteins produced in cell culture undergo extensive purification, virus inactivation, and viral particle clearance before becoming a drug substance for filling. However, impurities inevitably exist in the purified protein (Panel 4.4). The impurities can be from medium components or from molecules released from host cells or lysed host cells. The impurities may be compounds that are co-purified with the product protein or that are not sufficiently discriminated from the product in various product purification operations. They may also arise because of their physical association with the protein. Characterizing and quantifying these impurities is important in assuring the safety of the product. Impurities

Panel 4.4. Impurities

- Host cell protein (HCP)
  - Poses immunogenic concerns
  - May include enzymes that can cause changes in formulated drug products over time (even if the host cells are of human origin)
- Other co-purified media components, e.g., methotrexate (MTX)
- Host cell DNA
- Virus particles
- Endotoxins

that are commonly checked include host cell protein, host cell DNA, and endotoxins.

Host cell protein

The vast majority of therapeutic proteins are produced in non-human mammalian cells. The amount of native proteins secreted by host cells into the culture supernatant is small. However, with the high cell density and long duration of fed-batch cultures, these proteins may reach levels that affect final product purity. As the cell viability decreases in the late stage of culture, cell lysis inevitably leads to the release of host cell proteins. The host cell proteins co-purified with the product protein during recovery pose a risk of eliciting an immunoresponse in patients (Panel 4.4). Additionally, the contaminating host proteins may include some that have enzymatic activities or are prone to spontaneous chemical modification that, when present in the final drug product over a long period of time, may cause a change in product quality. The concern of host cell protein (HCP) contamination is magnified when the dose of the therapeutic protein is high, because the tolerance level of HCP is based on a dose, not the per-unit weight of the product. The detection and quantification of host cell protein contamination is thus required by regulatory agencies. A common analytical method relies on immunoassays, like ELISA, using the antiserum developed against the lysate of the host cell culture. The host cell lysate is used to establish a calibration curve for quantitation.

Structural Homogeneity of the Product

PROTEIN PRODUCTS MADE in cell culture may not be entirely uniform with respect to their structure (Panel 4.5). Glycans on different molecules of the same protein, or on different glycosylation sites of the same molecule, are somewhat different. In contrast, inhomogeneity may arise from the presence of a small fraction of molecules with variant structures, though the vast majority of the protein molecules have a "normal" structure, including primary, secondary, tertiary, and quaternary structures. For example, even though the fidelity of protein translation is very high, misincorporation of amino acids into the primary sequence occurs at low frequencies.[2] Under some culture conditions, such as shortage of a particular amino acid, the frequency of amino acid misincorporation may increase, leading to detectable quantities of variant proteins.[3] An enzymatic or post-translational modification reaction, such as signal peptide cleavage or C-terminal lysine cleavage, may not be completed, leading to a subpopulation of the product with variant structures.

In addition to the variants caused by the cells' protein processing variability, structural variants may also arise non-enzymatically after proteins are secreted into the culture medium, or occur in the course of downstream product recovery. For example, disulfide bonds in a protein may be reduced, further oxidized, scrambled, or cross-linked between different molecules.[4]

Many of the protein variants seen in cell culture processes are also seen in the human body. Oftentimes, structural heterogeneity or variability of proteins occur naturally. Nevertheless, the extent of their variability in the final product must be controlled within a specified range.

Panel 4.5. Protein Structural Heterogeneity

- Glycan heterogeneity occurs naturally through biosynthesis
- Protein structural variants arise from:
  - Errors in translation (amino acid misincorporation)
  - Mutations in a copy of the product gene in the host cell
  - Cellular biochemical reactions
  - Chemical or enzymatic modifications after protein secretion in the bioreactor
  - Chemical modifications in downstream processing
  - Structural changes in formulation and storage

Glycosylation Profile

As discussed above, the heterogeneity of glycan structures on proteins bears an important difference with other structural variations of proteins. Intrinsically, all product protein molecules should have the same primary, secondary, tertiary, and quaternary structure. Deviations or variants are anomalies and must be controlled to fall below certain levels. In contrast, heterogeneity in glycan structure is a natural phenomenon that occurs even for glycoproteins synthesized in the human body. When the glycans on a product protein are stripped off the protein and structurally analyzed, different structural classes of glycans are found to have different abundance levels for that protein, giving rise to a glycan distribution profile (Panel 4.6). Such distributions on a protein may vary with respect to the tissue in which it is expressed, the patient's health and age, etc.

Panel 4.6. Glycan Heterogeneity

- Glycan structure on different molecules is heterogeneous
- The glycan profile (i.e., the structural distribution pattern) is affected by:
  - The expression of the enzymes involved in glycosylation
  - The cells' metabolic state
  - Cultural conditions

For glycoproteins produced in cell culture, the glycan profile may vary with the host cell line or even the particular cell clone used, the cultural conditions, and the cells' metabolic state. Glycan structure may also be modified through degradation. For example, the presence of sialidases and other glycosidases, likely released from dead cells, causes the trimming of sialic acid or other sugar moieties from the glycan. The final glycan profile of the product is further affected by the isolation and purification process, as the selectivity of proteins with different types of glycans may differ across separation unit operations. Controlling glycan profiles within a prescribed bound throughout the manufacturing process is important for product release.

Glycan assays

The presence of glycans of different compositions and sizes on protein molecules cause small differences in their motility through gels in electrophoresis (Panel 4.7). Since sialic acid carries a negative charge, glycoproteins with differing sialic acid levels can be separated in isoelectric focusing gel, cation exchange chromatography, or capillary electrophoresis. These methods can be used in characterizing glycans, especially when the glycan structure is not expected to exert a major effect on its clinical efficacy, and the characterization primarily serves the purpose of assuring consistency of product quality.

Panel 4.7. Glycan Assays

- The sialic acid content of glycans affects the pI of the protein and can be assayed by isoelectric focusing or by anion exchange chromatography
- Binding assays using different lectins generate different binding patterns depending on the glycan structure
- N-glycans can be enzymatically stripped off the protein and analyzed in HPLC after being fluorescently labeled with 2AB

The glycan profile can also be determined by utilizing the varying binding affinities of different glycans to different lectins. The extent of a glycosylated protein binding to a lectin can be detected by biotin or digoxigenin-labeled lectins. Using a combination of lectins, often made into array format, one can gather information about a glycoprotein's carbohydrate profile. However, its resolution power is limited and it does not give complete structural information.

To structurally characterize the glycosylation profile on a protein, glycans are detached from the protein and subjected to chromatographic separation. For N-glycans, the cleavage of the glycan is achieved through a treatment with enzyme peptide-N-glycosidase F (PNGase F). The free glycans are then fluorescently labeled with 2-aminobenzamide (2AB) at the reducing terminal N-acetylglucosamine (GlcNAc). Following a cleanup step to remove protein and excess labeling reagent, the fluorescently labeled glycans are separated by liquid chromatography such as hydrophilic interaction chromatography (HIC). The relative % peak areas of the glycans are calculated and provide relative abundance of the various glycan species. However, there is no equivalent enzymatic treatment that releases O-glycans from serine or threonine. The release of intact glycan remains a major challenge in structural analysis of O-glycans. Chemical treatment can release these glycans, but it also alters the sugar structure, making it difficult to label and increase the sensitivity of detection. As a result, O-glycan analysis has lagged behind that of N-glycans. Structurally, O-glycans on a protein can be much more complex than N-glycans. The degree of complexity of O-glycans is dictated by the glycosylation gene expressed in the cell or tissue. Many of the O-glycosylated proteins are on the cell surface. Some expressed proteins in CHO cells are O-glycosylated, but, despite the large repertoire of enzymes involved in O-glycosylation, the number expressed in CHO cells is relatively small and the corresponding O-glycans are less structurally complex.

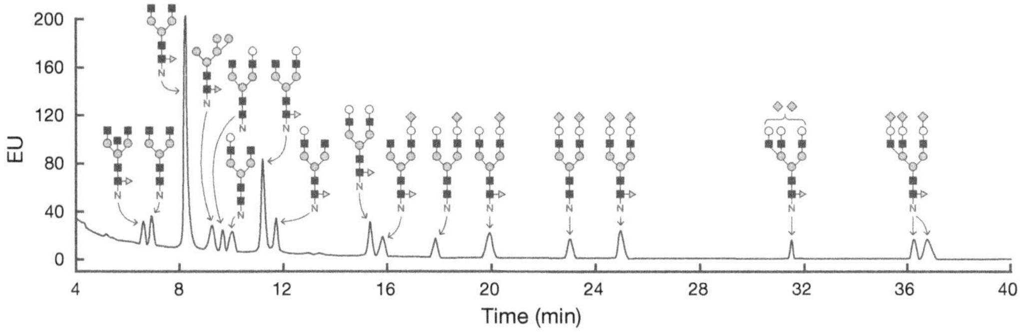

Figure 4.4. A chromatogram of eluted N-glycans.

When marked with a fluorescent label, the relative area/peak height of the glycan gives the relative abundance level of different glycans in the overall protein (Figure 4.4). However, when multiple glycosylation sites are present in a protein, it does not provide information on the distribution among the sites. It is possible to use protease to digest the protein to peptides, isolate and purify the glycosylated peptide, and then use mass spectrometry to determine the *m/z* ratio and determine the glycan distribution for each glycosylation site. However, if there are more than two sites the analysis can be rather complex. And even then, such analysis does not provide information on which glycans on one site are co-expressed with other glycans at another site of the same protein molecule. Such linked-glycan distribution on protein molecules will be readily available if a single-molecule-based analytical method is available.

Protein Structural Variants

Some protein variants bear alterations of their amino acid sequence. These alterations in the primary sequence may occur biologically through rare errors in protein or transcript processing or through chemical modification (Panel 4.8).[1] Some events, mostly occurring non-biologically, alter the higher-order protein structure, such as disulfide bond scrambling or the formation of aggregates.

Often, variants of a protein product are grouped based on routinely used protein characterization methods. Size (or aggregate) variants are those which appear as a large molecular weight peak based on size-exclusion chromatography. Charge variants are the molecules which are eluted from ion exchange chromatography and appear as separate and smaller peaks, or which form a separate band

Panel 4.8. Protein Structural Variants

Primary Sequence Variants

- Errors in protein expression
  - Amino acid misincorporation
  - Uncleaved leader sequence
  - C-terminal or N-terminal truncation
  - N-terminal pyroglutamate
- Abnormalities in biological events
  - Uncleaved C-terminal lysine
- Chemical modifications
  - Lysine succination
  - Deamidation of asparagine
  - Lysine glycation
  - Methionine oxidation
  - Aspartate isomerization
- Higher-order structural variants
  - Disulfide bond management
  - Aggregation

in isoelectric focusing.[5] As discussed above, the glycosylation pattern of a glycoprotein is normally heterogeneous but may be variable during the manufacturing process. Some consider glycan profile deviation from the norm or the prescribed range to be a structural variation. On the whole, structural variants are present as a relatively small fraction of the protein products. These variants may arise intracellularly before the protein molecule is secreted, or may be modified chemically or biochemically after secretion. The impact of these variants on the immunogenicity, efficacy, and potency of the protein has been described in literature.

Primary sequence variants

Primary sequence variants of a product protein may arise from a mutation in the DNA sequence which encodes the product gene in the producing cell. The mutation rate in the protein coding sequence during somatic cell replication is rather low. If any mutation is present in the product gene, it is likely to have originated from a single cell and occur in only one of the multiple copies of sequences that encode the product. A test to ensure the absence of such anomalies in the producing cell is typically carried out before the cell bank of a cell line is established.

Mistakes in the primary sequence are unlikely to be caused by transcription errors, as the frequency of such errors is low. Unspliced RNA species are seen in RNAseq at low frequencies, and the probability of a similar occurrence in the product transcript is therefore also very low. Such a transcript, if translated, will most likely generate a misfolded protein and be degraded. A product with a primary sequence variant is more likely to arise from errors in translation. If an amino acid is deficient in the medium, a mismatched amino acid may be used to charge the tRNA, leading the incorrectly charged amino acyl-tRNA to carry the wrong amino acid into the elongating protein molecule.

Other amino acid sequence variants may occur post-translationally. At a low frequency, the leader sequence that directs the nascent protein to translocate into the endoplasmic reticulum may not be cleaved, producing a variant. Many proteins have lysine or arginine in their C-terminus. The basic amino acids in the C-terminus are liable to cleavage by carboxypeptidase. The cleavage of lysine at the C-terminus of the heavy chain of many IgG molecules by carboxypeptidase is often not complete, leaving some molecules with uncleaved C-terminus lysine.

Chemically modified structural variants

Many structural variants arise through chemical modifications of amino acids without the involvement of cellular enzymes. These chemical modifications occur mostly after protein molecules are secreted into the extracellular environment. However, an amino group in lysine or a N-terminus amino acid may react with succinic acid or fumaric acid in the intracellular environment, although this does not appear to occur at high frequency.

Deamination, or the removal of an amide group, may occur in asparagine to form a succinimide or be further hydrolyzed to aspartate or iso-aspartate (Figure 4.5a). C-terminus glutamine can be cyclized, losing its amide group and forming pyroglutamate (Figure 4.5b). Glycation, the condensation of glucose or other reducing sugars to the ε-amino group of a lysine or the amino group of an N-terminus amino acid, occurs spontaneously in culture fluid (Figure 4.5c). Methionine residue may be oxidized by reacting with reactive oxygen species to form methionine sulfoxide (Figure 4.5d).

Higher-order structural variants

In additional to various chemical modifications that alter the amino acids in the primary structure, the disulfide bonds in a protein may be structurally modified in different ways. Disulfide bonds have a relatively low dissociation energy. Some disulfide bonds are located in the structurally flexible regions of a protein and are susceptible to chemical modifications that may lead to cleavage, trisulfide formation, etc. Disulfide bonds may also scramble their arrangement, often causing major conformational changes in a protein molecule and inducing aggregation. Protein aggregation may also arise from other mechanisms. For example, β-sheet secondary structures in a protein exert intra- or inter-molecular interactions that contribute to the tertiary and quaternary structure. Misalignment of intermolecular interactions may occur in the cell culture environment, especially in the sometimes extreme cultural conditions of late stage fed-batch culture. Some downstream processing conditions, such as low pH, hasten aggregate formation. Chemical modifications of disulfide bonds can be detected by peptide fingerprinting using mass spectrometry. Aggregation or protein size variants are discernible by size exclusion chromatography. This kind of chromatography separates proteins based on differences in their hydrodynamic volumes. Molecules with larger hydrodynamic protein volumes elute earlier than molecules with smaller volumes. The protein sample is loaded onto a size exclusion column, separated isocratically and the eluent is monitored by UV absorbance. Calculating the percentage of each separated component as compared to the total integrated area provides the relative abundance.

Charge Variants

The presence and abundance levels of various structural variants can be characterized by peptide fingerprinting in mass spectrometry or other instrumental analysis. However, for routine assessment, they are often assessed by more readily accessible assay methods, such as size exclusion chromatography and ion exchange chromatography.

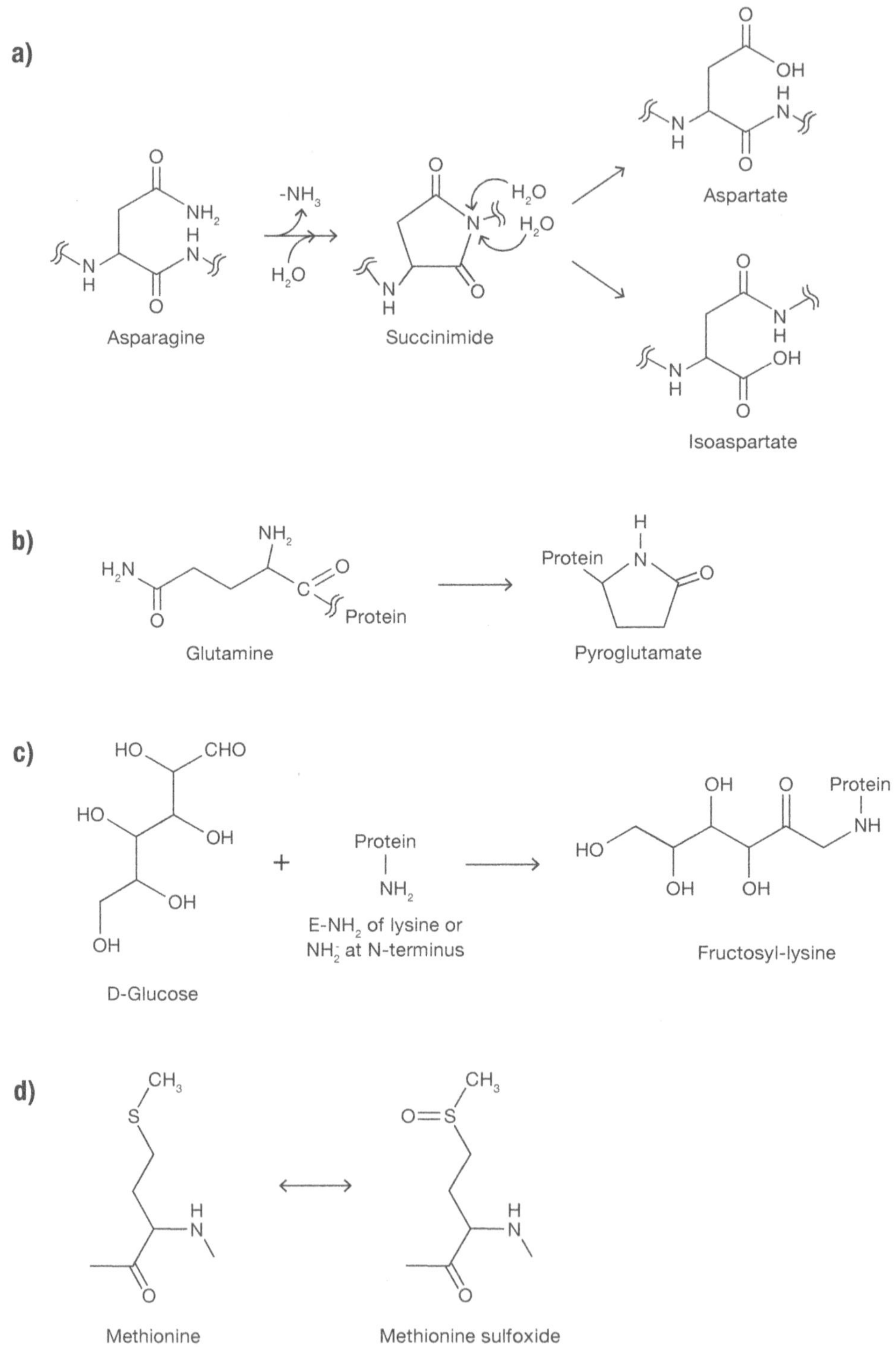

Figure 4.5. Chemical modifications that generate protein structural variants. (a) Deamidation of asparagine. (b) Pyroglutamate formation from glutamine. (c) Glycation of lysine. (d) Methionine oxidation.

Proteins carry many charged side chains in their amino acids. At a biological pH, most proteins are weakly charged. Changes in the net charge of a protein molecule may cause changes in its stability and activity (Panel 4.9). A charged variant may have an altered binding affinity to its target receptor or antigen.[6] It is thus important that the basic or acidic charge variants of a product protein be characterized and controlled within a specified range. Charge variants can be divided into acidic and basic charge variants. The former is eluted ahead of the main product peak in cation exchange chromatography (CEX), or afterwards in anion exchange chromatography (AEX), meaning that they have higher levels of a negatively charged acidic group (Figure 4.6). Basic charge variants are those which are eluted after the main product group in CEX or before in AEX.

Operationally, charge variants are defined by their elution profile in AEX or CEX. Typically, at a set pH, charged variants are separated on a cation or anion-exchange column and eluted using a salt gradient. The eluent is monitored by UV absorbance. The charged variant distribution is evaluated by determining the peak area of each variant as a percent of the total peak area. It is important to note that the elution profile of a protein in AEX and CEX is determined by a molecule's binding affinity in chromatography. Binding affinity is affected not only by the charge of a molecule but also by protein conformation and many other factors. This is in contrast to isoelectric focusing (IEF). IEF probes a protein's isoelectric point (pI), or the pH at which the protein has no net charge. A variant with a different isoelectric point forms a distinct band that differs from the standard product protein in the IEF gel. Some changes that do not affect the pI of a protein, such as the oxidation of methionine and the rearrangement of disulfide bonds, may change the binding affinity of the protein and cause it to appear as a separate minor peak in an ion exchange chromatogram. Different structural variants that appear as acidic and basic charge variants are listed in Panel 4.10.

Panel 4.9. Protein Charge Variants

- Proteins consist of many different amino acids comprising weak acidic and basic groups
- Identification and monitoring of charge variants throughout manufacturing is critical for the production of safe and effective drugs
- Charge variants impact the stability and activity of the protein, which may also cause immunogenic responses

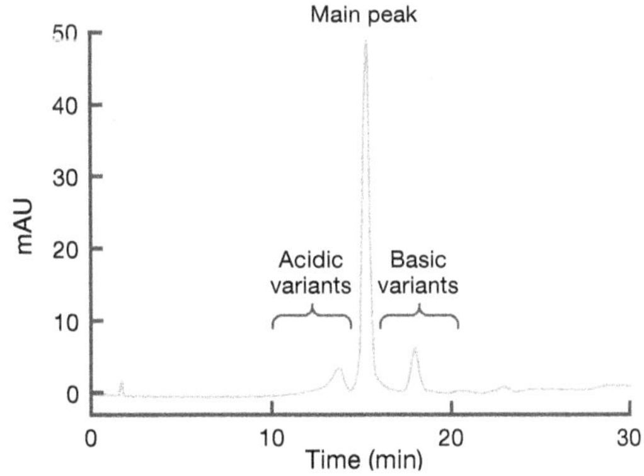

Figure 4.6. Elution of protein charge variants in cation exchange chromatography.

Panel 4.10. Effect of Structural Variation on Charge Variants

Structural Variants Contribute to:

- Increase of acidic species
  - Deamidation of asparagine
  - Leader sequence removal
  - Glycation of amino group in the N-terminus or ε-NH_2 of lysine
  - Glycosylation, sialic acid content
  - Amino acid misincorporation/mutation
- Increase of basic species
  - C-terminal lysine/arginine uncleavage
  - Succinic acid modification of NH_2 group
  - Amino acid misincorporation/mutation
  - In some cases, high molecular weight species
  - Some glycan attributes, such as high mannose
- No effect on charge variants
  - Methionine oxidation
  - N-terminal pyroglutamine
  - Aggregation
  - Disulfide bond rearrangement

Biological Activities

Functional Assay

STRUCTURAL CHARACTERIZATION as discussed above evaluates the quantity of various anomalies in the product. Some of the changes pose safety concerns or may affect the efficacy of the product. For example, disulfide bond scrambling is likely to elicit an immunogenic response. Some other changes, such as glycation and uncleaved C-terminus lysine, are seen in different proteins and at various levels in the blood circulation, and may not pose a major safety concern. In product characterizations, different structural variations are assayed separately and reported as a percentage of the whole product. When different alterations occur concurrently in a molecule, the effect on safety or efficacy may be different from when those changes occur in different molecules. However, the current methods of protein characterization do not reveal the combination of different structural variants and their distribution profile. There are large gaps in completely relating structural changes to functional changes. Hence, full characterization of a protein product must include the measurement of functional activities in a physiologically relevant manner.

A protein product may function through its catalytic activities, such as glucocerebrosidase's main function is to cleave the specific glycosidic bond. A product protein may also function through binding to its ligand in the body fluid, binding to receptors on the cell surface that trigger a downstream signaling response, or by blocking the competitive binding by its biological counterpart. A functional assay may involve quantifying the enzymatic activity of the product, measuring the binding strength of the product to the ligand, or using effector cells for a bioassay. In all cases, the appropriate assay to be used is chosen based on the understanding of the mechanism of action (MOA) of the molecule.

Binding assay

Since mAbs are the largest class of therapeutic proteins, binding assays between an antigen and an antibody are a commonly practiced functional assay. The binding kinetics are measured by surface plasmon resonance (SPR) (Figure 4.7). First, the ligand proteins are immobilized

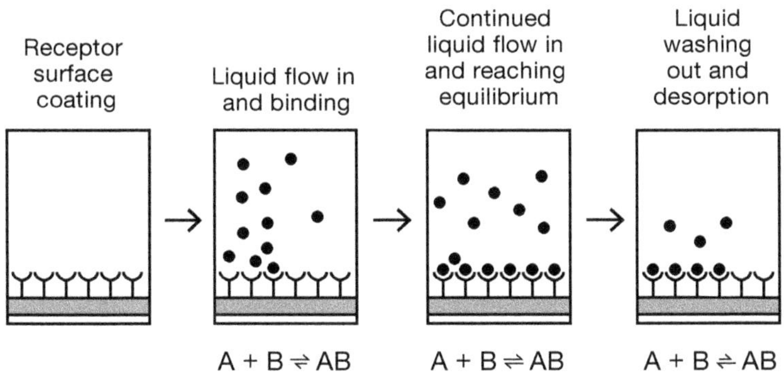

Figure 4.7. Adsorption and desorption kinetics of ligands in surface plasma resonance (SPR).

onto the sensor chip. The antibody protein is then flowed into the ligand-coated binding reaction chamber to allow for the measurement of antibody association to the ligand. This is followed by a period of dissociation in which a solution without the antibody is flowed through to measure the kinetics of dissociation. The process is repeated after regeneration of the surface immobilization of the ligand. The surface resonance increases as antibodies and ligands form complexes in the association phase and decreases in the dissociation phase. Using the binding kinetic equations, the rate constant of binding and dissociation can be determined.

Bioassays

Bioassays that quantify the biological activity of the protein using cultured cells are the most relevant evaluation of the product short of using animals. It enables the effect of changes in the protein on its biological activities to be quantitatively evaluated. But an effective bioassay is also difficult and time-consuming to develop. It relies on a good understanding of the mechanism of actions of the drug. An example is shown in Figure 4.8a and b that depicts the assay of antibody-dependent cellular cytotoxicity (ADCC) and complement-mediated cytotoxicity (CMC). In ADCC, an assay of two types of cells, effector cell and target cell, are used. The target cell expresses the antigen that is recognized by the antibody. The binding of the antibody to the target cell recruits the killer cell, which exerts cytotoxic action on the target cells. In CMC, the antibody binding to the antigen on the surface of the target cell recruits complements. The binding of the complement to the target cell then elicits complement-mediated cell lysis. With an effective bioassay, one can evaluate the activity at different doses of the antibody (Figure 4.8c). Consistent kinetic behavior in a bioassay assures the quality of the product. If a product batch is shown to have an altered behavior as shown in the response curve of a bioassay, further investigation is called for.

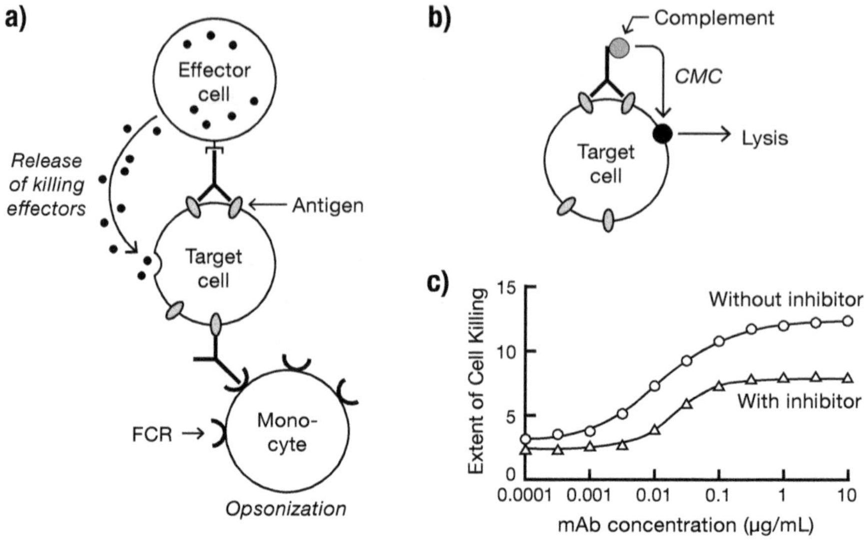

Figure 4.8. Bioassay of antibody-mediated target cell killing. (a) Antibody-dependent cellular cytotoxicity assay. (b) Complement-mediated cytotoxicity assay. (c) Killing results at different doses with and without an inhibitor that binds to target cells.

A common requirement in binding assays and bioassays is the availability of reference products and assay materials. For binding assays, the reference materials are the product and ligand molecules. For bioassays, the assay cells must be standardized, banked, and made readily available. Biological assays are often subject to a higher degree of variability. These additional challenges make bioassays less frequently used than chemical and structural characterizations.

Quality by Design (QbD) of Cell Culture Products

An important objective of process and product development is generating an understanding of process and product quality, and defining a strategy to monitor and control product quality during manufacturing. The variability of protein structures underlines the importance of identifying the properties whose variability affects the product's safety and efficacy, and the need to control those properties within a defined range. In the early 2000s, the FDA published a series of documents on cGMP in the twenty-first century and began a drive to put QbD into drug development and manufacturing (Figure 4.9).[7] Many of these documents, including ICH Q8 (R2) (Pharmaceutical Development), ICH Q9 (Quality Risk Management), ICH Q10 (Pharmaceutical Quality System), and

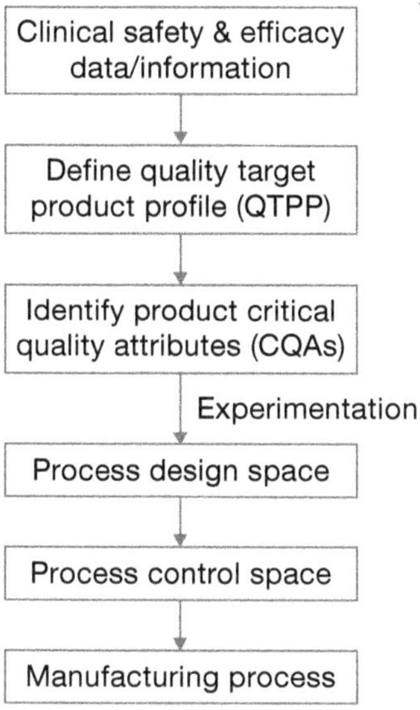

Figure 4.9. Incorporation of QbD in process development and manufacturing.

Panel 4.11. Elements of Pharmaceutical Development

1. **Quality Target Product Profile (QTPP)**
   A prospective summary of the ideal quality characteristics of a drug product, taking into account the safety and efficacy of the product
2. **Critical Quality Attribute (CQA)**
   A physical, chemical, biological, or microbiological property or characteristic that should be within an appropriate limit, range, or distribution to ensure the desired product quality (safety, identity, purity, biological activity, and stability). Ideally, this considers the mechanism of action of the protein.
3. **Design Space**
   The multidimensional combination and interaction of input variables and process parameters that have been demonstrated to provide assurance of quality
4. **Control Strategy**
   A planned set of controls, derived from current product and process understandings, that ensures process performance and product quality

ICH Q11 (Development and Manufacture of Drug Substance), provide high-level direction with respect to the scope and definition of QbD as it applies to the biologics industry. These documents also introduced a number of items that are listed in Panel 4.11.

QbD is an approach to process development wherein the manufacturing process and controls are proactively designed to deliver the intended product. QbD elements include the following: a Quality Target Product Profile (QTPP) that identifies the Critical Quality Attributes (CQAs) of the drug product; process design and understanding, including identification of Critical Process Parameters (CPPs), linking CPPs to CQAs; a control strategy that includes specifications for the drug substance and drug product as well as controls for each step of the manufacturing process; and process capability and continual improvement. QbD tools include prior knowledge, risk assessment, mechanistic models, design of experiments (DoE), data analysis, and process analytical technology.

The implementation of a QbD-based pharmaceutical development starts with defining the characteristics of the drug product that are required to achieve safety and clinical efficacy. This set of product characteristics is referred to as the Quality Target Product Profile (QTPP). Next, the physical, chemical, biological, or microbiological features of the drug product and drug substance that must be controlled within a bound in order to ensure the QTPP is met are identified. These features, called Critical Quality Attributes (CQAs), are variables that can be measured and quantified. How these CQAs are affected by process conditions in different operations of manufacturing is established from knowledge of the process or is determined experimentally. The information is used to define a design space of process variables within which CQAs of the product can be expected to fall into the permissible range. A control space is then established within which the process will be operated, thus ensuring that the CQAs of the resulting product meet the specifications. With a QbD-based approach, risk analysis is employed to assess various process aberrations that can cause CQAs to stray out of bounds, and mitigating measures are taken to reduce the risk. The implementation of QbD starts from the beginning of process development and continues through characterization, establishment, and commercial manufacturing. It is continually practiced throughout the product's life cycle so as to continually improve the process and enhance product quality.

The first task in implementing QbD is the development of knowledge of the relationship between process parameters (e.g., raw materials, cell inoculum), culture conditions, and product quality. Such relationships are inevitably complex, with multiple interactive input variables and process parameters that affect both productivity and product quality. Critical to understanding the relationship between process parameters and product quality are analytical tools which quantify various quality variables. Advances in analytical techniques have enabled sophisticated evaluation of physicochemical properties, biological activity, immunochemical properties, purity, and impurities.

Regulatory documentations (e.g., ICH Q6B) stipulate test procedures and acceptance criteria for biotechnological/biological products. Shown in Table 4.2 are some quality attributes with the category of quality as defined by regulatory agencies, the analytical methods often used to characterize the clinical product, and the material generated during process development studies.

QTPP and CQAs

QbD-based product and process development first considers the desired or target quality characteristics of the final drug product (QTPP) according to how it will be administered to a patient and what properties it must have to be safe and effective. An example of a QTPP for a recombinant antibody drug product is shown Table 4.3. The QTPP will be used to determine the CQAs of the product. For each characteristic, a qualitative target description or a quantitative measure is given. At

Table 4.2. Protein Structural Variants and Categories of Quality Attributes Affected

Product quality attribute	*Method(s) of characterization*	*Attribute category (ICH Q6B)*
Aggregation	SEC-HPLC	Identity, purity, stability
	Gel electrophoresis	Identity, purity, stability
	Analytical ultracentrifugation	Identity, purity
	SEC-HPLC with MALS	Identity, purity
C-terminal lysine	Ion exchange chromatography	Identity, purity, stability
	Isoelectric focusing	Identity, purity, stability
	Peptide mapping with MS	Identity, purity, stability
Deamidated isoforms	Ion exchange chromatography	Identity, purity, stability
	Isoelectric focusing	Identity, purity, stability
	Peptide mapping with MS	Identity, purity, stability
Glycosylation	Monosaccharide composition analysis	Identity
	Oligosaccharide profile	Identity
	Sialic acid content	Identity
	Galactose content	Identity
	Fucose content	Identity
Oxidation	Peptide mapping with MS	Identity, purity, impurity

Table 4.3. QTPP Listed in A-Mab Example from ICH Q9

Product attribute	*Target*
Dosage form	Liquid, single use
Protein content per vial	500 mg
Dose	10 mg/kg
Concentration	25 mg/mL
Mode of administration	IV, diluted with isotonic saline or dextrose
Viscosity	Acceptable for manufacturing, storage and delivery without the use of special devices (for example, less than 10 cP at room temperature)
Container	20R type 1 borosilicate glass vials, fluoro-resin laminated stopper
Shelf life	≥ 2 years at 2–8 °C
Compatibility with manufacturing processes	Minimum 14 days at 25 °C and subsequent 2 years at 2–8 °C, soluble at higher concentrations during UF/DF
Biocompatibility	Acceptable toleration on infusion
Degradants and impurities	Below safety threshold, or qualified
Pharmacopoeial compliance	Meets pharmacopoeial requirements for parenteral dosage forms, colorless to slightly yellow, practically free of visible particles and meets USP criteria for sub-visible particles

first glance, the list does not seem to present much information on the characteristics of the drug substance that are related to safety and efficacy. A closer examination will reveal the quality requirements of the product. The drug product must be stable as a liquid under refrigerated conditions at a concentration of 25 mg/mL for at least two years. The drug substance must also be stable for 14 days at room temperature and at higher concentrations during processing. A long stability is required since the degradants and impurities are below the safety threshold at the end of the shelf life. These product specifications point to the importance of protein stability. Hence the propensity of the drug product to aggregate, and the level of charge variants will likely impact the QTPP. The dose and concentration to be administered intravenously is high. This suggests that the level of host cell proteins and other impurities will likely impact the safety of the product.

Table 4.4 is a partial list of the quality attributes of a recombinant antibody product. Some attributes, such as pH and excipient concentration, are related to the composition and strength of the drug product. A number of attributes related to impurities or structural variations must be controlled at low levels to ensure the degradants and impurities aspects of the QTPP are met.

The list of quality attributes derived from the QTPP can be rather long, as seen in the example shown in Table 4.4. However, not all the attributes identified are critical and must be controlled within a range to minimize the risk of negatively affecting clinical outcomes (Panels 4.12 and 4.13). The task of identifying the CQAs among all the attributes typically takes place during product and process development, and before the manufacturing process is established and process and clinical data is available. The selection relies heavily on the knowledge of past products and processes, as well as the understanding of the mechanism

Table 4.4. Examples of Product Quality Attributes

Product variants	*Process-related impurities*	*Adventitious agents*	*Composition and strength*
• Aggregation • Conformation • C-terminal lysine • Deamidated isoforms • Disulfide bonds • Fragmentation • Glycation • Glycosylation • Oxidation • Thioether link	• Host cell DNA • Host cell proteins • Leached protein A • Selective agent (e.g., MTX) • Cell culture medium components (e.g., insulin) • Purification buffer components • Leachables	• Viral purity • Microbial purity • Endotoxins	• Product concentration • Potency • Osmolality • pH • Particulates • Clarity • Color • Volume • Excipient concentration

Panel 4.12. Identifying CQAs from QTPP-Derived Attributes

- CQAs are attributes that may affect the clinical outcome and must be controlled within bounds
- CQAs are identified before commercial manufacturing starts
- Identification relies on knowledge of past processes and the MOA of the product
    - Some molecular features of the product may be important for a particular MOA
- May use scoring or semi-quantitative mechanisms to rank CQAs

Panel 4.13. Establishing Quantitative Bounds for CQAs

- Deciding the value of CQAs is initially quite difficult, as process and clinical data is often insufficient
- May use functional assays (binding assay, bioassay)
- May use external conditions or purified variants to enrich variants under test conditions
- Employ various lots during testing
- Important to maintain a comprehensive library of samples generated, especially at larger scales as the process evolves
- Significant changes in attribute levels may occur *in vivo*

of action (MOA) of the product. A literature review on the MOA of the new candidate drug is performed and a knowledge database on experience from similar products is surveyed. For example, if the mechanism of a monoclonal antibody product is to block receptor sites, attributes that enhance antibody-dependent cellular cytotoxicity (ADCC) and complement-dependent cytotoxicity (CDC) may not be important for efficacy. As a result, fucosylation at the Fc region may not be considered a CQA. On the other hand, if the MOA is ADCC, the level of fucosylation of the glycan in the Fc region might be selected as a CQA.

There has been an increasing effort to employ a scoring system to select CQAs. Each member of a multifunctional team scores the attributes. Each attribute is given an impact score based on its perceived importance to the fulfillment of the QTPP and an "uncertainty" score that accounts for the extent of the variability of the attribute. The selection of CQAs is then based on the overall score. As the product advances and experiences from clinical and *in vitro* studies accumulate, knowledge of the degradation pathways and structure activity relationships of the product are used to refine CQAs. A prudent approach is to adopt a risk assessment procedure, as will be discussed later in this chapter. In general the identification of CQAs and definition of their bounds is an iterative process that is refined as the product matures through commercialization life cycle.

CQA range and validation

After the selection of CQAs, the appropriate value or value range must be specified for each CQA. Deciding the value of CQAs is a difficult task, especially in the early stage of product development when sufficient clinical and manufacturing data is unavailable. In the early stages of process development and QbD establishment, data from *in vitro* binding assays and bioassays are used to evaluate the effects of different ranges of

CQAs on the performance of the product. Since the drug substance production process typically produces various variants within a range, in order to test the conditions out of typical range, extreme stress conditions are used to accelerate the generation of some structural variants, such as degradation products and aggregates. Some variants, including charged variants, can be enriched or purified for testing *in vitro* or in small animals. The data from *in vitro* and *in vivo* studies are augmented with historical data from the established drug product in the initial definition of CQA values. When refining CQAs for manufacturing, it is important to incorporate a large number of product lots produced under different settings so as to better identify CQAs and set release specifications. In addition to development and clinical manufacturing lots, lots with "extreme" variant levels generated under extreme conditions may also be included.

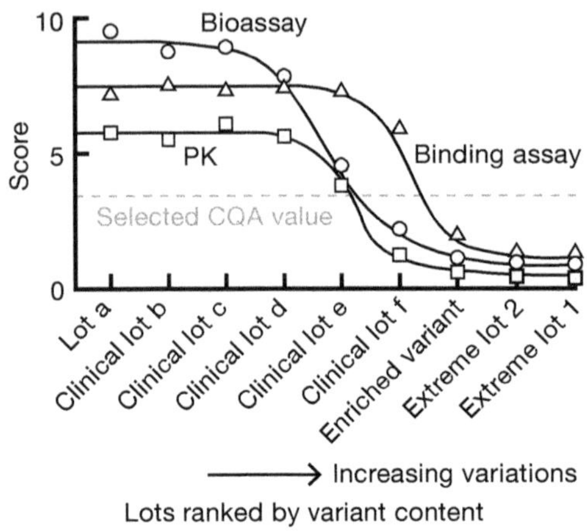

Figure 4.10. A hypothetical instance of using quality scores of different assays to select CQA specifications.

A hypothetical scenario using binding assays, bioassays, and PK to understand the impact of variants on the efficacy of the product is shown in Figure 4.10. In the early stage of product development, various assays can be used to set the ranges of CQAs. As clinical lots for clinical trials become available and clinical studies proceed, the data can be used to support the ranges of CQAs selected for clinical manufacturing lots. It is important to generate a comprehensive library of samples generated, especially at larger scales as the process evolves, so that the final validated bioassay can be used to characterize as many development and clinical lots as possible before final CQAs and specifications are set.

A consideration in evaluating acceptable ranges for product attributes is that significant modification of the product may occur when it is administered to the patient. For example, deamidation of proteins is known to occur in the blood stream. Predicting such attribute changes *in vivo* using a model or by measurement will generate data that supports the selection of a wider range of CQA values.

CQAs in Different Stages of Manufacturing

A manufacturing process of therapeutic proteins can be largely divided into four stages: drug substance manufacturing (consisting of production, downstream recovery, and purification); drug substance handling; drug product manufacturing; and drug product handling (Figure 4.11). Each stage further consists of different unit operations. The bulk of production occurs in the drug substance manufacturing stage. However, it

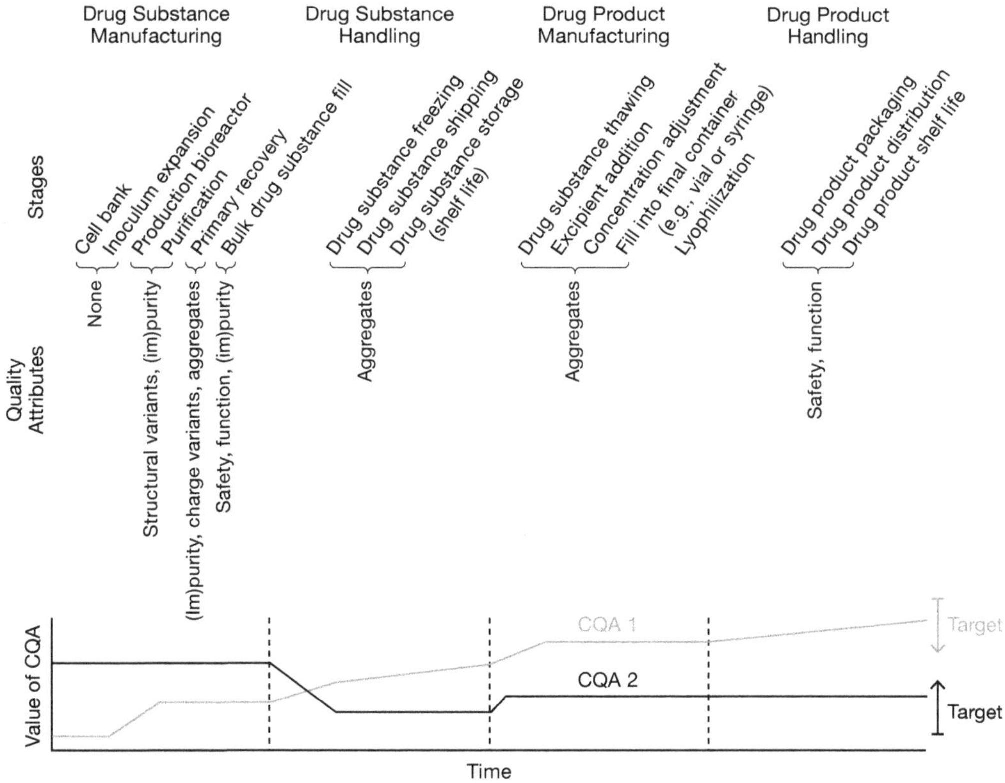

Figure 4.11. Key quality attributes vary across stages of manufacturing.

is the final drug product that directly impacts safety and efficacy in patients, and the final drug product that the QTPP is designed for. The value of the CQAs may not be constant throughout the manufacturing process. As the product stream advances through different stages of manufacturing, the purity and possibly the homogeneity of the product will increase, and the CQA value should approach the final value of the drug product. For example, if protein aggregation is a CQA and is set to below a threshold value, as the product goes through harvest and purification the fraction of aggregates should decrease. A CQA may become significant only after a certain stage of the manufacturing process. For example, a particular type of product variant may arise that is due to exposure to low pH conditions; it thus appears in downstream processing but not in the production bioreactor. A comprehensive understanding of CQAs in different stages of manufacturing facilitates the setting of design targets or specifications around CQAs which the product needs to meet.

Quality Considerations before Process Development

Increasingly, quality assessment is being incorporated into decision-making on various matters related to the production of the drug, sometimes even before process development starts, since some decisions will affect product quality robustness in future manufacturing (Panel 4.14). For example, sometimes multiple candidate molecules are considered for the same drug, such as in the case of antibodies with somewhat different primary sequences in the complementarity determining region (CDR) that bind to the same antigen. Product quality, stability, and platform fit are considered when designing and selecting lead monoclonal antibodies and other protein therapeutic candidates. For example, the propensity of these molecules to form protein aggregates may differ. Selecting one with a low propensity for aggregation will reduce yield loss in process and possibly impact long-term product stability. Another example is the viscosity of the protein solution, which may impact the delivery for the protein therapeutic.

Panel 4.14. CQAs in Early Development and Manufacturing

- CQAs considered even before process development
  - Selection of candidate molecules (e.g., propensity to aggregate)
  - Selection of cell line (e.g., glycosylation stability)
  - Can also adopt a scoring system in decision-making
- CQA values change in different stages of manufacturing

Producer cell lines used in biomanufacturing have all been clonally selected. Different cell clones are often somewhat variable not only in terms of productivity but also in many properties that affect product quality. Products produced by different clones can have different features that are considered CQAs, such as fucosylation or propensity to aggregation. These quality factors are considered in the selection of clones for manufacturing. Shown in Table 4.5 is an example of a scoring table that assesses factors that affect the stability of the cell line (integration site, copy number, epigenetic stability), productivity (growth rate, titer, specific productivity), and attributes of the product quality (sequence variants, aggregation, structural variants, glycosylation, modifications to the CDR). Different weights are assigned to different characteristics and the overall score is used to select the best clone to ensure "manufacturability" and reduce the risk of CQAs falling out of range.

Quality Risk Management

Quality risk management (QRM) is a systematic process for the assessment, control, communication, and review of risks to the quality of the drug product across the product's life cycle. Although the quantitative risk may be calculated in different ways, the evaluation fundamentally includes an identification and assessment of hazards throughout the totality of product characterization, process development, and clinical experience. To initiate a QRM process when developing a product, a cross-functional team which assembles background information and data on the product and process and identifies available tools and timelines is typically formed (Figure 4.12). Risk assessment is then performed to identify the hazards, or process failure modes, that may impact the

Table 4.5. A Hypothetical Example of Cell Line Screening by Quality Score

Clone	***Stability Attributes***				***Productivity Attributes***					
	Transgene		*Integration site*		*Specific productivity*		*Titer*		*Growth rate*	
	Copy # / integrity	*Score*	*Quality*	*Score*	*pg/cell/ day*	*Score*	*g/L*	*Score*	h^{-1}	*Score*
1	1/+	10	+	8	25	7	4.1	6	0.03	8
2	3/++-	7	+--	3	17	5	4.5	7	0.028	8
3	5/+----	5	++---	4	15	4	3.9	6	0.032	8
4	3/+++	8	+--	3	21	6	2.9	5	0.026	5
5	1/-	4	+	2	12	2	2.5	5	0.025	5

Clone (cont.)	***Quality Attributes***						***Total Score***
	Aggregation		*Structural variants*		*Afucosylated glycans*		
		Score		*Score*		*Score*	
1	++	4	++	3	2.1	5	7
2	+	5	-	7	1.5	6	9
3	+++	2	-	7	1.7	6	8
4	-	6	+	4	3.5	2	7
5	-	6	+	4	3.2	2	6

CQAs. The probability that a given failure mode will occur is then evaluated. Finally, risks are evaluated against criteria and categorized, often using a semi-quantitative scoring system or by categorizing them as low, medium, and high risk. Many tools and types of risk assessments can be used in QRM.

Once risks are identified and classified, the team determines if the identified and analyzed risks are acceptable. If not, the likelihood and impact of a hazard are investigated and better understood, and then actions are identified to reduce or eliminate the risk. These actions to reduce the severity and probability of harm often include increasing the detectability of hazards. In some cases, residual low-level risks are deemed acceptable.

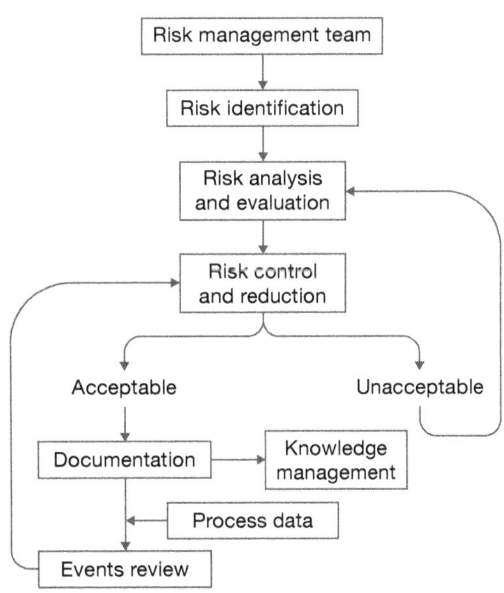

Figure 4.12. A general flow chart of quality risk management.

The output of the QRM process is typically a document that summarizes the risk assessment analysis as well as the risk control actions taken. However, this document is only a snapshot of the knowledge available at the time of its completion. As risks deemed unacceptable are mitigated, another round of risk identification, analysis, and evaluation is completed to ensure evaluation of any new risks introduced or increased significance of existing risks. In addition, risks should be periodically reviewed to ensure new knowledge and experience are considered.

Risk Mitigation

Upon identification of risks, means of mitigation are devised (Table 4.6). One may take a rigorous risk management approach using tools such as Ishikawa (fishbone) diagrams, cause-and-effect analyses, or failure modes and effects analysis (FMEA).

When investigating the relationship between the process variable in question and the outcome of the quality attributes, it is always wise to start with a review of the historical data. For example, the duration of the process may be identified as a factor that can potentially affect the CQA host cell protein content. An investigation of the historical data can be initiated to determine the incidence frequency that the HCP exceeds the set value for the process time periods in question. To facilitate risk and benefit analyses, data on productivity and other process variables are also analyzed. A decision regarding a risk management plan for the identified process variables can then be made. After the investigation, a smaller number of process parameters with a high impact on the CQAs is

Table 4.6. Example of Risk Assessment of Process Parameters

Process Parameter in Production Bioreactor	***Failure Mode***	***Risk Level****							***Risk Mitigation***
		Quality attributes affected				*Product attributes affected*			
		Aggregation	*α-Fucosylation*	*HCP*	*DNA*	*Product yield*	*Viability at harvest*	*Turbidity at harvest*	
Inoculum density	Low density	L	M	M	M	H	M	M	DOE
Inoculum viability	Low viability	M	M	H	H	H	M	M	DOE
Inoculum age	Old inoculum culture	M	M	M	L	H	L	L	Mining knowledge database
Medium lot	Mixed medium lots	L	M	L	L	L	L	L	Not required
⋮	⋮	⋮				⋮			⋮

*H: high, M: moderate, L: low

selected for more detailed experimental characterization, and to generate a model that quantifies their effects, singularly and interactively, on the CQAs. This is performed on the unit operations that contribute to the variability of the CQAs.

Risk assessments are initiated during the product and process development stages, and are used in defining the operating (control) space of the process parameters. The assessment continues and is updated throughout the process characterization and performance verification stages. Even after routine manufacturing, it continues throughout the product life cycle management stages.

Knowledge Management

An important component of QRM is the creation of knowledge from data and the subsequent utilization and dissemination of that knowledge (Figure 4.13). Starting from the beginning, prior knowledge of the product and process from public literature, internal institutional documents,

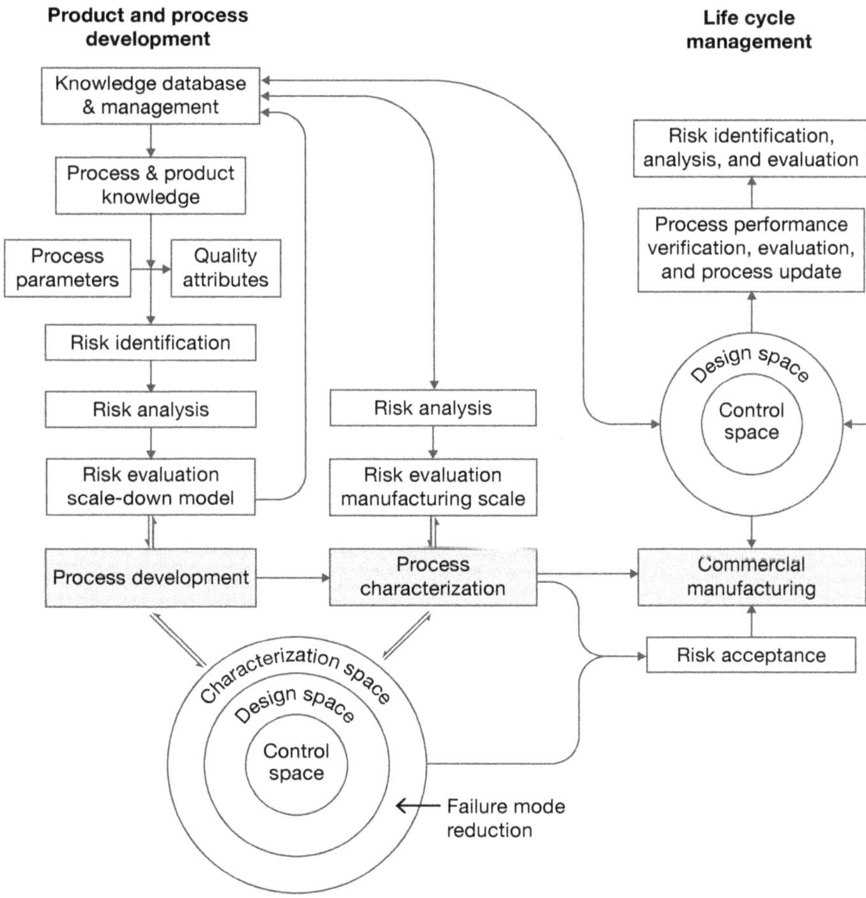

Figure 4.13. Incorporating risk management into process development, manufacturing, and life cycle management.

including prior regulatory submissions, and knowledge databases are captured for QRM. As product development progresses, information on the safety and efficacy of the product is generated from non-clinical and clinical studies, and CQA risk assessments are updated. Process development studies and clinical manufacturing also generate additional knowledge prior to commercialization. Once the product is launched, experience in commercial manufacturing and technology transfers to alternate manufacturing facilities and creates much insight into the relationship between process parameters and CQAs. Compiling, analyzing, storing, and disseminating the knowledge gained from all of the above, a process known as knowledge management, is critical to continued product quality enhancement (Panel 4.15). Many companies utilize spreadsheets, databases, and specialty software to ensure that that knowledge is managed in an efficient way. Some create process libraries that include standardized risk assessments and compilations of data from past process development experiences. Information technology (IT) tools such as Biovia Discoverant® and Tibco Spotfire® can be used as knowledge management tools to aggregate large datasets.

Design Space and Control Strategy

The ultimate objective of QbD is to establish a manufacturing process that generates products which meet the QTPP and CQAs. In the jargon of QbD for pharmaceutical development, this is achieved through defining the process design space. Mapping and defining the design space is an important task in process development (Panel 4.16). However, due to the high attrition rate of protein drug candidates, design space mapping is often not initiated until a candidate has shown promise in clinical studies.

A manufacturing process, and each unit operation in the process, can be described in general terms as having two types of inputs, controlled and uncontrolled, and each of these can have multiple inputs. The inputs include materials inputs such as inoculum and raw materials, as well as process parameter values such as a pH set point. The process then converts the inputs into outputs. Similarly, outputs include materials such as the product, cells, and contaminants, as well as non-material parameters such as product quality attributes (Figure 4.14a). The range of process inputs, including operating parameters, that gives the highest yield and product quality and results in the "optimal" process outcome is called the design space.

When defining a process design space, information from past process development and knowledge of processes can be called upon; the initial operating range of each process parameter obtained from such knowledge is called the knowledge space (Figure 4.14b). The knowledge

Panel 4.15. Knowledge Management in QbD

- Important part of QRM
- Types of knowledge:
  - Literature and internal documents on products and processes
    - Safety, efficacy from clinical/non-clinical studies
  - Manufacturing experiences with CQAs
- Entails data compilation and results analysis

Panel 4.16. Defining the Design and Control Spaces

- The design space is designed for each unit operation
- Defining spaces is time- and effort-intensive, and is performed only on promising products
- Determines the relationship between controllable inputs (material and parametric) and quality output
- Works from past experience to explore process quality
- Augmented by experimentation to refine the process-quality relationship
- Identifies the "optimal range" as the design space
- Spaces intersect among different CQAs
- Define a control within which routine operations will be performed

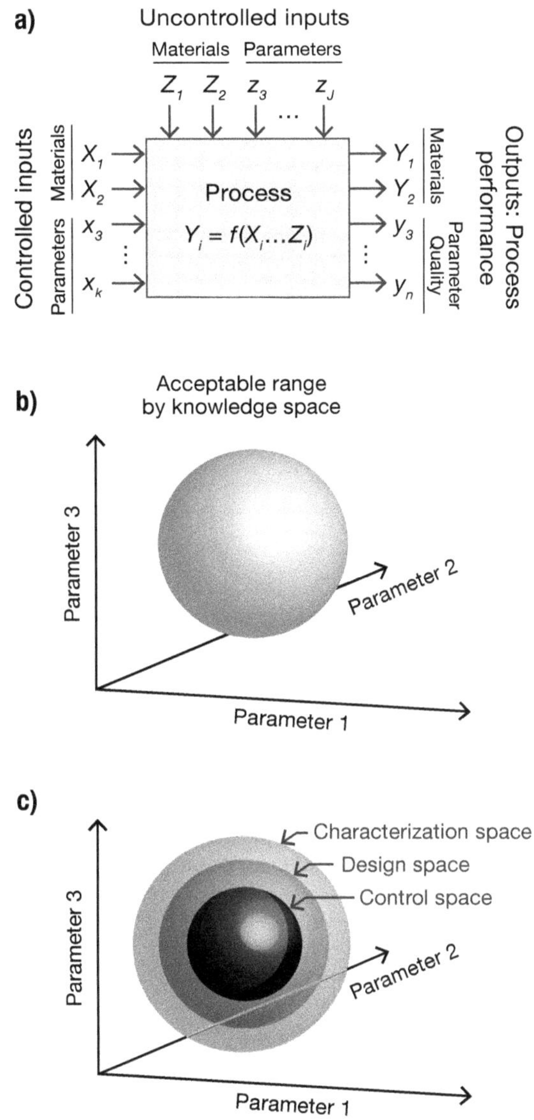

Figure 4.14. (a) Inputs and outputs of a unit operation of process. Some inputs are critical process parameters. (b, c) The knowledge space of critical process parameters facilitates the definition of the design and control spaces. The characterization space is typically larger than the design space.

space is not sufficient to define the design space. In order to map the design space, one has to understand the space beyond its boundary. This space, called the characterization space, spans a wide range of process parameters that reach regions not necessarily optimal for productivity or product quality. Furthermore, the characterization process often includes a thorough assessment of parameter interactions. Upon completion of characterization, the boundary of the design space is defined such that any combination of inputs within the boundary will produce a product of acceptable quality. The control space, within which manufacturing operations are practiced, is typically within and smaller than the design space, thus assuring optimal process performance. The behavior of CQAs in the design space and control space is carefully studied for individual unit operations of the manufacturing process.

Critical Process Parameters (CPPs)

Design of experiments (DOE) studies are used to identify the criticality of process parameters. A critical process parameter is defined in ICH Q8 as a process parameter whose variability has an impact on a CQA and should therefore be monitored or controlled to ensure that the product has the desired quality (Panel 4.17). A process parameter that does not impact a CQA but may affect process consistency is a non-CPP.

The goal of the evaluation of the impact of process parameters on quality attributes, and the subsequent definition of the design space, is to identify high-risk inputs and include them in the process characterization studies. If necessary, low-resolution screening experiments including several parameters can be first performed to eliminate non-critical inputs from further experimentation.

Panel 4.17. Critical Process Parameters

- CPPs affect CQAs and must be controlled within a range
- The intersection of optimal spaces for different CQAs gives the "optimal" range for each CPP
- The relationship between CPPs and CQAs is used to develop control strategies
- This is unique to each unit operation and the results are linked for overall manufacturing

The design space for a unit operation can be determined by overlaying the permissible regions of each CQA (Figure 4.15). From surface response experiments in DOE studies, the boundary between permissible and impermissible regions can be described by an equation or a model of the process. Next, models for all CQAs are overlaid to identify the region that is permissible to all CQAs. In practice, productivity is also taken into consideration. From such a plot, the normal operating range of the critical process parameters can be determined.[9] Normal operating ranges and acceptable ranges (e.g., action limits) are described in batch records and automation recipes in manufacturing.

Scale-Down Model and Design Space

The experimental characterization of the effect of process parameters on CQAs is carried out in a multiplex manner in a scale-down model (SDM) that is predictive of the manufacturing-scale process (Panel 4.18). It is therefore crucial to understand the capabilities and limitations of the scale-down model for evaluating the process parameters of interest in the manufacturing scale. For the bioreactor step of cell culture process,

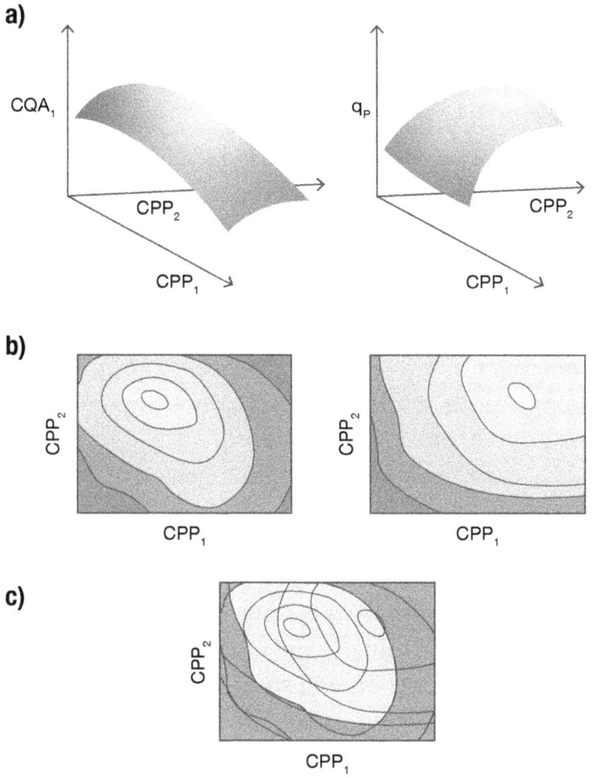

Figure 4.15. (a) Illustration using surface design to superficially evaluate the effect of CPPs on different CQAs and q_p. (b, c) Defining the "optimal" range of CPPs by their intersection.

Panel 4.18. Scale-Down Models for Process Characterization

- Usually carried out in a multiplex manner
- In scale translation, all parameters cannot be kept constant
- Requires a strong understanding of the capability and limitations of scale-down models
- The most important physical parameters are selected to be kept constant between the production scale and the scale-down model
- A successful scale-down model should reproduce the time profiles of growth, key nutrients, and metabolites, and have the same product quality as the manufacturing scale under standard culture conditions
- Test parameters can be varied over a range according to DOE protocol

the comparison of the scale-down and the manufacturing-scale models should include cell growth, metabolism, productivity, and product quality. Longitudinal analysis of the time series data of all runs in the scale-down model and all runs in the manufacturing scale should be performed, and statistically acceptable criteria defined.

It is important to keep in mind that in scale translation all scale-sensitive parameters cannot be kept constant. If one chooses a critical parameter to be constant when comparing different scales, some other parameters will change. For example, if the aeration rate per reactor volume is kept constant, then the superficial gas velocity will be higher in the larger scale. Furthermore, the hydrostatic pressure at the large scale will be higher, as well as the oxygen transfer driving force at the bottom of the reactor. As a result, the capacity of CO_2 stripping and oxygen transfer differs between the scale-down model and the manufacturing bioreactor. The scale-down model study should focus on the critical process parameters and take into account other uncontrolled parameters that are also affected by scale. In the definition of the design space, other relevant datasets (such as pilot-scale runs) should also be considered.

Process characterization in scale-down models typically employ design of experiments (DOE) principles to investigate the impact of multiple variables on cell growth, metabolism, productivity, and, most importantly, product quality. The information learned from the experiments depends on how the experiments are designed. Therefore, careful analysis of preliminary data is the first step in planning such experiments.

Process characterization in scale-down models for design space mapping can be carried out first in lower-resolution screening studies to explore several process parameters and identify the most impactful. The tradeoff of such screening studies is that the interactions of parameters may be confounded, or primary effects may be confounded with interactions. The most impactful variables are then studied again, in a more extensive DOE design, to establish a model of the effects and interactions of these parameters using response surface designs (Figure 4.15a).[9]

The determination of CPPs and the design space is performed for each unit operation as well as for the entire drug substance and drug product manufacturing process. The final steps involve connecting the design spaces of all unit operations via linking studies to ensure that the overall control strategy is effective. A table can be constructed that lists all the critical data of the CQAs and CPPs involved in each unit operation for better dissimilation of the information (Table 4.7). For product variants, such as glycosylation, deamidation, oxidation, glycation, C-terminal lysine, and N-terminal pyroglutamate, the production bioreactor has the dominant effect. The purification process can do little to modulate these attributes. The hold steps can potentially increase the levels of some variants, such as deamidation. Therefore, CPPs of bioreactor operation must be established to control the bioreactor output for these CQA(s). pH, temperature, duration of culture, feed rates, timing and components, dissolved oxygen levels, medium age, cell age can impact these CQAs. Not all of these may simultaneously be CPPs for a given product. Additionally protein recovery processes can also impact certain attributes, especially those relating to aggregation or fragmentation. For aggregates, fragments, host cell proteins, and host cell DNA, the cell culture process is the primary source of quality variability, but purification steps can remove these product- and process-related impurities. Therefore, CPPs in downstream process can be leveraged, and linkage studies are used to identify critical upstream and downstream process parameters.

Different unit operations in a process are thus interactive. The material input into a unit operation is affected by its upstream step and may require adjustment of its CPPs in order to control its CQAs. Linkage studies to examine the combined effects on CQAs are performed to ensure that the combination of worst-case inputs allowed by the design

Table 4.7. Tabulation of the Effects of Process Steps on Quality Attributes

Product Quality Attribute	*CQA*	*Production bioreactor*	*Protein A chromatography*	*Low pH virus inactivation*	*CEX*	*AEX*
Aggregation	Yes	Form	Removal	Risk to form	Removal	Removal
Deamidated isoforms	No	Form				
Oligosaccharide	Yes	Form				
Process-related impurities						
HCP	Yes	Form	Removal	Removal	Removal	Removal
DNA	No	Form				Removal
Protein A	No		Form		Removal	Removal
Viral Safety	Yes			Inactivation		Clearance

Cont.

Product Quality Attribute	*Nanofiltration for viral particle removal*	*UF/DF*	*Compounding*	*Filtration*	*Filling, stopper, cap*
Aggregation			Risk to form		Risk to form
Deamidated isoforms					
Oligosaccharide					
Process-related impurities					
HCP					
DNA					
Protein A					
Viral Safety	Clearance				

space for each CQA would produce a product that meets its specifications. Worst-case conditions are run in bioreactors in the laboratory or at pilot scale and purified at the worst-case boundary of the design space.

QbD in Different Stages of the Product's Life Cycle

In the past decade, QbD has taken root in the manufacturing of biopharmaceuticals. The road map it established has facilitated streamlining product development and compressing the process development timeline. Key to this success was the adoption of a risk analysis-based QbD approach, the integration of knowledge management, and the utilization of platform processes. Nowadays, process characterization at small scales makes up the bulk of process development data. Modern-day,

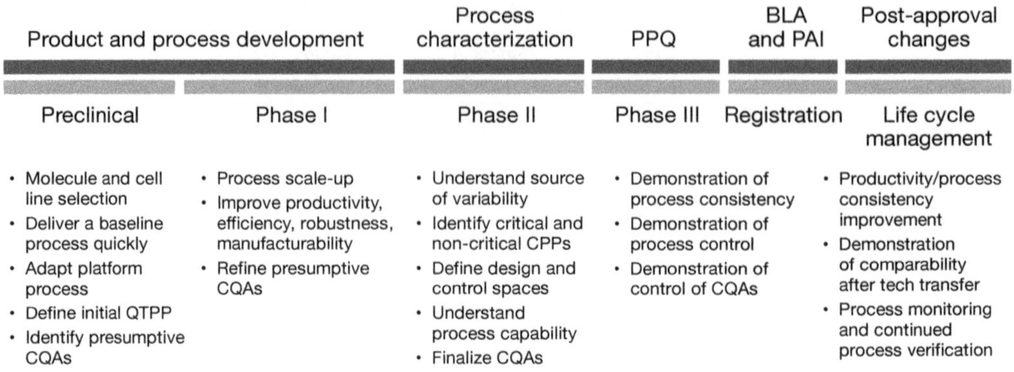

Figure 4.16. QbD-related tasks across the process and product development life cycle.

high-yielding processes result in fewer clinical manufacturing runs, and process performance qualification (PPQ) acceptance criteria are often developed with statistical analyses that include small-scale data.

Figure 4.16 depicts the activities related to QbD during the product's life cycle. In the early stages of product and process development, activities focus on understanding the QTPP of the product and establishing a process to move quickly and safely to clinical trials. As more experience is gained in the process and more knowledge acquired from *in vitro* and *in vivo* studies, understanding of CQAs improves. At the point when, based upon positive clinical trial data, a product is deemed commercialization-ready, the design space mapping studies are completed, critical process parameters are identified, and the control strategy is finalized. The final stage of process development includes Process Performance Qualification (PPQ), which demonstrates that the process is capable of consistently meeting CQA acceptance criteria.

Through the identification of CPPs and their relationship to CQAs, the QbD approach aims to harness the power of Process Analytical Technologies (PAT). The breadth of knowledge generated with multivariate explorations and the development of the design space can be leveraged through use in real-time monitoring and control for enhanced product robustness.

Continued process verification and process changes

During the commercial life cycle of the product, process improvements and technology transfers to new manufacturing facilities may require demonstrations that the product manufactured after such changes is comparable to that produced with the process approved at launch. These types of activities leverage the comprehensive understanding of CQAs and the design space.

The QbD principles used in process and product development are utilized throughout the life cycle of the product. Continued process verification (CPV) is an important continuation of process validation, and generates additional process knowledge beyond the original design space mapping undertaken during process development. CPV can be applied to monitor the robustness of the manufacturing process and allow action to be taken to improve consistency and product quality within the design space.

A CPV may involve identifying trends during process monitoring. Control limits of CQAs are calculated from historical manufacturing data (~30 batches), including the lower control limit (LCL), upper control limit (UCL), and lower and upper specification limits (LSL and USL, respectively) (Figure 4.17). Western Electric rules can be used to determine when the process is trending out of control. For example, when batch data are outside the control limit, six consecutive descending points occur, or nine consecutive points above or below the average occur, action would be taken to move the process back into a state of control. Points that are outside the specification limits are investigated via the Quality System. With continued verification and identification of the cause(s) of incidents which result in points falling outside the specification limit, corrective measures can be taken to narrow the process variability.

The concept of CPV is shown in Figure 4.18. Ideally, batch data is monitored in real time and interfaces with electronic batch records and data aggregation software. Multivariate analysis or data mining of production datasets is then conducted to identify opportunities and propose hypotheses for improvement. Laboratory studies may be conducted to prove these hypotheses via controlled experimentation for possible implementation in manufacturing. Diligence in process monitoring and subsequent action can result in year-over-year improvements to robustness and yield. Importantly, with a QbD-driven process understanding,

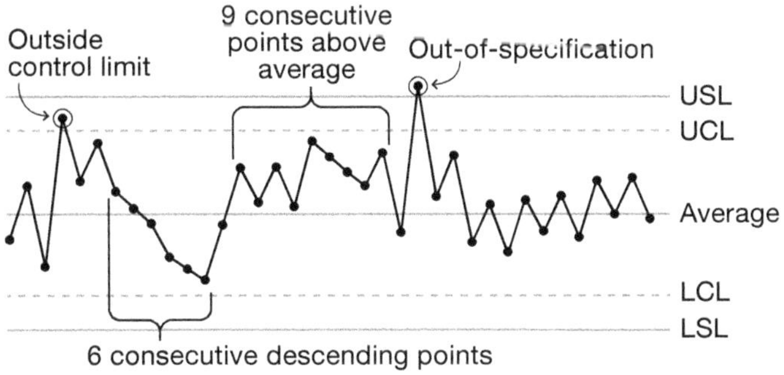

Figure 4.17. Example control chart of CQA records for continual process verification.

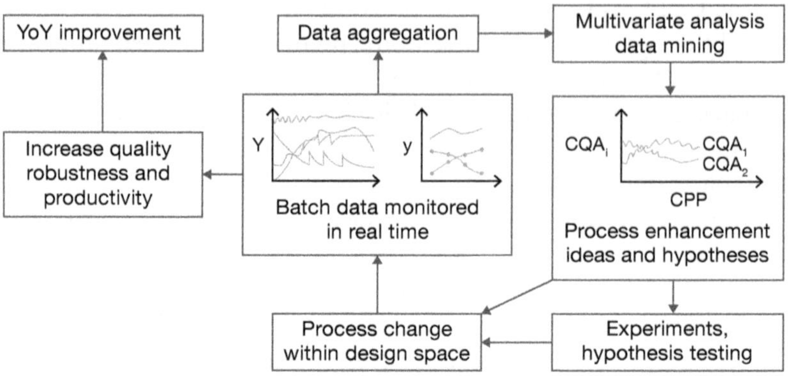

Figure 4.18. Continued process verification provides assurance that the process remains in a state of control.

post-approval process improvements can be performed with parameter values within the design space and potentially without prior health authority approval. This can greatly accelerate the speed at which process robustness adjustments can be implemented in manufacturing. This is fairly straightforward when the proposed changes are within the design space originally defined. However, the industry has been striving to include more complex changes, such as manufacturing site changes, within the scope of a QbD filing. With certain post-approval commitments, this has been possible on occasion, especially when there is sufficient knowledge and experience with legacy products and facilities.

Concluding Remarks

In this chapter we discussed the properties of protein products that contribute to their quality. A sound understanding of these products is important for controlling the product quality during the manufacturing process. For better control of product quality, current product and process development has implemented QbD. The goal of QbD is to reduce product variability, thereby enhancing process development and manufacturing efficiency. This is achieved by designing a robust manufacturing process and establishing clinically relevant specifications. The key elements of QbD include the QTPP, process design and understanding, the control strategy, and continued life cycle improvement.

As the concept of QbD matures, there will be a more consistent approach to QbD and a clearer vision of what its implementation means for all audiences. It is important to understand that QbD is not for reduced regulatory oversight. Rather, it is a broad paradigm of life cycle risk management and process understanding. It provides the opportunity to be transparent about uncertainty and how it will be managed, and to accept that process understanding and knowledge will grow throughout the product's commercial lifetime.

References

1. Beck, A., et al., *Characterization of therapeutic antibodies and related products.* Anal. Chem., 2013. **85**(2): p. 715–36.
2. Harris, R.P. and P.M. Kilby, *Amino acid misincorporation in recombinant biopharmaceutical products.* Curr. Opin. Biotechnol., 2014. **30**: p. 45–50.
3. Khetan, A., et al., *Control of misincorporation of serine for asparagine during antibody production using CHO cells.* Biotechnol. Bioeng., 2010. **107**: p. 116–23.
4. Moritz, B. and J.O. Stracke, *Assessment of disulfide and hinge modifications in monoclonal antibodies.* Electrophoresis, 2017. **38**(6): p. 769–85.
5. Du, Y., et al., *Chromatographic analysis of the acidic and basic species of recombinant monoclonal antibodies.* MAbs, 2012. **4**(5): p. 578–85.
6. Khawli, L.A., et al., *Charge variants in IgG1: Isolation, characterization, in vitro binding properties and pharmacokinetics in rats.* MAbs, 2010. **2**(6): p. 613–24.
7. Yu, L.X., et al., *Understanding pharmaceutical quality by design.* AAPS J., 2014. **16**(4): p. 771–83.
8. Abu-Absi, S.F., et al., *Defining process design space for monoclonal antibody cell culture.* Biotechnol. Bioeng., 2010. **106**(6): p. 894–905.
9. Rouiller, Y., et al., *Application of Quality by Design to the characterization of the cell culture process of an Fc-Fusion protein.* Eur. J. Pharm. Biopharm., 2012. **81**(2): p. 426–37.

Additional Materials

- **ICH Quality Guidelines**
  www.ich.org/products/guidelines/quality/article/quality-guidelines.html
- **FDA Guidelines**
  https://www.fda.gov/vaccines-blood-biologics/guidance-compliance-regulatory-information-biologics
- **A-mAb Case Study in Bioprocess Development (CMC-Biotech Working Group)**
  https://www.casss.org/page/286

Contents

5

Stoichiometry and the Kinetics of Cell Cultivation

Introduction

IN A BIOMANUFACTURING PROCESS, cells take up nutrients which generate energy and are converted to more cells, metabolites, and product. For many microbial biochemicals, the cost of those raw materials constitutes a major portion of the total production cost. It is therefore important to maximize the efficiency of converting the raw materials to the product. Most cell culture products are not raw-material intensive. However, efficient nutrient utilization is still important. Over-abundance and over-consumption of nutrients leads to inhibitory metabolite accumulation, which affects productivity and even product quality.

Cells convert materials through their biochemical reaction network. Depending on the chemical environment, nutrients are taken through different routes as they are converted to cells, metabolites, and product. These routes have different material utilization efficiencies and, more importantly, different effects on overall process productivity. During conversion, the principle of material balance cannot be violated. Using the material balance principle to establish the stoichiometric relationship between inputs and process outputs is the first step for process design and enhancement.

When optimizing a process, in addition to the information on the stoichiometric relationships between materials, it is important to know how to appropriate those materials over time and how the outcome variables will behave in response. Quantitative descriptions of how the important material variables (e.g., cell biomass, nutrients, metabolites) change with other process variables and time are key tools when developing a plan to guide the process along a desired path. This chapter discusses the stoichiometry and kinetics of cell growth in a bioprocess.

Material Conversion in Cell Culture

CELLS OBTAIN THEIR MATERIAL RESOURCES from their medium (Panel 5.1), including water, which is the most abundant component in a cell. Additionally, cells consume oxygen so they can derive energy from the nutrients they consume. The three nutrients consumed in the greatest quantity by most cells in culture are oxygen, glucose, and glutamine. Some adapted or engineered cells do not require glutamine. Inside the cell, the material inputs are converted in the cell's biochemical reaction network to metabolites in the course of generating energy and synthesizing cell mass and product. To generate energy, glucose is converted to lactate, CO_2, and H_2O through glycolysis, the TCA cycle, and the pentose phosphate pathway. Glutamine is deaminated, releasing NH_3, before its carbon skeleton enters the TCA cycle through α-ketoglutarate for energy metabolism. The major metabolites generated from energy metabolism are thus lactate, NH_3, CO_2, and H_2O. The amino group from the metabolized glutamine and other amino acids may also be exported as non-essential amino acids (e.g., alanine, asparagine, proline) in addition to being excreted as NH_3. The energy metabolism generates ATP, NADH, and NADPH to meet the energetic needs of many biosynthetic reactions.

Panel 5.1. Inputs and Outputs of Cell Growth Processes

Inputs
Glucose
Amino acids
Other: lipids, nucleotides, etc.
Bulk salts
Trace minerals
Oxygen

↓
CELLS
↓

Outputs
Biomass
Product
CO_2
Water
Lactate, NH_3
Excreted amino acids

In the course of metabolic material conversion, many reaction intermediates are generated, most of which continue to be catabolized or are used in biosynthesis to make other cellular components. Some will be secreted as metabolites. The line between intermediates is a blurred one. Many reaction intermediates, such as components of the TCA cycle, are also secreted, albeit in small proportions to the amount synthesized. For example, virtually all of the lactate synthesized during cell growth is secreted as metabolites, while only a small portion of the TCA cycle intermediate citrate may be secreted under some culture conditions.

From a material balance perspective, we can treat growing cells as a system on which the material balance is to be performed. The inputs include all materials that are taken up by cells, i.e., the consumed nutrients (say, over a time period or, if one can measure it so quickly, at an instant of time). The outputs include the new biomass that is synthesized and the metabolites, including lactate, NH_3, CO_2, and H_2O that are secreted. The balance thus focuses on the exchange of materials between the system (cells) and its environment. In the following sections, we will discuss the major items of inputs and outputs in a cell culture system. Material balance is also performed on the intracellular bioreactor network based on the material exchange with the environment. The analysis, referred to as metabolic flux analysis, will be discussed in Appendix A.

Cell Biomass

Before material balance on cell growth can be performed, one has to determine the material content of the cell. The size of a cell varies widely from microorganisms to those of mammals. Order of magnitude-wise, the dry biomass of an average cell of bacteria, yeast, and animal cells is 10^{-12}, 10^{-11}, and 5×10^{-10} g per cell, respectively (Table 5.1). The most abundant chemical species in a cell is water, accounting for up to 90% of the volume of plant cells, ~80% of animal cells (Table 5.2), and 70% of bacterial cells. Since cell cultivation is carried out in aqueous environments and the amount of water taken up by cells during growth is difficult to measure, the material balance on cell culture is typically performed only on "dry" matter, thereby excluding water.

Macroscopically, cells are made of a few classes of macromolecules (proteins, carbohydrates, DNA, and RNA) or macromolecular assemblies (primarily lipids). These organic matters constitute the vast majority of the dry mass in a cell. Protein molecules constitute the largest portion among them, providing the machinery for DNA, RNA, and protein synthesis. Protein molecules also serve as the structural components of the cell and execute all of the catalytic, transport, and communication functions. The lipid content of an animal cell is higher than in a bacterium. The abundance of organelles and the presence of a lipid bilayer membrane in an animal cell contributes to their higher numbers.

Intracellular carbohydrates exist as oligosaccharides on many proteins and lipids. Carbohydrate

Table 5.1. Typical Dry Weight of Cells

Type	*Dry weight*
Bacterial	10^{-12} g/cell
Yeast	10^{-11} g/cell
Average Animal Cell	$3–6 \times 10^{-10}$ g/cell

Table 5.2. Average Composition of an Animal Cell

	pg/cell	*Range*	*%*	*% of dry biomass*
Wet weight	3500	3000–8000		
Dry weight	600	300–1200		
Protein	250	200–300	10–15	~50–70
Carbohydrate	150	40–200	~1–5	~5
Lipid	120	100–200	~1–2	~5
DNA	10	8–17	~0.3	~2
RNA	25	20–40	~0.7	~4
Water			55–80	
Volume	4×10^{-9} cm^3			

also exists as five carbon carbohydrates (ribose and deoxyribose) in RNA, DNA, and nucleotides (e.g., ATP, GTP). Only a small fraction exists in a free (or phosphorylated) form. The cellular content of carbohydrates is harder to estimate because most carbohydrates are conjugated to lipids, proteins, or nucleosides.

Variation in cell size

Cells isolated from different tissues or of different differentiation types vary in size (Table 5.3, Panel 5.2). The volume of a typical mammalian cell is in the order of a few picoliters (about 1,000 times larger than a bacterium). The average diameter of a population of cells in culture ranges from 10 to 20 μm. Cells of different differentiation types that were isolated from the same tissue often differ in size. For example, megakaryocytes are much larger than T lymphocytes and natural killer cells. Even hepatocytes isolated from different regions (zones) of the liver differ in size. Pluripotent stem cells and other more primitive stem cells are fairly small. Their nucleus spans more than 70% of the cell diameter and their cytoplasm is relatively small. Liver cells and antibody-secreting plasma cells are at the other end of the cell size spectrum, containing a significant amount of cytoplasm for protein secretion.

Even cells in a homogeneous population of the same cell type or cell line have different sizes. Cells are concurrently undergoing different stages of the cell cycle; those immediately before mitosis are twice as large as those after cell division. Like many other properties, such as DNA content per cell, the cell size of a population distributes over a range with more than a 2-fold difference between the highest and lowest subpopulations. For aneuploid cells, the distribution of size is typically greater than for normal diploid cells. Some aneuploid cell lines have a wide distribution of chromosome numbers, and some even have a tetraploid karyotype. Those cells are larger than diploid cells.

The distribution of cell size changes with the growth stage. Rapidly growing and quiescent cells may have different sizes. Furthermore, in a culture, cells that lose viability often become visibly smaller, as measured by flow cytometry.

Table 5.3. Size of Animal Cells

Cell type	*Volume* (μm^3)	*Diameter* (μm)
Hybridomas	900–4000	12–20
Endothelial cells	1400–2500	14–17
Chinese hamster ovary cells (suspension)	1200–1800	~14
Human foreskin fibroblasts (FS-4)	3000–6000	18–22

Panel 5.2. Size of Cells in Culture

- Cell volume distributes over a range
- Dead cells are often smaller
- Cell size varies with culture stage

Cell number is traditionally used for the quantification of cell concentration, even though it does not sufficiently capture the metabolic and production capacity of the culture since cell size may change during the culture and among different cultures. It is instructive to remember that the volume of a sphere (which is a reasonable approximation of a cell) is proportional to its diameter raised to the third power. Therefore, cells that are twice as large in diameter are eight times larger in cell volume. In addition to cell size variation, the density, or the water content, of a cell may also vary in different stages of a culture. In a fed-batch culture, the osmolality increases over culture time, possibly causing the density of cellular content to vary. Thus, under some conditions, such as investigating cell growth or productivity change, it is necessary to quantify cell size distribution and cellular density in addition to cell number.

Amino acid composition of cellular proteins

Microbial and plant cells often grow on simple carbon sources supplemented with an inorganic nitrogen source, such as ammonium or urea. These cells convert inorganic nitrogen to all 20 natural amino acids that are used to make proteins. Animal cells lack the capability to make 11 to 12 of those 20 natural amino acids. These essential amino acids must be supplied for animal cell culture to enable them to grow and make products.

The protein content and composition of cells change under different growth conditions; however, they are seldom measured. Table 5.4 lists the approximate amino acid composition of a cell and IgG. Given target levels of cell biomass and product to be produced, a stoichiometric amount of all essential amino acids must be supplied. In addition to essential amino acids, non-essential amino acids are usually also supplied in a cell culture medium.

Table 5.4. Approximate Amino Acid Composition of Cells and IgG

	Cell composition mean (%)	*Standard deviation*	*IgG composition*
Ala	9.03	0.32	5.31
Arg	4.74	0.32	2.43
Asn }	10.08	0.59	3.49
Asp }			3.95
Cys	0.26	0.04	2.43
Gln }	12.62	0.63	5.01
Glu }			5.16
Gly	9.14	0.57	6.98
His	2.22	0.07	1.67
Ile	5.73	0.35	2.43
Leu	9.00	0.68	6.83
Lys	6.85	0.49	6.98
Met	2.27	0.14	1.37
Phe	3.73	0.31	3.49
Pro	5.51	0.58	7.13
Ser	6.19	0.16	12.90
Thr	5.42	0.22	7.74
Tyr	2.73	0.14	4.10
Val	6.54	0.27	9.10

Intracellular fluid

Water constitutes 70–80% of total cell volume. The soluble low molecular components (excluding polymeric materials and lipid membrane assemblies) in water also make up a large fraction of the biomass. The intracellular fluid contains electrolytes, carbohydrates, amino acids, metabolic intermediates, nucleotides (ATP, ADP, etc.), and many other components. The vast majority of intracellular amino acids reside in cellular proteins and only a small fraction exist as free amino acids in intracellular fluid. Most amino acids are present at the 0.05–0.5 mM range in intracellular fluid. Their normal concentration is sufficient to sustain a high fidelity in charging their corresponding tRNA.

Typical concentrations of amino acids and other major soluble components in intracellular fluids are listed in Table 5.5 and Table 5.6, respectively. The intracellular concentration of amino acids can vary widely among different cell lines and under different growth stages and culture conditions. Also shown in Table 5.6 are the typical concentrations of solutes in the cell's natural environment (i.e., the interstitial fluid). Potassium, magnesium, and phosphate are present at high concentrations (millimolar range). A large fraction of cellular Mg^{2+} is associated with ATP, which is typically present at the 1–3 mM range. In addition to free phosphate, phosphate is also present in phosphorylated compounds (DNA, RNA, nucleotides, phosphorylated sugars, lipids, etc.).

Many inorganics, including phosphate, potassium, and magnesium, are present at much higher concentrations intracellularly than they are in extracellular fluid or culture medium. As cells grow, they take up nutrients in large enough quantities that they accumulate in the intracellular fluid. It is important to ensure that those inorganic nutrients are also supplied in sufficient quantities. By knowing their intracellular content, the stoichiometric amount required to produce the target biomass can be estimated.

In addition to the major inorganic species (K^+, Na^+, phosphate, Mg^{2+}, and Cl^-), many minute inorganic elements are also constituents of cellular components, including iron, copper, selenium, zinc, and cobalt. Many primarily exist as a prosthetic group of proteins. These elements must also be supplied in enough quantities to generate biomass. The cellular content of those elements is seldom reported and may vary widely among different cell types or even under different culture conditions. For example, the zinc content

Table 5.5. Intracellular Concentrations of Amino Acids

Name	*mM*	*Name (cont.)*	*mM*
Ala	0.2–2.0	Lys	0.1–0.6
Arg	<.05	Met	0.01
Asp	0.4–0.8	Ornithine	0.120
Asn	0.4–0.8	Phe	0.3–0.5
Citrulline	0.036	Proline	0.137
Gln	0.05–4	Ser	0.149
Glu	0.3–12	Thr	0.1–4
His	<0.05–0.09	Tyrosine	0.059
Ile	0.3–0.5	Valine	0.171
Leu	0.1–0.4		

Table 5.6. Approximate Concentrations of Components in a Cellular Environment

	Interstitial fluid (mM)	*Intracellular fluid (mM)*
Na^+	140	6–14
K^+	4.0	100–140
Ca^{2+}	1.2	0.01
Mg^{2+}	0.7	3–20
Cl^-	108	4
HCO_3^-	28.3	10
HPO_4^{2-} $H_2PO_4^-$	2	11
SO_4^{2-}	0.5	1
ATP		1.5
Amino acids	2	8
Carnosine		14
Creatine	0.2	9
Glucose	5	0.05
Lactate	1.2	1.5
Phosphocreatine		40
Protein	0.2	3–4

is much higher in pancreatic cells than in other cells. Similarly, iron is rich in muscle and red blood cells. Without quantitative data on the cellular contents of those elements, one resorts to titration experiments under defined culture conditions to see whether the supply of those elements is limiting cell growth.

Metabolites

Cells produce and excrete metabolites as they grow. Some metabolites are produced as the product of oxidation of the main nutrient for energy generation (typically glucose), such as lactate, CO_2, and H_2O. Some are produced by metabolic reactions undertaken to balance the amino acids (e.g., alanine, glutamic acid, and asparagine). Others are degradation products of excess amino acids; for example, hydroxybutyrate is produced due to the degradation of isoleucine. The excretion rate of a metabolite is affected by the abundance level of the enzyme of the synthetic pathway, that of the transporter for its excretion, and the concentration of the substrate and the metabolite.

It is virtually impossible to provide cells with every nutrient in a "correct" stoichiometric amount and maintain it at an "optimal intracellular" level so that cell metabolism is at a completely homeostatic state with minimal metabolite excretion except those derived from energy generation (i.e., lactate, CO_2, and H_2O). In other words, the excretion of other metabolites is unavoidable, especially those associated with amino acid conversion and degradation. To close material balance, one has to determine the exact types and quantities of metabolites produced. However, many metabolites are not easily measured as a routine laboratory analysis. Some degradation products are growth inhibitory to cells, so taking a metabolomic approach to identify the potential presence and accumulation of metabolites is certainly worthwhile. By performing material balance on the cell culture system and examining how closely the carbon and nitrogen elements are closed, one can also evaluate the likelihood that additional metabolites may be accumulating that are yet unidentified.

Material Balance on Cell Growth

THE PROCESS OF GROWING CELLS and producing a product can be formulated into a stoichiometric equation of "overall biomass synthesis" (Panel 5.3). This equation can be viewed as the composite of all reactions involved in generating energy and synthesizing biomass and product. At the center of the reaction is the biomass. The cell mass is expressed as a chemical formula of elements C, H, N, O, P, S, etc., based upon its elemental composition. Usually, we use a formula of only carbon (C), hydrogen (H), nitrogen (N), and oxygen (O), neglecting the other mineral

Panel 5.3. Example of an "Equation" for Cell Growth

$$C_6H_{12}O_6 \; + \; 0.15\, C_{6.14}H_{12.36}N_{1.50}O_{2.08} \; + \; 0.34\, C_5H_{10}N_2O_2 \; + 1.39\, O_2 \rightarrow$$

(glucose) *(weighted average of amino acids)* *(glutamine)*

$$2.37\, CH_{1.97}N_{0.26}O_{0.49} + \; 0.0058\, CH_{1.83}N_{0.14}O_{2.06} + 1.53\, CO_2 + 1.28\, H_2O +$$

(cell mass) *(antibody)*

$$1.44\, C_3H_6O_3 + \; 0.16\, NH_3 + 0.13\, C_4H_7NO_2 \qquad \textbf{(Eq. 5-1)}$$

(lactate) *(alanine)*

elements. This formula presents the mass ratio of those elements. One can arbitrarily assign the stoichiometric numbers to give them different "formula weights." An example of a biomass formula is shown in the cell growth stoichiometric equation (Eq. 5-1, Panel 5.3), in which the stoichiometric number of carbon is chosen to be 1. Others may prefer to assign the formula mass as 100.

The inputs in a cell growth process, i.e., the reactants of the stoichiometric reaction equation of cell growth, include all nutrients consumed by cells and used to proliferate. Since only C, H, N, and O are considered in the stoichiometric equation, we consider only glucose, glutamine, other amino acids, and oxygen. Other minute media components containing C and N, such as vitamins, are neglected. Instead of writing down all amino acids separately, one may use a weighted average derived from the stoichiometric ratio of their consumption.

The outputs, the product side of the stoichiometric equation, include "new" cell mass that is generated as well as the metabolites and product that are excreted. One can write a molecular formula describing the formation of the protein product from its amino acid sequence. The metabolites excreted include lactate, ammonia, some non-essential amino acids, and other excreted minor metabolites.

Growth Stages of Mammalian Cells in Culture

The data of a cell culture process is typically presented as time profiles of the concentrations of cell biomass, nutrients, metabolites, and product. A typical cell growth curve can be divided into a number of growth stages: lag phase, exponential growth phase, stationary phase, and death (or the decline phase) (Figure 5.1). The exponential growth phase is characterized by an exponential increase in biomass, or a period of constant slope in biomass increase on a semi-logarithmic plot over time. For mammalian cells, the period of exponential cell expansion in a culture is often relatively short (typically no more than a 10-fold increase in cell concentration). In some cases, a period of stagnant cell concentration

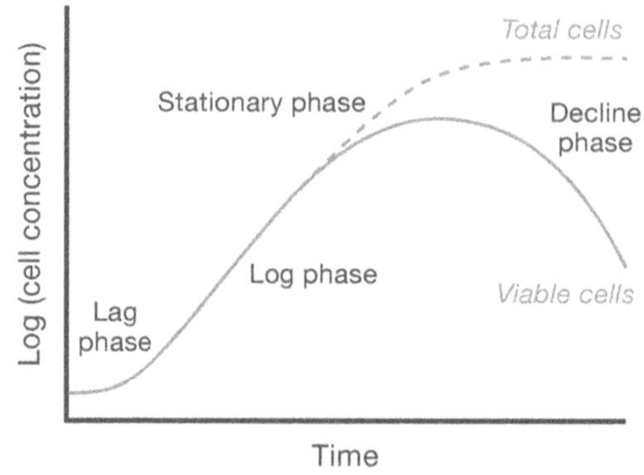

Figure 5.1. A typical growth curve of mammalian cells and growth stages in a suspension culture.

Table 5.7. Doubling Time of Some Culture Cells

Cell type / conditions	*Doubling time*
Mouse embryonic stem cell	~12 h
Human diploid fibroblast (10% FBS)	24–40 h
Human diploid fibroblast (2% FBS)	40–60 h
CHO-K1 (5% FBS or in rich medium)	16–24 h
NS0 cell	16–24 h

may precede the exponential phase, called a lag phase. A lag phase may be caused by using an inoculum from a culture that has already reached a stationary phase, or by using an inoculum grown previously in a vastly different medium. Initiating a culture with a suboptimal cell concentration may also cause poor initial growth. In general, anchorage-dependent cells should be inoculated at a minimum density of 10^4 cells/cm^2 while also maintaining a cell concentration of 10^5 cells/mL. Cells grown in suspension are generally started at about 10^5 cells/mL, but at a higher level for very small blood cells such as natural killer cells.

The exponential phase is marked by a constant cell growth rate, or doubling time (the time duration that cell biomass or cell number doubles). The growth rate of a cell line is dependent on the medium, temperature, pH, and many other physical and chemical conditions. Additionally, most cell lines, cell strains, or tissue cells require a set of growth factors or cytokines for growth. The growth rate is affected by the levels of the required growth factors. The maximum growth possible under most favorable conditions for different cell lines and cell types spans over a wide range, from a fast rate of 12 h for mouse embryonic stem cells to ~24 h for human diploid fibroblasts (Table 5.7). Cells commonly used in therapeutic protein production double every 16–30 h.

After the rapid growth period, due to nutrient depletion or the accumulation of metabolites, the growth rate diminishes and the cell concentration stops increasing. This is referred to as the stationary phase. A stationary phase may also reflect a balance between cell growth and death. In the latter case, the growth curve is characterized by a constant viable cell concentration, along with an increasing concentration of dead cells.

Once cell viability begins to decrease, the culture is considered to be in a decline phase. The growth behavior at a late stage of culture often varies, depending on whether cells are adherent or in suspension. Many anchorage-dependent cells grow substantially slower once cell density on the surface approaches confluence. They can also sustain a confluent cell density over a period of days without entering the death phase.

Conversely, cells grown in suspension, especially those with an inherently strong receptor-mediated apoptotic mechanism, often enter the death phase soon after the viable cell concentration peaks.

Kinetic Description of Cell Growth & Product Formation

Specific Rates

A CELL CULTURE PROCESS can be characterized by how the concentrations of cell biomass, nutrients, and product change over time. From these data of time profiles of concentrations, one can determine the rates of change of those concentrations, how those rates relate to each other, and how active or productive the process is.

The rate of concentration change of a culture component is determined from the slope of its concentration profile over time and is referred to as the volumetric rate. Eq. 5-2, Panel 5.4 shows the volumetric rate for cell growth. One can similarly express the volumetric rates of nutrient consumption or product formation, although they are not frequently used. A culture that is quickly accumulating a large number of cells per culture volume (i.e., has a high volumetric growth rate) may have started from a high concentration of cells or from cells that are more actively growing. Hence a more useful quantity is the specific rate, which describes how active cells are.

Usually, cell volume constitutes only a very small fraction of the volume in a suspension cell culture. The volumetric rate is essentially based on the liquid volume. In some cases, such as with high-density solid microcarrier culture, the volume occupied by solid beads is large; thus, liquid volume and culture volume are not equal. Usually the concentrations of nutrients, metabolites, and product are measured based on liquid volume, not on culture or reactor volume. In such cases, the "volume" used for different kinetic parameters must be clearly stated.

Another set of descriptors, called the specific rate, is normalized to a per cell basis. These activity parameters describe how active each unit of the cell is for activities like making new biomass (specific growth rate), consuming glucose (specific glucose consumption rate), or producing lactate (specific lactate production rate).

The speed at which cells generate new biomass is linearly dependent on the number of cells in culture, as shown in Eq. 5-3 (Panel 5.4). The proportionality (or the rate constant) is called the specific growth rate, μ. This quantitative description allows one to compare the speed of growth among different cells or cells under different culture conditions. It has a

Panel 5.4. Specific Rates Describing Growth Kinetics

Growth Rate (G): Change in cell concentration per unit time (cell number/L-h; or g cell/L-h)

$$G = \frac{dx}{dt} \quad \textbf{(Eq. 5-2)}$$

Specific Growth Rate (μ): Cell number/cell-h or g cell/g cell-h

$$\frac{dx}{dt} = \mu x \quad \textbf{(Eq. 5-3)}$$

$$\mu dt = \frac{1}{x} dx \quad \textbf{(Eq. 5-4)}$$

Doubling Time

$$\ln x = \mu t \quad \textbf{(Eq. 5-5)}$$

$$x = x_0 e^{\mu t} \quad \textbf{(Eq. 5-6)}$$

$$\ln \frac{x_2}{x_1} = \mu (t_2 - t_1) \quad \textbf{(Eq. 5-7)}$$

$$t_d = \frac{\ln 2}{\mu} = \frac{.693}{\mu} \quad \textbf{(Eq. 5-8)}$$

Specific Nutrient Consumption Rate: g nutrient/g cell-h

$$q_s = \frac{-1}{x} \frac{ds}{dt} \quad \textbf{(Eq. 5-9)}$$

Specific Product Formation Rate

$$q_p = \frac{1}{x} \frac{dp}{dt} \quad \textbf{(Eq. 5-10)}$$

unit of the inverse of time. For mammalian cells and microorganisms in culture, it is normally expressed in h^{-1}. This is in contrast to the volumetric cell growth rate, which has units of cells-$L^{-1}h^{-1}$. Note that the quantity of cells may be described in terms of cell mass, cell number, or some cellular component, depending on the method used to measure cell concentration. In the specific growth rate, the cell quantity descriptor is not present. However, a specific growth rate determined using different cell measurements may give somewhat different values, because cell size, cell mass, and cellular content may not be related to one another in the same proportion under different culture conditions. For example, under some conditions cells may not divide and increase in number, but expand in cell volume and biomass. So the specific growth rate based on the cell number measurement is different from that obtained with cell mass.

The cell doubling time is used frequently to describe how quickly cells grow. By separating the variables x and t in Eq. 5-3 into two sides of the equation (Eq. 5-4) and integrating with respect to time and cell concentration, respectively, one obtains the relationship between cell concentration and time, given a constant specific growth rate (Eqs. 5-5 and 5-6). The doubling time is the time it takes to develop a 2-fold increase in cell concentration (Eq. 5-8).

A descriptor for the activity of nutrient (often also called substrate) consumption, or the specific rate of nutrient consumption, can be similarly defined as shown in Eq. 5-9. Note that a negative sign is given since the nutrient concentration decreases over time. The specific nutrient consumption rate is obtained by dividing the volumetric nutrient consumption rate (*ds/dt*) by the cell concentration. Similarly, a specific product (or metabolite) formation rate can be defined (Eq. 5-9).

In some cases, non-viable cells constitute a significant portion of the cell population. A death term is added to the cell concentration balance (Eq. 5-11, Panel 5.5). Note that x represents viable cells, and x_d denotes dead cells. The total cell concentration is the sum of x and x_d (Eq. 5-12). The dead cells arise from viable cells at a frequency described by the specific death rate, μ_d. Using the definition shown in Panel 5.5, the specific growth rate is calculated from the rate of change of the total cell concentration. If one uses the viable cell concentration to calculate the growth rate (*dx/dt*), the resulting specific rate (Eq. 5-12) is $\mu-\mu_d$.

Cell lysis, in which dead cells further lyse and become invisible in cell counting, may occur in culture. Their numbers are sometimes estimated from the measurement of the release of cellular content such as DNA or an intracellular enzyme. Cell lysis is assumed to occur from dead cells (Eq. 5-14). To describe the growth kinetics of such a culture, one can also include a cell lysis term (Eq. 5-15).

Panel 5.5. Specific Rates in the Presence of Cell Death

Cell Growth in Culture

$$\frac{dx_d}{dt} = \mu_d x \qquad \text{(Eq. 5-11)}$$

$$\frac{dx}{dt} = \mu x - \mu_d x \qquad \text{(Eq. 5-12)}$$

$$x_t = x + x_d \qquad \text{(Eq. 5-13)}$$

Specific Rate

$$\mu = \frac{1}{x}\frac{dx_t}{dt} \qquad \text{(Eq. 5-14)}$$

$$q_s = \frac{-1}{x}\frac{ds}{dt} \qquad \text{(Eq. 5-15)}$$

With Cell Lysis

$$\frac{dx_d}{dt} = \mu_d x - \lambda x_d \qquad \text{(Eq. 5-16)}$$

$$\frac{dx}{dt} = \mu x - \mu_d x - \lambda x_d \qquad \text{(Eq. 5-17)}$$

Panel 5.6. Yield Coefficient and Stoichiometric Ratios

Yield of Biomass on Substrate

$$Y_{x/s} = \frac{dx}{ds} \approx \frac{\Delta x}{\Delta s} \qquad \text{(Eq. 5-18)}$$

Yield of Product on Substrate

$$Y_{p/s} = \frac{dp}{ds} \approx \frac{\Delta p}{\Delta s} \qquad \text{(Eq. 5-19)}$$

Stoichiometric Ratio of Two Substrates

$$\alpha_{s_1/s_2} = \frac{\Delta s_1}{\Delta s_2} \qquad \text{(Eq. 5-20)}$$

Stoichiometric Ratio of Lactate to Glucose

$$\alpha_{L/G} = \frac{\Delta L}{\Delta G} \qquad \text{(Eq. 5-21)}$$

The calculation of the other specific rate must also account for cell death (Eq. 5-16). The cell concentration used to calculate the specific nutrient consumption or product formation rate is the viable cell concentration (Eq. 5-17). In other words, one assumes that dead cells are metabolically inactive.

Stoichiometric Ratio and Yield Coefficient

In evaluating the performance of a process, we assess the efficiency of material conversion, i.e., how much of a raw material is converted into the product. The performance is then compared to a target value for further optimization.

A yield coefficient is frequently used to describe the efficiency of raw material conversion to cell mass or product in microbial fermentation, where the cost of raw materials is often a very significant fraction of the production cost. The yield coefficient is given a symbol $Y_{x/s}$ or $Y_{p/s}$, for cell mass or product, respectively (Eqs. 5-18 and 5-19 in Panel 5.6), and is based on the consumption of a particular substrate. It is the ratio of the quantity of the product or cell produced to that of the substrate consumed. A yield coefficient for a cell can be based on different input materials, e.g., glucose or other nutrients.

Yield coefficients are rarely used in animal cell culture processes. Cell concentration in a typical animal cell culture process is seldom measured in mass. Furthermore, the medium components that are used in large quantities by cells (glucose and amino acids) are not the top contributors to the cost of goods.

The stoichiometric ratio of various nutrients and metabolic products is more frequently used in cell culture processes. They are the ratio of the amount of different nutrients consumed or metabolites produced over a time period in culture (Eqs. 5-20 and 5-21). Under different metabolic conditions or in different growth stages, cells utilize various nutrients differently and change the number of different metabolites produced. Changes in the stoichiometric ratios of various nutrients and metabolites are indicators of alterations in metabolism (Panel 5.7). Furthermore, knowing the stoichiometric ratio of nutrient consumption is essential in devising a sound nutrient feeding scheme in fed-batch and perfusion cultures.

For example, the stoichiometric ratio of lactate to glucose (e.g., the ratio of the amount of lactate produced to that of glucose consumed) is a strong indication of whether the energy metabolism is at a highly glycolytic or highly oxidative state. If most

Panel 5.7. Description of the "State" of the Cultures

- Physiological state
  - There is no unique or universal definition for the physiological state
  - In general, the kinetic parameters described above are sufficient to describe it
- Metabolic state
  - Can be described by the stoichiometric ratios

glucose is channeled through glycolysis to lactate, the ratio is close to two moles of lactate per mole of glucose. Conversely, if most glucose is directed toward the TCA cycle for aerobic oxidation, the ratio will be close to zero, while the stoichiometric ratio of oxygen to glucose will be closer to six moles of oxygen per mole of glucose.

The stoichiometric ratio and yield coefficient, based on a given pair of compounds, can be expressed in different units, e.g., mol/mol or g/g.

Integral Cell Concentration

In the early days of therapeutic protein development, the quantitative assay of product was often performed by specialized scientists. The measurement usually took a substantial amount of time that delayed process development. To facilitate the assessment of various process conditions for production, integral cell concentration was used. The product accumulation rate in a culture can be described by multiplying the specific product formation rate by cell concentration (Eq. 5-22 in Panel 5.8). Integration over the culture period from t_0 to t_f obtains the product concentration p_f at t_f (the starting product concentration is 0) (Eqs. 5-23 and 5-24). If q_p is constant, one can take it out of the integral. One can see that the final product concentration is proportional to the integral of cell concentration (Eq. 5-25). In a plot of cell concentration (x) vs. time, the integral cell concentration is the area under the curve of the cell concentration curve (Figure 5.2). Hence, integral cell concentration was used as an indication of the product accumulated in culture. This is still in practice widely. However, one should be aware that q_p is often not constant. It is a better practice to examine q_p at different stages of the culture for process optimization.

Panel 5.8. Integral Viable Cell Concentration

$$\frac{dp}{dt} = q_p x \qquad \textbf{(Eq. 5-22)}$$

$$dp = q_p x\, dt \qquad \textbf{(Eq. 5-23)}$$

$$\int_0^{p_f} dp = \int_{t_0}^{t_f} q_p x\, dt \qquad \textbf{(Eq. 5-24)}$$

$$p_f = q_p \int_0^{t_f} x\, dt \qquad \textbf{(Eq. 5-25)}$$

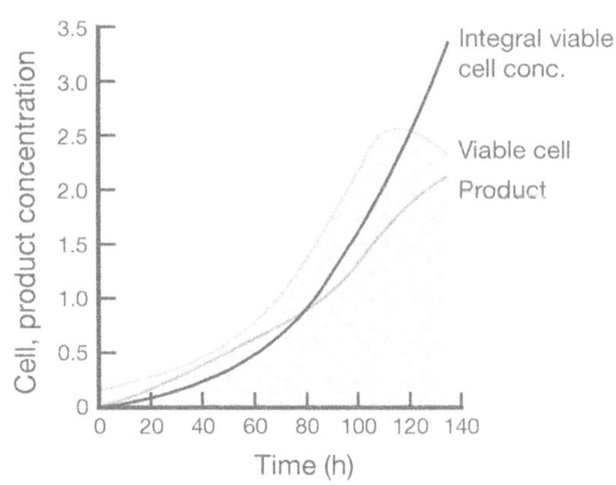

Figure 5.2. Cell and IgG product concentration in a batch culture and the viable integral cell concentration over time.

Kinetic Model of Cell Growth

IN THIS SECTION WE WILL DISCUSS using mathematical models to conduct computer experiments to evaluate process strategies. The topic would have seemed unfeasible only a few years ago. There are different levels of using a mathematical model for cell culture process development.

First, a model can be used to summarize a large volume of experimental data. In describing a process, we plot the concentrations of cells, key substrates/metabolites, and product (the core components of cell culture processes) over time. The data or their plots are extremely difficult to describe in words. By converting the data or plots into a set of equations or a mathematical model, regardless of whether the model is empirical or mechanistic, the behavior of the data can then be recreated by anyone who is given the model and value of the parameters.

A model can also be used to explore ideas and test hypotheses. When we study a physical system and its behavior, we first develop a verbal description of the system. We may then propose a hypothesis about the system's behavior, also verbally. The verbal description of the system and hypothesis can be translated into a mathematical form. The mathematical model can then be used to explore the system's possible behavior under different conditions. One may be able to compare the model simulation and experimental results and use the insight to refine the hypothesis. Ultimately, if one has a well-constructed model, one can use the model to predict the behavior of the systems. If the model is developed based on a mechanistic understanding of the system, it may be used to predict regions of parameter space that have not been experimentally tested previously. It can also be used to optimize or control the dynamics of the system.

Monod Model

At the core of any model for a cell culture system is the description of how the cell growth rate changes with a changing environment. A number of mathematical models have been used to describe mammalian cell growth in culture. Most are based on Monod-type models that were traditionally used to describe microbial growth. Verbally, the Monod model states that the growth rate of microorganisms increases with increasing limiting nutrient concentration, but eventually reaches saturation (maximum growth) as the nutrient level keeps increasing. The model was first introduced by Jacques Monod to describe the growth of *E. coli* cells on glucose. Monod then expressed the statement in a mathematical form.

The Monod model uses two parameters, μ_{max} and K_s, to describe the relationship between the specific growth rate, μ, and the limiting substrate concentration, s (Eq. 5-26, Panel 5.9, Figure 5.3). The equation gives a saturation type of behavior; μ increases with the limiting substrate

Panel 5.9. Monod Model

$$\mu = \frac{\mu_{max} s}{K_s + s} \qquad \textbf{(Eq. 5-26)}$$

when $s >> K_s$, $\mu \rightarrow \mu_{max}$

when $s << K_s$, $\mu \rightarrow \mu_{max} s/K_s$

- An empirical model for first-order approximation of cell growth
- Under most process conditions, growth is not limited by nutrient availability

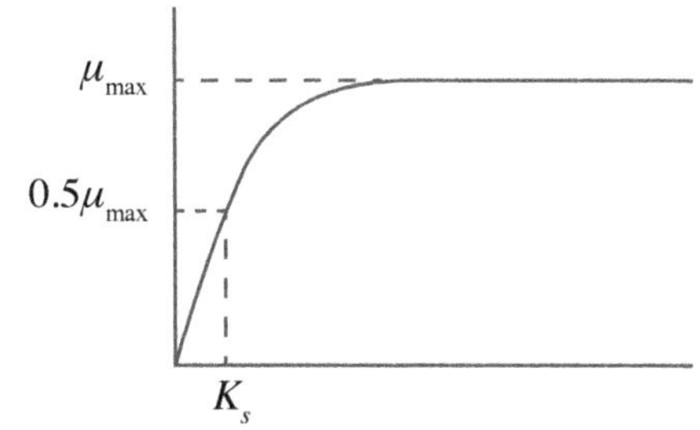

Figure 5.3. The specific growth rate as a function of growth-rate-limiting substrate concentration as described by the Monod growth model.

concentration until it reaches a maximal value, μ_{max}, asymptotically. At low concentrations, increasing the rate-limiting nutrient concentration increases the specific growth rate linearly. At very high concentrations, the specific growth rate is constant at μ_{max}.

The Monod model is an empirical model, where one uses experimental data to fit the model and determine the value of μ_{max} and K_s. Nevertheless, it depicts the growth curve of microbes well. When glucose is imposed as the limiting nutrient for mammalian cells in culture, it adequately describes the growth rate dependent on glucose concentration. However, under most culture conditions as well as *in vivo*, glucose is not the limiting nutrient. Instead, the growth rate is often dependent on the concentration of growth factors and subject to the regulation of signaling pathways. In industrial cell culture, the growth is more likely to be influenced by the accumulation of metabolites. Nevertheless, the Monod model is simple and has its merits for exploring different culture conditions.

Models for Multiple Nutrients and Adherent Cells

The Monod model describes growth dependence on a single limiting nutrient. Mammalian cells require multiple nutrients for proliferation. In addition to glucose, cell growth also requires glutamine, other amino acids, and many other nutrients. Although most are provided in excess and are not limiting, under some conditions the growth or metabolic behavior is affected by more than one nutrient simultaneously. Monod models are modified to describe the dependence of growth rate on multiple nutrients. Most common modifications incorporate multiplicative terms involving multiple substrates. Some also include the inhibitory effect of metabolites on the growth rate.

Panel 5.10. Multiplicative Saturation Kinetics Model

$$\mu = \mu_{\max} \cdot \frac{s_1}{K_1 + s_1} \cdot \frac{s_2}{K_2 + s_2} \qquad \textbf{(Eq. 5-27)}$$

$$\mu = \mu_{\max} \cdot \frac{s_1}{K_1 + s_1} \cdot \frac{s_2}{K_2 + s_2} \cdot \frac{K_{iA}}{K_{iA} + A} \qquad \textbf{(Eq. 5-28)}$$

In the multiplicative model for cell growth, two substrates, s_1 and s_2, are considered to influence the growth rate (Eq. 5-27, Panel 5.10). Each substrate has its corresponding half-saturation constant K_s. In most common applications, the two substrates are glucose and glutamine. The model can be extended to consider metabolite inhibition (Eq. 5-28). In the example shown in Eq. 5-28, A is an inhibitory metabolite, like lactate.

Panel 5.11. Growth Rate Dependence on Cell Density of Adherent Cells

$$\mu = \mu_{\max} \left(1 - \frac{y}{y_{\max}}\right) \qquad \textbf{(Eq. 5-29)}$$

Many cell types, including some used in viral vaccine manufacturing, are adherent. Their growth rate is not only affected by nutrient and metabolite concentrations, but also by the cell density on the surface to which they adhere. At low cell densities, cells grow at a maximum rate; as the cell density approaches "confluence" (i.e., the maximum density), the growth rate decreases (Eq. 5-29, Panel 5.11).

Modeling Productivity

THE KINETICS OF PRODUCT FORMATION in microbial systems are typically described using an empirical formula of proportionality to the specific growth rate (Eq. 5-30, Panel 5.12). The specific productivity, q_p, is considered to be influenced by two factors: a "growth-associated" term, α, which describes dependence on the specific growth rate; and a "non-growth associated" term, β. Depending on the relative magnitude of α and β, the production can be positive-, negative-, or non-growth associated.

Panel 5.12. Specific Product Formation Rate: Dependent on Growth Rate

$$q_p = \alpha\mu + \beta \qquad \textbf{(Eq. 5-30)}$$

Such classifications of specific productivity have limited applicability to mammalian cell culture systems. In general, the specific productivity of cell culture processes is relatively insensitive to the growth rate. For some recombinant proteins, the productivity is somewhat higher in the stationary phase of fed-batch cultures. At that slow-growing stage, many factors, including osmolality, lactate, and CO_2 concentration, have all deviated from optimal growth conditions. However, it is generally thought that the dominating factors that increase the productivity in the late stage of a culture are related to the stress conditions but not the growth rate. With a sufficient amount of data, one may use an empirical model to describe the specific productivity as a function of the stress index that considers different factors contributing to stress, such as osmolality and the concentration of some metabolites, like butyrate, hydroxybutyrate, and lactate.

Modeling a Cell Culture System

An important objective of developing a mathematical model for a cell culture system is being able to use it to predict culture behavior by testing different process conditions *in silico*, especially for conditions that cannot be easily carried out experimentally. In the section below, we will discuss the basic considerations in establishing a model for a cell culture system.

Elements of a System Model

The first step to building a model is identifying the pivotal variables of the system (Panel 5.13). In general, they include the concentrations of cells, glucose, lactate, and other factors affecting the specific growth rate and specific productivity. One then wants to develop a description of how those variables change over culture time due to their dynamic interactions with one another. Because changes in both those variables and their interactions are dynamic, one uses a set of differential equations to describe their dynamics. The second step is thus to establish the model or the system of equations of the cell culture system. Next, one specifies the interactions of the variables using mathematical descriptions and uses the experimental data to quantify the parameters used in the model so one can solve the equations and simulate the system behavior under different conditions.

Panel 5.13. Necessary Components for a Mathematical Description of a Cell Culture Process

- A description for:
  - Cell growth (and death)
  - Product formation
  - Nutrient utilization
- Growth model
  - The dependence of the specific consumption rate on the "controlling variable" (growth rate, nutrient concentration, etc.)
- Product formation
  - The dependence of the specific production rate on the "controlling variable"
- Experimental data
  - The model will have some parameters (e.g., half-saturation constants). The experimental data are used to determine the value of those parameters.
- Material balance equations
  - For "state variables" (the concentrations of cells, nutrients, product, inhibitors, etc.)

Balance equations on pivotal variables

After the pivotal variables for the system being considered are identified, a material balance equation for each variable is written. Shown in Panel 5.14 are example equations for a batch system. The rate of change of cell concentration (left-hand

Panel 5.14. A Cell Growth Model Incorporating Interactions, Growth, and Metabolism

Cell Balance

$$\frac{dx}{dt} = \mu x = \frac{\mu_{\max}}{(K_L + L)} \quad \textbf{(Eq. 5-31)}$$

Glucose Balance

$$\frac{dG}{dt} = -q_G x = -q_G(\mu, L)x \quad \textbf{(Eq. 5-32)}$$

Product Balance

$$\frac{dp}{dt} = q_p x \quad \textbf{(Eq. 5-33)}$$

Lactate Balance

$$\frac{dL}{dt} = q_L x = q_L(\mu, G, L)x \quad \textbf{(Eq. 5-34)}$$

side of Eq. 5-31) is the consequence of only cell growth (i.e., the specific growth rate multiplied by cell concentration) since there is no input or output for a batch culture. For a fed-batch or continuous culture, the balance equations will include additional terms to account for inputs and outputs. This will be discussed in Chapters 9 and 10. Equations are shown for the balance of glucose, lactate, and a product (Eqs. 5-32, 5-33, and 5-34). Again, because neither input nor output are present in a batch system, the balance is only attributed to the consumption of glucose and the production of lactate.

The reason that one includes all those variables is their interactions. In other words, the concentration of one variable affects the rates of change of other variables. For example, the specific growth rate may be affected by the concentrations of glucose and lactate. The cell concentration affects the rates of change of both glucose and lactate; hence the three variables (concentrations of cells, glucose, and lactate) are mutually interacting. If the specific production rate is affected by glucose and lactate, but the product production does not affect the specific rates of growth, glucose consumption, and lactate production, then the balance equation for the product does not need to be included in the system. One can first solve the time profiles of cells, glucose, and lactate, and then compute the product profile. On the other hand, if the product is a cytokine that has growth inhibitory effects, then the balance equation needs to be included in the system of differential equations.

Kinetic model for growth and metabolism

A model that describes the growth kinetics of cells in a culture therefore consists of material balance equations of the pivotal variables, including cell mass, nutrients, and metabolites (glucose and lactate equations). Embedded in the system of equations are quantitative descriptions of the relationship between the specific rate of growth, nutrient consumption, metabolite production, and nutrient and metabolite concentrations, or other factors affecting those rates. Once those relationships and the appropriate initial conditions (i.e., the initial concentrations of cell mass, glucose, lactate, and product) are specified, the system of equations can be solved and the behavior of the culture simulated.

The key to developing a model of a cell culture system is thus the description of the relationships between growth and metabolism. Approaches largely fall into two types: one employs empirical relationships determined by fitting experimental data to the model, and the other uses a mechanism-based metabolism model but relies on empirical modeling to describe the relationship between the growth rate and the environmental variables.

Empirical Models

Our current understanding of growth rate control does not permit us to develop a mechanistic model for cell growth that is responsive to environmental chemical and physical conditions. A Monod model or a modified Monod model is commonly used to tie the balance equations of cell and substrate together. The model typically describes substrate consumption as being related to growth rate empirically, for example, by using a proportional relationship. Such a model, when used within the range of experimental observation, can describe cell growth and the substrate profile. One may extend the model to describe the effect of inhibitory metabolites and osmolality on the growth rate. An empirical model is useful in simulating culture performance under conditions that do not deviate hugely from those previously observed, but its predictive value is limited.

Hybrid System of Mechanistic/Empirical Models

In the past two decades, scientists have developed extensive understanding of the kinetic behaviors of nearly all the enzymes involved in energy metabolism. Additionally, our knowledge of the regulation of those enzymes at transcription, translation, and activity levels is rapidly expanding. It is now possible to develop a kinetic model that describes a cell's energy metabolism. Material balance equations for reactions intermediate of energy metabolism, complete with enzymes' kinetic expressions, can be established. The uncertainty in estimating the level of each enzyme involved can be dealt with by using the transcriptome data as a first-order estimation followed by a parameter estimation using experimental data. A mechanistic model was used to describe metabolic shift from a high glycolytic flux state to a low flux state, a phenomenon frequently seen in fed-batch culture and which is tied to a high productivity in recombinant protein production.[1] Such an intracellular metabolic model can be linked to an extracellular model that describes the concentrations of cells, glucose, and lactate as described above. The linkage is mediated by the transporters for glucose, lactate, etc. Such an integrated intracellular and reactor model invariably deals with an intracellular model of metabolism in a fast time scale and a reactor model that is on a much slower time scale (Figure 5.4). The integrated model, hybrid model

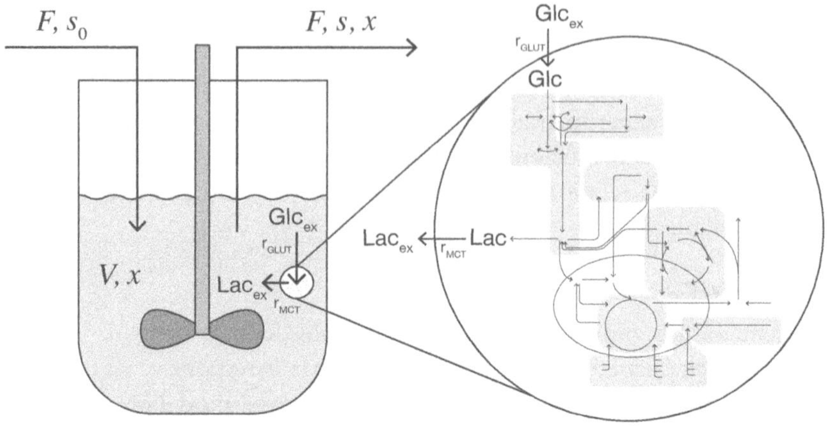

Figure 5.4. Depiction of a hybrid multiscale model for cell growth in culture.

of empirical growth, and mechanistic metabolic model will be multiscale, dealing with both the fast time scales of the metabolic reactions and the slow time scales related to the growth and change of chemical environments in the reactor. A multiscale hybrid model has been applied to explore the multiple steady state behavior in continuous culture.[2]

Concluding Remarks

Stoichiometric and kinetic relationships among process variables are important in describing the behavior of cell culture processes. They provide a framework for quantitative comparison of different processes. Empirical growth kinetic models allow for simulation of a culture's kinetic behavior under different growth conditions. Advances in metabolic science have enabled the development of mechanistic metabolic models. A hybrid multiscale model integrating empirical growth kinetic descriptions and a mechanistic metabolic computation framework will enhance our ability to optimize the cell culture process through *in silico* experiments. It is a goal that is attainable and will gain more momentum in the coming years.

References

1. Mulukutla, B.C., et al., *Bistability in glycolysis pathway as a physiological switch in energy metabolism.* PLoS ONE [Electronic Resource], 2014. **9**(6): p. e98756.

2. Yongky, A., et al., *Mechanism for multiplicity of steady states with distinct cell concentration in continuous culture of mammalian cells.* Biotechnology & Bioengineering, 2015. **112**(7): p. 1437–45.

Nomenclature

$\alpha_{s1/s2}$	Stoichiometric coefficient of substrates s_1 and s_2
α, β	Growth associated (α) and non-growth associated (β) constant of specific productivity
λ	Specific cell lysis rate
μ	Specific growth rate
μ_d	Specific death rate
A	Metabolite concentration
G	Glucose concentration
K_s	Half-saturation constant in Monod model
L	Lactate concentration
p	Product concentration
q	Specific rate of substrate consumption or product formation
s	Substrate (nutrient) concentration
t	Time
t_d	Doubling time
x	Viable cell concentration
x_d	Dead cells
Y	Yield coefficient
y	Cell density on growth surface

Contents

6

Cell Line Development

Introduction

MAMMALIAN CELL LINES are used to produce therapeutic and diagnostic proteins as well as viruses for gene therapy and vaccines. In all applications, the cell line used for the production has a large effect on productivity and product quality. Viruses have diverse tissue tropism and host species range. The infectivity of a virus is often affected by the tissue and species that the host cell is derived from. Virus infection elicits various responses from host cells that suppress its replication (Panel 6.1). The selection of a host cell with temperate responses to virus infection is thus important for high virus productivity. Most virus products do not undergo the extensive purification seen in the production of protein therapeutics. Their recovery often involves only the capture and concentration of the virus, the removal of major contaminants, and buffer exchange. Extensive purification causes virus titer loss. Selecting a host cell with a minimum of contaminants can ease the demand on downstream processing. Because of the possibility of incorporation of genetic elements from the host cell into the viral genome, the use of tumorigenic host cells for virus production has traditionally been shunned.

Panel 6.1. Factors Affecting Host Cell Selection for Industrial Applications

Viruses (Vaccines, Gene Therapy)

- Tropism of virus for tissues and species
- Virus productivity
  - Host cell response
- Host cell response to virus infection
- Genetic stability (tumorigenicity)
- Processability
- Regulatory acceptability

Heterologous Gene Products

- Ease of genetic manipulation
- Genetic stability
- Secretory plasticity
- Processability
- Regulatory acceptability (endogenous retrovirus, cell history)

Selecting the right host cell line or cell strain for virus production is thus of paramount importance.

For the manufacturing of most viruses, once a host cell line is selected the development effort focuses on the process rather than on cell line development. In some cases, the host cell line is engineered to endow it with some "helper" functions for virus replication. However, with the increasingly easy access to genome engineering, the use of cellular engineering to alter host cell response or even a synthetic biology approach to engineer virus production systems will likely be exploited.

For the biomanufacturing of therapeutic proteins, subunit vaccines, and virus-like particles (VLPs), the production cell line is among the most important factors affecting productivity and product quality. Having a cell line with the ability to synthesize a high-quality product at a sufficiently high specific productivity is the prerequisite to a high productivity process. For the production of recombinant proteins, the factors affecting the selection of the production cell line are different from those for virus production (Panel 6.1). This chapter will focus on the development of cell lines for therapeutic proteins.

Transient Production vs. Stable Expression Cell Line

Transient Expression

RECOMBINANT PROTEINS are produced in mammalian cells through a transient expression of the gene of interest (GOI), or by establishing a cell line that produces the GOI product permanently (Panels 6.2–6.4). Transient expression is frequently used for the generation of small quantities of proteins (up to a few grams) for protein characterization or toxicity testing. First, cells are transfected with a high concentration of plasmids that include the GOI. A large number of plasmids are taken up by cells into the cytoplasm. A fraction of those plasmids then enter the nucleus, where the GOI is transcribed into mRNA. The mRNA molecules are exported back to the cytosol, translated into proteins, and secreted into the medium. Due to the large number of plasmids transfected and the strong promoter used, a large amount of product

Panel 6.2.
Protein Production in Mammalian Cells

Transient Expression

- The transgene is encoded in a plasmid or non-integrating virus vector
- The DNA-based vector must enter the nucleus for transfection, after which the mRNA of the transgene is translated in the cytoplasm
- The vector is not integrated into the genome; product synthesis stops after the vector is degraded or diluted due to cell division
- Useful for rapid production of research quantities of proteins

Stable Cell Line Expression

- The heterologous gene is integrated in the chromosome and expressed
- The heterologous gene can stay in the genome and be expressed stably for a large number of generations

Panel 6.3. Host Cells Frequently Used for Transient Expression

- HEK293 (human embryonic kidney fibroblasts)
- COS (from green monkey kidney cells)
- BHK cells
- CHO cells

Panel 6.4. Host Cells for Stable Expression of Heterologous Proteins

- CHO, NS0, SP2/0, HEK293, BHK, PER.C6
- Different host cell lines have different post-translational capabilities
- Different mammalian host cells produce somewhat different glycan structures
  - CHO and NS0 have different sialylated glycans (NANA and NGNA, respectively)

is produced. The plasmids in the cell are diluted as the cells divide and degraded over time. Thus, the transient production of the protein is usually limited to a few days. Nevertheless, the process is easily carried out (requiring only plasmid preparation) and is frequently used to produce a small quantity of proteins for characterization and testing.

Stable Expression

Stable expression involves the integration of the GOI into the genome of the host cell and the isolation of a cell line that has the capability of expressing and producing the GOI over a long period of time. Establishing a stable cell line for a given product has been the core of cell culture-based biomanufacturing. The cell line must be capable of producing product of the same quality in different batches, and in different locations, over the years. Once a production line is selected, cells are expanded to establish master and working cell banks. The stockpiled cells, typically stored in liquid nitrogen, are used for manufacturing purposes throughout the life cycle of the product.

Host Cells

An ideal production cell line has a high productivity, is able to synthesize the product to the desired quality, and is genetically stable for consistent product synthesis over the product's life cycle. Many factors affect the ease or difficulty of establishing a production cell line, including the choice of host cell, the vector carrying the GOI, and the method of inserting the GOI into the genome of the host cell. Of all the factors, the most important is the selection of the host cell.

Factors Affecting Host Cell Choice

In the early days of recombinant therapeutic protein technology, many studies explored the possibility of using various cell lines of rodent, monkey, and human origins for production, including mouse C127 and myeloma (NS0), Baby Hamster Kidney (BHK), and Human Embryonic

Kidney (HEK293) (Panel 6.4). It did not take long for CHO cells to dominate the scene.

When developing a production cell line, the host cell line is transformed from cells producing low levels of secreted proteins to highly secretory cells. Many different cell lines can be transfected with the GOI and made to produce the protein. However, only a few are able to have their genetic circuitry reprogrammed to expand their protein folding and other processing and secretory capabilities to the level necessary to become a hyperproducer. Different host cells have different degrees of "plasticity" that enable them to undergo such a transformation. The extraordinary capability of CHO cell lines to undergo the still poorly characterized transformation process is a major factor that made them the host cell line of choice for stable expression of recombinant proteins.

Different host cell lines have different capabilities and specialized post-translational modifications, such as gamma-carboxylation, phosphorylation, and lipidation. Cell lines derived from different species or different tissues can have different compositions of glycosyltransferases that catalyze different glycosylation reactions, or have different isoforms with different substrate specificity. Glycans on proteins produced in CHO, NS0, and other cell lines are somewhat different from each other because of their different sialylation.[1] Murine cells like NS0 and SP2/0 express galactose-α1,3-galactosyl transferase and can potentially add α-galactose to the glycan. As discussed in Chapter 3, non-human mammalian cells produce N-glycolylneuraminic acid (Neu5Gc/NGNA) that is absent in human glycan. Human cells express only Neu5Ac/NANA (N-acetylneuraminic acid). CHO cells express only α2,3-sialyltransferase thus do not synthesize the α2,6-sialic acid that is seen in human glycans.

Nowadays, most new therapeutic protein products are produced in CHO cells. The CHO cell line was derived from an inbred strain of Chinese hamster.[2,3] Many sublines and mutants were subsequently derived.[4] The early passing-on of those sublines was not well documented. Among the CHO cell lines commonly used in industry, DXB11 and CHO-K1 were derived from the same ancestor line, while CHO-S and DG44 share the same lineage.[4] It is common practice for a company to select a CHO cell line as the host cell platform for virtually all future products.

The choice of host cell also determines the possible host cell components that may contaminate the product. Non-human host cell proteins that are released into the medium upon cell lysis may become residual contaminants in the final product and elicit an immunogenic response upon administration to the patient. In the manufacturing process, the contaminating host cell proteins are removed and reduced to an acceptable level in downstream processing.

Most mammalian genomes harbor sequences of endogenous retroviruses. Sequences of endogenous retroviruses constitute about 1% of the human genome. Many of these sequences are heavily mutated and are inactive. However, some viral proteins are expressed, and viral particles or even retroviruses can be produced from some host cells. The release of these viral particles by a host cell may contaminate the product and pose a safety concern. The level of viral particles released varies among host cell species and cell lines. Murine cell lines tend to produce high levels of viral particles. Both host cell proteins and viral particle contamination issues are mitigated through the manufacturing process. Nevertheless, the issues are factors that must be considered in host cell selection.

A human host cell line has the advantage of synthesizing "human form" glycans and eliminating the problem of non-human host cell protein contamination. The issue of human retrovirus particle inactivation and removal can be mitigated with the unit operation steps already implemented in contemporary product purification. However, in the past three decades the host cell line employed for biologic manufacturing has converged on CHO cells, rather than on one of the human cell lines established as a producer. CHO cells have a high level of genetic plasticity and can be readily reprogrammed into high protein secretory cells. They are also relatively easy to "adapt" to different culture conditions.

Host Cell Adaptation and Engineering

Many cell lines commonly used for recombinant protein production, including CHO, BHK, and HEK293 cells, are derived from adherent cells that proliferate by attaching to a compatible surface. They can be adapted to acquire the ability to grow in suspension (Panel 6.5). The adaptation to suspension growth is often achieved by eliminating the surface for attachment and waiting for adapted cells to grow. This is sometimes aided by adding a dispersing agent like heparan sulfate or dextran sulfate to the culture medium to reduce cell aggregation. After a couple of weeks, cell growth gradually resumes and cell concentration increases to allow for subculturing.

In addition to suspension growth, host cells can also be adapted to other culture conditions favorable for manufacturing. For example, the requirement of complex lipid additives and growth factors may be reduced or even eliminated by gradually forcing cells to grow under decreasing concentrations of the additives. The molecular mechanism of cell adaptation is not well understood. One case that is known is the adaption of NS0 cells to become cholesterol-independent. In the auxotrophic NS0 cells, a gene in the cholesterol biosynthetic pathway is silenced due to extensive methylation in the CpG island region of its promoter.[5] The adapted cells have a reduced methylation in the promoter region to allow for the expression of the gene, thus alleviating the cholesterol dependency. Whether the adaptation to other culture conditions is similarly affected by epigenetic changes is not known.

Many traits are highly desirable in a producing cell line, including favorable metabolic characteristics, enhanced protein secretion capacity, and the capability to synthesize specific glycan features. However, many of these traits are highly complex and are most likely under the influence of an array of genes instead of a single gene. For example, the capability of growth in suspension likely involves alterations in the expression of a plethora of genes in the signaling pathway of growth control and in the organization of the cytoskeleton. Nevertheless, many desirable traits, such as the cholesterol independence described above, are under the control of a single gene or a small number of genes that are amenable to genetic manipulation.

Panel 6.5. Alteration of Host Cell Lines for Protein Production

Host Cell Adaptation to Acquire the Capability of:

- Suspension growth
- Serum-free, defined medium growth
- Altered nutritional requirements
  - e.g., glutamine- or cholesterol-free growth
- The genetic/epigenetic changes incurred in adaptation are likely complex and are not well understood

Host Engineering

- Effective when the gene-trait relationship is known
- Used to achieve:
  - Favorable metabolic characteristics
  - Apoptosis resistance
  - Improved glycosylation or other post-translational modifications

Increasingly, the host cell line is being genetically altered to harbor favorable traits (Panel 6.5).[6] Host cells can be engineered to overexpress anti-apoptotic genes that suppress apoptosis in the late stage of fed-batch culture and prolong the production phase. Cellular metabolism may be engineered to reduce waste metabolite formation and to facilitate the switch to a lactate-consumption type of metabolism. Host cells can be engineered to alter their glycosylation enzyme expression profile.[7] Expression of heterologous 2,6-sialyl transferase allows for the addition of sialic acid through the 2,6-glycosidic linkage that is absent in the glycan synthesized by CHO cells.[8] Knock-out of endogenous fucosyl transferase eliminates the fucose residue in the glycan-attached Asn_{297} of IgG heavy chain.[9] The afucosylated form of IgG has been reported to enhance antibody-dependent cellular cytotoxicity.[9]

Generating a Recombinant Protein High-Producing Cell Line

The first step to constructing a producing cell line from a host cell line is establishing a vector that contains the expression unit of the GOI coding for the product (Panel 6.6). The sequence encoding the GOI, often along with the vector sequence, is then inserted into the genome of the host cells so that a production cell line can be established (Figure 6.1). Some methods of gene insertion insert a defined

Panel 6.6. Steps in Cell Line Generation

- Vector/GOI transfection and integration into host cell genome
- Selection
- Amplification (may be omitted)
- Single-cell cloning
- Screening/productivity assessment/isolation of clones
- Cell stability evaluation
- Cell/product characterization

DNA sequence into either a predefined locus or random locations in the host cell genome. Most commonly, multiple copies of the vector, often in somewhat modified vector structure, are inserted into random locations in the genome. However, targeting a defined sequence into a prescribed site is increasingly being explored for creating a production cell line.

Typically, the vector is propagated in *E. coli* to generate sufficient quantities for transfection into the host cell. The vector typically carries a selectable marker gene so that the cells that have acquired the vector can be enriched after transfection using a selection agent (Figure 6.1). *In vitro* screening of post-selection cells is then performed to screen for clones in this population that produce the product at high levels.

Until recently, it was common practice to amplify the copy number of the integrated vector in the isolated clones by subjecting them to a high concentration of selection agent. This kills the vast majority of cells, sparing only those with multiple copies of the vector; some of the amplified copies will have both the selectable marker gene as well as the GOI. Screening of the amplified cell population is then carried out to identify the candidate clones that produce the highest levels of product.

After vector integration and amplification, the cell population is rather heterogeneous. Different cells have different genomic loci of vector integration. They may also have different regions of their genome that have been rearranged, amplified, or mutated. Their growth behavior may also be different. Such a genetically heterogeneous cell population

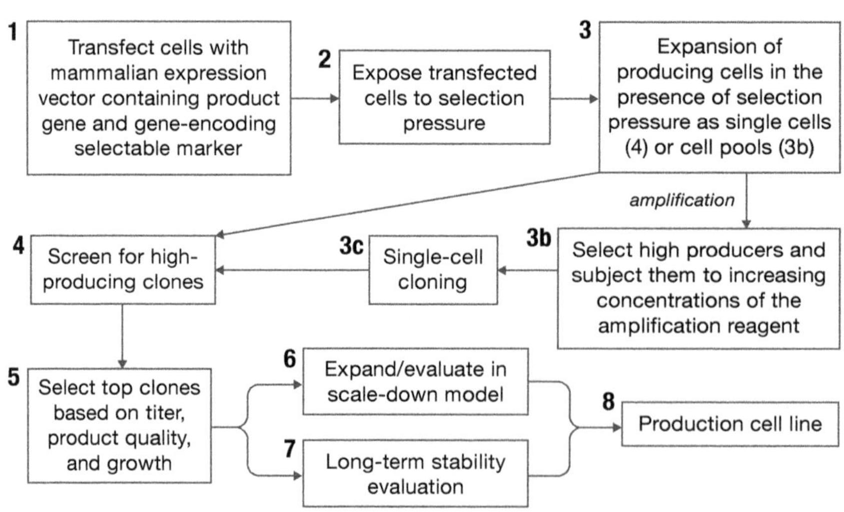

Figure 6.1. Typical steps in developing a high-producing cell line.

is called a "cell pool." Subsequently, single-cell cloning is performed by sorting single cells into culture wells using flow cytometry or by culturing each individual cell into separated colonies in a plate. After single-cell cloning, the productivity of each clone is assessed. Those with high productivity are isolated and further expanded for growth characterization and product quality assessment for final production cell line selection. In some cases, the producing cells are then adapted to growth conditions more amenable to manufacturing conditions.

Vectors

A number of different types of vectors are used to insert a GOI into the host cell genome. Most commonly used is a plasmid vector that contains a complete GOI expression unit (Figure 6.2, Panel 6.7). The vector typically includes two sets of sequences: one for expressing the GOI and for selection of the host cell, and one consisting of bacterial sequences that enable the plasmid to replicate in *E. coli* so that a large quantity of plasmids can be obtained. The bacterial elements include a bacterial selectable marker for plasmid-carrying *E. coli* and the bacterial origin of replication for propagation in *E. coli*. These sequences, while essential in plasmid propagation, are not needed for GOI expression in the host cell.

Often, the plasmid is cut with a restriction enzyme to "linearize" the plasmid at a site that is not essential for the expression of the GOI before its delivery to the host cell. One may also use a specialized plasmid replication system that makes "minicircles," thus eliminating most of the bacterial vector sequences from the final minicircle. The minicircle plasmid has a smaller size than the complete plasmid and gives a better efficiency in transfection.

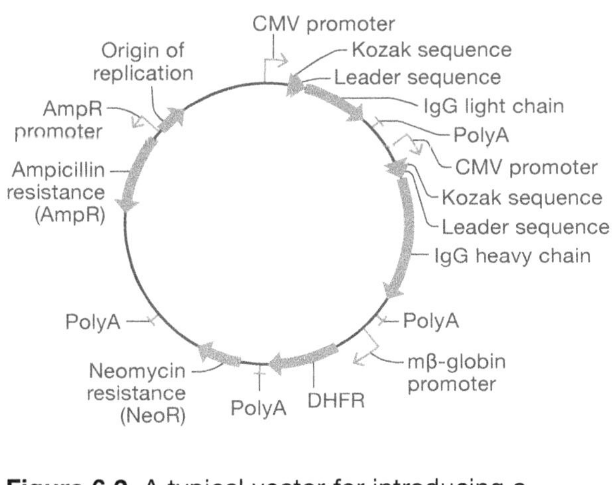

Figure 6.2. A typical vector for introducing a transgene into a host cell for recombinant protein production.

Panel 6.7.
Elements in a Plasmid Vector

- Promoter
- Coding sequence of the gene of interest (GOI)
- PolyA signal
- May have integration signal/ homology arms to facilitate targeted integration
- Selectable marker/ amplification marker
- *E. coli* plasmid element (origin of replication, selectable marker)

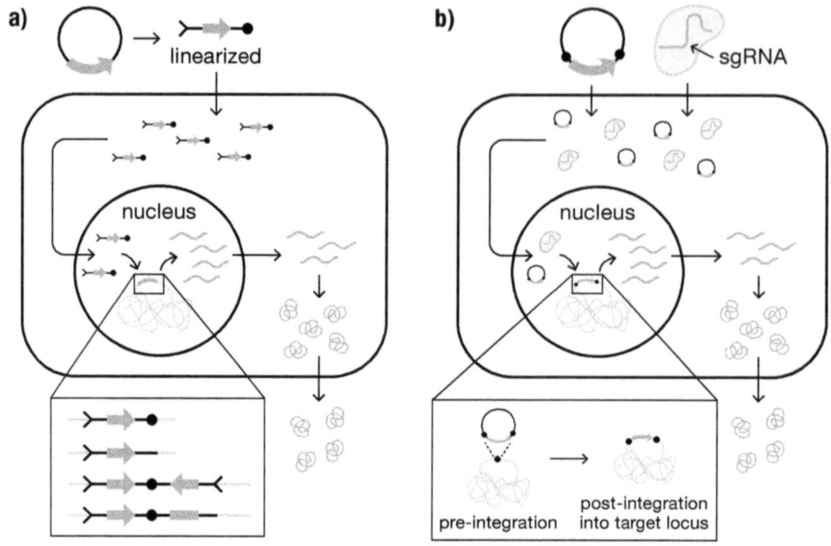

Figure 6.3. Integration of the GOI into the host cell genome. (a) Random integration using plasmids. (b) Targeted integration. In random integration, the GOI may be structurally modified before integration.

The plasmid is then transfected into the host cell. The plasmid is often subjected to intracellular degradation, and the truncated or intact plasmid may be ligated in various orientations. Through random events during host DNA replication and repair, the vector is integrated into the host genome without additional elements provided exogenously. The integrated sequence is thus highly variable in terms of its length and vector-host genome junction, and so is the locus on the host cell genome that the vector is integrated into (Figure 6.3a).

Increasingly, the GOI in a vector is directed to a specific locus and integrated as a defined sequence (Figure 6.3b). This is usually done by using DNA integration methods like CRISPR, transposon, and recombinase-mediated cassette exchange (RMCE) (such as LoxP or Flp systems). These methods all involve guiding the GOI to the locus using matching sequences that are homologous, or complementary (inset in Figure 6.3b). For CRISPR, the destination is a locus in the genome as defined by the DNA sequence (Figure 6.4a). A guide RNA directs Cas9 nuclease to the locus to cut the DNA. Through the cell's DNA repair system, the DNA segment containing the GOI is then inserted into the cut site. Next, the GOI is integrated into the genome at the target site. For LoxP and Flp, the destination is a pre-inserted LoxP or Flp sequence in the genome (Figure 6.4b). In the case of transposon-mediated gene insertion, the sequence marking the insertion loci is shorter and less defined. The integration is often numerous in the genome (Figure 6.4c). In all these methods, an enzyme (Cas9 for CRISPR, recombinase for LoxP/Flp, and transposase for transposon) is provided in the form of a plasmid component,

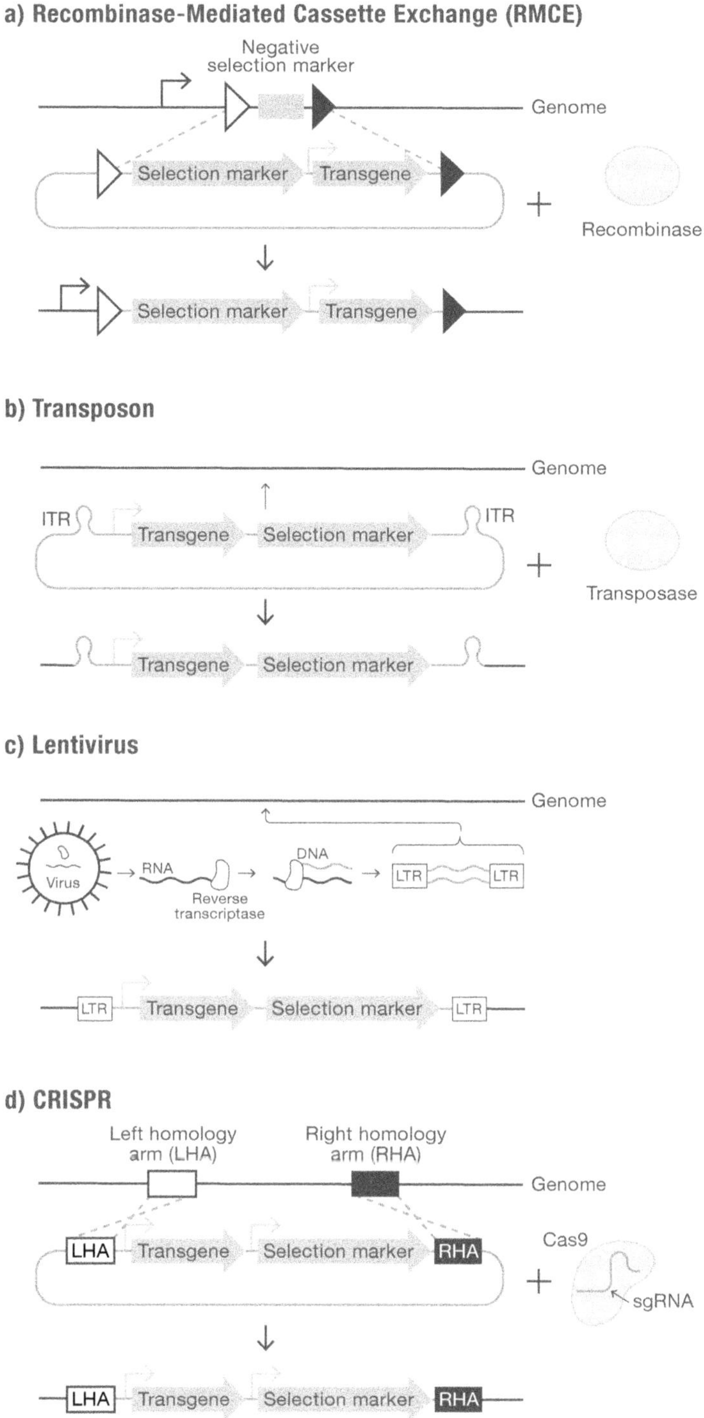

Figure 6.4. Schematic of GOI integration by (a) recombinase-mediated cassette exchange (RMCE), (b) transposase, (c) lentiviral, and (d) CRISPR systems. In CRISPR and RMCE, the GOI integrates into a defined locus. In transposon and lentiviral vectors, the integration is random.

mRNA, or protein, and co-transfected with the GOI and other components to facilitate the cutting of the genomic DNA and the integration of the insertion sequence.

Viral vectors are also used to deliver the GOI to host cells. Viruses enter the cell by binding to specific cell surface receptors, and then integrate into a chromosome to generate cells expressing the GOI. Using a lentivirus, the number of vectors needed for an infection is rather small compared to that for a plasmid vector. Because of the traditional safety concerns, viral vectors have not been used extensively for introducing GOIs for recombinant protein production until recent years. Lentivirus has been used as an *ex vivo* gene delivery system to cells for gene therapy. The viruses are modified to be devoid of viral protein sequences in the genome and are replication deficient. In place of viral genes, the lentiviral vector carries a payload that includes the GOI (Figure 6.4d). Upon integration into the host cell genome, the GOI is inserted along with the viral vector. Lentiviral vectors have a propensity for integrating into actively transcribed regions of the host cell's genome. They insert a defined DNA segment consisting of the virus genome into the host cell genome. Viral vectors have a very high efficiency of infection and integration. Using low multiplicity of infection (MOI), it is possible to isolate cells that have a very small number or even just one copy of the virus integrated into the host cell genome, and which have a defined viral genome sequence. Retroviral vectors are also frequently used in gene therapy; however, they do not transduce CHO cells as efficiently as lentivirals do. Both lentiviruses and retroviruses are RNA viruses, and are reverse transcribed to double-stranded DNA for replication upon infecting host cells. Reverse transcription has a higher error frequency than DNA replication. Thus, sequence verification after cell lines are isolated is necessary, just as in the case of other vectors.

Promoters and other regulatory elements

Many commonly used promoters that drive the transcription of the GOI are derived from mouse or human viruses, for example the CMV (cytomegalovirus), SV40, and HSV (herpes simplex virus) thymidine kinase (TK) promoters (Panel 6.8). Promoters for highly expressed genes from the host cell have also been used. Additionally, synthetic promoters have been isolated with the aim of increasing transcript levels.[10] The promoters used for driving GOI expression are mostly constitutive and are expressed all the time.

The promoter used for product protein expression is also subjected to epigenetic modification of the host cells. Silencing of the promoter will affect the expression level of the GOI and therefore productivity. For example, the methylation of CpG dinucleotides in the CMV promoter/enhancer in a recombinant CHO cell line can cause a decrease in productivity.[11]

Panel 6.8. Driving Transcription of a GOI

Promoters for Transgene Expression

- Mostly constitutive promoters, virus origin
  - SV40, CMV, thymidine kinase (TK)
- Strong promoters from host cells
- Synthetic promoters
- Sometimes used with an enhancer, flanking control regions

Epigenetic Effects on Promoter Expression

- Some promoters are subject to CpG silencing
- Transcription affected by local context of integration site
- The position effect of heterochromatin may be minimized by using insulators

In addition to the strength of the promoter, other elements affect the transcript level of the GOI. The human CMV, in the presence of intron A, has been shown to enhance the protein expression levels of the GOI.[12] By including the 5′ and 3′ flanking control regions of EF-1α, the expression of the GOI driven by CMV or the human EF-1α promoter in CHO cells can be increased.[13] Enhancers are DNA sequences that increase gene transcription and may not necessarily be located close to the promoter sequence. In some cases, a promoter such as CMV is used in conjunction with its associated enhancer sequences in the vector construct.

The expression of the inserted GOI is affected not only by the promoter and enhancer used, but also by the regional context of the genome near the integration locus. Some regions on the genome may be unfavorable to transcription. For example, the presence of CpG islands in close vicinity to a heterochromatin region may negatively affect gene expression. To minimize the probability that a randomly integrated GOI falls under the influence of negative position effect, a cis-acting barrier element called an insulator can be placed in front of the promoter of the GOI.

GOI-coding sequence and leader sequence

The cDNA of a GOI isolated from a source organism may not be the most efficient for expression in a host cell from a different species. The coding sequence of the GOI is often modified to enhance the efficiency and fidelity of its translation (Panel 6.9). The codon usage can be changed according to the abundance profile of tRNAs in the host cell. Codons may also be changed to remove cryptic splice sites, optimize GC content, and remove direct repeats.

Secreted proteins have a leader sequence for translocation into the endoplasmic reticulum during translation. The leader sequence differs in different proteins and species. A strong signal sequence, such as one derived from human albumin, may give a higher productivity. However, the effectiveness of a signal sequence may also be influenced by the downstream protein sequence. Furthermore, the leader sequence needs to be cleaved upon the entry of the nascent protein molecule into the ER. Incomplete or nonspecific post-translational cleavage of the leader

Panel 6.9. Expression Unit of a GOI

- Coding sequence is often codon-optimized
- A leader sequence for efficient secretion (e.g., human albumin leader)
  - The leader sequence selection affects secretion and cleavage
- 5′ UTR and 3′ UTR (untranslated region) influence transcript stability, transcription termination

sequence can result in heterogeneity in the mature protein sequence at the N terminus. The efficiency of cleavage of the leader signal is thus an additional consideration in its selection.

The 3′ UTR, i.e., the region between the stop codon and the polyA tail on the mRNA, is important for proper termination of transcription and also plays a role in mRNA stability. It contains the signal for the addition of polyA to the transcript that allows for its transport from the nucleus to the cytoplasm. Commonly used polyA signals for recombinant protein production include SV40 and BGH (bovine growth hormone) polyA.

Selectable marker and reporter

Upon the introduction of the plasmid vector into the host cells, some cells will receive the vector. A portion of those vectors will enter the nucleus and express the encoded genes. Among the genes encoded in the plasmid is a selectable marker such as antibiotic resistance. The expression of the selectable marker allows the cell to survive in the presence of the selection reagent. Eventually, some plasmids that have entered the nucleus will be integrated into the genome and become a permanent part of the genome. Plasmids are not capable of self-replication, and free plasmids in the cell are gradually degraded. Only those cells that express the selectable marker will survive under selective conditions. Those that survive and grow under selection over time therefore have the vector, including the selectable marker gene, integrated into the genome. Many of the integrated vectors will also contain the GOI. The selectable marker gene and the selective condition, typically the presence or absence of the selective agent, are thus used in pairs. Some commonly used selectable markers and their mode of action in mammalian cells are listed in Table 6.1.

Selectable markers fall into two categories: dominant and recessive (Panel 6.10). A recessive marker resides in cells with a particular genetic background and causes a growth deficiency. The introduction of a compensatory gene leads to overcoming the deficiency. For example, cells without a functional dihydrofolate reductase (DHFR) gene will require the additional supplementation of thymidine and glycine in the culture medium for growth. The introduction of a functional DHFR enables them to grow without the supplements. Therefore, after the transfection of a plasmid containing DHFR, only the transfectants that express the newly acquired DHFR will grow in the absence of thymidine/glycine. Similarly, thymidine kinase (TK)-defective mutants require thymidine to be included to the culture medium. The introduction of a functional TK

Table 6.1. Commonly Used Antibiotics for a Selection of Stably Transfected Mammalian Cells

Antibiotic	*Mode of selection*	*Resistance gene*	*Mode of resistance*
Geneticin (G418)	Blocks protein synthesis by inhibiting elongation step	Neomycin phosphotransferase (npt)	Phosphorylation of Geneticin
Hygromycin B	Inhibits protein synthesis by disrupting translocation and promoting mistranslation	Hygromycin phosphotransferase (hpt)	Phosphorylation of Hygromycin B
Puromycin	Blocks protein synthesis by causing premature chain termination	Puromycin N-acetyltransferase (pac)	Acetylation of Puromycin
Blasticidin S	Inhibits protein synthesis by interfering with peptide bond formation	Blasticidin S deaminase (bsr)	Deamination of Blasticidin S
Zeocin (Bleomycin)	Intercalates into and cleaves DNA	Bleomycin resistance protein (ble)	Binds stoichiometrically and prevents Zeocin from binding to DNA

gene allows for cell growth in the absence of thymidine. Conversely, the presence of a dominant selectable marker is protective to the cell under selective conditions. By introducing the selectable gene to the cell, the cell is endowed with a resistance to a given lethal condition.

A resistance gene typically encodes an enzyme that modifies the selective chemical agent to destroy its activity. The phosphorylation and acetylation reactions employed for inactivating antibiotics require intracellular reactants (ATP, acetyl group donor). Those enzymes are therefore only effective in destroying the selective agents intracellularly. The hydrolysis enzyme, on the other hand, may be active even when released into the medium after cell lysis. In the selection process, the concentration of the selective chemical agent decreases with time and the rate of decrease is dependent on the type of resistance and the concentration of transfected cells. Thus, the optimal concentration for the selection for each agent is not only dependent on cell line but also on cell concentration. Clonal selection and population selection may have rather different optimal concentrations of the selective agent.

Panel 6.10. Selective Markers for Transgene Expression

Recessive Selection

- DHFR (dihydrofolate reductase)
  - Effective on DHFR-deficient background
- TK (thymidine kinase)
  - Effective on a TK-deficient background

Dominant Selection

- Antibiotic resistance
  - Neomycin, hygromycin, blasticidin
- Use a mutated weaker selectable marker and a weak promoter to select the integrant into a transcriptionally active site

Another class of selective agents interferes with the uptake of a toxic selective agent. The multidrug resistance gene (MDR) confers cells with resistance by increasing their ability to pump toxic substances, such as colchicine, out of cells. Its overexpression allows for selection from a background of cells not expressing MDR.

The expression of the selectable marker is usually driven by a weaker promoter than the promoter driving the expression of the protein of interest. Sometimes a mutated selectable marker with a lower activity to rescue the missing enzyme function is used. With such a "weakened" expression of the selectable marker, the integration site must be transcriptionally active and allow for a higher transcript level of the selectable marker in order to counter the selection pressure. This may lead to increased expression of the protein of interest.

Not all cells that have received transfected plasmids and express the selectable marker protein will also express the GOI. The frequency of the false positives (those expressing the selectable marker but not the protein of interest) varies with the selectable marker used and the vector design. Some false positives may arise through failure to co-integrate the GOI, or through mutations that develop alternate mechanisms to neutralize the effect of the selectable marker. Thus, choice of the selectable marker can be important for reducing false positive clones.

The selection of transfectants that harbor the GOI is performed at single-cell cloning stage to isolate clones, as well as at a population level to generate the transfected cell pool. The isolation of GOI-transfected cells may also be performed without using a selectable marker and selective pressure. Instead of a selectable marker gene, a reporter gene may be used in the vector. The expression of the reporter protein gives the transfected cell fluorescence upon excitation with light of the appropriate wavelength. The transfected cells can be sorted using fluorescence-activated cell sorting (FACS) into single cells for single-cell cloning or into a pool. Commonly used reporter proteins include GFP (green fluorescent protein), mCherry, tdTomato, and RFP (red fluorescent protein). Such reporter-gene-based transfectant isolation is frequently used in research, but less often for industrial cell lines.

Delivery of Foreign DNA and Its Integration into the Host Cell Genome

Successful transfection of a plasmid vector requires that the host cell membrane first become permeable or else that a mechanism is in place for the plasmid uptake. Calcium phosphate precipitation, cationic polymers, and liposomes of the DNA molecules are commonly used to deliver plasmids to the cytoplasm of the host cell (Panel 6.11). The plasmids are first prepared as particles or encapsulated into lipid vesicles. Then, through particle interactions with the cellular membrane and an endocytosis-like mechanism, plasmids are taken up by cells. Electroporation is another frequently used method. It delivers the DNA molecules to the cytoplasm through the openings in the membrane generated by an electric field. All of these methods use a very high plasmid-to-cell ratio, but only a moderate DNA concentration. While thousands of the plasmid molecules may enter each cell, only a small portion translocate to the nucleus where they can be transcribed to express the gene(s) encoded in the plasmid, including the selectable marker gene.

Typically, the transfection efficiency is evaluated by the expression of the resistance marker gene or another reporter gene. This is in turn dependent on the plasmids' entry into the nucleus and subsequent expression. Because of this, the efficiency rarely approaches 100%.

Panel 6.11. Methods of Gene Transfer

Lipofection/Lipid-Mediated Gene Transfer

- A mixture of DNA with an amphipathic compound (DOTMA, DOPE, etc.) that simultaneously interacts with DNA and hydrophobic portions of the membrane, allowing passage of DNA into the cell

Electroporation

- Exposes cells to a high-voltage electro-pulse in the presence of the DNA solution
- Introduces pores in the plasma membrane, allowing entry of DNA
- Duration of pulse and strength of the electric field varies with cell type

DNA-Calcium Phosphate Co-Precipitation

- DNA and calcium chloride are added dropwise into a HEPES buffer with sodium phosphate (1 mM)
- A fine precipitate forms in 5–30 min, and is added directly to the cells
- Cells take up DNA-containing particles via an endocytosis-type mechanism

The plasmid, in its circular and supercoil form, may be used for transfection directly. Often, it is first linearized using a restriction enzyme whose recognition sequence is outside of the GOI and selectable marker. Linearized plasmid DNA is thought to result in a higher efficiency of genome integration. While some transfected cells may have a single copy of the integrated vector carrying the GOI, others may have multiple copies that integrated into different loci upon transfection. Furthermore, the vector integrated into the host cell genome may not be intact, but rather be truncated or fused by multiple copies (Figure 6.1).

Gene Amplification and Integration of the GOI

For three decades, an effective and commonly practiced step in cell line development has been to subject the vector-transfected cells to very high levels of selective pressure to increase the number of copies of the GOI in the host cell genome. The rationale is that more copies of the GOI give a high transcript level and a high productivity. Upon exposure to a very high level of selective pressure, most cells will die, leaving only those with many copies of the selectable marker gene. Since the GOI is nearby the selectable marker gene, it is highly probable that it will be co-amplified during the amplification and produce the product at high levels. Structurally, the vector sequence may be amplified alone, or it may be co-amplified with adjacent segments of the genome. The amplicons may be in tandem or combined in different orientations. Some amplified segments may be translocated to different loci nearby or far away in the same or different chromosome.

DHFR amplification system

Two gene amplification markers are commonly used in mammalian cells: DHFR and glutamine synthetase (GS). The DHFR gene as a selectable marker for gene amplification is used in conjunction with its chemical antagonist, methotrexate (MTX) (Figure 6.2, Panel 6.12). DHFR is an enzyme that catalyzes the conversion of folate to tetrahydrofolate, a compound required for the biosynthesis of glycine, thymidine monophosphate, and purine. MTX is a folate analogue that inhibits DHFR, thereby leading to cell death in the absence of thymidine and purine in the medium.

After exposure to a high concentration of MTX for a period of time, the surviving cells have multiple copies of DHFR. The size of amplicons surrounding DHFR varies widely, from 10 kbp to 10 Mbp. The GOI and DHFR genes are co-amplified. All copies of the GOI may not be transcriptionally active. Some may be structurally compromised due to DNA truncation, fusion, or other modifications. It is difficult to assess which of the intact copies of the GOI are transcriptionally more active. The amplification process can be performed as a single step of MTX exposure over one to two weeks, or in multiple step-wise increases in MTX concentration. Highly MTX-resistant cells may contain several thousand copies of DHFR.

DHFR-based gene amplification is more efficient in a DHFR-defective genetic background to minimize or eliminate the possibility that the resistance arises from the amplification of the endogenous DHFR without concurrent amplification of the GOI. CHO mutants (DSB-11, DG44) that are deficient in DHFR are commonly used for amplification-based cell line development. These DHFR-deficient cells require the addition of thymidine, glycine, and hypoxanthine to the medium. These cells do not grow in the absence of added nucleosides unless they acquire a functional DHFR gene.

GS amplification system

The GS selection system is based on the reaction that synthesizes glutamine from the substrates glutamate and ammonia. Most mammalian cell lines require glutamine to grow, since their endogenous GS activity is low (Panel 6.13). The GS expression vector contains a GS gene, along with the GOI, to allow for the selection by growth in glutamine-free cell culture media. GS is usually driven by a weaker promoter than the GOI.

Panel 6.12. DHFR Amplification System

- Use $DHFR^{-/-}$ (DG44) or $DHFR^{+/-}$ (DXB11) background, vector contains coding sequences of DHFR and the linked GOI
- DHFR catalyzes conversion of dihydrofolate to tetrahydrofolate for purine biosynthesis, and the reaction is inhibited by methotrexate (MTX)
- Treated with a high concentration of MTX, the surviving cells attain multiple copies of the DHFR gene and the GOI

Panel 6.13. GS Amplification System

- Cultured cells have very low glutamine synthetase activity and require glutamine for growth
- GS catalyzes conversion of glutamate/NH_3 to glutamine, can alleviate glutamine requirement
- The reaction is inhibited by methionine sulfoximine (MSX)
- After exposure to a high concentration of MSX, the surviving cells have multiple copies of the GS gene and GOI

By supplementing a high concentration of GS inhibitor methionine sulfoximine (MSX) in the medium, it is possible to select for transfectants that have multiple copies of GS.

Low multiplicity of GOI integration

In the development of B lymphocytes, only one copy of the IgG heavy gene is actively transcribed, while the copy in the other allele is inactivated so as to render it untranscribable, a phenomenon called allele exclusion. Yet, B cells continue to develop into high antibody-producing plasma cells. The phenomenon suggests that, with the right cellular machinery, a single copy of the GOI is sufficient for high protein expression and secretion. This was realized in a high-producing NS0 cell line that was engineered to have only one copy of IgG heavy chain and light chain genes.[14] Having multiple copies of the GOI certainly facilitates high transcription, but a high level of expression is achievable even with a single copy of the GOI.

A high degree of gene amplification through the application of a high concentration of selective pressure often gives rise to phenotypically and genotypically unstable producing cells. It is advantageous to accomplish the same high level of transcription using a lower degree of amplification. A number of methods aim to attain high expression levels with a low multiplicity of the GOI. DHFR-selectable markers are used in conjunction with an impaired neomycin resistance gene. After plasmid transfection and antibiotic G418 selection, only cells to whose genome the vector has integrated in a transcriptionally active region and that give a high level of neomycin resistant transcripts can survive. The method thus allows for the selection of cells that the GOI has integrated into in a very transcriptionally active region known as a "hot spot" region. The amplification is then performed at a lower level of the DHFR inhibitor MTX. Since the locus of integration is transcriptionally active, the high-expressing clones isolated have only a few copies of the GOI.

A low multiplicity of the GOI can also be obtained by omitting the amplification step. After transfection, a large number of plasmids enter the cell and only a small fraction of those integrate into the genome. The number of copies of the vector (and the GOI) integrated into a cell distributes over a range. By titrating the selective agent concentration, transfectants above different thresholds of multiplicity of the GOI are selectively enriched.

Integration site of the GOI

The locus of GOI integration influences how actively it is transcribed and how stable it is over time (Panel 6.14). The genome region of GOI integration may also affect the stability of the GOI structurally. The genome of aneuploid cells constantly acquires structural variants due to the reduced DNA replication fidelity. Some regions in the genome may be prone to breakage or gain or loss of copies. The data on genome structural variants for industrial cell lines, and even for research cell lines, is scarce, but can be readily acquired through genome sequencing.

Panel 6.14. GOI Integration Site

- Genome regions of integration may have varying structural stabilities
  - Unstable if near a telomere
  - Some regions may be prone to segmental deletion, insertion, or translocation
- Local context of genome may affect transcription
  - Vicinity to heterochromatin may encounter heterochromatin invasion and silencing
  - Integration in transcriptionally active, highly accessible, or super-enhancer regions may favor transcription

Some genome segments are transcriptionally more active than others, or are considered to be 'hot spots' because of their high accessibility to transcription factors. The genome regions that have an abundance of such highly accessible segments are considered to be super-enhancer regions. Although data is still lacking, it is possible that the insertion of the GOI into such a highly accessible region will facilitate high transcription of the GOI. Local context may negatively affect transcription. The genome consists of euchromatin regions and heterochromatin regions. In heterochromatin regions, DNA is densely packed, making it less accessible to transcription factors. Super-enhancer and heterochromatin regions are both lineage-specific and are inherited in cell division. However, variation does occur and the boundary of a heterochromatin region may extend, causing a decrease in transcription of nearby genes.

Hence, in targeted integration of GOI by CRISPR or RMCE, the ideal target site on the genome should reside in a structurally stable region where the transcription activity is very high. In the case of random integration, the integration locus of the GOI can be identified by genome sequencing or focused integration site sequencing, and a candidate cell line that has a favorable integration can be selected for production.

Isolating a Production Cell Line

Hyperproductivity traits

Classical cell line development largely revolves around means that increase the transcript level of the GOI. For example, GOI amplification and its hot spot integration primarily affect the transcript levels of the GOI. Although highly abundant transcription is necessary for high productivity, by itself it is hardly sufficient for converting a host cell into a hyperproducer.

In our body, B cells differentiate to antibody secretory plasma cells upon antigen stimulation over the course of a few days. The changes in cell properties in such a transition may give hints to the traits that a hyperproducing cell must acquire in addition to achieving highly active transcription of the GOI (Figure 6.5).

After antigen stimulation, B cells show the characteristics of unfolded protein response (UPR). UPR is typically induced in cells overloaded with secretory protein molecules and is marked by an expanded ER. Over the course of about four days, B cells increase their size significantly and expand their ER by at least 15-fold. The expansion appears to involve the whole content of the ER, not merely the selective increase of some ER proteins. The expansion starts before mass antibody production. The enzyme disulfide isomerase (PDI), which catalyzes disulfide bond formation, increases in the ER as the B cells differentiate. Many of the redox balance enzymes in the cytosol and mitochondria are also upregulated. There is evidence that the Golgi increases along with the ER. Along with these changes, the metabolic capacity also increases significantly. The transcription factor XBP-1, which activates transcription of many ER proteins, is upregulated, coinciding with the increase in antibody production around a day post-stimulation.

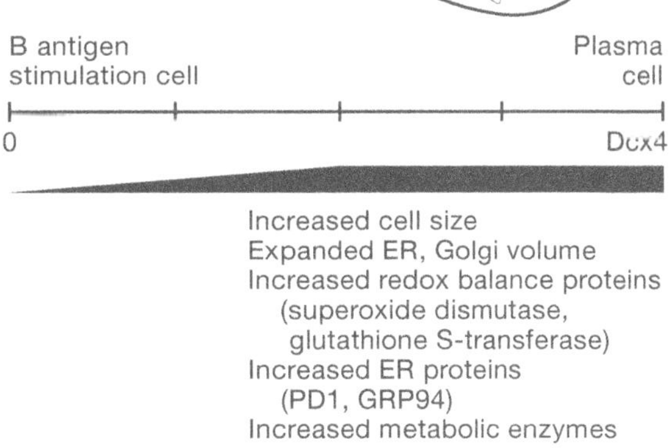

Figure 6.5. Cellular changes occur in the differentiation of a B cell to a plasma cell.

Taken together, the traits that give rise to the hyperproductivity of plasma cells include high secretory capacity, elevated energy generation,

Panel 6.15. Cellular Properties Contributing to Hyperproductivity

- High transcript level of the GOI
- High translation capacity
- High secretory and post-translational processing capacity
- High activity for energy metabolism
- High redox balance capacity
- Ability to grow to high cell concentration and sustain a high viability in stationary phase

and redox balance (Panel 6.15). For glycoprotein products, elevated post-translational processing capacity, including high mannose glycan trimming in the ER, glycan extension in the Golgi, and membrane vesicle trafficking for membrane homeostasis, also contribute to a high productivity.

In addition to an increased product synthesis capacity, a hyperproducer should have superior growth kinetics. A higher tolerance to adverse growth conditions in the late culture will enhance product accumulation by sustaining a high cell viability over a longer duration in the stationary growth stage. Hyperproductivity is thus a complex combination of many cellular properties that each by itself is a complex trait, including high secretory and post-translational modification capacity, redox balance, and increased antiapoptotic activities. Each complex property is likely to be subjected to the control of multiple regulatory networks rather than that of a single master regulation gene. The path that leads a transfected host cell to becoming a hyperproducer is not unique. It is likely that many different combinations of cellular changes can lead to similar levels of high productivity. In other words, different hyperproducing cell lines are likely to have different changes in their gene expression that make each of them a hyperproducer. A more complex and comprehensive combination of important traits may lead to higher productivity. The possible involvement of many alternative combinations of traits that can give rise to hyperproductivity may make the identification of "marker" genes for screening or engineering a hyperproducer elusive.

Single-cell cloning and cell characterization

After the selection of the transfectants, the resulting cell population is heterogeneous in their integration sites, copy number of the GOI, growth behavior, product secretion kinetics, etc. In a heterogeneous population, the fastest growing subpopulation will become predominant over time. The population shift may cause the product quality to shift as well. Single-cell cloning is performed on the cell pool by limiting dilution or by FACS to ensure that the production cell line starts with a homogeneous genetic and epigenetic background (Panel 6.16). As will be discussed next, it is inevitable that the population will become heterogeneous again. Nevertheless, barring further genetic change to the GOI, the ensuing producing cell line will have identical genetic information and local gene expression context.

It takes roughly 10–14 days for a single cell to grow into a colony of about 500–1000 cells. The duration is constrained by the cell growth rate. The low cell concentration used in single-cell cloning (~1 cell/100 μL in a culture well, or ~10 cell/mL) likely incurs a long lag phase and extends the duration of single-cell cloning.

Panel 6.16.
Isolating Hyperproducing Cell Clones

- Single-cell cloning
  - To ensure producing line starts with a homogeneous population
- Improve efficiency by
  - Using conditioned medium
  - Clone-picking instrumentation
    - Cell colony-gel system, entrapment, or 3-D printing
    - Robotic microfluidic nano-well system (increases cell concentration by using nano-wells)
- The increased capacity of screening enables the isolation of rare hyperproducers

When using well plates, the number of clones that can be screened is often limited. The adoption of high-throughput methods using colony-in-gel systems allows thousands of colonies to be screened, drastically increasing the probability of isolating hyperproducers. Typically, an in-gel product concentration assay is integrated into an automated colony-in-gel system to identify and isolate cell colonies that produce more product. Fluorescence-labeled antibodies against the product are added to the cell-gel, where they bind to the secreted product molecules that accumulate around the colony. The radius and the total fluorescence around the colony is a quantitative indicator of the amount of product produced, while the size of the colony is used to estimate the growth rate. The increased screening capability has allowed very high secretory cell clones to be isolated.

After single-cell cloning, clones with a high productivity are isolated. Those which also exhibit robust growth and production characteristics are further characterized for product quality and production stability before the final production line is selected.

Single-cell cloning and population heterogeneity

Aneuploid cells have abnormal chromosomal counts, and many of their chromosomes have macroscopic structural aberrations. Such chromosomal alterations accumulate over time, generating a heterogeneous cell population created by different cells carrying different sets of chromosomes with different structural abnormalities. Some cell lines are highly heterogenous, with a wide range of chromosome abnormality and numbers. For example, some CHO cell lines have a wide range of structural variation.[4, 15] Others, such as Vero cells,[16] have a relatively low level of heterogeneity, with a narrow distribution of chromosome numbers and variations of karyotypes. At a detailed molecular level, base-pair mutations, varying lengths of segmental DNA duplications, deletions, inversions, and translocations frequently occur in aneuploid cells. These structural variations occur during DNA replication at the single-cell level. Some genomic structural variants may carry phenotypic or physiological variations as well. Cells with some types of genome or karyotype variations may grow to become a noticeable subpopulation. The population of an aneuploid cell line is thus heterogeneous in genomic variants.

In single-cell cloning, the progeny population starts with a uniform integration locus and genotype. However, the karyotype in the cell population inevitably becomes heterogeneous again. In CHO cell lines, the heterogeneous distribution in the karyotype is re-established within a couple dozen generations after single-cell cloning.[15] What single-cell cloning from a cell pool achieves is thus the homogenization of the GOI integration loci and other genetic and epigenetic alterations derived during cell line development. The population heterogeneity at the genome level is re-established just like in a typical aneuploid cell line.

Assessing Production Performance of Candidate Cell Lines

After the single-cell cloned candidate cells are expanded, several cell characteristics are evaluated for the selection of the production cell line, including productivity, product gene integrity, the GOI transcription level, and product quality. The productivity of the candidate clones determined in cell-gel or well plates is under conditions that are rather different from those during manufacturing. Before selecting the finalist candidates, small-scale cultures ought to be carried out under simulated process-scale conditions. The selection is based on productivity and protein quality parameters such as structural variants and the glycosylation profile (Panel 6.17). A variety of culture devices are used for high-throughput small-scale culture, from culture tubes and shaker flasks to automated miniaturized systems that mimic stirred tank bioreactors.

Panel 6.17. Characterization of Candidate Cell Lines for Production

Process Performance Evaluation

- Use high-throughput systems, parallel evaluation of multiple candidates, and replicated experiments for statistical confidence

Cell Characterization

- Integrity of integrated GOI copies
  - DNA sequencing of the GOI
  - RNAseq, analyze GOI transcripts
  - Mutation may be present only in one GOI copy and only in some cells if the mutation occurs after GOI integration or single-cell cloning
- Integration site determined by sequencing
  - Structural stability of the GOI integration region
  - Distance to heterochromatins

Product Characterization

- Protein integrity
- Glycan profile

The integrity of all of the expressed copies of the product gene(s) integrated in the genome can be verified by genome sequencing or by PCR amplification of the integrated GOI followed by amplicon sequencing. Mutations or structural variants in one or more copies of the GOI must be absent in the cell line. RNAseq may also be used to detect mutations and other structural variants.

The transcript of the GOI is usually expressed at a very high level, and is covered at a very great depth in RNAseq. The depth of coverage allows for the identification of rare base changes and insertions or deletions (indels). Therefore, a sequence change is detectable even if it occurs only in a small fraction of mutant cells.

Ultimately, the structural integrity is evaluated by structural analysis at protein levels. Some glycan structure features affect the biological functions of some classes of therapeutic proteins with the same mode of action. If the preferred glycan feature is known, the glycan profile of the product is an important parameter in cell line selection. For example, for an IgG product that mediates ADCC, the preferred cell line is one that produces the afucosylated glycan as the dominant form.

Evaluation of Productivity Stability

From a single cell at the beginning of single-cell cloning until the progeny cells reach the final stage of the manufacturing culture, a production cell line is likely to undergo more than 60 population doublings (Panel 6.18). The number of cell divisions exceeds that of a fertilized egg growing into a hamster, a mouse, or even a human adult. DNA replication, both *in vivo* and in cultured diploid cells (before senescence), has very high fidelity, maintaining sequence identity and structural integrity at a chromosomal level. However, aberrant changes occur constantly in aneuploid cells. Some of these changes may lead to a decrease in productivity. Hence, the finalist cell lines are serially cultured for 60 doublings, or more than the number of doublings required to reach the manufacturing scale, and their productivity examined over time to verify their stability over the product life cycle (Panel 6.19).

Besides the structural instability and transcriptional silencing of the GOI, changes unrelated to the cell's capability to synthesize the product may impact the product titer accumulated in the culture (Panel 6.20). For example, the transition

Panel 6.18. Stability of Producing Cell Lines

- Diploid cells used for vaccine production are stable within their allowable passage number
- Recombinant producing lines are aneuploid, have an unstable karyotype
- From cell cloning to product, the life span may take 60 cell doublings
- From cell thawing and expansion to manufacturing, may take 40 cell doublings
- Production cell line must remain productive throughout its life cycle

Panel 6.19. Evaluation of Productivity Stability

- Serial expansion of cells for a large number of doublings (e.g., 40–60), evaluate titer at the end
  - Labor intensive, long time duration
  - Can detect only significant decrease in product titer
- Detect change in clonal distribution of productivity over 10–20 cell doublings
  - Productivity will decrease only if the low-producing subpopulation increases, or if (unlikely) the entire population becomes less productive
  - Examine the proportion of low-producing clones over time
    - By single-cell cloning and determining the percentage of clones whose titer is below the threshold
    - By single-cell sorting followed by single-cell quantitative PCR to quantify the transcript of the GOI to check transcriptional stability

Panel 6.20. Productivity Instability

- Transgene alterations cause instability
  - Loss of GOI copy
  - Transcriptional silencing: promoter silencing, heterochromatin invasion
- To enhance stability
  - Robust promoter, native promoter
  - Low transgene copy number
  - Integrate to a structurally and transcriptionally stable genomic region
- Cell property alterations may cause secondary effects on productivity
  - Metabolic behavior change
  - Secretory pathway change
  - Change in growth kinetics

from the production of lactate to its consumption in the stationary phase of growth has an impact on growth and production kinetics. A change in this metabolic shift behavior of the cell line will cause the productivity to decrease. It is also plausible that changes that broadly affect cell physiology, such as protein processing capacity or altered redox balance, may cause productivity to decrease. In those cases, the drop in productivity is real, but its cause is most likely unrelated to the instability of the productivity. In the following section, we will focus on that instability.

Possible causes of lost productivity

Loss of functional copy of the GOI

In cell division, a cell may lose a copy of the GOI that makes a large contribution to transcription and become a low- or non-producer. In spite of their propensity to develop genomic structural variations, the vast majority of genes in aneuploid cells are stable and not easily lost. GOIs, if integrated into the genome like an endogenous gene, should also be stable. However, GOI copies that are amplified in tandem, integrated near a telomere, or integrated in a region that is vulnerable to structural change may have a higher frequency of loss.[15]

Transcriptional silencing

Mutation or epigenetic changes in the regulatory region of the GOI may diminish its transcription. The methylation of the CMV promoter at its CpG residues causes a reduced transcription of the GOI.[11] The invasion or spread of the neighboring heterochromatin region into the locus of the GOI insertion may cause transcription silencing of the GOI.

Population heterogeneity and low producer dominance

Genomic structural variations and epigenetic silencing occur at a low frequency in every cell division. At a low frequency, some low- or non-producing cells may arise through such genomic or epigenetic changes. The low-frequency occurrence of low- or non-producing cells is not deleterious by itself. A decrease in the product titer is detectable only when the mutant population becomes large. Two mechanisms can lead to an increase in the representation of the mutant: (1) an accumulation of mutants through continued mutations of the parent population, or (2) a faster growth rate of the mutants than of the parent. As shown in Figure 6.6, if the mutant does not grow faster, continued mutations alone will not lead to a high enough non-producer abundance level to cause a detectable drop (20%) in productivity over 60 generations unless the mutation rate is at an exceedingly high level of 1/1000 in a generation. On the other hand, if the non-/low-producing cell grows 20% faster than the parent, even at a moderate mutation rate of 1 in 10^5 per generation, there is a significant drop in the percentage of parent cells (and in productivity) over time. In a continuous culture study where non-producing hybridoma cells became a significant fraction of the population over time, the non-producing cells were found to grow nearly 20% faster than the parent producing cells.[17]

It is highly probable that a mutant cell with a low productivity grows faster. The energetic cost of product synthesis in a hyperproducing cell line is high. Free of that energetic load, a mutant cell may grow faster, although this may be dependent on culture conditions.

One way to evaluate cell line stability is thus to perform single-cell cloning on two populations that are 10–20 generations apart to quantify the change in the percentage of low- and non-producing clones (Panel 6.19). Whether the percentage increases or not over generations will give a measure of the stability of the productivity. Alternatively, one can perform single-cell quantitative PCR on the product transcript level after cell sorting. An increase in a subpopulation with a reduced transcript level will suggest productivity instability.

With targeted integration, one can guide the integration of the GOI to a stable genomic locus that has a very low

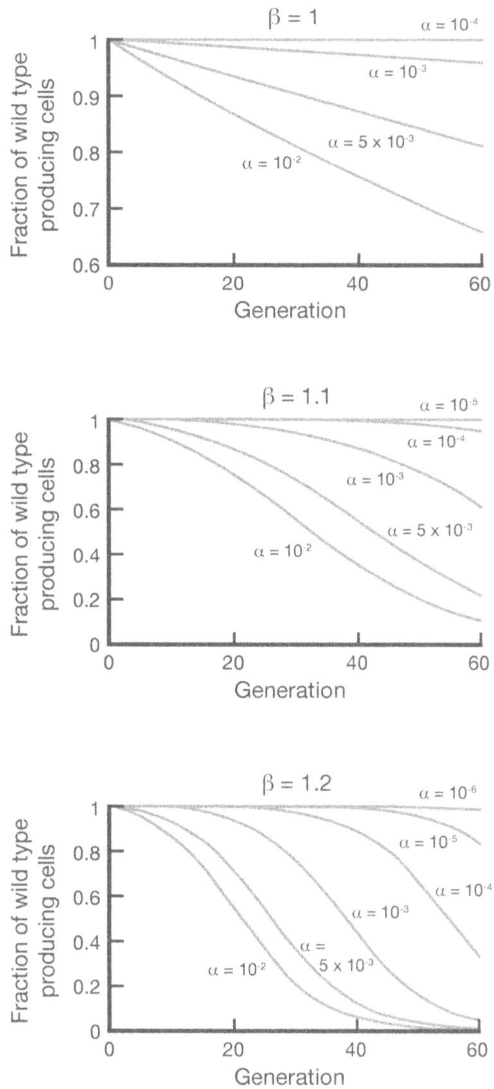

Figure 6.6. An increasing mutant subpopulation through continued mutation or a faster growth rate. α is the frequency that a low-/non-producing mutant cell arises in each generation. β is the ratio of the growth rate of the mutant to the parent. At β = 1, mutants arise only from mutation. At β = 1.1 and 1.2, mutant cells grow 10% and 20% faster, respectively.

frequency of structural variation in the host cell and which is distal to heterochromatin regions. Hence, an advantage of targeted integration is a high degree of confidence that the resulting production cell line will be stable.

Streamlined Cell Line Development

Cell line development is a long process that is repeated for each new drug candidate. Shortening the timeline of cell line development will facilitate new drug development (Figure 6.7, Panel 6.21). A bottleneck of cell line development is the single-cell cloning step. Tens of thousands of producing cells can be generated after the transfection of the GOI. The single-cell cloning step reverts the cell number back to 1, adding at least 25 cell doublings, or more than three weeks to the timeline.

Single-cell cloning may increase the risk of generating a population that has a skewed distribution of genomic content and cellular properties. When applied to a normal and genetically nearly uniform population, single-cell cloning serves to "purify" the population. However, cell lines used in bioprocessing are highly heterogeneous in many aspects. A cloned single cell may not give back the same cell distribution as the original cell population. For example, a mutant cell lacking both alleles of gene a ($a^{-/-}$) isolated from a mixed population of $a^{-/+}$, $a^{-/-}$, and $a^{+/+}$ will not give back genotype $a^{+/+}$ in its progeny population. In order to streamline cell line development and preserve the properties of the population, single-cell cloning should be practiced to the extent of ensuring the initial homogeneity of the production cell line, but not unnecessarily repeated afterwards.

Directed integration of the GOI into a stable and accessible locus is being increasingly adopted in cell line development. After integration, the resulting producing cells consist of three types of GOI insertion status

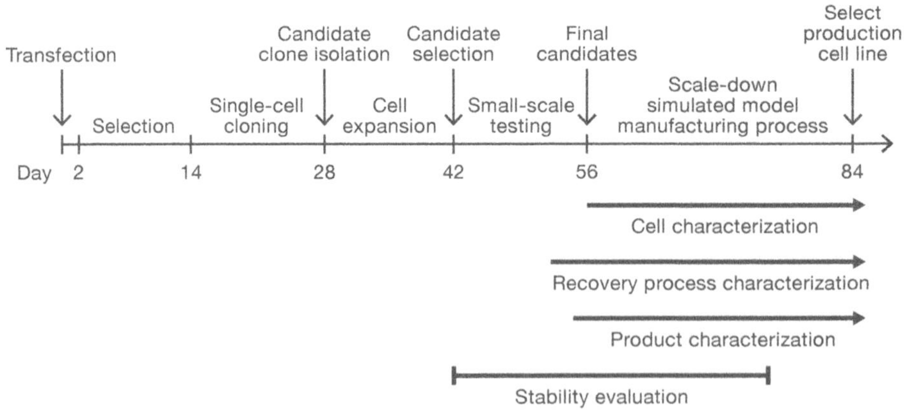

Figure 6.7. A timeline for generating an industrial producing cell line.

at the locus: $^{+/-}$, $^{-/+}$, and $^{+/+}$ (assuming the cell line has two sets of chromosomes). Note that a pair of sister chromosomes may not be identical in aneuploid cells, as one or both may be abnormal, fragmented, or translocated, and given different regional contexts. Insertion of the GOI into two alleles may give different phenotypes. Thus, directed GOI integration will likely still be followed by single-cell cloning. However, directed GOI integration may increase the frequency of obtaining stable hyperproducers.

To avoid single-cell cloning, one may attempt to integrate the GOI in all alleles in the cell. By eliminating the occurrence of off-target integration and then collecting cells of $^{+/+}$ with respect to GOI integration, one can generate a population of cells that have the same distribution of genomic contents as the host cell population except for the homogenous insertion of the GOI.

Panel 6.21. Streamlining Cell Line Development

- Single-cell cloning is a bottleneck in making cell line development more efficient
- Single-cell cloning homogenizes GOI integration sites and genomic structural variations
  - However, karyotype heterogeneity and structural variability quickly return to the single-cell cloning "purified" production cell line
  - Repeated single-cell cloning may give rise to a skewed cell genomic distribution that is different from the original early cell line
- Targeted integration may generate (GOI-integration locus)$^{+/+,\ +/-,\ -/+}$ cells
  - ($^{+/-}$) and ($^{-/+}$) may not be identical
  - A pool of ($^{+/+}$) with no off-site integration would have retained the distribution of the genomic structure of the host cell population. Perhaps this can accomplish what single-cell cloning is intended to achieve?

Concluding Remarks: Genomic Technology & Cell Line Development

THE PAST TWO DECADES have seen a tremendous increase in the productivity of cell culture process. Much of the productivity increase can be attributed to the generation of better-producing cell lines. Although the cell line development process remains largely empirical, nowadays industrial producing cell lines can secrete proteins at levels that rival the professional secretors in our bodies. The increased ease and versatility in performing genetic manipulations has played a major role in better cell line development. Perhaps even more important is the enhanced capability to identify and isolate high-producing clones that perform well at manufacturing scales. Automation of single-cell cloning and high-producing clone identification and isolation allows more clones to be evaluated and greatly increases the probability of isolating a high producer that is at the far high end of the distribution curve. Gene amplification, a main step in cell line development for three decades, is giving way to controlled low- or single-copy GOI for improved cell stability. The advances in genomics are shifting the paradigm in cell line development. The targeting of a product gene to specific loci that give high expression levels and are stable over time is facilitating cell line construction. As our understanding of the role of epigenetics in conferring the trait of hyper-productivity improves, epigenetic intervention to enhance those traits will become an inviting area of study.

References

1. Goh, J.B. and S.K. Ng, *Impact of host cell line choice on glycan profile.* Crit. Rev. Biotechnol., 2018. **38**(6): p. 851–67.
2. Puck, T.T., *Development of the Chinese hamster ovary (CHO) cell.*, in *Molecular cell genetics.*, G. M. M., Editor. 1985, John Wiley and Sons: New York. p. 37–64.
3. Yerganian, G., *The biology and genetics of the Chinese hamster,* in *Molecular cell genetics.*, M.M. Gottesman, Editor. 1985, John Wiley and Sons: New York. p. 3–36.
4. Wurm, F. and M. Wurm, *Cloning of CHO cells, productivity and genetic stability—A discussion.* Processes, 2017. **5**(4).
5. Seth, G., R.S. McIvor, and W.S. Hu, *17Beta-hydroxysteroid dehydrogenase type 7 (Hsd17b7) reverts cholesterol auxotrophy in NS0 cells.* Journal of Biotechnology, 2006. **121**(2): p. 241–52.
6. Lim, Y., et al., *Engineering mammalian cells in bioprocessing—current achievements and future perspectives.* Biotechnol. Appl. Biochem., 2010. **55**(4): p. 175–89.
7. Butler, M. and M. Spearman, *The choice of mammalian cell host and possibilities for glycosylation engineering.* Curr. Opin. Biotechnol., 2014. **30**: p. 107–12.
8. Lee, E.U., J. Roth, and J.C. Paulson, *Alteration of terminal glycosylation sequences on N-linked oligosaccharides of Chinese hamster ovary cells by expression of beta-galactoside alpha 2,6-sialyltransferase.* Journal of Biological Chemistry, 1989. **264**(23): p. 13848–55.
9. Yamane-Ohnuki, N., et al., *Establishment of FUT8 knockout Chinese hamster ovary cells: an ideal host cell line for producing completely defucosylated antibodies with enhanced antibody-dependent cellular cytotoxicity.* Biotechnology & Bioengineering, 2004. **87**(5): p. 614–22.
10. Brown, A.J., et al., *Synthetic promoters for CHO cell engineering.* Biotechnol. Bioeng., 2014. **111**(8): p. 1638–47.
11. Yang, Y., et al., *DNA methylation contributes to loss in productivity of monoclonal antibody-producing CHO cell lines.* J. Biotechnol., 2010. **147**: p. 180–85.
12. Xia, W., et al., *High levels of protein expression using different mammalian CMV promoters in several cell lines.* Protein Expression and Purification, 2006. **45**(1): p. 115–24.
13. Running Deer, J. and D.S. Allison, *High-level expression of proteins in mammalian cells using transcription regulatory sequences from the Chinese hamster EF-1alpha gene.* Biotechnology Progress, 2004. **20**(3): p. 880–89.

14. Hartman, T.E., et al., *Derivation and characterization of cholesterol-independent non-GS NS0 cell lines for production of recombinant antibodies.* Biotechnol. Bioeng., 2007. **96**: p. 294–306.

15. Bandyopadhyay, A.A., et al., *Recurring genomic structural variation leads to clonal instability and loss of productivity.* Biotechnol. Bioeng., 2019. **116:** p. 41–53.

16. Osada, N., et al., *The genome landscape of the African green monkey kidney-derived vero cell line.* DNA Res., 2014. **21**(6): p. 673–83.

17. Frame, K.K. and W.-S. Hu, *Kinetic study of hybridoma cell growth in continuous culture: II. Behavior of producers and comparison to nonproducers.* Biotechnol. Bioeng., 1991. **38**: p. 1020–28.

Contents

7

Medium Design for Cell Culture Processing

Introduction

CELLS ARE THE CORE of cell technology. Medium to cells is like food to humans. It is the most important factor that impacts a cell's growth, metabolism, and other biological activities. For industrial processes, it dictates the productivity and often the product quality. Over the past three decades, we have drastically expanded our understanding of the chemical environment that enables cells that have been isolated from different tissues to grow, as well as the environment that allows cells to differentiate from one state to another. Cells which have been isolated from different tissues or species, or which are at different differentiation state, have different needs in terms of growth factors and cytokines. But the basic nutritional requirements of mammalian cells in culture are strikingly similar among cells of different tissue origins and species.

Early in the industrialization of cell culture, media design efforts focused on enhancing cell proliferation and on eliminating animal serum or other animal-derived media components. The work was mostly spearheaded by bioscientists using a mechanistic understanding of cells' nutritional needs (Panel 7.1). The arrival of the antibody era in the second half of the 1990s and the sudden demand for large quantities of antibody product called for increased cell concentration and productivity and changed the face of cell culture processes. Fed-batch culture and concentrated feed medium were used more frequently. Instead of providing cells with optimal growth conditions as had been done during the previous decade, cells were increasingly cultured under conditions suboptimal for growth, or even under stressed conditions that favored product formation. The intense emphasis on maximizing the productivity and modulating product quality demanded that medium development work take a stoichiometric and kinetic perspective. The philosophy of medium optimization changed from pampering cells with inviting growth conditions to maximizing cell density and stressing cells in order to squeeze out every last bit of productivity. Much of the success in increasing product titer by at least an order of magnitude over the course of a decade can be attributed to better media formulation. As optimizing cell growth was de-emphasized, the media that was optimized for production was also used for cell expansion in seed cultures.

Over the life cycle of a product, cells spend far more time expanding their population—from single-cell cloning and cell banking to seed train—than in the final production bioreactor. An optimized production medium maximizes productivity. A well-balanced cell expansion medium optimizes cell growth and minimizes deleterious metabolite formation, and sometimes even enhances the cells' structural and phenotypical stability. A holistic view should be taken in medium development. The industrial cell culture medium should not be monolithic; on the contrary, it should be a multitude of media that enable cells to proliferate optimally at a fast rate while maintaining cellular, structural, and genomic stability during the expansion stage, while also allowing cells to achieve conditions favorable to maximum productivity and high product quality (Panel 7.2).

The renewed emphasis on cell expansion in media design is critical for cell therapy and regenerative medicine. In those applications, cells are the product that is administered to patients. They must have high viability and functional activity. In the past decade, genomic science has profoundly changed our approach to biomedical and biochemical exploration. Modern medium design will increasingly exploit the genomic data of the cell of interest and design the medium

Panel 7.1. Historical Approaches in Industrial Cell Culture Medium Design

Classic Medium Design

- Optimized for cell growth
- Eliminates serum and animal-derived components
- Relies on a metabolic understanding of cells' nutritional requirements and meticulous experimentation

Antibody Era

- Maximizes productivity even at the expense of the growth rate
- Emphasizes stoichiometric and kinetic perspectives

Integrated Approach

- Combines mechanistic, stoichiometric, and kinetic perspectives
- Optimized for both cell expansion and production
- Employs genomic and transcriptome information on cells' biological capabilities

Panel 7.2. Industrial Cell Culture Media for Cell Expansion and Production

Industrial Cell Culture Process

- A long period of cell expansion followed by a relatively short production/differentiation period

Medium Design Objectives

- Cell expansion medium
  - Optimized for growth
  - Minimizes deleterious metabolites and reactive oxygen species (ROS)
  - Enhances cells' structural integrity and production stability
- Production medium
  - Maximizes cells' activity, productivity, or product quality

Table 7.1. Cellular Chemical Environment *in Vivo**

Approximate Concentrations in a Cellular Environment	*Interstitial (mM)*	*Intracellular (mM)*
Na^+	140	14
K^+	4.0	140
Ca^{2+}	1.2	0.01
Mg^{2+}	0.7	20
Cl^-	108	4
HCO_3^-	28.3	10
HPO_4^{3-}, $H_2PO_4^{2-}$	2	11
SO_4^{3-}	0.5	1
Amino acids	2	8
Lactate	1.2	1.5
Glucose	56	
Protein	0.2	4
Total chemical species (mmole/L)	301.8	302.2
Corrected osmolar activity (mM)	281.3	281.3

*Adapted from *Textbook of Medical Physiology*, A.C. Guyton.

Panel 7.3. Some Characteristics of Interstitial and Intracellular Fluid

- Interstitial fluid represents cells' native niche environment
- Total osmolality is ~280–300 mOsm
- Na^+ and Cl^- are the most abundant in interstitial fluid
- There is a large concentration difference between extracellular and intracellular fluid in the concentration of:
  - Na^+, Cl^-, and Ca^{2+} (high outside the cell)
  - K^+, Mg^{2+}, and PO_4^{3-} (high inside the cell)

based on the cell's intrinsic capabilities as gleaned from it genetic and epigenetic features.

A Guide for Medium Design: Body Fluids

Originally, all cells in culture were isolated from animal tissues. The composition of the chemical environment of their natural niche is a relevant reference for designing an optimal medium for their cultivation. The vast majority of cells in the body are not in direct contact with blood, and are instead surrounded by interstitial fluid. The protein and hormone content of interstitial fluid differs in various tissues, but the general chemical composition of small molecular weight solutes in interstitial vs. intracellular fluid is similar among cells of different tissues. A representative composition of interstitial fluid is shown in Table 7.1, along with that of intracellular fluid.

The low molecular weight solute composition of interstitial and intracellular fluid bears a few important characteristics. The total osmolality is around 280–300 mOsm, or about a total of 280–300 mM of dissociated solutes in both the intracellular and extracellular environments (Panel 7.3). The media for cells used in research has largely kept the total osmolality in the range of 280–300 mOsm, close to that of interstitial fluids. Bear in mind that a dissociable ionic compound contributes its total dissociated ions to the osmolality, whereas an undissociated molecule contributes an equivalent of its concentration. Thus 1 mM of $MgCl_2$ contributes a total 3 mOsm; while 1 mOsm of glucose contributes 1 mM in an ideal solution. A couple of percentage points of error notwithstanding, the osmolarity can be taken as the sum of the molarity of all dissolved species in the fluid. The largest contributor to the final osmolarity of the interstitial fluid is Na^+, followed by Cl^-. A number of other inorganic species including K^+, Mg^{2+}, and Ca^{2+} are present at low concentrations in the millimeter range.

The major anions are Cl^- and the bicarbonate ion (HCO_3^-) (10–30 mM) that serves as a pH buffer.

A number of ion species are present at very different concentrations between the interstitial space and the cytoplasm. Both Na^+ and Cl^- are present outside the cell at a 10-fold higher concentration than inside the cell, as is Ca^{2+}. In contrast, K^+, Mg^{2+}, and PO_4^{3-} concentrations are much higher on the intracellular side.

Cells' tolerance to deviations from "optimal" conditions varies with the physical and chemical conditions. The sublethal but non-lethal range of different compounds also varies considerably. For example, cells in culture can tolerate low (80 mM) or high (140 mM) sodium concentrations, or high osmolarity (up to 400 mOsm), for a period of days. Hamster cells may tolerate or even grow with a reduced growth rate in hypothermic conditions of 33 °C or a low pH of 6.8, but not at the higher temperature of 40 °C at a pH of 7.8.

Types of Media and Classes of Medium Components

A complete cell culture medium has two major categories of components: basal medium and growth supplements (Panel 7.4). The basal medium consists of the small molecular weight components such as sugar, amino acids, vitamins, nucleosides, and inorganic salts. The basal medium provides sugar as an energy source and various nutrients for making new cell mass and product. It also provides balanced salt concentrations and osmolarity to allow for cell growth. In general, the basal medium can be prepared as a mixture of chemicals in powder form and is commercially available. Upon dissolution in pure water and sterile filtration, it is ready for use. In practice, a number of components that are unstable (i.e., spontaneously degraded) in solution or are hygroscopic if present in the power mixture are excluded from the powder medium. Many amino acids and vitamins are heat labile, so the basal medium is typically sterilized by membrane filtration.

Many cell lines and primary cells explanted from tissues will not proliferate if provided with basal medium alone, as basal medium does not contain growth factors or other factors necessary for growth. Growth supplements that may be added to basal medium include growth factors, phospholipids, soy hydrolysate, and serum. These supplements may promote cell growth by providing molecules that modulate

Panel 7.4. Major Classes of Medium Components

Basal Medium
- Sugar
- Amino acids
- Fatty acids, lipids, and their precursors
- Vitamins
- Nucleosides
- Bulk salts
- Trace elements
- pH buffer

Supplements
- Serum
- Hydrolysates
- Growth factors
- Carrier proteins

specific signaling pathways, or by filling some nutritional needs (such as delivering cholesterol). Some signaling molecules, instead of stimulating cell growth, trigger a signaling pathway that directs cellular differentiation or maintains cells at a particular differentiation state.

Types of Media

Serum-free and animal component-free medium

Animal serum has been a regular component of cell culture medium, used at 5–20% (by volume) levels in addition to basal medium. Serum is highly complex in its chemical composition. It contains a large mixture of proteins, from serum albumin, immunoglobulins, clotting proteins, and lipids to small molecular weight compounds including sugar and amino acids. The outbreak of the prion disease bovine sponge form encephalopathy in the United Kingdom in the late 1980s compelled the cell culture industry and the regulatory agencies to undertake large efforts to eliminate serum from media. Since then, the use of serum-free and animal component-free media has become the industrial manufacturing norm (Panel 7.5). Removing raw materials of animal origin minimizes animal virus or prion contamination. It also eliminates the possibility of some antibodies present in the serum reacting with a cell antigen or the product (especially important for viruses produced for use as vaccines), which would reduce productivity. The use of serum-free and animal component-free media has been extended to the cell line development stage to minimize the exposure of cells to adventitious agents throughout the entire cell banking and manufacturing process. In the early days of serum-free medium, this effort was thwarted by the requirement of growth factors or other materials that were isolated from animal tissue, such as serum albumin and transferrin. Most of those supplements are now produced by recombinant DNA technology in microbial or mammalian cells. However, animal serum is still widely used in manufacturing viral vaccines. In the cultivation of human primary differentiated cells for clinical applications, inactive human serum is often used.

Panel 7.5. Serum-Free, Animal Component-Free Medium

- Serum-free media consists of a nutritionally complete basal medium supplemented with an empirically determined mixture of hormones, growth factors, cell attachment proteins, and binding proteins
- Many serum-free media contain a complex mixture of undefined components, such as soybean-meal, another plant hydrolysate, peptone, or beef hydrolysate
- Animal component-free media eliminates all materials derived from animals, such as growth factors, transferrin, and beef peptone
- The growth rate in serum-free media is often slower than in serum-containing media

Complex vs. chemically defined media

Many medium components that have been traditionally used in cell culture are a complex mixture of different components, such as animal

Panel 7.6. Chemically Defined Medium

- Contains no complex nutrient mixtures
- All components are completely characterized in terms of chemical composition
- Its adoption was accelerated by the drive to exert control over all aspects of production and downstream processing for licensing by the FDA
- Its wide use is facilitated by
  - High-throughput screening of media components
  - The availability of growth factors produced by rDNA technology
  - Adaptation of cells to leaner growth environments
  - The development of small molecules or synthetic peptides that are biologically active in stimulating cell proliferation or directing differentiation

serum, plant hydrolysate, soybean phospholipids, and yeast extract. The inclusion of any of those materials in the medium make its composition incompletely characterized. Any medium that has an undefined chemical composition is a complex medium. In addition to a complex mixture like beef peptone, some seemingly "pure" components may readily associate with impurities that are not easily characterized; for example, serum albumin is a carrier of lipids and many metabolites that are not easily removed during purification. Those components are also regarded as having a complex composition.

A chemically defined medium contains only components whose chemical composition is known and characterized, and has all of its chemical species specified (Panel 7.6). It does not contain any mixture of components with an unknown or undefined composition. For example, "lipids" or "phospholipids" are not well-defined compounds, but are mixtures of a class of compounds that are not chemically specified. A chemically defined medium may contain purified growth factors, cytokines, and carrier proteins. A chemically defined medium is not necessarily protein-free.

A large number of manufacturing processes for protein therapeutics employ chemically defined media. Many cell cultures for cell therapy or regenerative medical applications also employ chemically defined media. The application of chemically defined media in industrial manufacturing has become more widespread largely due to the drive of regulatory requirements, the availability of better-characterized raw materials, and the increasing ease in adapting cells to different culture environments (Panel 7.6).

Protein-free medium

By definition, a protein-free medium contains no protein. Most protein-free media are chemically defined. However, a protein-free medium may contain undefined lipids or fatty acids and may therefore not be chemically defined. Many cells that grow well in protein-free media are highly adapted or transformed. Most normal diploid cells, including stem cells and various differentiated cells, require some growth factors or cytokines in order to proliferate in culture.

Classes of Medium Components

Stoichiometric vs. unconsumed medium components

Medium components fall into two general categories: stoichiometric and habitable (also called "unconsumed"). Stoichiometric components are taken up by cells, metabolized, and used to make more cell mass and product (Panel 7.7). Virtually all organic components are stoichiometric components, except for some antioxidants. Some inorganic salts are also consumed to become a significant portion of cellular materials, like phosphate is a constituent of nucleic acids. A stoichiometric relationship between the biomass and the medium components that are consumed can be established for stoichiometric medium components. If one wishes to grow more cells or produce more product, these components must be supplied in larger quantities.

Medium components of the second category, mostly inorganic compounds, play the critical role of providing a chemical environment that is suitable for cell growth, but they are barely consumed by cells. They are called habitable (or habitat-enabling) components. Many of these salts are also components of the intracellular chemical environment (e.g., Na^+ and Cl^-). Even water is taken up by cells; as the total cell volume expands, water must be taken up along with other components that constitute cellular materials. However, their intracellular concentrations are low and the quantity taken up by growing cells from the medium is unappreciable. In medium design, the level of unconsumed components is to be maintained in a range that sustains the habitable environment.

Panel 7.7. Two General Types of Medium Components

Stoichiometric Components

- Medium components that are consumed in appreciable amounts
- e.g., glucose, amino acids, fatty acids, some growth factors, and some salts like phosphate
- Should be supplied in some proportion to the cell and product concentrations to be produced

Habitable (Unconsumed) Components

- Components that are hardly consumed by cells but are needed in some level to provide conditions necessary for cell proliferation
- Many such components are taken up by cells because they are constituents of the intracellular environment. However, their consumption rate is very low and almost unmeasurable.
  - e.g., Cl^- and Na^+
- Should be maintained at relatively constant levels

The classification of stoichiometric and unconsumed nutrients is therefore operational. A normally unconsumed component may be taken up in a substantial quantity at a high cell concentration. Thus, unconsumed medium components may become stoichiometric components under high-density cultivation conditions; this is illustrated with phosphate in Panel 7.8.

Macromolecules as consumed and unconsumed medium components

Cell culture medium often contains macromolecules such as insulin, fibroblast growth factor (FGF), and serum albumin. These constituents serve a variety of purposes; some are carrier proteins that ferry ligands into the cell, while others

Panel 7.8. Phosphate as a Stoichiometric Medium Component

- PO_4^{3-} concentration in medium is typically ~1 mM, while its cellular concentration is 11 mM. The intracellular ATP/ADP concentration is about 2mM, or ~6mM of phosphate. Total DNA and RNA contain another 35 mM equivalent of phosphate. A culture with cells of an average volume of 2 x 10^{-12} L and at a cell concentration of 10^{10} cells / L takes up about 1 mmol / L of PO_4^{3-} [(11 + 6 + 35) mmol / L x 2 x 10^{-12} L / cell x 10^{10} cell / L].
- Cell consumption of PO_4^{3-} will drastically reduce its concentration at such a cell concentration

are growth factors that bind to receptors on the cell's surface. Molecules that are taken up by cells may need to be replenished in order to maintain their concentration, whereas molecules that transmit signals by binding to cell surface receptors (and are therefore not internalized and degraded) may not need to be replenished.

Panel 7.9. Macromolecular Medium Components

- Some growth factors and carriers are unconsumable
  - e.g., fibroblast growth factor, transferrin
- Some are consumed or degraded
  - e.g., insulin at high levels, low-density lipoproteins (LDL)

Many macromolecules are internalized, some are degraded (consumed), and others are recycled (Panel 7.9). For example, the ferric ion carrier, transferrin, binds to a transferrin receptor and is internalized. It is then translocated to the lysosome, where it releases ferric ions in the lysosome's low pH environment. Lastly, the transferrin is recycled to the extracellular medium. As long as ferric ions are available (i.e., replenished), transferrin can continue its role as a ferric ion carrier.

Insulin, another commonly used growth factor, binds to the insulin receptor and triggers a signaling event that does not involve its own internalization. However, when present at a high concentration (a few micrograms per mL), insulin is internalized by adipocytes and degraded.

An example of a consumable macromolecule is low-density lipoprotein (LDL). LDL is the carrier for lipids and cholesterol. After an LDL particle binds to an LDL receptor on the plasma membrane, the receptor-ligand complex is internalized in a clathrin-coated pit that pinches off intracellularly to become a coated vesicle. Subsequently, the clathrin coat depolymerizes, resulting in an uncoated (smooth-surfaced) vesicle, often called an endosome. The endosome then fuses with an uncoupling vesicle that has an internal pH of about 5.0, which causes the LDL particles to dissociate from the LDL receptors. The LDL receptors are then recycled back to the plasma membrane. The vesicles containing the LDL particles fuse with lysosomes, in which the cholesterol esters are hydrolyzed to fatty acids and cholesterol. Cholesterol is then incorporated into cell membranes.

Process Medium Design: Kinetics, Stoichiometry, Omics & Optimization

HISTORIC MEDIA, such as MEM, DMEM, and RPMI, were used in conjunction with animal serum. The development of media in those early days of cell culture largely focused on compounds that were present in high concentrations and are stoichiometrically important (e.g., sugar, amino acids and vitamins) in order to reduce the reliance on serum. Bulk salts were added to balance osmotic pressure and pH.

The drive to develop serum-free media in the 1970s created a new generation of media in which cells' nutritional needs as well as their growth/hormonal factor requirements are experimentally identified. The work in that period largely defined the general nutritional needs of mammalian cells in culture. The baseline of nutritional needs is generally quite similar among cells isolated from different tissues and different species. Some variability does occur, as a gene essential for the transport or catabolism of a particular nutrient may not be expressed in some tissues or in some species, but overall such variabilities are exceptions.

Optimal Growth Media via Clonal Growth Assay

A technique commonly used in the early period of cell culture enlightenment is clonal growth (Panel 7.10, Figure 7.1). Cells of interest were plated at the very low density of a few dozen per cm^2 in Petri dishes containing media of different nutrient compositions. At such a low density, each cell is well separated from the others. The dish was then set aside in an incubator and left undisturbed for about 10 days. During that time, some cells grew into colonies on the dishes' surface. At the end of the experiment, each dish was stained with a dye that colored the cell colonies. The extent of the color stain then gave a semi-quantitative measure of cell growth. Since the cell density used was so low, and the number of nutrients consumed through the entire period very small, the medium composition was relatively constant and the number of metabolites accumulated very small. This allowed the growth-promoting capability and the concentration of each medium component to be determined, as well as enabling simultaneous evaluation of many different chemical combinations. When this method is applied to cells grown in suspension, the cells are embedded in agar or another type of gel.

Panel 7.10. Determining Nutritional Requirements by Clonal Growth

- The test cell is plated on dishes at low density in a test medium
- Cells grow and become well-separated colonies (~10–15 days)
- The dish is stained with a dye; the degree of staining reflects the extent of cell growth
- The growth extent reflects the capability of the test nutrient at that particular level to support cell growth
- This method defined the general nutritional needs of cells in culture and paved the way for the industrialization of cell culture process

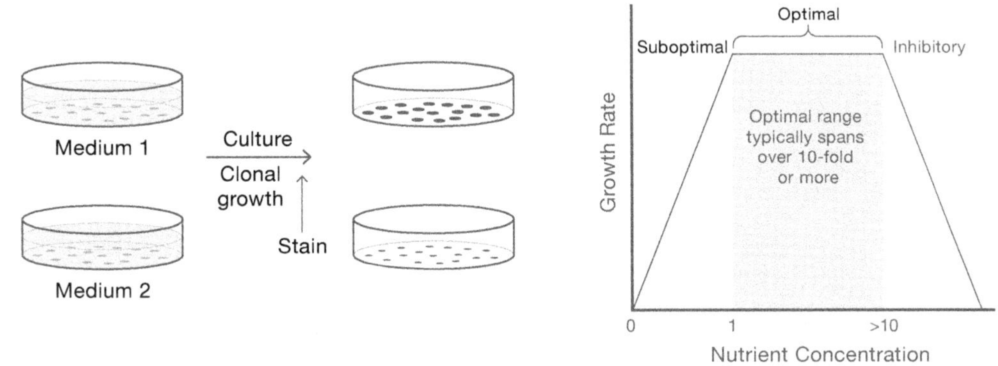

Figure 7.1. Clonal growth and optimal range of nutrients for cell growth.

By the mid-1980s, various serum-free and chemically defined media (e.g., F-12, MCDB104) had been developed, and insulin and PDGF were shown to be required for the growth of human diploid fibroblasts.[1] Clonal growth is still being performed today in small well plates to determine the growth factor/cytokine requirements of many types of cells. Since different nutrients and growth factors may have an interactive effect on cell growth, the design of experiments (DOE) approach is often taken in choosing combinations of nutrient levels to be tested (Figure 7.2).

Many cell strains or cell lines are now grown with serum-free chemically defined or semi-defined (sometimes containing serum-albumin, as will be discussed later in this chapter) media, including pluripotent and many other stem cells, and many NS0 and CHO cell lines. The discovery was made around this time that the concentration range of most nutrients that give cells their maximum growth rate (i.e., a growth-optimal concentration) spans at least 10-fold (Figure 7.1). This suggests that, in a media optimization experiment, the stepping of concentrations must not be too close to each other.

The first generation of serum-free media defined the general nutritional requirements of cells in culture. These achievements greatly facilitated media development for many cell lines, as well as for industrial processes.

Medium for Industrial Cell Culture

Before the arrival of therapeutic proteins, cell culture media were developed for biomedical research and sustained only relatively low levels of cell density. For industrial processes, which require higher cell and product concentrations, the medium's nutrient content must be increased (Panel 7.11, Figure 7.2). The general approach is to provide a more complete nutritional content at a higher concentration while minimizing the increase of osmolality, or simply by using a mixture of DMEM (which has higher nutrient levels and supports higher cell concentration) and

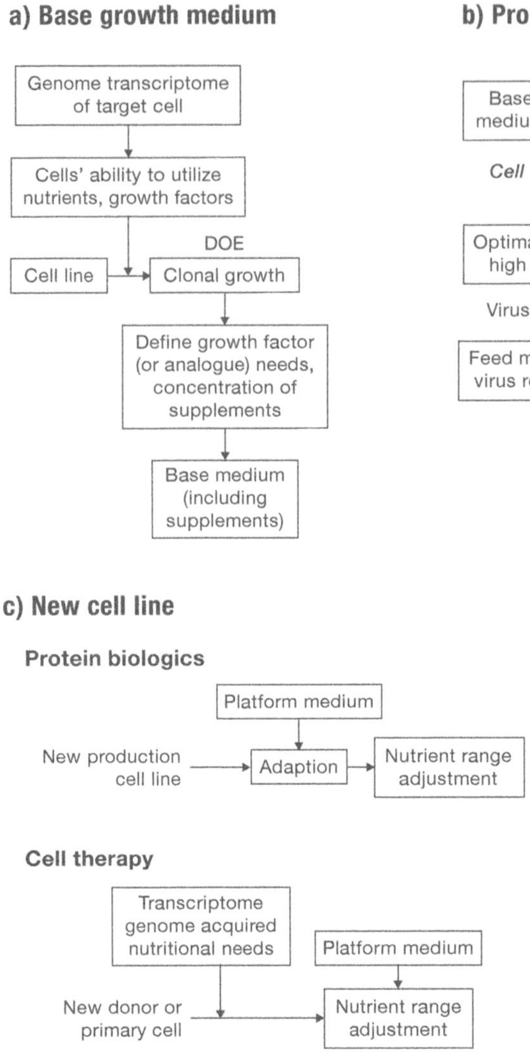

Figure 7.2. Steps in medium development for (a) base growth medium, (b) production platform medium, and (c) medium for new cell lines.

F-12 (which has a more complete nutritional content). Nowadays, most industrial media have higher levels of nutritional components, including amino acids, other organics, and a small number of salts (such as phosphate and magnesium, which will be discussed later), in order to meet the stoichiometric needs of generating more biomass. The medium composition, including the feed medium for fed-batch cultures, is typically determined empirically. In industrial settings, this is often performed using high-throughput instrumentation with pH monitoring and control and automatic sampling. The fortified nutrient levels can give many-fold higher cell and product concentrations. Inevitably, metabolites accumulate as cell concentration increases and cell viability decreases in the late stage of the process.

The increased nutrient level and osmolality takes a toll on cells' growth rate. Furthermore, for cost reasons growth factors are often eliminated when possible or used at suboptimal levels for growth in industrial manufacturing. Hence, the cell growth rate in industrial production is often lower than in a medium optimized for growth. Nevertheless, the high cell density and product titer achieved through the employment of these industrial media contributed to the rapid growth of antibody therapeutics at the turn of the millennium. In many cases, the same medium is used for both cell expansion and production.

Panel 7.11. Developments of Industrial Cell Culture Media

- Basal media include more complete nutrients (e.g., include all nutrients in Ham's F-12) at higher levels (like in DMEM or higher)
- Adapts cells to proliferate in leaner media with low or no growth factor supplement
- Adjusts the levels of key stoichiometric medium components to enable cells to grow to a high concentration
- Small flask cultures or high-throughput automated instrumentation are often used to conduct DOE-based experimentation to determine the nutrient range
- May result in higher osmolality, higher metabolite accumulation, and slower growth rate, but higher cell and product concentration
- Establishes a platform medium that is used as the basis for rapid development of new media for cell lines producing new products

Platform Medium and Extension to Additional Cell Lines

To streamline the process development of a new product, the medium that gives rise to high productivity for a given product is adopted as a platform medium for future media development for cell lines (Panel 7.11, Figure 7.2). The platform medium provides the basis for the fine-tuning of medium composition for a new cell line. Some cell lines have a high variability at the genome and transcriptome levels, causing different sublines to have different metabolic behavior. One may first adapt the new cell line to grow in the platform medium. Subsequent adjustment of nutrient concentrations can then be undertaken to optimize process performance.

Integrating Physiological Insight and Genomic Data into Medium Design

Work in the past few decades has established the nutritional needs of cells in culture. Medium design efforts today often concern culturing a new cell type, or a new cell isolate of a particular differentiation lineage. Different tissues in mammals have different specialized metabolic functions, and different tissue cells respond differently to growth regulators. Cells derived from a particular tissue generally inherit the characteristics of the tissue from which they are derived. However, upon cell isolation, gene expression changes may occur to the isolated cells; this is partly caused by changes in cell-tissue interactions and local chemical cues. Importantly, the acquisition of the capability to grow in culture is accompanied by major alterations in cellular growth regulatory architecture.

Hence, even primary cells in culture often bear significant differences in gene expression from the tissue cells of their origin. Such differences in gene expression may affect the cells' nutritional needs. Furthermore, even though all mammals share highly similar nutritional needs, there are many variabilities among different species.

Further contributing to the variability of nutritional needs are the genetic and epigenetic changes that occur in the process of deriving cell clones from a progenitor cell line. This is especially true in the case that clones are isolated under selective pressure following genetic manipulation. Through genomic, transcriptomic, proteomic, and even metabolomic measurement, or by gleaning omics data available in the public domain, much insight on cells' metabolic capability in utilizing specific components in the medium can be gained to facilitate medium design (Panel 7.12).

Modern medium design should thus return to a mechanism-based approach while integrating omics information. Figure 7.3 illustrates the supply and utilization of typical nutrients in a medium. Glucose is always the most abundant organic nutrient. Sometimes, alternative sugars such as fructose, galactose, and mannose are included with the aim of modulating energy metabolism or to influence the glycosylation profile of the product. Additional components include amino acids and other organics. Every nutrient must have a route (a transporter or transport system), to enter the cell. Many transporters have multiple isoforms with different transport kinetics that are expressed differently in different tissues and cell lines.

Panel 7.12. Physiological Considerations in Medium Design

- A cell's ability to use a medium component is dictated by its expression of the relevant genes
- Depending on the tissue source, species, and epigenetic events, cells have varying metabolic capabilities
- The effective utilization of a nutrient requires the expression of a transporter, utilization genes, and degradation/catabolic enzymes
- "Alternative" substrates (such as fructose, mannose, or galactose in the place of glucose) also require an efficient route of entering the main pathway
- Reactive nutrients that may otherwise cause damage to cells or other nutrients additionally require "sequesters" or carriers both extracellularly and intracellularly
- To respond to a growth regulator and signaling molecule, cells must have the corresponding receptor
- Transcriptome and proteome data can reveal a cell's nutrient utilization capability and can be used to guide medium development

Upon entering the cell, some nutrients (e.g., glucose and most amino acids) enter their metabolic pathways (e.g., glycolysis and tRNA loading) readily. Some other nutrients, such as galactose and mannose, may need to be converted to another intermediate in order to enter the main metabolic artery. For example, galactose and fructose are first converted to UDP-galactose and fructose 1-phosphate, respectively, to enter glycolysis. The proteins for the entry must be expressed at a sufficiently high level to sustain a flux that is high enough for the desired biochemical objective.

Some nutrients are highly damaging to cells at high levels. *In vivo*, these nutrients

a)

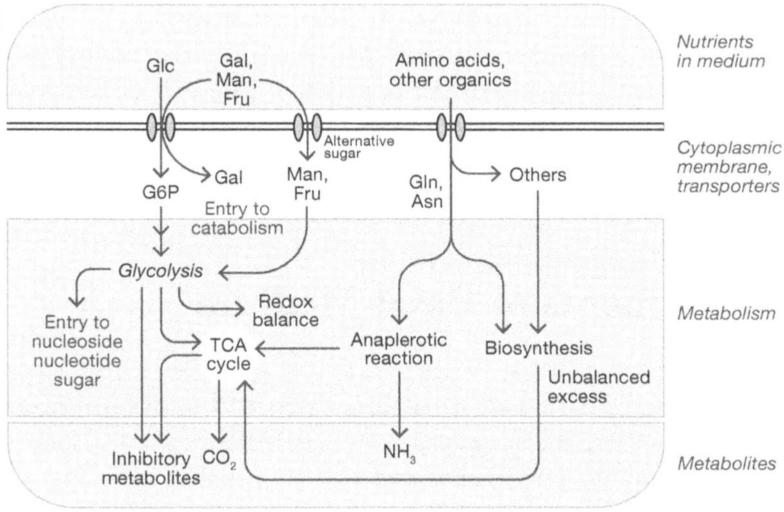

b)

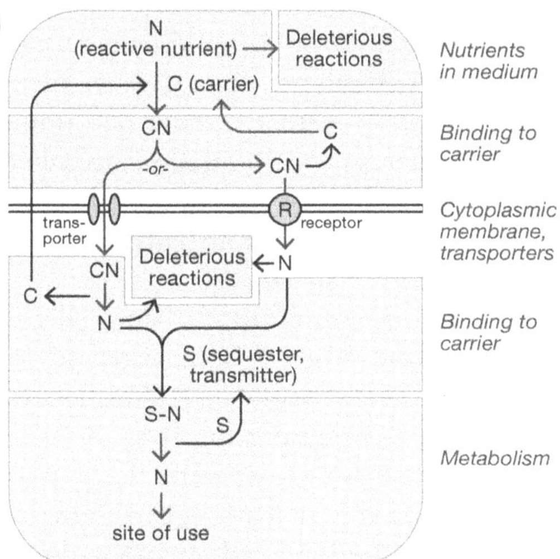

Figure 7.3. a) The general scheme of utilization of medium components. The nutrient present in the medium enters the cell through the transporter/receptor on the plasma membrane where it is metabolized and, if in excess, degraded to generate metabolites. b) Nutrients that can potentially cause deleterious effects are often bound to a carrier that acts as a sequester. Once entering the intracellular environment, another sequester is present to minimize its deleterious effect. The presence of the machinery for the utilization of a nutrient can be evaluated using a cell's transcriptome.

are sequestered outside and inside the cell; in the latter case, they conjugate to a binding protein. For example, both copper and ferric ions are highly oxidative. Ferric is bound to transferrin in blood circulation to prevent it causing damage to cells. Carrier or sequester compounds may need to be included in culture medium to safeguard the extracellular environment. For highly reactive species, such as ferric ion, carrier or sequester proteins are present in the cell to mitigate their deleterious effect on cellular constituents.

Nutrients that are taken up by cells in excess of what is needed for cell growth and other functional activities will enter pathways for their degradation and excretion. Nutrient levels in industrial media typically far exceed the optimal for growth. However, the somewhat reduced growth rate results in a high product titer and productivity. At a high concentration, some nutrients are taken up by cells far in excess and are shunted to degradation pathways. Some metabolites from the degradation are growth inhibitory or even toxic at high levels. The expression level of enzymes involved in the degradation pathways of compounds leading to inhibitory metabolites will therefore affect the performance of the culture. Hence, medium design may need to take the presence or absence of degradation pathways into consideration.

Components of Basal Medium

Water

Mammalian cells are exceedingly sensitive to the quality of water used for media preparation. City water, the usual source of water for medium preparation, contains particulates, including bacteria that can introduce endotoxins, trace organics, and various inorganic ions such as potentially harmful heavy metals (Panel 7.13). Those contaminants must be removed before water is used in cell culture. Typical water preparation processes include deionization through ion exchange, microfiltration to remove particulates and bacteria, and finally reverse osmosis to reduce conductivity (or increase resistance) to >20 MΩcm.

In some applications, especially cell therapy, the product is subjected to minimal purification before final filling as a drug product; to minimize the entry of any pyrogenic contaminants, water for injection (WFI) is used in preparing cell culture medium. WFI is prepared by low evaporation rate distillation, which minimizes the chance of a water droplet in the evaporating steam carrying a solute or particle from the water.

Panel 7.13. Water for Cell Culture Media

- Types of contaminants in city water
  - Inorganics: heavy metals, iron, calcium, chlorine
  - Organics: byproducts of plant decay, detergents
  - Bacteria: endotoxins, pyrogens
  - Particles: colloids, particles
- A typical water preparation process for cell culture involves deionization, microfiltration, and reverse osmosis. An ultrafiltration step is sometimes included.
- WFI, which is prepared by water vapor compression or by multiple-effect distillation, is also used in cell culture manufacturing

Hexoses

Glucose and glutamine are the primary nutrients that supply a cell's energy needs in culture (Panel 7.14). The physiological concentration of glucose in blood is ~0.8 g/L (4.4 mM). In culture, glucose is typically present from 1–5 g/L (5.5–27.5 mM). In the production reactor, sometimes a high level of glucose, as much as 15 g/L (82.5 mM), is used. In this case, glucose is a large contributor to the osmolarity of the medium, and adjustments to the medium's composition must be made (by reducing sodium and chloride concentrations) to maintain osmolarity in a growth-permissible range.

Other sugars, especially galactose, mannose, and fructose, may also be used as alternative sugars. All cultured cells express the GLUT1 transporter at a significant level, and take up glucose readily under a normal glucose level in medium. Galactose is also transported by GLUT1, and can thus be used as an alternative sugar to glucose. The K_M for galactose uptake is higher than for glucose. In the concentration range commonly used for glucose, galactose is taken up by cells at a lower rate, resulting in lower lactic acid production in the culture. Fructose is transported by the GLUT5 transporter. The K_M for fructose transport by GLUT5 is also high. Thus, similar to galactose, the uptake rate for fructose is lower than for glucose unless a high concentration of fructose is used. However, not all cells can utilize fructose as some do not express GLUT5.

Panel 7.14. Six-Carbon Sugars

- Glucose is typically used in 1-5 g/L
- Physiological level is ~0.8 g/L
- Fructose and galactose may also be used
- Galactose and glucose can both be transported by the GLUT1 transporter, which is present in most cells
- Fructose is transported by a different transporter (GLUT5); unless the transporter is expressed, the cell may not be able to use fructose efficiently
- An alternative sugar is often taken up by cells at a slower rate, which may reduce lactate production and help cell maintenance
- Pyruvate and ribose are sometimes supplied in small quantities that are insufficient to supply a cell's energy needs

Galactose and fructose do not get phosphorylated at their C6 position and enter the glycolysis pathway directly. They can serve as the main hexose source for cells through alternative entry points to glycolysis. Galactose is phosphorylated at C1 to become galactose 1-phosphate, and then through a transferase and an epimerase catalyzed reaction it becomes glucose 1-phosphate, which is converted to glucose 6-phosphate. Fructose, once in the cytosol, is converted to fructose 1-phosphate and split into dihydroxyacetone 3-phosphate and glyceraldehyde by aldolase. Both are then converted to glyceraldehyde 3-phosphate and enters glycolysis (Figure 7.4).

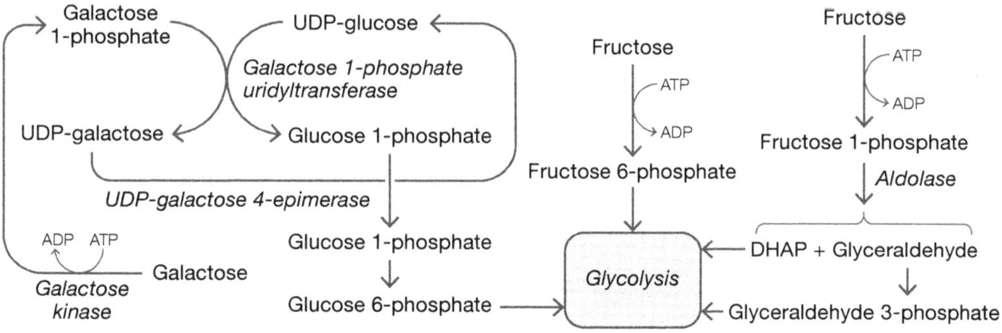

Figure 7.4. Entry of galactose and fructose to glycolysis.

With their slower uptake rates, galactose and fructose give a low flux type of metabolism when used as the main hexose. The alternative sugar can also be used in conjunction with glucose. As glucose concentration decreases to a low level, the consumption rate of galactose or fructose declines, facilitating the switch of the glycolysis to a low flux state. Galactose and mannose have also been used to increase galactosylation of N-glycans. However, when used together with glucose, a high concentration may be necessary in order to compete with glucose for transport through GLUT1. It is important to keep in mind that when using galactose, fructose, or other sugars as the sole course of sugar, the carbon flux distribution and the intracellular nucleotide sugar concentrations are different than if glucose had been used. Since nucleotide sugars are the precursors to glycan, the effects of alternative sugar on the glycan profile must be examined if glycosylation is a product quality attribute.

Amino Acids

Glutamine is an essential amino acid for cells in culture. Most tissues express glutamine synthase and make glutamine from glutamic acid. Glutamine is thus not an essential amino acid (Panel 7.15). However, glutamine becomes an essential nutrient in culture. Among all the amino acids, glutamine is consumed in the highest quantity and is present in the medium at the highest concentration. It is the second most abundant organic component in the medium after glucose and is consumed at roughly 1/3 to 1/10 of the molar consumption rate of glucose. The exceptions are cell lines that have been engineered to express glutamine synthetase, which enables cells to synthesize glutamine. Glutamine supplies the amino group for incorporation into proteins and for nucleotide biosynthesis. It also supplies the carbon backbone of TCA cycle intermediates by converting to α-ketoglutarate in anaplerotic reactions (see Chapter 3). Glutamine spontaneously degrades in aqueous solutions and releases ammonium. Consequently, in order to avoid degradation glutamine is typically added to culture medium immediately prior to use.

Panel 7.15. Glutamine as a Nutrient

- Glutamine is an essential amino acid for cultured cells, but not for mammals
- Its main roles as a nutrient:
  - Used in protein and nucleotide synthesis
  - Enters the TCA cycle after deamination for anaplerotic metabolism
  - Consumed at a high level, second only to glucose, at about 1/3 to 1/10 of glucose
- Spontaneously degrades in aqueous solutions

Amino acids are classified as essential or nonessential based on nutritional studies using animals or tissue culture cells (Table 7.2). Mammals lack the synthetic pathway of aspartate family amino acids that produce lysine, isoleucine, methionine, and threonine. The synthetic pathways for the aromatic amino acid family (phenylalanine and tryptophan), and for leucine, valine, and histidine are also incomplete. Mammals thus acquire these amino acids through food to meet growth needs. Some amino acids are essential only for cells in culture, but not for animals (indicated by a * in Table 7.2). In animals, different tissues may cross feed each other, such that amino acids synthesized in one tissue (especially liver) may be transported to cells in other tissues. The expression levels of enzymes involved in some amino acid biosyntheses may decrease in cultured cells to make the amino acid become essential. Glutamine is nonessential for animals; it is synthesized from glutamic acid through glutamine synthase. Proline becomes an essential amino acid for some cell lines.

Cell culture media developed in the 1960s and 1970s contained at least the 14 essential amino acids. Those media were designed to be used with serum supplementation, which supplies some additional amino acids. Media designed for serum-free culture typically include all amino acids.

Table 7.2. Essential and Nonessential Amino Acids

Essential	*Nonessential*
L-arginine*	L-alanine
L-cysteine*	L-asparagine
L-histidine	L-aspartic acid
L-isoleucine	L-glutamic acid
L-leucine	L-glycine
L-lysine	L-proline
L-methionine	L-serine
L-phenylalanine	
L-threonine	
L-tryptophan	
L-tyrosine*	
L-valine	
L-glutamine*	

*Essential for cells in culture, but not for animals.

Some nonessential amino acids may be secreted by cells and accumulate in culture medium, including alanine, asparagine, glutamic acid, and proline. In principle, nonessential amino acids can be synthesized by cells, and thus the depletion of a nonessential amino acid is not detrimental. However, it has been reported that the depletion of asparagine in medium resulted in some degree of its substitution by serine in the therapeutic protein produced.[2] Caution should be taken in medium development to monitor possible side effects of nutrient depletion.

Vitamins

Many vitamins serve critical biochemical roles in metabolism. Their nutritional deficiency in humans is the root cause of numerous diseases and led to their discovery in the early twentieth century. Even though they are grouped as a common class of nutrients, their chemical natures and biological roles are diverse. They are all essential for the vitality of humans, but are needed only in minute quantities compared to glucose and amino acids. Some vitamins are cofactors of enzyme reactions, while others are a prosthetic group (i.e., bound to the protein) of an enzyme. These vitamins are essential for the catalytic activity of those enzymes.

Not all vitamins are needed by cells because of species or tissue variability. For example, vitamin C is essential for humans but not for mice. Vitamin K is nonessential only for some differentiated cells (or genetically engineered cells) that synthesize proteins that require glutamic acid γ-carboxylation. The nutritional deficiency of these vitamins leads to various diseases in humans, and their deficiency in cell culture can impair energy metabolism or biosynthetic functions (for review see reference 3). The vitamins that are commonly included in cell culture media are listed in Table 7.3.

Biotin (vitamin B7) is a prosthetic group tightly bound to many enzymes involved in carboxylation reactions. It functions as a CO_2 carrier in enzymes such as pyruvate carboxylase that are involved in anaplerotic reactions, and acetyl-CoA carboxylase in fatty acid synthesis.

Thiamine (vitamin B1) participates in biochemical reactions as thiamine pyrophosphate. It forms a transient C–C bond with the α-carbon of α-keto acid in α-keto acid dehydrogenase reactions to facilitate the departure of CO_2 in the decarboxylation reaction. It is also a coenzyme of transketolase in the pentose phosphate pathway.

Pantothenate (vitamin B5) is a precursor of CoA and a critical participant in a large number of reactions in energy metabolism, lipid catabolism, and biosynthesis. Plants, yeast, and many bacteria can synthesize CoA *de novo* from aspartate and ketovalerate through pantothenate. Mammals lack the capacity to synthesize pantothenate and must take it up from a food source.

Table 7.3. Vitamins

Compound	***Vitamin***	***Biological Roles***
Participate in vital biochemical reactions required by all mammalian cells		
Biotin	Vitamin H	A prosthetic group as a CO_2 carrier in carboxylase reactions
Ca-pantothenate	Vitamin B5	Precursor of CoA
Cyanocobalamin	Vitamin B12	Catalyzes reactions involving the exchange of an alkyl group with hydrogen
Folic acid	Vitamin B9	C1 (methyl, methylene, formyl) carrier
Niacinamide niacin, nicotinic acid	Vitamin B3	A precursor of NAD, NADP
Pyridoxine HCl	Vitamin B6	A prosthetic group in amino group transfer reactions
Riboflavin	Vitamin B20	Cofactor in many redox reactions (as FMN, FAD)
Thiamine HCl	Vitamin B1	Thiamine pyrophosphate participates in decarboxylation reactions
Antioxidants		
Ascorbic acid	Vitamin C	Electron donor, reduces free radicals
D-alpha-tocopheryl acetate	Vitamin E	Antioxidant, protects polyunsaturated fatty acids
Required only in cells carrying out certain reactions or undergoing specific differentiation		
Phylloquinone, menadione, menaquinone, sodium bisulfite	Vitamin K	Cofactor of vitamin K-dependent γ-carboxylation
Cholecalciferol	Vitamin D	Involved in calcium homeostasis, supplemented to culture media of some differentiated cells

Folic acid plays the role of a C1 (methyl, methylene, formyl) carrier. It is involved in glycine synthesis through the use of serine as a methylene donor. Its derivative, dihydrofolate, participates in the nucleotide salvage pathway. Dihydrofolate reductase is a target enzyme used in the DHFR gene amplification system.

Niacin, or nicotinamide (vitamin B3), is a precursor of NAD and NADP. Riboflavin (vitamin B2) or flavin mononucleotide (FMN) and flavin adenosine dinucleotide (FAD) are important cofactors in many biochemical reactions.

Ascorbate (vitamin C) is a water-soluble reducing agent and electron donor that is involved in many antioxidant reactions. It is a cofactor in the proline hydroxylation reaction in the biosynthesis of collagen I and III. It also participates in dioxygenase catalyzed reactions and histone demethylation. Plants and most mammals can synthesize ascorbate, but the enzyme for the last step of its biosynthesis, gulonolactone oxidase, is not functional in humans, other primates, and some other species. Hence, ascorbate is a vitamin necessary for humans. In the dioxygenase reaction, ferrous ion is oxidized to ferric. Ascorbate reduces the ferric ion back to ferrous and allows dioxygenase to return to its original state (Figure 7.5).[4] The oxidized ascorbate (dehydroascorbate) is regenerated back to ascorbate enzymatically by consuming NADH or NADPH. It can also revert to ascorbate by oxidizing Fe^{2+} to Fe^{3+} while generating free radicals, which can further react with oxygen to form hydrogen peroxide, especially at a high concentration of free Fe^{3+}. *In vivo*, most iron is sequestered in transferrin in plasma or is bound to ferrin inside the cell; thus, the prooxidant effect of ascorbate is not prominent. Under cell culture conditions of high ascorbate and free Fe^{3+} concentrations, the possible role of ascorbate as a prooxidant,

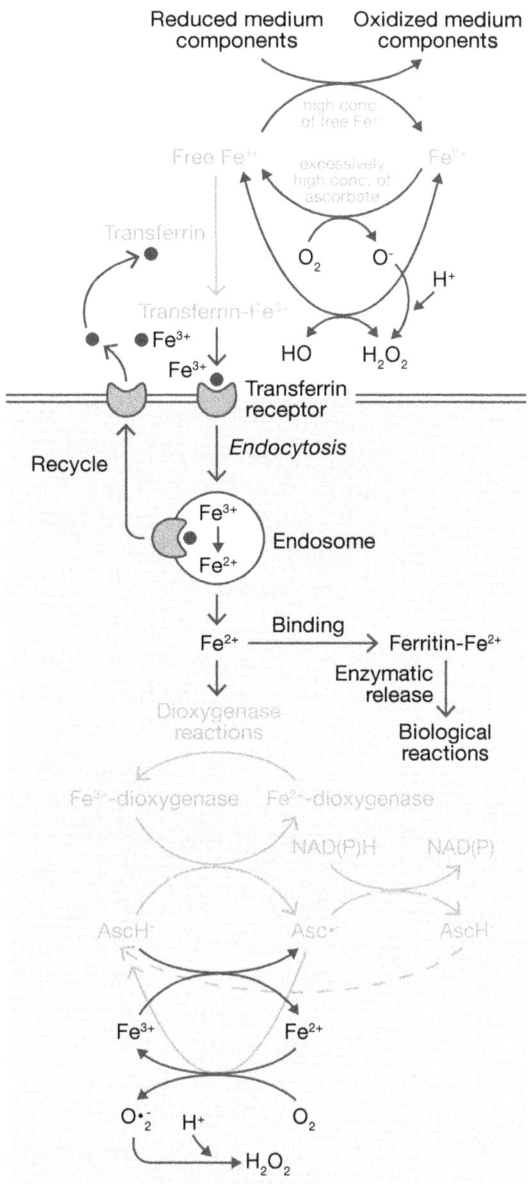

Figure 7.5. The interactions of iron, transferrin, ferritin, ascorbate, and the iron metabolism.

instead of an antioxidant, should not be ignored. To minimize the possible prooxidant effect of ascorbate, ascorbate 2-phosphate is sometimes used. Ascorbate 2-phosphate is not oxidized as rapidly as ascorbate in solution, and thus has a low prooxidant effect. It is taken up by cells at a slower rate, giving it a longer availability in cell culture media.

The vitamins discussed above are all highly soluble in water except for biotin, which has a somewhat lower solubility. A few vitamins (vitamins A, D, E, and K) have a low solubility in water and are fat-soluble.

Vitamin E (tocopherol) is an antioxidant that prevents polyunsaturated fatty acids from being oxidized and forming peroxides. Its presence thus reduces the production of free radicals and reactive oxygen species. Most of the vitamin E associated with cells resides in membranes, especially the Golgi membrane. The oxidation of polyunsaturated fatty acids generates C-centered free radicals that can propagate and generate more oxidized polyunsaturated fatty acids in membranes. The presence of vitamin E in the membrane stops such chain reactions. Vitamin E and ascorbic acid are thus complementary to each other; vitamin E has an antioxidant effect on unsaturated fatty acids in cellular membranes, while ascorbic acid prevents the formation of reactive oxygen species in the intracellular soluble environment. Both vitamin E and ascorbic acid are regenerated by glutathione.

Vitamin K is a cofactor in the γ-carboxylation of glutamate that occurs in the synthesis of many blood coagulation proteins, including prothrombin, factors VII, IX, and X, protein C, and non-coagulation proteins such as osteocalcin and transthyretin. Vitamin K is supplemented to cultures of cells carrying out γ-carboxylation reactions.

Vitamin D is a group of related compounds naturally synthesized in humans and other mammals from cholesterol (called cholecalciferol). It becomes deficient under conditions of low exposure to sunlight in the UV spectra range. The photoactivation reaction completes the last step of vitamin D3 synthesis. Individuals with low UV exposure must obtain it from dietary sources. Vitamin D3 is a classical hormone that regulates the transcription of a variety of genes in a tissue-specific manner. At the whole-body level, vitamin D regulates the homeostasis of calcium and phosphorus through its effects on their tissue uptake. It is supplemented to cultures for the purpose of modulating the differentiation process or cells' differentiated properties. At high doses, vitamin D inhibits cell growth.

A group of compounds generally referred to as retinoids possess vitamin A activities. Their forms are retinol, retinal, retinoic acids, and ester. Their most important role is the vision function, with retinal serving as the photosensitive chromophore through a covalent link to the photoreceptor protein rhodopsin. In addition to operating as a vitamin, the retinoic acid derived from retinol plays a role in the regulation of transcriptional regulation in embryo development and in stem cell differentiation. Vitamin B is used to guide the differentiation process of specific lineages.

Nucleosides

Nucleosides and nucleic acid make up a very significant portion of a cell's content, as they constitute the genome. Most cultured cells are diploid, although many industrial cell lines are multiploid. Nucleosides constitute all RNAs whose cellular content is higher than DNA. Additionally, free ribonucleotides, like ATP, ADP, and AMP, are present in relatively high abundance in the cell. In fast-growing cells, deoxyribonucleotides (deoxyadenosine, deoxyguanidine, deoxythymidine, deoxycytidine) and ribonucleotides (adenosine, guanidine, uridine, cytidine) must be synthesized sufficiently quickly or otherwise be supplied exogenously to sustain growth. Mammals have *de novo* synthesis pathways for making purines (adenosine and guanidine) and pyrimidines (uracil, cytidine, thymidine), although the major site of such synthesis is in the liver (Panel 7.16). Both bases (purine and pyrimidine) and nucleosides (base + ribose or deoxyribose, without phosphate) can be transported into cells from tissue fluid or culture medium. Cultured cells, notably cancer cells, can develop the capability to synthesize purines and pyrimidines. Nucleic acids, especially mRNAs, are subjected to degradation due to gene expression regulation and turned over rapidly in the cell. These degradation products, mostly nucleosides or nucleoside monophosphates, are recycled to nucleoside triphosphates through salvage pathways for incorporation into nucleic acids again.

Panel 7.16. Nucleosides and Precursors

- DNA, various RNAs, and nucleotide monomers constitute a major portion of cellular materials
- *De novo* synthetic pathways exist, but most synthesis occurs in the liver in mammals
- They are subject to degradation and turnover; salvage pathways recycle (and interconvert) the nucleosides, but do not provide what is needed for growth
- Bases and nucleosides can be transported into the cell
- Hypoxanthine and thymidine are typically added to the medium, and their stoichiometric quantity can be calculated

Thymidine and hypoxanthine are commonly used as pyrimidine and purine sources in contemporary media. They are readily linked to a ribose 5-phosphate in the cell to become inosine monophosphate (IMP) and TMP. IMP enters the synthetic pathway to make AMP and GMP, which are further phosphorylated into ATP and GTP. TMP is converted to UMP and CMP. If hypoxanthine is not available in the medium, *de novo* synthesis is the primary source and must be sufficiently active to sustain cell growth since the salvage pathway only recycles the base "salvaged" from nucleic acid degradation, but does not generate the nucleosides needed for proliferation. In *de novo* synthesis, the four nitrogen atoms in each purine ring are derived from glutamine, aspartate, and glycine, while obtaining carbons requires CO_2, formate, glycine, and folic acid. The *de novo* synthesis of pyrimidine similarly requires glutamine, aspartic acid, and CO_2. One can see that without a sufficient supply of hypoxanthine, many precursors will be needed for nucleoside biosynthesis.

As discussed in Chapter 6, dihydrofolate reductase mutants of Chinese hamster ovary cells are frequently used for co-amplification of the DHFR gene and transgene. Hypoxanthine and thymidine are supplemented for the culture of those mutants but are removed to facilitate the process of gene amplification.

Fatty Acids and Lipids

Lipids constitute a significant portion of animal biomass (Panel 7.17). Phospholipids, cholesterol, and sphingolipids make up the cell membrane. The composition of lipids in a membrane affects its fluidity and dynamic behavior. The lipid bilayer membrane forms vesicles which play key roles in protein secretion and envelop virus replication. Mammals synthesize different lipids in different tissues. Unsaturated fatty acids of different carbon lengths constitute about half of the fatty acids in phospholipids. However, mammals cannot add double bonds to w-3 and w-6 carbons. Linoleic acid (18:2 cis-$\Delta^{9,12}$, a w-6 fatty acid with an 18-carbon total and 2 double bonds at positions 9 and 12 from the carboxylic carbon) and linolenic acid (18:3 cis-$\Delta^{9,12,15}$, a w-3 fatty acid) are thus essential fatty acids for mammals. They acquire these w-3 fatty acids from diet. From linoleic acid and linolenic acid, other long-chain fatty acids with double bonds in w-3 and w-6 carbons can be synthesized. Hence, a complete medium for serum-free culture usually contains linoleic acid and sometimes also arachidonic acid (24:4 cis-$\Delta^{5,8,11,14}$) and oleic acid (18:1 cis-Δ^{9}).

Panel 7.17. Lipids and Fatty Acids

- Lipids constitute a few percent of biomass in most cells
- Phospholipids and cholesterol form lipid bilayer membranes; about half of the fatty acids in the lipid bilayer membrane are unsaturated
- Linoleic acid (18:2 cis-$\Delta^{9,12}$) and linolenic acid (18:3 cis-$\Delta^{9,12,15}$) are essential fatty acids
- Lipids have a low solubility in media, but carrier proteins (e.g., serum albumin) or synthetic carriers may increase their bioavailability
- Phospholipid precursors (ethanolamine, choline) are often present
- Most cells can synthesize cholesterol, but for some it must be provided

Although linoleic and linolenic acids are often referred to as essential fatty acids, their effects on the growth of cultured cells is subtle. The depletion of essential amino acids has an almost immediate retarding effect on cell growth, but the failure to supplement cells with essential fatty acids is not as profound and does not cause immediate cessation of growth.

The vast majority of fatty acids supplemented to media are incorporated into phospholipids and reside in cell membranes. Phospholipid supplementation to culture medium is a logical approach in medium design. Supplementing tissue extracts rich in phospholipids, cholesterol, and lipoproteins (such as Ex-Cyte derived from bovine or lecithins isolated from various sources) promotes cell growth. However, the demands to employ media components that are non-xeno and the drive to use chemically defined medium has thwarted the use of those extracts except in special applications.

Instead of tissue extracts, precursors of phospholipids (choline, ethanolamine, inositol) are supplied. Cholesterol is supplied to its auxotrophic mutants, such as NS0 cells. The frequent addition of chemically defined phospholipids, phosphatidyl -choline, -ethanolamine, -inositol, and dipalmitoyl phosphatidic acid to culture medium had been shown to have a growth-promoting effect on human diploid fibroblasts.[1]

Fatty acids, cholesterol, and phospholipids all have low solubility in aqueous environments. In mammals or in culture media containing animal serum, lipids are bound in serum albumin. *In vivo*, cholesterol is carried in high-density and low-density lipoproteins in circulation. When lipids are used in disposable plastic bioreactors, the adsorption to the plastic surface and possibly reduced bioavailability should be determined. Ways to increase the solubility have been explored. Cholesterol can be derivatized with a dicarboxylic acid (e.g., succinic acid) to form an ester and increase its solubility, or be supplied as a cyclodextrin conjugated complex. Phospholipids may be prepared as liposomes[1] or dissolved in DMSO for delivery to laboratory cell culture; however, its industrial applications may not be imminent.

Bulk Salts

Many mineral elements, in the form of ions (e.g., Na^+ or Cl^-) or polyatomic ions (e.g., PO_4^{3-}), are essential constituents of cell mass and must be provided in the medium. A list of the ions contributed by bulk salts and their concentrations in several media are shown in Table 7.4. Among all ions, Na^+ is present in the highest concentration, followed by Cl^- by a rather large margin because Na^+ is also contributed by other anionic species, especially HCO_3^-, which is added as $NaHCO_3$ for buffering

Table 7.4. Concentrations of Bulk Ions in Basal Medium (mM)

Ion	*DMEM*	*F-12*	*RPMI*	*William's E*	*E8*
Na^+	155.29	145.02	138.52	144.44	103.53
K^+	5.33	2.98	5.33	5.33	4.16
Mg^{2+}	0.81	0.60	0.41	0.81	0.60
Ca^{2+}	1.80	0.30	0.42	1.80	1.05
Cl^-	119.28	135.80	108.78	126.18	181.24
PO_4^{3-}	0.91	1.04	5.63	1.01	0.95
HCO_3^-	44.04	14.00	23.81	26.19	40.52
SO_4^{2-}	0.81	0.07	0.41	0.81	0.41
NO_3^-	0.00	0.00	0.85	0.00	0.00

pH. Potassium concentration in the medium is typically only 1/50 to 1/25 that of Na^+. Its intracellular level is 10–20 times higher than Na^+ (Table 7.1).

Phosphate makes up part of the nucleic acids and nucleotides of cell biomass. It is also present as various phosphorylated metabolic intermediates and as inorganic phosphate. A high level of phosphate in the medium is necessary for carrying out the stoichiometric need to synthesize new biomass. Magnesium plays key metabolic roles in energy metabolism and is conjugated to many metabolites. The intracellular level of Mg^{2+} is high, but free Mg^{2+} is only at about ~.25–1 mM, as the majority is complexed to ATP and other organic acids. Phosphate, K^+, and Mg^{2+} are thus three inorganic components that appear at higher concentrations intracellularly than in the medium. Calcium is essential for signaling in some differentiated cells, and is present in high concentrations in the endoplasmic reticulum (ER) and the sarcoplasmic reticulum in muscle cells. The concentrations of the most abundant bulk ions (Na^+, Cl^-, and K^+) in commonly used media spans a small range. In contrast, HCO_3^{2+} spans a much wider range.

An important role of bulk ions is to balance the osmolarity to be in, or at least near, the physiological range of 270–300 mOsm. The most important contributors to osmolarity are Na^+, Cl^-, and HCO_3^{2+}. While the composition of bulk ions may vary in different medium formulations, the total osmolarity of most industrial media is maintained in the range of 270–330 mOsm. When bicarbonate is not used in culture medium, or when glucose is added at a very high concentration (such as 50 mM, contributing nearly 50 mOsm of osmolality), the osmolarity of the culture medium needs to be rebalanced by adding or removing bulk salts. A common practice is to add a mixture of NaCl and KCl to maintain the molar ratio at about 30.

Trace Elements

Five transition metals—iron, copper, manganese, zinc, and cobalt—play key roles in the biological functions of all mammalian cells (for review, see reference 5). They are naturally present in serum, but must be included in serum-free media (Table 7.5). Additionally, some other heavy metal ions, including molybdenum (Mo), vanadium (V), strontium (Sr), and selenium (Se), also appear to participate in biological reactions. Among these metals, iron and zinc are present in cells and human bodies at much higher levels than other trace elements (Panel 7.18).

The zinc ion has only one 2+ valence state (Zn^{2+}), is not redox active, and is less toxic to cells. All the other metals have multiple oxidation states; for example, iron has Fe^{2+} and Fe^{3+}, copper has Cu^+ and Cu^{2+}, and manganese has II, III, and IV states. These ions incur cytotoxicity when present at high concentrations, by oxidizing components of cells or the medium, or by generating reactive oxygen species. This is especially true for copper, which has a redox potential of +0.2 to +0.8 V, and iron, which has a redox potential of 0.77 to ~1.0 V. In media and in free forms in a typical solution (not oxygen-depleted), they exist as Cu^{2+} and Fe^{3+} and are highly oxidative. In circulation, Fe^{3+} is bound to the ~200 binding sites on transferrin (Figure 7.5). At a bound state in transferrin, Fe^{3+} is not available for reactions involving Fe^{3+}/Fe^{2+} transition. Intracellularly, Fe^{3+} is bound to protein ferritin and is delivered to its site of usage through carrier proteins. The binding to carrier proteins sequesters the oxidizing effect of free Fe^{3+}. Although it is beyond the scope of this chapter to discuss this matter in detail, the other trace metals that exhibit cytotoxicity at high levels likely have similar mechanisms: extracellular and intracellular carriers that sequester the cytotoxic effect, a transportation system that allows for intake across the cytoplasmic membrane, and a protected means of delivery to reach the target site.

Table 7.5. Trace Elements in MCDB 104 (a serum-free medium for human diploid cells)

Element	*Compound*	*Concentration (μM)*
Fe	$FeSO_4 \cdot 7\ H_2O$	5.0
Cu	$CuSO_4 \cdot 5\ H_2O$	1.0×10^{-3}
Zn	$ZnSO_4 \cdot 7\ H_2O$	5.0×10^{-1}
Mn	$MnSO_4 \cdot 5\ H_2O$	1.0×10^{-3}
Mo	$(NH_4)_6Mo_7O_{24} \cdot 4\ H_2O$	1.0×10^{-3}
Ni	$NiCl_2 \cdot 6\ H_2O$	5.0×10^{-4}
Se	SeO_2	3.0×10^{-2}
V	NH_4VO_3	5.0×10^{-3}
Si	$Na_2SiO_3 \cdot 9\ H_2O$	5.0×10^{-1}
Sn	$SnCl_2 \cdot 2\ H_2O$	5.0×10^{-3}

These trace metals play key biological roles. Their multiple valence states make them easy participants in electron transfer reactions. They can form a prosthetic group, function as a cofactor of an enzyme, or be part of a complex that is integrated into a protein. Iron is responsible for the key activity of heme in hemoglobulin, which allows red blood cells, myoglobulin in muscle cells, and cytochromes in virtually every cell to carry oxygen molecules. Iron also plays a key role in electron transfer complexes in the mitochondria. Furthermore, iron is a metal ion component of many important proteins expressed at high levels in the cell. Among the trace elements, iron's cellular demand is the highest and is therefore provided in the medium at a relatively high concentration.

In comparison, the copper ion, which has an even higher oxidative potential than Fe^{3+}, is present only in small numbers of cellular proteins, many of which are extracellular or are involved in prevention of copper toxicity.[5] Copper is provided at low levels in most traditional media, about three orders of magnitude lower than iron. Nevertheless, it plays a critical role as the cofactor in cytochrome C oxidase that catalyzes the last step of the electron transfer chain by reducing oxygen to water and pumping four protons across the mitochondrial inner membrane. Increasing copper levels in traditional media has been shown to facilitate the transition of glucose metabolism from a high flux state to low flux state in the late exponential growth phase. Since copper is highly oxidative, caution should be taken when raising copper to very high levels.

Panel 7.18. Trace Elements in Serum-Free Media

- Trace elements play a variety of biological roles and must be supplied in a chemically defined medium
  - Fe^{3+} is the part of heme that carries oxygen
  - Fe^{3+} and Cu^{2+} are both involved in electron transfer in the mitochondria
  - Mn^{2+} is a component of many enzymes, including DNA/RNA polymerases and glycosyl transferases
- Fe, Zn, and Si are present in higher concentrations than all others
- Except for Zn, these elements have multiple oxidation states, are redox active, and are harmful at high concentrations
- Fe and Cu are especially reactive and can generate ROS. Their concentration is very low in body fluids.
- Fe^{3+} binds to transferrin in blood and is sequestered in cells by binding to ferritin

Zinc is present at high concentrations in the pancreas and is conjugated with insulin. Its distribution in tissues is somewhat uneven. Zinc does not participate in redox reactions, but is a cofactor of a large number of enzymes. Many metalloproteases are zinc proteins, including carboxypeptidases and matrix metalloproteases. Other zinc proteins include many zinc finger proteins that bind to DNA and play a regulatory role, as well as carbonic anhydrase, which is localized on the mitochondria membrane and facilitates the solubilization of carbon dioxide to bicarbonate.

Manganese is involved in photosynthesis in the oxidation of H_2O to O_2 in plants. It is a component of many enzymes including DNA/RNA polymerases and glycosyl transferases (involved in glycosylation). Manganese is also associated with enzymes involved in reversing oxidative stress such as catalases and peroxidases. It is an essential element, although it is required at a low level.

Cobalt is a component of vitamin B12. Selenite serves as an antioxidant. It was not realized as an essential nutrient in the early days of serum-free medium development because it was supplied as a minute contaminant in another media component. The serum-free medium F-12 developed in the 1970s had only the trace elements described above. In later development of MCDB, the list was expanded to that shown in Table 7.5.

Non-Nutritional Medium Components

Some medium components are added primarily for operational purposes instead of nutritional ones. They can be omitted without harmful effects. For example, sodium bicarbonate is a pH buffer and Pluronic F-68 is used as a protective agent from mechanical stress. Sodium bicarbonate can be replaced by another pH buffer such as HEPES. Pluronic F-68 may be removed when cells are grown in flasks without mechanical agitation.

Sodium bicarbonate buffer

Cellular respiration generates a large quantity of CO_2 that becomes bicarbonate after dissolving in water. Bicarbonate provides a natural pH buffer in a human's body fluid. In the early days of cell culture media development, sodium bicarbonate was adopted as a pH buffering reagent.

Consider that CO_2 is present in a system consisting of a gas phase and an aqueous phase. Over a period of time, CO_2 distributes between the gas phase and the liquid phase and reaches equilibrium (Eq. 7-1, Panel 7.19). At equilibrium, the concentration of CO_2 between the two phases is described by Henry's law (Eq. 7-2). Upon association with water, the dissolved $CO_{2(aq)}$ is dissociated into bicarbonate (HCO_3^-) and H^+ (Eq. 7-3). At equilibrium, the relationship between the $CO_{2(g)}$ in the gas phase and the bicarbonate and H^+ in the liquid phase is described by Eqs. 7-4 and

Panel 7.19. CO_2 as a pH Buffer

- CO_2 is the pH buffer in the human body's fluid
- Media that contains bicarbonate becomes alkaline very rapidly due to the loss of CO_2 when removed from the incubator
- The pK_a of bicarbonate is low (6.3)
- The $NaHCO_3$ buffer requires appropriate CO_2 concentrations in the gas phase. The reactions are:

CO_2 dissolves in an aqueous solution to reach equilibrium:

$$CO_{2(g)} \rightleftharpoons CO_{2(aq)} \quad \textbf{(Eq. 7-1)}$$

The CO_2 concentration in liquid is described by Henry's Law:

$$P_{CO_2} = H\left[CO_{2(aq)}\right] \quad \textbf{(Eq. 7-2)}$$

CO_2 in an aqueous solution forms a bicarbonate ion:

$$CO_{2(aq)} + H_2O \rightleftharpoons H_2CO_3 \rightleftharpoons HCO_3^- + H^+ \quad \textbf{(Eq. 7-3)}$$

The equilibrium is described as:

$$K_{eq} = \frac{\left[HCO_3^-\right]\left[H^+\right]}{\left[CO_{2(aq)}\right]} \quad \textbf{(Eq. 7-4)}$$

Applying Eq. 7-2 to the equilibrium relationship:

$$K_{eq} = \frac{\left[HCO_3^-\right]\left[H^+\right]}{P_{CO_2}/H} \quad \textbf{(Eq. 7-5)}$$

pH = $-\log_{10}[H^+]$ and pK_a = $-\log(K_{eq})$, so:

$$pH = pK_a + \log\frac{\left[HCO_3^-\right]}{P_{CO_2}/H} \quad \textbf{(Eq. 7-6)}$$

The pH of the solution is affected by P_{CO_2} and HCO_3^-.

7-5. By taking a logarithmic transformation and applying the definition of pH ($\log_{10}[H^+]$) and pK_a ($\log_{10}K_{eq}$), one obtains the relationship between the pH and the gas phase level of CO_2 and HCO_3^-.

In a cell culture medium, HCO_3^- is supplied as sodium bicarbonate ($NaHCO_3$). In order to reach the target pH, some amount of CO_2 must be provided in the gas phase depending on the sodium bicarbonate concentration in the medium. The buffering capacity in the medium is dependent on the concentration of the bicarbonate in the medium, and the buffering action of bicarbonate requires the presence of CO_2. At a given concentration of bicarbonate, the pH of the medium is dependent on the CO_2 level in the gas phase. Similarly, at a constant CO_2 level the pH at equilibrium is affected by the bicarbonate concentration. The relationship can be plotted from Eq. 7-6 as shown in Figure 7.6.

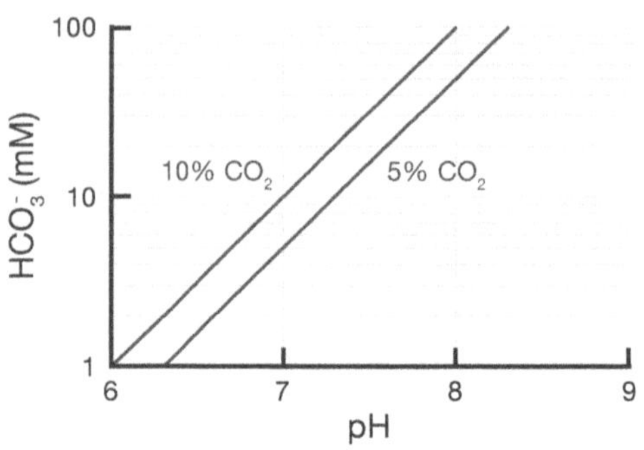

Figure 7.6. The relationship between the sodium bicarbonate concentration, pH, and atmospheric carbon dioxide level.

When selecting a buffer, one normally chooses a compound whose pK_a is near the target pH. The optimal pH for most cells is in the neutral range, or even 7.2–7.4, except in bioprocess, where the pH is often controlled at the low side, around 6.9–7.0, to reduce lactate production. The pK_a of bicarbonate is about 6.3, making it not an ideal buffer for neutral pH. A typical cell culture medium also contains sodium phosphate and HEPES, which has a pK_a in the neutral range, to increase the pH buffer capacity in neutral pH.

Buffering effect of bicarbonate, lactate, and CO_2

Typical cell culture medium contains 14–44 mM $NaHCO_3$. In a cell culture incubator with 10% CO_2 in ambient air, the pH in a DME medium will be initially ~7.4. As cells grow, lactate and CO_2 are produced. Each mole of lactate excreted by cells is accompanied by a mole of H^+. The pK_a of lactic acid is 3.7. Hence, all lactic acid produced from cell metabolism is dissociated to lactate and H^+ under cell culture conditions. Similarly, $CO_{2(aq)}$ produced from metabolism associates with water to generate a bicarbonate and a H^+. Because the pK_a of bicarbonate is pH 6.3, when in neutral pH the ~40 mM of bicarbonate in a DME medium has only ~20 mM of buffer capacity.

The bicarbonate level in a human's blood is about 25 mM, while the normal lactate level is about 2 mM. The buffer capacity of bicarbonate is sufficient to hold blood pH neutral. However, the buffer system provided by CO_2 in our blood is a dynamic system. CO_2 is continuously produced by cells and ventilated in the lungs, creating a balance of production and ventilation that maintains the CO_2 level in blood at ~25

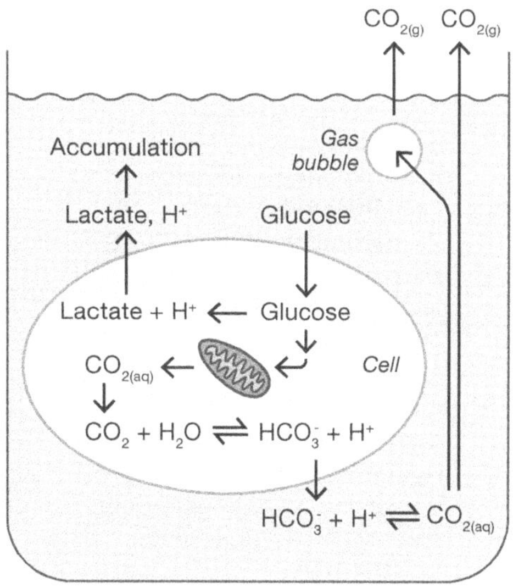

Figure 7.7. Stripping of CO_2 from the bioreactor.

mM. The ventilation is necessary to maintain the blood's pH. The other metabolite produced in very large quantities that influences blood pH is lactate. Lactate, once it is generated through muscle exercise or other activities, is consumed by some tissues like muscle and liver (Figure 7.7).

The buffering of pH in a cell culture process is not only provided by the sodium bicarbonate initially present in the medium, but also by the CO_2 produced by cells. Like the dynamic system present in the blood, there should be a balance between production and removal through aeration. At the beginning of a culture process, CO_2 is present at a level that is appropriate for the sodium bicarbonate concentration in the medium and for the maintenance of the pH. As cells grow and lactate accumulates, the CO_2 level in the gas phase should be reduced to strip $CO_{2(aq)}$ from the medium. This neutralizes H^+ stoichiometrically (Eq. 7-3 and Figure 7.7) and requires that the CO_2 stripping capacity in the reactor is sufficient. However, the CO_2 stripping capacity of most industrial bioreactors is insufficient, as will be discussed in Chapter 11.

Alternative buffers

Under some conditions, such as when a CO_2 incubator is not available, sodium bicarbonate is unsuitable for buffering pH. A number of alternative pH buffers are listed in Table 7.6. Most cells have a relatively high tolerance to HCO_3^- concentration (over 50 mM), but the tolerance to alternative buffers is lower. At excessively high concentrations, these pH buffers can exert a negative effect on cell growth (Panel 7.20). Among these buffers, HEPES is the most commonly used, as its pK_a is in the neutral range and most cells have a moderate tolerance to its presence.

It should be noted that CO_2 is also an essential nutrient for cell growth. It is a substrate in many biochemical carboxylation reactions. Normally, the demand for CO_2 is met by its production in cell metabolism. If cells are cultured in a bicarbonate-free medium and the ambient air is aerated vigorously through the medium to strip off the CO_2 produced by the metabolism, one may see cell growth inhibition until the aeration is stopped, thus allowing the CO_2 produced in the metabolism to accumulate. When a sodium

Table 7.6. Non-Bicarbonate Buffers Used in Cell Culture

	pK_a value at 37 °C	*Anhydrous mol. wt.*	*Working concentration (mM)*
HEPES	7.31	238.3	10–28
TRICINE	7.80	179.2	<50
MOPS	7.01	209.3	10–20
Glycylglycine	7.95	132.1	10–20

Panel 7.20. Alternative Buffers

- Sodium β-glycerophosphate (20 mM) functions as a detoxifier of ferric chloride hydroxo compounds (i.e., Fe^{3+} chelator)
- Zwitterionic buffers: HEPES (N-2-hydroxyethylpiperazine-N-2-ethane) is commonly used between 10–30 mM
- Alternative buffers can be growth inhibitory at high concentrations
- Requires balance of osmolality by adjusting bulk salt levels by the addition of NaCl/KCl

bicarbonate-free medium is used to grow cells, it is common practice to maintain the CO_2 in the gas phase at 0.2–0.5%. Many media have low concentrations of sodium bicarbonate and HEPES to provide a neutral pH at the beginning of a cell culture process. The low level of bicarbonate provides the CO_2 required for cell growth.

Antioxidants

Panel 7.21. Physiologically Relevant Antioxidants

- Glutathione (reduced)
- Vitamin E
- Vitamin C
- Taurine
- Selenium
- Catalase
- Superoxide Dismutase
- β-mercaptoethanol

During cell metabolism, reactive oxygen species (ROS), free radicals, and various superoxide radicals arise and accumulate both inside and outside the cell. Additionally, some medium components such as Fe^{3+} and Cu^{+} are oxidative. As well as vitamins E and C for species that do not synthesize vitamin C, physiologically relevant compounds such as glutathione, enzymes that neutralize oxidized and potentially damaging compounds, and even chemicals that are not naturally occurring in the cell (e.g., mercaptoethanol) are sometimes included in the medium (Panel 7.21). Among them, glutathione and mercaptoethanol are the most commonly seen.

Glutathione is a key player that cells use to reduce ROS. Glutathione is a tripeptide of glutamic acid, cysteine, and glycine. In mammals, it is synthesized primarily in the liver and transported through blood circulation to other tissue cells, where it is further transported across the cytoplasmic membrane into the cytoplasm and certain organelles. Glutathione is often included in cell culture media, especially when cultivating differentiated tissue cells.

In addition to physiological antioxidants like vitamins C and E and glutathione, non-physiological antioxidant compounds are often included in culture media to minimize oxidation of labile medium components in culture fluid. β-mercaptoethanol is routinely used in the cultivation of stem cells and differentiated cells. Caution should be taken when using reducing agents in a medium if secreted proteins containing disulfide bonds are being produced. Mercaptoethanol can reduce the disulfide bridge to free thiols, which may then become reoxidized and form incorrectly paired disulfide bonds.

Agents modulating cells' physical interactions with the environment

Some medium components are added with the intention of modulating the physiochemical interactions between cells and neighboring

Panel 7.22. Media Components that Modulate Physical Interactions

- Heparan sulfate, heparin, and dextran sulfate can be used to facilitate the adaptation of cells to suspension growth
- Pluronic F-68 is used to protect cells from mechanical damage caused by agitation and aeration

cells or cells and their environment. The viscosity of the medium was speculated to be important for sustaining cell growth in culture in the early days of serum-free culture medium development. Some compounds, such as polyethylene glycol and carboxymethylcellulose, were tested in the early development of serum-free media, but are seldom used now (Panel 7.22).

Many cells used in bioprocessing are adherent cells. To adapt them to grow in suspension, sometimes heparan sulfate or dextran sulfate is added to the medium to reduce cell aggregation and adhesion to surfaces and facilitate the adaptation to suspension growth. The effectiveness of the compound is affected by the charge and the molecular weight.

Most cells are derived from tissue cells that are not normally exposed to high fluid shear conditions like those found in a bioreactor. When adherent cells are grown on microcarriers or suspension-adapted cells are grown under agitated conditions, cell damage caused by aeration and mechanical agitation is frequently seen. To protect cells grown in stirred tank bioreactors from damages caused by extensive fluid mixing and aeration, a protective agent, Pluronic F-68, was used in the cultivation of BHK cells in 1960s. Many Pluronic surfactants are available; all are block copolymers of polyoxyethylene (POE) and polyoxypropylene (POP). The larger the POE group, the more hydrophilic the molecule is and the greater its detergent-like activity and cell cushioning effects. The larger the POP group, the greater the toxicity and the anti-foaming ability. Pluronic F-68, in a concentration range of 0.01–0.2%, provides a protective effect under agitation and aeration conditions. The degree of foaming is high. Therefore, it is desirable to determine if a suitable replacement is available. Pluronic F-88 and F-77 may also provide suitable protective effects. As a surface-active agent, Pluronic F-68 alters the interfacial properties of the cell–fluid and gas bubble–fluid interfaces. Despite this, the mechanism of its protective effect is not well understood.

Antibiotics and selective agents

Penicillin and streptomycin have been frequently used in cell maintenance in research laboratories. The practice was relatively routine in the earlier days, when medium preparation equipment and practice were not as robust. For manufacturing processes, antibiotics are rarely used, although in vaccine production neomycin, polymyxin B, streptomycin, and gentamicin are sometimes used. For the cultivation of primary cells, which are not subjected to a long duration of quality control prior to processing, the use of gentamicin reduces the rate of contamination significantly. Some antibiotics are only moderately more inhibitory to bacteria than to cultured cells. Toxicity testing is necessary before use. Many cell lines have been subjected to selection using antibiotics or metabolic inhibitors, such as methotrexate, as discussed in Chapter 6. In some cases, a low level of selective agent is present in the cell maintenance medium to reduce the risk of gene copy loss and the subsequent reduction of productivity.

High Molecular Weight and Complex Supplements

Some cells are capable of rapid growth using only a basal medium without any further supplements. However, such an extraordinary capability is the exception rather than the norm of cells in culture. Most cells in culture require supplementation of a number of growth factors and carrier proteins. Some industrial cell lines can be grown with only a basal medium supplemented with a few non-protein compounds (thus a protein-free medium), but at a substantially lower growth rate. When an enriched medium is used, cells grow at a faster rate. Many stem cells and other normal diploid cell lines require high concentrations of serum to grow. Serum is still commonly used in the isolation of primary cells, at least in the early stage of cell cultivation. In this section we will discuss the role of a few key high molecular weight supplements in cell culture.

Serum or Biological Fluids

Serum is the fluid left behind after blood coagulates, free of blood cells and most coagulation proteins. It is a complex mixture containing nutrient substances, metabolites, hormones, substances released from damaged cells (e.g., hemoglobin and growth factors from platelets), and various plasma proteins. Fetal bovine serum (FBS) is the most widely used serum in animal cell culture because it contains high concentrations of growth stimulatory factors and low concentrations of growth inhibitory factors. Other common sources of sera are human, bovine, newborn bovine, donor bovine, and donor horse.

Serum serves many different and important roles in cell culture (Panel 7.23). In addition to providing nutrients not sufficiently present in basal media (e.g., cholesterol), serum provides factors for the attachment of adherent cells (e.g., vitronectin, fibronectin) and modulates colloid osmolarity. Serum contains protease inhibitors and neutralizes trypsin used in cell detachment and other enzymes released by dead cells. Serum contains the carrier proteins transferrin and serum albumin. Carrier proteins chaperone components that are in very low concentrations, are poorly soluble (e.g., fatty acids carried by albumin), or are unstable (e.g., ferric ions carried by transferrin).

Serum is rich in "bulk" proteins (e.g., serum albumin) that can prevent nonspecific adsorption of critical factors to culture vessels. Serum also plays an important role as a scavenger. In cell cultivation, various contaminants may arise from numerous sources. For example, minute chemical components may leach out from parts of the reactor or the filter used in medium preparation. Some of these chemicals may be detrimental to cell growth or have other negative effects on cells. Due to the presence of serum in the medium, those compounds may be sequestered by adsorption to serum proteins before they can act on cells, thus minimizing potential damage.

Animal serum, however, has numerous disadvantages in addition to cost and the difficulty of maintaining consistent quality (Panel 7.24). The most serious concern is the possibility of contamination with animal viruses or prions that are infectious to cells used in the production. The presence of serum in culture medium makes downstream processing more complicated. It also makes the task of final product characterization more complex, as the number of heterologous proteins from the serum must be lowered to beneath an acceptable threshold. Serum also carries antibodies against various antigens to which the animals have been exposed. One has to check each lot of serum to ensure no reaction to the product protein or virus. For virus production processes, if serum antibodies cross-react with the product virus the production will be drastically affected.

Panel 7.23. Roles of Serum in Cell Culture Media

- Provides hormones and growth factors
- Protease inhibitors (e.g., α-2-macroglobulin) neutralize proteases used in trypsinization or produced by dead cells
- Provides carrier proteins
  - For low molecular weight substances (e.g., transferrin)
  - For nutrients which dissolve poorly (e.g., fatty acids, cholesterol, apolipoprotein)
- Binds nonspecifically and releases slowly the medium components that are inhibitory when present in excess
- Binds and/or neutralizes toxic substances (e.g., detergents)

Panel 7.24. Disadvantages of Serum in Cell Culture Media

- Potentially introduces animal viruses and other undesirable contaminants into cell culture (e.g., adventitious agents, antibiotics, proteases)
- Variability in quality
- High running costs and unnecessary capital outlay
- Normally purchased in large lot sizes, resulting in costly storage
- Serum lot testing can be tedious and costly
- Increases the complexity of downstream processing and final product characterization

Insulin and Insulin-Like Growth Factor (IGF-1)

Insulin plays a key role in regulating glucose uptake for many cell types. In addition to modulating glucose metabolism, insulin also exhibits mitogenic effects and stimulates cell growth through an overlapping pathway with IGF-1 (Panel 7.25). IGF-1 has an acute effect on protein and carbohydrate anabolism by increasing cellular uptake of amino acids and glucose, and by stimulating glycogen and protein synthesis. IGF-1 also affects cell proliferation, differentiation, and apoptosis. It is

Panel 7.25. Insulin and Insulin-Like Growth Factor

- Insulin stimulates glucose uptake by adipocytes and other cells, and also has a mitogenic effect at high concentrations
- Insulin is used in culture at the 1–10 μg/mL range. The blood insulin level is 4μU/mL or 1.3 μg/mL (1 μU = 0.33 μg); the IGF-1 level is 100–200 ng/mL.
- Insulin and IGF-1 have an overlapping signaling pathway through IR and 1GF1R
- IGF-1 can replace insulin in cell culture at a lower concentration

a potent mitogen that increases DNA synthesis and stimulates the expression of cyclin D in a wide variety of cells.

Both insulin and IGF-1 bind to the insulin receptor (IR) and IGF receptor (IGF1R), but with different affinities. After insulin binding, IR or IGF1R is phosphorylated, leading to activation of an insulin receptor substrate (IRS). There are multiple isoforms of IRS that are distributed differently in cells of different tissues. The signal is then relayed to downstream signaling pathways.

The response of the cell to insulin and IGF is dependent on the abundance level of the different IRS isoforms. Most cells, including CHO cells, express both IR and IGFR. However, NS0 cells express only IGFR. Differential binding to IR and IGF1R, as well as differential activation of various IRS isoforms, leads to different responses to insulin and IGF-1.

Insulin is used in cell culture at concentrations that are nearly 100-fold higher than that found in blood. At such a high concentration, insulin can trigger a mitogenic response. IGF has a much stronger affinity for IGF1R and is used at a much lower concentration than insulin.

Industrial NS0 cells or other myeloma lines are grown without insulin supplementation, and many CHO cells have been adapted to grow without insulin. Some differentiated cells derived from cancer, such as PC12 cells derived from pheochromocytoma, can also grow without insulin supplements. The changes in the signal transduction pathway that enable those cells to rapidly proliferate without insulin or IGF-1 stimulation is not yet clear. IGF-1 has a short half-life and is comparatively costlier than insulin. Also available is a protein-engineered IGF-1 called Long R3-IGF-1, which has a slower degradation rate and longer bioavailability in culture. Insulin, as one of the first recombinant proteins to be made available for widespread use in the treatment of diabetes, is a commodity chemical. For the manufacturing of recombinant therapeutics, insulin is commonly used instead of IGF-1.

Transferrin

Transferrin is the iron carrier glycoprotein in blood circulation in mammals. As we discussed in the trace metal section, ferric ions are highly oxidative. In the human body, it exists primarily in bound form to heme and other proteins with iron centers, or in bound form to transferrin in blood circulation (Figure 7.5) (Panel 7.26). Transferrin has a large number of binding sites with very high binding constant for ferric ion. When bound to transferrin, Fe^{3+} is unavailable for oxidizing other chemical species. Transferrin binds to the transferrin receptor on a cell's surface to form a complex which is internalized into an endosome. Upon internalization, Fe^{3+} is released from the transferrin into the low pH environment of the endosome, where it is reduced to Fe^{2+} by ferrireductase and transported into the cytoplasm. Transferrin and its receptor are recycled to the extracellular environment and cytoplasmic membrane, respectively. Once in the cytoplasm, Fe^{2+} is stored in ferritin or is transported into sites of use, such as the mitochondria or target proteins.

Panel 7.26. Transferrin

- Typically used at 1–30 μg/mL (MW 80kDalton, 10 μg/mL = 0.1 μM)
- Binds to iron very strongly
- May be replaced by an iron-chelating agent, such as citrate

Recombinant human transferrin is commercially available. In general, transferrin will bind to receptors from many different species, although the binding affinity may be affected by the species origin of the transferrin–transferrin receptor pair. The concentrations required for cells of different species may differ and must be empirically determined. Furthermore, the cost of recombinant transferrin is a factor in its widespread application.

In addition to transferrin-mediated iron uptake, another route, non-transferrin-bound iron uptake, is used by cells.[6] A significant portion of Fe^{3+} in blood is bound to citrate instead of transferrin. In lieu of transferrin, Fe^{3+} can be chelated to citrate in a freshly prepared stock solution for addition to cell culture medium. Depending on the chelating agent used, the resulting iron-chelator complex may sequester iron availability to cells if a membrane transport mechanism is unavailable, or may make the complex available if an uptake system exists. Evidence suggests that the iron-citrate complex is taken up by cells.

Serum Albumin and Other Carrier Proteins

Serum albumin is the most abundant protein in human plasma. It is a nonglycosylated, negatively charged protein of 585 amino acids and serves as a carrier for many compounds that may have low solubility in aqueous solutions (Panel 7.27). Most notably, serum albumin is a carrier for fatty acids, bilirubin, heavy metal ions, and other agents that may harm cells. Serum albumin is probably the most important protein that mediates scavenger functions of serum in cell culture medium.

Panel 7.27. Serum Albumin

- Most abundant protein in blood (~3.5–5 g/L), MW 66.5 Kda, used in medium at 0.1–5 g/L
- Carrier protein for many different molecules, fatty acids, lipids, Fe^{3+}, and inhibitory metabolites like bilirubin
- Highly versatile molecule, but difficult to completely chemically characterize the compounds bound to it

Recombinant forms of human albumin are available. Most recombinant proteins are easily characterized in a chemically defined medium with respect to their purity and minute contaminating species. In comparison, the diverse binding capacity of serum albumin to a wide variety of species makes determining its complete chemical properties difficult. Serum albumin from different preparations may be bound to different amounts and varieties of fatty acids or other compounds. Hence, serum albumin may not be used in a chemically defined medium.

Protein Hydrolysates

Hydrolysates, or extracts from yeast, animal, or plant tissues, have been commonly used in cell culture processes to reduce the need for animal serum. Beef extract, such as the commercial Ex-Cyte, served as a ready-to-use source of phospholipids and other nutrients until the use of animal extracts was mostly discontinued in industrial cell culture. Some processes employ hydrolysates derived from soy, rice, and other plants by enzymatic or acid hydrolysis. Plant hydrolysate is a complex mixture consisting of various oligopeptides of different compositions and lengths, oligosaccharides, other organics, and even various metal ions including heavy metals. The composition varies with the method and conditions of hydrolysis, the source of the raw materials, and the season of harvest. Variation between lots is inevitable and difficult to quantify. The roles of hydrolysates are not completely understood. They may provide some nutrients and minerals, and may also act as scavengers through undefined molecular interactions with possible contaminants.

Concluding Remarks

OUR UNDERSTANDING of the nutritional requirements of cells in culture has improved drastically over the years, aided by the advances in chemical and biochemical analytical tools. The aim of the development of industrial cell culture media is to harness these understandings and incorporate stoichiometric and kinetic considerations. This chapter gives succinct descriptions of the roles of key medium components in supporting cell growth. While general nutritional needs are common to all cells, certain needs arise that are specific to particular cell lines or cell types. In the coming years, industrial medium development will take a genomics-assisted approach, employing epigenomic or global gene expression profiles of the target cell to design the medium based on the cell's capability to utilize certain nutrients or its need for specific growth factors. Industrial medium development will also seek to optimize cell growth during the cell expansion stage and maximize productivity in the production bioreactor. This chapter has not discussed the potential effects of medium design on cell stability, since little is known and few systematic studies have been done. A number of medium components affect the generation of cell-damaging compounds such as ROS and free radicals. In conclusion, the use of contemporary methods of medium design based on biochemical mechanisms, genomic and transcriptomic insights, and high-throughput experimental exploration will facilitate further advances in cell culture biomanufacturing.

References

1. Bettger, W., S.T. Boyce, and B.J.H. Walthall, R.G., *Rapid clonal growth and serial passage of human diploid fibroblasts in a lipid-enriched synthetic medium supplemented with epidermal growth factor, insulin and dexamethasone.* Proc. Natl. Acad. Sci. U.S.A., 1981. **78**: p. 5588–92.
2. Khetan, A., et al., *Control of misincorporation of serine for asparagine during antibody production using CHO cells.* Biotechnol. Bioeng., 2010. **107**: p. 116–23.
3. Combs, G.F.J. and J.P. McClung, *Vitamin E.*, in *The vitamins: fundamental aspects in nutrition and health, fifth edition.* 2017: Elsevier. 207–42.
4. Du, J., J.J. Cullen, and G.R. Buettner, *Ascorbic acid: Chemistry, biology and the treatment of cancer.* Biochimica et Biophysica Acta (BBA) - Reviews on Cancer, 2012. **1826**(2): p. 443–57.
5. Bleackley, M.R. and R.T.A. MacGillivray, *Transition metal homeostasis: From yeast to human disease.* Biometals, 2011. **24**: p. 785–809.
6. Brissot, P., et al., *Non-transferrin bound iron: A key role in iron overload and iron toxicity.* Biochim. Biophys. Acta, 2012. **1820**(3): p. 403–10.

Nomenclature

H	Henry's law constant
K_{eq}	Equilibrium constant
K_M	Half-saturation constant
P	Pressure or partial pressure
pK_a	Dissociation constant

Contents

8

Cell Culture Bioreactors

Introduction

AT THE DAWN of recombinant DNA technology, cell culture was largely a research tool. The industrial use of animal cells was limited to the production of viral vaccines for human and veterinary applications. Some cell lines used for virus production could be grown in suspension cells and were cultured in flasks or stirred tanks. The majority of cell lines used were adherent and were mostly grown in roller bottles. Nowadays, industrial cell culture is operated in scales up to tens of thousands of liters. In spite of the dramatic increase in its industrial output over the last four decades, the basic design and operation of the bioreactor in cell culture processes have remained remarkably similar. The stirred tank bioreactor that has been the workhorse of bioprocess for microbial fermentation is also the reactor of choice for cell culture bioprocess. However, the adoption of the stirred tank for cell culture was not uniformly accepted in the early stage of cell culture bioprocess development. For nearly a decade until the early 1990s, much research and commercial development focused on exploring new bioreactors to overcome a number of hurdles cell culturing faced. The complex nutritional requirements of mammalian cells, especially their dependence on animal serum for growth and the resulting very high protein content in the medium, posed challenges in air sparging. The mechanical damage caused by agitation was another concern. What was unanticipated was the extraordinary capability of cells to adapt to suspension growth in highly turbulent flow conditions, and to adapt to serum-free, and even protein-free, culture media. The adoption of much leaner and simplified media helped expand the bioreactor's operational parameter space, especially the reduction of protein content in the medium, which eased the use of air sparging for oxygen supply. By the early 1990s, the stirred tank that had been the workhorse for biochemical manufacturing for products such as antibiotics and amino acids became the bioreactor of choice. What has not changed since is the fact that the culture medium, constrained by cells' growth needs and intolerance to high osmolality, can support only relatively low cell concentrations and resultant low product concentrations, and that the accumulation of metabolites causes growth inhibition and limits productivity. These constraints are major factors in selecting the operating mode of a process.

This chapter will discuss the fundamentals of bioreactors. Despite virtually all industrial bioreactors being of the stirred tank type, some auxiliary equipment used in the process utilize a different type of reactor, and should be analyzed as such. Hence, an understanding of the basics of reactor performance is useful. For more in-depth discussion, readers are referred to a biochemical engineering textbook.[1] This chapter will also highlight a number of bioreactors that were developed in the exploration stage of cell culture processing. These examples are useful for their conceptual novelty and help to illustrate the limitations on their use as industrial bioreactors. Although these reactors are not commonly used in current processes, they might find specialized applications in other areas, such as tissue engineering and gene or cellular therapy.

Basic Types of Bioreactors

BIOREACTORS CAN BE GENERALLY CATEGORIZED according to their mixing characteristics. The two extremes of mixing are complete, instantaneous mixing and a solid-like complete absence of mixing. The differences are best illustrated using a continuous flow reactor, although many bioreactors are used in batch fashion.

The two extreme mixing patterns characterize two types of idealized continuous reactors: the plug-flow (tubular) reactor (Figure 8.1) and the well-mixed stirred tank (Figure 8.2). In an ideal well-mixed bioreactor, the mixing is assumed to be so intensive that the fluid is homogeneous

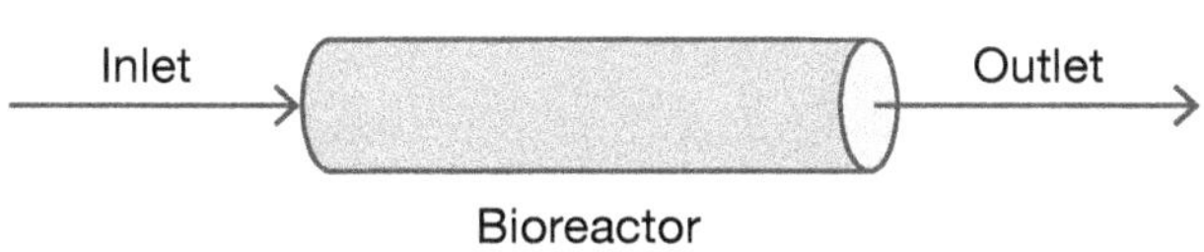

Figure 8.1. Schematic of a tubular flow reactor.

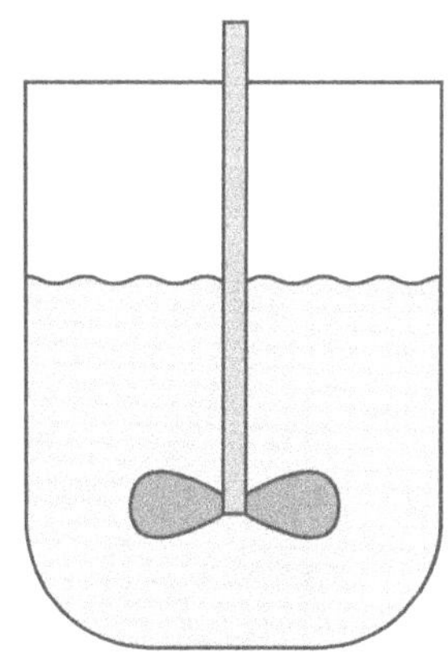

Figure 8.2. Schematic of a stirred tank reactor.

Panel 8.1. Idealized Bioreactor

- Plug flow reactor (no mixing)
- Well mixed reactor (instantaneously perfectly mixed)

throughout the reactor. When a new solution is added to the reactor, the solute is instantaneously and uniformly distributed.

A tubular reactor is the other extreme of an idealized bioreactor; it has absolutely no mixing (Panel 8.1). It is also called a plug flow reactor or piston flow bioreactor. As its name implies, a solution entering the bioreactor from one end moves downstream like a wall or a plug, and continues steadily forward until it reaches the exit.

Stirred Tank (Well-Mixed) vs. Tubular Reactor (Plug Flow)

The distinction between a well-mixed continuous stirred tank reactor (CSTR) and plug flow reactor (PFR) is best illustrated by a comparison of their behavior after a step change in feed concentration. Consider a continuous reactor that has an inlet stream (feed) and an outlet stream (effluent) that are equal in volumetric flow rate; thus, the volume of the reactor is constant. Initially, both the fluid in the reactor and the feed stream are colorless. At time = 0, the feed stream is switched to a fluid containing a red dye at a concentration of c_i. We then observe how the concentration of the dye has changed at the outlet.

In a PFR, the red dye moves downstream like a sharp band, since there is no back mixing or diffusion to blur the sharp boundary between the colorful and colorless streams. After switching the feed stream to the red stream, it takes one holding time (V/F, see Figure 8.3b) for the red color stream to fill the reactor and reach the exit. When the dye begins to leave the reactor, its concentration is identical to that at the feed (Figure 8.3).

In the case that the reactor is well mixed (i.e., a CSTR), the color is distributed uniformly throughout the reactor as soon as the dye is added to the feed stream. Because the dye is uniformly distributed, the effluent will have the same concentration of dye as the reactor, regardless of where it is drawn from. Of course, for the purposes of our discussion, we

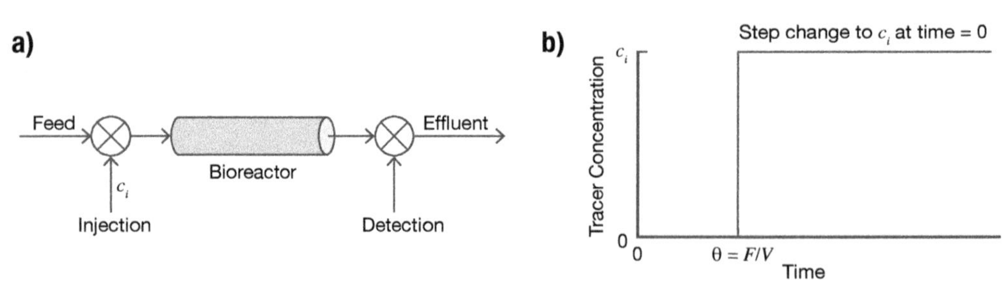

Figure 8.3. (a) A tubular reactor with a switch for tracer injection, and (b) the concentration profile of the tracer after the switch is turned on.

neglect the time delay caused by flow in the pipe leading into and out of the reactor. As more colored feed enters the reactor, the concentration of the color seen in the outlet will gradually increase. The balance equation for the concentration of the dye is shown in Panel 8.2 (Eqs. 8-1 and 8-2). Note that in this case, the concentration in the outlet flow is the same as in the reactor, because of the assumption of being well mixed.

By setting initial condition $c = 0$, one can solve the equation and plot the profile of the normalized concentration of the dye (c/c_i) over time (Eq. 8-3, Panel 8.2). V/F has units of time and represents the holding time (θ) (Eq. 8-4), the time it takes one reactor volume of feed to pass through the reactor. One can see that, after one holding time, the concentration of the dye in the reactor is 0.63 of that in the feed. It takes three holding times (3θ) for the concentration to approach (~98%) that in the feed (Figure 8.4). It is clear that a PFR has a very different mixing behavior from a CSTR. Of course, the reactors that are actually in use do not exhibit such ideal behavior, but many will have a mixing profile closer to one type than the other.

CSTR and PFR with Reactions

The discussion above considers only mixing behavior in the reactor, essentially just mixing tanks. In a bioreactor,

Panel 8.2. Material Balance of a Continuous, Well-Mixed Stirred Tank Bioreactor with a Step Change of Tracer Concentration

$$\frac{dV}{dt} = F_{in} - F_{out} = 0 \qquad \textbf{(Eq. 8-1)}$$

$$F_{in} = F_{out} = F$$

$$V\frac{dc}{dt} = F(c_0 - c) \qquad \textbf{(Eq. 8-2)}$$

Initial condition:

$$t = 0 \quad c = 0$$

Solution:

$$c = c_0\left(1 - e^{\frac{-F_t}{V}}\right) \qquad \textbf{(Eq. 8-3)}$$

Define holding time: $\theta = \dfrac{V}{F}$ **(Eq. 8-4)**

$$c = c_0\left(1 - e^{\frac{-t}{\theta}}\right) \qquad \textbf{(Eq. 8-5)}$$

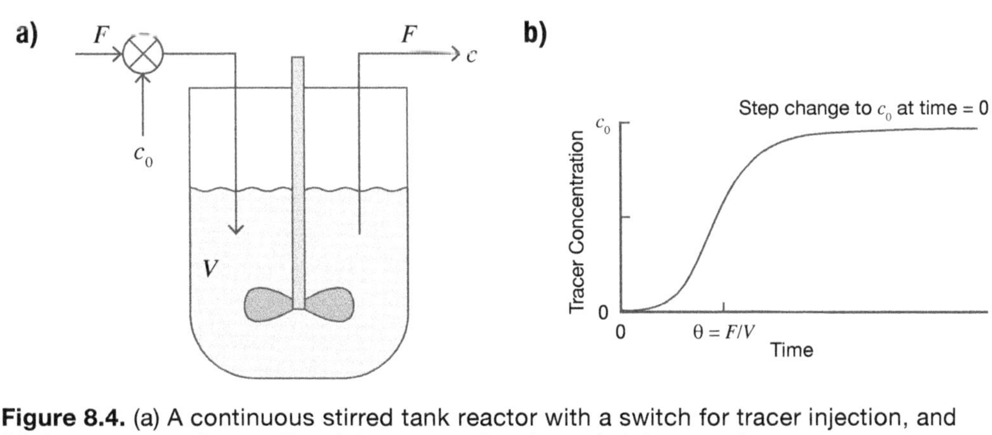

Figure 8.4. (a) A continuous stirred tank reactor with a switch for tracer injection, and (b) the concentration profile of the tracer after the switch is turned on.

Panel 8.3. Material Balance of a Stirred Tank Bioreactor with a Reaction at Steady State

$$V\frac{dc}{dt}=F\left(c_0-c\right)-rV \quad \textbf{(Eq. 8-6)}$$

$$V\frac{dc}{dt}=\frac{F}{V}\left(c_0-c\right)-r \quad \textbf{(Eq. 8-7)}$$

$$V\frac{dp}{dt}=\alpha_{p/c}r-\frac{F}{V}p \quad \textbf{(Eq. 8-8)}$$

At steady state $\frac{dc}{dt}=0$

$$\frac{F}{V}\left(c_0-c\right)=r \quad \textbf{(Eq. 8-9)}$$

$$\frac{F}{V}p=\alpha_{p/c}r \quad \textbf{(Eq. 8-10)}$$

Panel 8.4. Material Balance of a Plug Flow Reactor

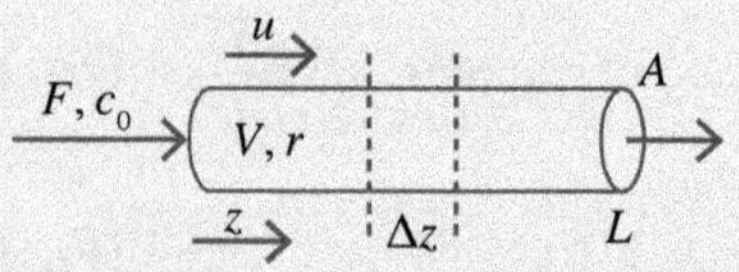

$$0=Fc\big|_{z+\Delta z}-Fc\big|_{z}+r\Delta zA \quad \textbf{(Eq. 8-11)}$$

$$Fc\frac{dc}{dz}=-rA \quad \textbf{(Eq. 8-12)}$$

$$\frac{F}{A}\frac{dc}{dz}=-rA \quad \textbf{(Eq. 8-13)}$$

$$u\frac{dc}{dz}=-r \quad \textbf{(Eq. 8-14)}$$

consumption of nutrients and generation caused by product formation and growth take place; these processes (reactions) will cause the concentration profile to differ from that in a simple mixing tank. For a CSTR, the assumption that the concentration at the outlet is identical to that in the reactor holds. The balance of a reactant, i.e., how the amount of a reactant in the reactor changes with time, is thus the input of the reactant into the reactor, minus the output and the amount of the reactant consumed during the reaction (Eq. 8-6, Panel 8.3). For the product, it is then the input (usually nil) minus the output, plus the amount generated from the reaction. The effects of the reaction on the material balance equations are dealt with by adding a term to the material balance equation to account for the reaction (Eqs. 8-6 and 8-8, Panel 8.3). One balance equation is written for each of the reactants and products.

The equation describing the overall balance of the dye in a PFR differs from that for a CSTR. In a PFR, the concentration is not constant throughout the reactor, but is dependent on the position within the reactor. As the stream flows down along the reactor, the reactant concentration reduces as it is converted into product. Thus, the product concentration increases because the reactant and product concentrations both change along the reactor length. For a reaction that follows first-order kinetics (i.e., the reaction rate is proportional to the reactant concentration), the reaction rate decreases exponentially along the reactor length. At a steady state, the concentration at different positions in the reactor becomes constant. To account for the effects of position, we carry out a material balance on the dye over a very small section along the reactor (Eq. 8-11, Panel 8.4). The equation becomes an ordinary differential equation with the position as the only independent variable (Eq. 8-14).

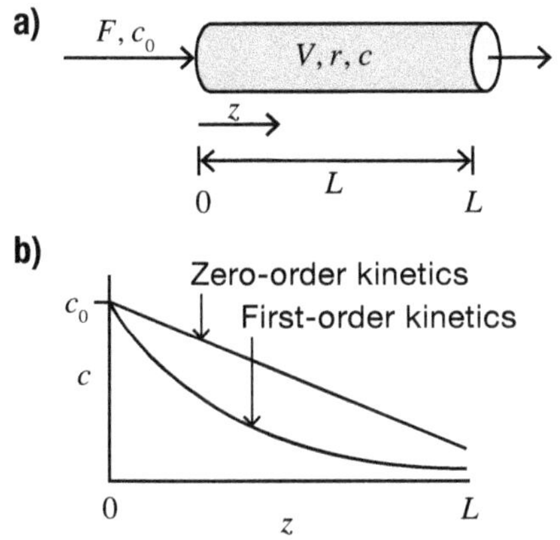

Figure 8.5. (a) Concentration profile of the reactant in a plug flow reactor at steady state. (b) The reactant concentration decreases along the reactor as it is converted to the reaction product.

Panel 8.5. Reaction Kinetics and Reactor Mixing Behavior

- The same reaction in a PFR or a CSTR has a different kinetic profile:
  - In a PFR, as the feed moves downstream, the reactant(s) is consumed and the product(s) is produced. The reactant and product concentrations change with the position in the flow direction.
  - In a CSTR, the concentrations of the reactant(s) and product(s) are homogeneous. The concentrations at the exit are the same as in the reactor.

It can be seen that, in the presence of a reaction, the concentration of the reactant decreases as the feed stream moves downstream, while the product concentration increases (Figure 8.5). For bioprocess applications using a PFR, the nutrient level will decrease while metabolite concentration increases along the bioreactor. At a certain point, the nutrients will become depleted, and a reactor longer than that length becomes unproductive.

Implications of Growth or Reactions Occurring in the Reactor

Plug flow bioreactors are intrinsically more difficult to scale up than mixing vessels, as the concentration gradient of essential nutrients will inevitably become limiting in the downstream region of the reactor (Panel 8.5). One way to overcome nutrient limitations is to increase the nutrient supply rate by using a higher concentration in the feed or by operating at a higher flow rate. There are, however, practical limits on both nutrient concentration and flow rate. For example, nutrient concentration is limited by its solubility, and a high flow rate requires a higher capacity pump to overcome a higher pressure drop across a larger reactor. These kinds of limitations are especially true for growth of aerobic organisms, because oxygen solubility in water is very low and becomes quickly depleted unless supplied continuously. Thus, the size and scalability of a PFR reactor is somewhat limited. In a CSTR, all cells in the reactor encounter the same environment. The nutrients that feed into the reactor are distributed uniformly, albeit at abundant or suboptimal levels. Since all animal cells grow aerobically, virtually all industrial reactors for biologic manufacturing are of stirred tank types. However, plug-flow type devices are used in various auxiliary operations such as the cell settler, where it is used for concentrating cells for recycle into the reactor.

Heterogeneous Reactor: High Solid Content

A bioreactor typically encompasses three phases: a liquid phase of dissolved nutrients and metabolites; a solid phase consisting of insoluble nutrients, cells (which can be regarded as "solid"), and sometimes cell support materials; and a gas phase (bubbles for aeration). Since culture samples taken from the liquid phase do not contain gas phase, and cell mass represents a very small fraction of the culture liquid volume, the content of a cell culture bioreactor is often treated as homogeneous, as if having a single liquid phase. Cells, like nutrients, are treated as part of the liquid phase, and no special consideration is given to the volume taken up by cells in establishing material balance equations.

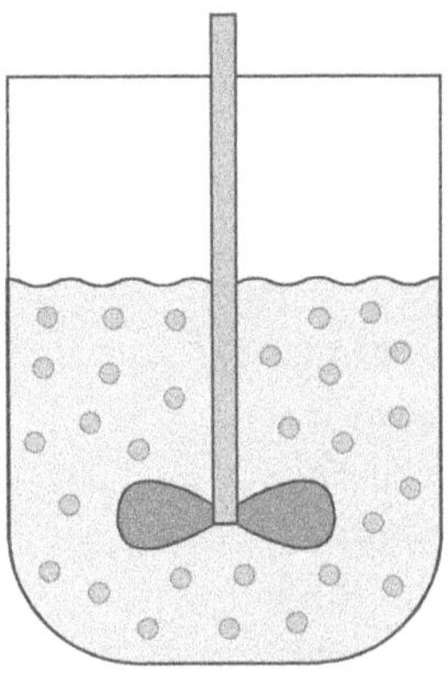

Figure 8.6. A microcarrier bioreactor. Cells attach to the surface of microcarriers, which are suspended in the culture medium. It is a heterogeneous reactor with a solid phase (microcarriers) and liquid phase.

In microbial fermentation and plant cell culture, cells may make up a large fraction (up to 50%) of total reactor volume. The reactor content is therefore partitioned into cell mass and culture medium, each occupying some volume. The cell and nutrient concentrations calculated from medium volume or total reactor volume have different values.

Some cell culture bioreactors contain a significant volume fraction of solid phase such as microcarrier beads (Figure 8.6). Such heterogeneous bioreactors must be treated as two separate phases in material balance. In a microcarrier bioreactor, for example, microcarrier beads often constitute 10–30% of the culture volume. In this circumstance, even cell concentration needs to be well defined. For example, it must be specified whether 10^7 cells per milliliter is with respect to total culture volume or to liquid volume.

Tissue Culture and Disposable Cell Culture Systems

Arguably, the oldest bioreactors for protein and virus production are animals themselves or their tissues. The application of virus vaccines produced in animals dates back two hundred years to when Edward Jenner used cow pox to inoculate humans for protection against small pox. Many viral vaccines developed in the first half of the twentieth century employed animal or animal tissues as production vehicles. Even today, many viruses, including the vast majority of influenza viruses used in vaccination, are produced in embryonated chicken eggs. The production of biologics is carried out in a wide variety of bioreactors, from tens of thousands of chicken eggs in a viral vaccine factory to multiple stirred tanks of tens of cubic meters in size. They cover applications from treating a single patient a limited number of times to providing medicine to tens of thousands of patients for the treatment of a chronic disease. In this section, we will describe reactors used for large-scale cultivation of mammalian cells.

Disposable Culture Systems

The classic Roux culture flask, the narrow-mouthed, rectangular glass flask, started to be replaced by disposable plastic T-flasks four decades ago. The takeover was complete by the early '90s. Today, virtually all laboratory cell culture in the western world today is carried out in disposable flasks, including T-flasks, roller bottles, and multiple flat panels. For small quantity production, such as in the case of growing cells for autologous therapy, flasks are still commonly used for both attachment-dependent and suspension cells.

In the last two decades, the disposable culture system has gone beyond the traditional bottle- and flask-based systems (Panel 8.6). Instead of relying on employing more units to increase production, the new disposable systems aim to increase reactor volume and reduce the requirements of labor and auxiliary equipment. These disposable systems have been widely used in the seed train or cell expansion stage after thawing from cryopreserved vials. As discussed in Chapter 1, even thousand-liter disposable reactors are available for manufacturing.

Panel 8.6. Single-Use Bioreactor

- Simple, low-capital investment in plant construction, fast to implement, ease of equipment validation
- A variety of systems available with vastly different mixing mechanisms
- Less well-characterized in fluid dynamics, mass transfer
- Suitable for seed culture, scales up to hundreds of liters
- Possible faster turnaround
- Lacks the strength of steel; some operations that are routinely performed in steel vessels, such as pressurized liquid transfer, are not easily undertaken in single-use vessels

Roller bottles, flasks

Roller bottles are cylindrical screw-capped bottles with a total volume ranging from 1 to 1.5 L, and are suitable for a culture volume of 0.1 to 0.3 L. The large surface that is exposed to air allows for oxygen diffusion into the medium. Adherent cells can attach to the inner wall of the bottle, while suspension cells simply grow unattached (Figure 8.7). Stacks of bottles are placed on a rack and rotated at 1 to 4 rpm. For small-scale operations, roller bottles provide many advantages for the cultivation of adherent cells. The system is relatively inexpensive to set up and allows for rapid adjustment of production throughput in response to changing needs. Furthermore, complete replacement of medium (e.g., from growth medium to product medium) is relatively straightforward.

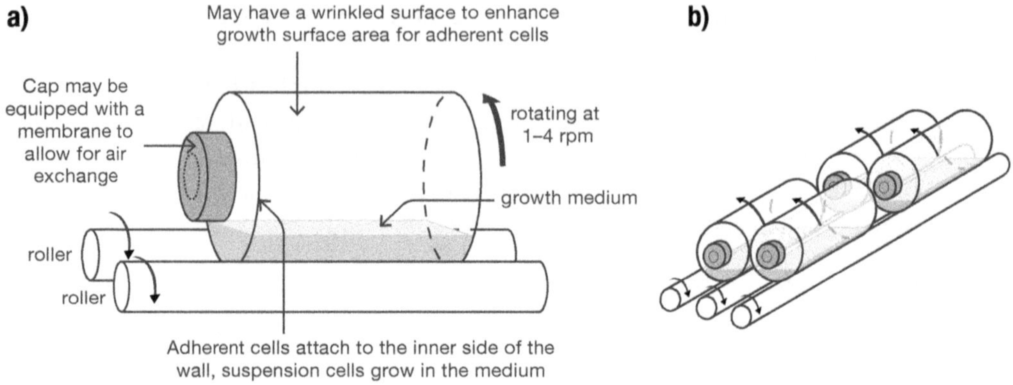

Figure 8.7. A roller bottle for cell culture.

Roller bottles are particularly useful if a serum-containing medium needs to be replaced by a serum-free or protein-free medium for the production phase. The transparent plastic wall allows visual or microscopic examination of the culture status. Most microbially contaminated bottles can be readily identified and discarded before mixing with others.

For large-scale production of biologics, however, there are numerous drawbacks to roller bottles. In-line and on-line environmental monitoring and control is virtually impossible or at least impractical. Aseptic bottle handling for inoculation, protease-treatment for cell detachment and expansion, medium exchange, and product harvest are labor intensive and must be performed by skilled technicians to ensure a low failure rate. A batch of a size suitable for manufacturing purposes may require hundreds or even thousands of bottles, and the large number of manual steps involved dramatically increases the risk of microbial contamination.

Despite these significant drawbacks, roller bottles are widely used in the production of viral vaccines. In some cases, roller bottles are still in use because the legacy product was approved by regulatory agencies decades ago and a process change would incur significant costs for regulatory approval and could even result in product comparability issues. In other cases, roller bottles are selected because a roller bottle process is easy to implement and the production capacity needed is manageable. However, when faced with the need to increase production, e.g., due to market expansion, scale-up of a roller bottle process is challenging.

Notable examples of industrial roller bottle processes include the production of erythropoietin (EPO) using recombinant Chinese hamster ovary cells and the production of live attenuated chickenpox (varicella) vaccine and herpes zoster (shingles) vaccine using the adherent secondary human lung fibroblast cell strain, MRC-5. In some manufacturing facilities, a fully automatic, robotic roller bottle handling system is available. However, a robotic system does not handle a large number of bottles in parallel, thus prolonging the duration of handling when a large number of roller bottles are to be processed. The virus inoculation step can be stretched for many hours, from starting the first bottle to finishing the last bottle, possibly making the production conditions less optimal.

Roller bottles are typically incubated in a temperature-controlled environment in an industrial process. Usually no air exchange is necessary in a days-long process. Some roller bottles have a gas-permeable membrane for gas exchange for use in a CO_2 cell culture incubator. Recently, variations of T-flasks and roller bottles have been used in the cell therapy community. In the cultivation of T cells and NK cells, stationary culture has been found to be preferred, possibly to allow for minute factors to accumulate in the cells' surrounding microenvironment. These culture devices install a gas-permeable membrane at the bottom that allows cells to settle on it. Even blood bag-like devices with high gas permeability are available commercially.

Multiple plate systems

Systems consisting of multiple parallel plates enclosed in a chamber are widely used in laboratories and in industrial production of viral vectors, vaccines, and cells for therapy. These parallel plates, which are stacked inside a transparent plastic box, increase the available surface area for adherent cells to attach to and grow on. A 40-tray Nunc Cell Factory® (NCF) has a capacity of 25,280 cm^2 and an ~8 L liquid volume. One notable example of NCF use is the production of the multivalent Rotavirus vaccine using Vero cells, a continuous African green monkey kidney cell line. Another multiple plate system, CellCube® Module, consists of nine-inch square polystyrene plates stacked vertically in a resin case with 1 mm spacing. The space between stacks is completely filled with medium. Continuous medium circulation is used for oxygen and nutrient supply. CellCube has been used in the cultivation of MRC-5 cells (human fibroblasts derived from the lung) for the production of both attenuated and inactivated hepatitis A virus (HAV) vaccines. These disposable bioreactors have also been used in the clinical production of cells for immunotherapy and gene therapy vectors.

Bags and moderately scalable disposable culture wares

Blood bags were used for the cultivation of cells for cell therapy three decades ago. Small bags are placed in incubators for temperature control. Typically, the bag is not filled up with medium; instead, an air space is left in the bag for oxygen supply. For larger, single-use bag systems, mechanical mechanisms provide mixing and oxygen transfer. Wave™ consists of pre-sterilized, disposable Cellbags resting on a special rocking platform.[2] The rocking motion of the platform induces waves in the culture fluid, which provides mixing and enhances oxygen transfer (Figure 8.8).

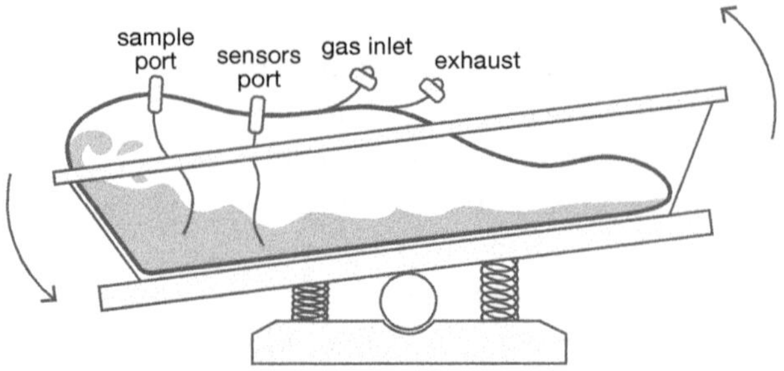

Figure 8.8. Schematic of a wave bag cell culture system.

This and other simple disposable systems are basically the combination of a culture medium container and an external mechanism to move the container and provide mixing. Their ability to increase in size is limited by the capability to supply oxygen and the ability to move a large mass of liquid.

Disposable systems provide opportunities for manufacturers to offload to specialized disposal suppliers the work related to equipment preparation, validation, and maintenance. In order to allow for further scale up, direct aeration into the liquid phase must somehow be introduced. Single-use plastic equivalents of existing stainless-steel bioreactor vessels emerged to fill this niche. Such a reactor consists of a reusable stainless steel outer support container and a single-use cell chamber with a working volume of up to 1000 L, which can be integrated into a bioreactor control system. A manufacturing plant consisting of these modest-scale, single-use reactors is becoming an alternative to a conventional stirred tank made of stainless steel with fixed piping, auxiliary equipment, etc. Because of the material of construction, its operation faces limitations. For example, it cannot be pressurized to quickly transfer its contents to other process units. It is also challenging to scale it up much beyond its current size, unless the mechanical strength of its materials is reinforced to withstand the higher power input at an affordable cost, while still mitigating the environmental issue of employing disposable plastic.

Adherent Cell Support Systems

Suspension vs. Adherent Culture

For processes whose production scale is relatively small, a disposable system provides some advantages. For protein therapeutics needed in large quantities, the use of disposable systems may be limited to inoculum preparation. Large-scale operations therefore continue to employ fermenters or other bioreactors.

The majority of cells used for protein production via rDNA are suspension cells grown in stirred tank bioreactors. However, many cell lines used for the production of viruses for vaccine and gene therapy applications are adherent. For the cultivation of these cells, a supporting surface, sometimes called a substrate, is needed. Many disposable systems have been commercially developed to provide a large amount of cell growth surface area in a single structure.[3] These systems are invariably limited in their scale of operation. For applications that require a very large quantity of cells, it is necessary to employ support for cells to adhere to and grow on in a bioreactor.

Microcarriers

Microcarriers are small particles that are large enough to support the growth of adherent cells. In a stirred tank bioreactor, cells can be cultivated on the surfaces of the suspended microcarriers, thus allowing the process to be scaled up.[3] They fall into three general types: conventional solid or microporous microcarriers that provide external surface area for cell attachment and growth, macroporous microcarriers that also allow cells to grow in pores in their interiors, and microspheres that facilitate the formation and growth of cells as aggregates.

An advantage of microcarrier culture is the ease of separating cells from the culture medium. Many microcarrier cultures require medium exchanges during cultivation to remove lactate, ammonia, and other metabolites and to replenish nutrients, or to remove serum in the medium before virus infection. This can be accomplished by simply allowing cell-laden microcarriers to settle, such that the spent medium can be withdrawn from the top and replenished. In large-scale operations, continuous or semi-continuous perfusion is accomplished by withdrawing the medium through a coarse screen or from a settling tank to retain microcarriers in the reactor.

Solid and microporous microcarriers

Microcarriers were first used for cell culture by Van Wezel in the 1960s. They changed the cultivation vessel of adherent cells from a two-dimensional surface to a three-dimensional, scalable reactor. To reduce the power required to suspend the cell-laden microcarriers and reduce the mechanical damage of agitation, microcarriers typically have a diameter range of 100–300 μm and a specific density of 1.02–1.05 g/cm^3 (Panel 8.7). This diameter range provides good growth surface area per reactor volume. Even at a moderate microcarrier concentration occupying

Panel 8.7. Desired Characteristics of Microcarriers

Density	~1.02 g/cm^3	Only slightly higher than water for easy suspension and settling
Diameter	150–200 μm	Carriers with a small diameter that still allows for cell spreading (more surface area per bead volume, slower settling velocity)
Porosity	From solid to nearly 90%	Prefer solid for use as inoculum (easier bead-to-bead transfer)
Surface properties	ECM coating, slight positive charge	Positive charge enhances initial cell attachment

8–15% of the culture volume, a significantly larger surface area per reactor volume can be achieved than is possible with roller bottles.

Microcarriers have been fabricated from many different materials, including dextran, gelatin, polystyrene, glass, and cellulose. Many that were once commercially available are no longer on the market because microcarriers constitute only a niche market for some vaccines and a limited number of recombinant proteins. In general, microcarriers have a wettable surface and are sometimes coated with collagen or other adhesion molecules to enhance cell adhesion and spreading. The most widely used microcarriers, which are based on dextran, are derivatized with charged molecules or denatured collagen. These carriers are microporous. In its dry form, the bead is very small. Upon hydration in the medium or isotonic buffer, the bead diameter increases many times and ~90% of the bead volume is water. Hence the relatively low specific density. Normal diploid fibroblasts can grow well on microcarriers and exhibit normal morphology or even the typical parallelly aligned pattern seen when grown on Petri dishes (Figure 8.9). Continuous cell lines grown to high densities sometimes even form multiple layers (Figure 8.10).

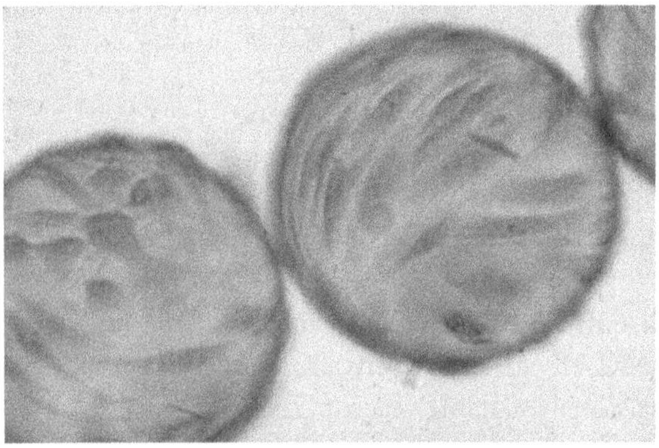

Figure 8.9. Human foreskin fibroblast FS-4 grown on charged dextran (Cytodex 1) microcarriers.

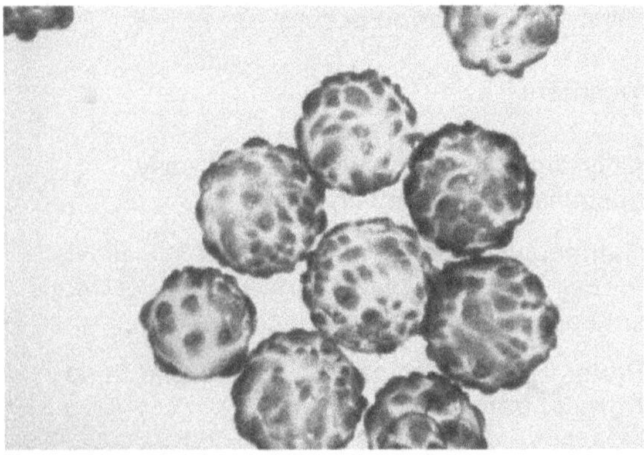

Figure 8.10. CHO cells grown on charged dextran (Cytodex 1) microcarriers.

Other types of commonly used microcarriers are made of polystyrene. Their specific density and settling velocity in water is higher than the dextran-based microcarriers. On a unit weight basis, polystyrene microcarriers have much less surface area than dextran carriers. While dextran beads are transparent under a microscope, easy for microscopic observation of cell morphology, polystyrene beads are refractive and therefore not convenient for cell observation. Additionally, the surface of polystyrene beads is hydrophobic, making them unsuitable for cell attachment. Thus they are subjected to irradiation, plasma treatment, or another chemical modification to increase the wettability of the surface, or are derivatized with cell adhesion peptides to facilitate cell attachment.

A wide variety of cell types have been grown on microcarriers, including fibroblasts, epithelial cells, hepatocytes, neuroblastoma cells, and endothelial cells from various species. Overall, microcarrier culture allows a very large number of adherent cells to be grown in a stirred tank bioreactor. Depending on the cell type, a cell concentration of 2–10 x 10^9 cells/L is achievable. However, in very large-scale operations it is not feasible to inoculate cells from conventional flask or roller bottle cultures; the inoculum will have to be from a seed microcarrier culture. For cell therapy applications, it is also necessary to detach and recover cells. Detaching adherent cells from microcarriers requires extensive optimization to achieve a high recovery yield. The selection of a microcarrier type will influence the ease of cell detachment.

Macroporous microcarriers

Macroporous microcarriers contain large interconnected internal pores (Figure 8.11). The void space inside allows cells to be cultivated on internal and external surfaces. Cells in the interior are less susceptible to mechanical damage caused by agitation and gas sparging. On the downside, cells in the interior of microcarriers are more likely to be subject to oxygen limitation due to the long diffusional distance, especially since most macroporous microcarriers have a larger diameter (500 μm–2 mm) than standard microcarriers.

Macroporous microcarriers are made of various materials, including gelatin, collagen, and plastic. Many cell lines have been successfully grown on macroporous microcarriers, such as Vero, HepG2, CHO, and HEK293 cells. The final cell concentration achieved tends to be higher than that obtained with an equivalent volumetric concentration of conventional microcarriers. In some cases, however, the growth kinetics are slower because the penetration of cells into the interior may be slowed or even retarded by the restrictive openings of the pores (Figure 8.12).

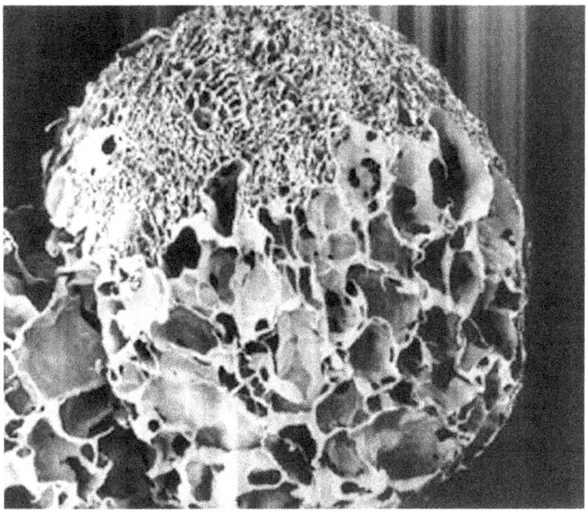

Figure 8.11. Scanning electron micrograph of a collagen-based macroporous microcarrier.

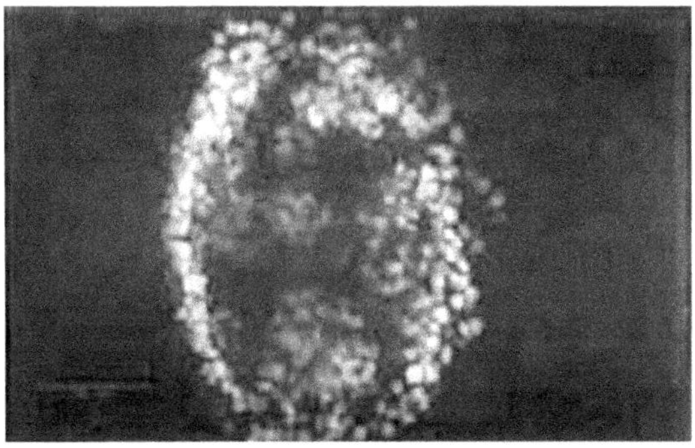

Figure 8.12. Confocal microscopic section of HepG2 cells grown in a macroporous microcarrier.

Microsphere-induced cell aggregates

Most anchorage-dependent cells do not develop normal morphology and do not multiply on surfaces with an excessively high curvature. When cultivated on the surface of a small cylindrical rod

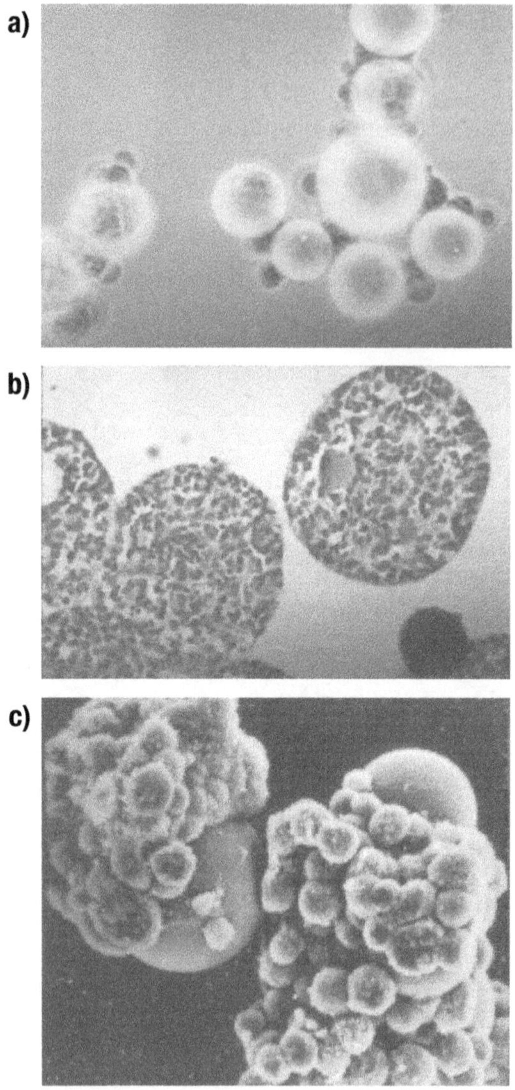

Figure 8.13. Microsphere-induced cell aggregates. (a) CHO cells attaching to a microsphere 1h after inoculation. (b) A thin section of HEK293 cell aggregates. (c) Scanning electron micrograph of HEK293 cell aggregates.

with a sufficiently large diameter, fibroblastic cells will grow in random orientations. If the diameter is reduced to a level with a high curvature, cells will begin to align in the longitudinal, rather than radial, direction. Similarly, if the diameter of a microcarrier is too small, many cells will not spread out on the surface and grow. However, many continuous cell lines can multiply even if they are not spread out like they would be on a flat surface. When these cells are grown on very small microcarriers, they agglomerate to form aggregates and continue to grow to a high density. These microspheres, usually possessing a diameter of about 50 μm, provide the initial surface for cell agglomeration and proliferation. These agglomerated microspheres become aggregates as the cells grow. If the aggregate diameters become too large, dead cells can be seen in the interior as a result of transport limitations. Many cell lines, including BHK, CHO, HEK293, and swine testicular cells, have been cultivated as aggregates using microspheres (Figure 8.13).

Cell Aggregates

Some cells can be grown as aggregates when cultivated in shaker flasks or stirred vessels without any surface support. These include not only continuous cell lines like BHK and HEK293, but also some primary tissue isolates and stem cells. For example, pluripotent stem cells can be grown as aggregates called embryoid bodies and increase in cell number and differentiate. Hepatocytes isolated from liver can form three-dimensional spheroids that retain differentiated properties but have limited capacity for cell expansion. These cells form cell-cell interactions in a short time when placed in a suspension, and can then grow without cell-substrate adhesion (Figure 8.14). It has been shown that high Ca^{2+} concentrations promote HEK293 cell aggregation, while low concentrations elicit partial dissociation. Some cells do not form aggregates quickly when placed in

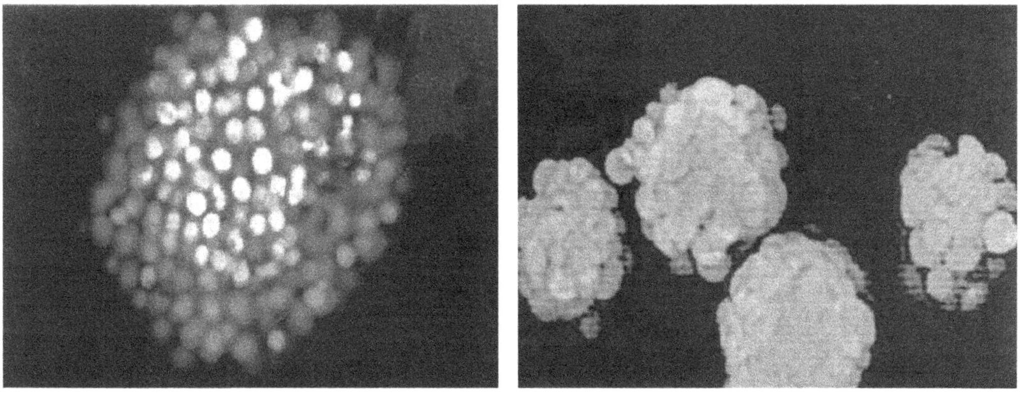

Figure 8.14. Confocal microscopic section of HEK293 cells grown as aggregates.

suspension. These cells, e.g., embryonic stem cells, can be forced to form aggregates by placing them in round-bottom, low-attachment wells or in suspended medium drops (hanging drops) for about two days. Upon forming aggregates, they can be cultured in stirred tanks.

Aggregate cultures have advantages similar to microcarrier cultures. They can be cultivated in conventional stirred tank reactors with environmental controls. Upon the cessation of agitation, particles will settle relatively rapidly, permitting easy medium replenishment or perfusion. A downside of aggregate cultures is the controllability of the aggregate size. Excessively large aggregates often have a lower viability that may influence product quality if the product is degraded or otherwise modified by enzymes released by lysed cells. For some applications that require dispersed cell suspension rather than cell aggregates, dissociation of highly viable cells from aggregates will require some optimization effort.

Bioreactors Developed for Cell Culture

At the dawn of recombinant DNA technology, cell culture faced great challenges. The traditional processes for virus vaccine production did not support the productivity and product quantities required for new protein biologics. Cells were grown in media containing high levels of protein, for which supplying oxygen by direct air bubbling was difficult. Cells were sensitive to mechanical stress incurred by impeller agitation, and they produced large amounts of lactate, ammonium, and other metabolites that inhibited cell growth. To overcome those hurdles, much effort was devoted to cell culture technology development and numerous innovations were developed. In the end, the problems were resolved by means other than adopting new types of manufacturing bioreactors. With the advances in our understanding of cells' nutritional needs, serum was eliminated from media and the protein content drastically reduced, making direct sparging of air manageable. Over time, cells were generally adapted to sustain the mechanical agitation of a stirred tank bioreactor. The problem of low cell concentration due to metabolite accumulation was overcome by adopting fed-batch or perfusion culture and managing the accumulation of inhibitory metabolites. In the end, the prevailing bioreactor used in cell culture was, and remains to this day, the stirred tank. It is instructional to briefly review some of those earlier innovations in order to glean the lessons learned. Furthermore, some bioreactors may find applications in tissue engineering or cell therapy applications.

Stirred Tank Bioreactors

Stirred tanks have been the workhorse for microbial fermentation since the 1950s and began to be used in cell culture for vaccine production using BHK cells in 1960s.[4] The basic configuration of the stirred tank bioreactor for mammalian cell culture is similar to that for microbial fermentation. While the Rushton type impeller is the norm in microbial fermenters, mammalian cell culture fermenters mostly employ axial flow type impellers (see Chapter 12). The difference reflects the different purposes of agitation in microbial fermentation versus cell culture. In microbial fermentation, agitation is needed at a high power input to disperse air bubbles and to increase oxygen transfer efficiency, whereas in mammalian cell culture the primary purpose of agitation is to maintain relatively uniform suspension of cells or microcarriers.[5] The power input per unit volume of bioreactors is also substantially lower. In general, the mixing time in a mammalian cell culture bioreactor is substantially longer than that in a microbial fermenter of similar scale. The oxygen transfer capacity in a cell culture bioreactor is also substantially lower than that in a microbial fermenter. However, the typical oxygen demand in a mammalian cell culture is 10 to 50 times lower than that in microbial fermentation.

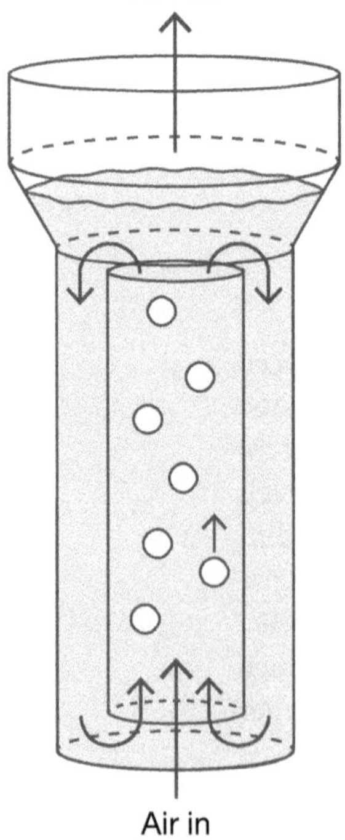

Figure 8.15. An airlift bioreactor.

Bioreactors with Alternative Mixing Mechanisms

Airlift bioreactor

An airlift reactor uses large air bubbles to drive the liquid in a segregated region in the bioreactor upward, which then induces the liquid to move downward in a second region, thus creating an internal circulation loop that suspends cells, provides mixing, and supplies oxygen. Airlift reactors are used to grow mammalian, insect, and plant cells. A common configuration has a cylindrical internal draft tube that is concentric to the reactor wall above the air sparger (Figure 8.15). The air bubbles drive the liquid upward because of the upward momentum generated by the air flow and the lower effective fluid density in the draft tube. Medium flows upward through the sparged section (riser) and downward in the bubble-free section (downcomer). This method of generating circulation has a low energy requirement compared with stirred-tank reactors. Airlift bioreactors for cell cultivation are considered to be low-shear devices because there is no mechanical agitation. They have been used successfully with suspension cultures of BHK21, human lymphoblastoids, CHO, hybridomas, and insect cells. Airlift bioreactors are effective when the cell concentration is low and a small air flow rate is sufficient to provide mixing and oxygen. Without mechanical agitation to break up air bubbles, the bubble size is large and oxygen transfer efficiency is lower at a given air flow rate per reactor volume. As cell concentration increases in a fed-batch culture, a higher aeration rate has to be used. Airlift bioreactors are primarily used in laboratories for insect or plant cell cultures.

Fluidized bed bioreactor

Fluidized bed reactors have long been used in chemical heterogeneous catalysis. The fluid stream carrying the reactant (often gas phase) enters through a flow distributor at the bottom of the reactor at a sufficiently high velocity to suspend the solid catalyst particles. The reactant enters the catalyst, and product diffuses out into the fluid, where it is carried out at the top of the reactor. A separator prevents the catalyst particles from being blown out. The main advantage of a fluidized bed is the high heat transfer efficiency between the fluid and the catalyst surface. When applied to cell culture, cells must be immobilized on particles, typically macroporous microcarriers, so that the particle can be separated from the exiting liquid phase and be retained in the reactor (Figure 8.16). However, unlike in chemical catalysis, mass transfer and heat transfer at the external surface of the microcarrier are not rate controlling. Rather, the limiting factor is often the accumulation of metabolites in the medium. Furthermore, the consumption of the nutrient is relatively low; one single pass of medium through the reactor would consume only a small fraction of the nutrient. Hence, the bulk of the medium is recirculated continuously to suspend the particles. The recirculating medium also supplies the oxygen by passing through an

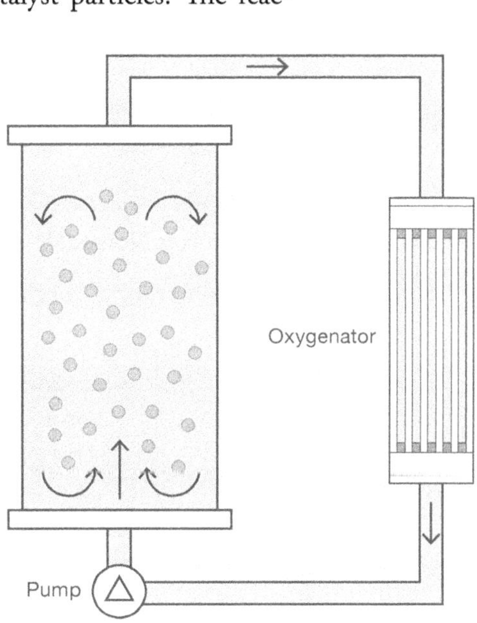

Figure 8.16. Schematic of a fluidized bed bioreactor. The carrier-free culture fluid is pumped into the fluidized bed. The heavy carriers recirculate in the reactor. The culture fluid exits from the reactor and is re-oxygenated before returning to the reactor.

oxygenator before returning to the reactor. In order to increase productivity, the microcarrier concentration is high, which in turn requires a high medium recirculation rate to supply a sufficient amount of oxygen. To sustain a high flow rate, the density difference between solid phase and liquid phase must be sufficiently large to allow for the retention of particles. In the fluidized bed bioreactors for cell culture commercialized in the 1980s, the carriers were weighted down by iron particles to increase the density difference between fluid and carrier. This allowed the particles to be retained in the reactor at the flow rates required to supply sufficient oxygen for cell growth.

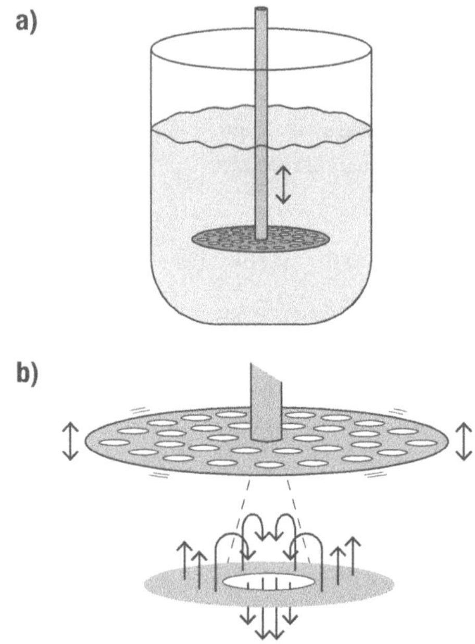

Figure 8.17. (a) A vibromixer bioreactor and (b) its fluid flow pattern.

Vibromixer

A vibromixer uses a perforated disk as the mixing mechanism rather than a conventional impeller. The disk vibrates in a vertical direction (perpendicular to the plane of the disk) at a high frequency, causing liquid to circulate through the perforated holes and provide mixing (Figure 8.17). The vibromixer was used for the cultivation of suspension cells for virus production. It has since been replaced by the stirred tank. However, it is still used to provide mechanical agitation to suspend microcarriers at high concentrations after trypsinization, in order to facilitate the detachment of cells from microcarriers.

Membrane stirred tank

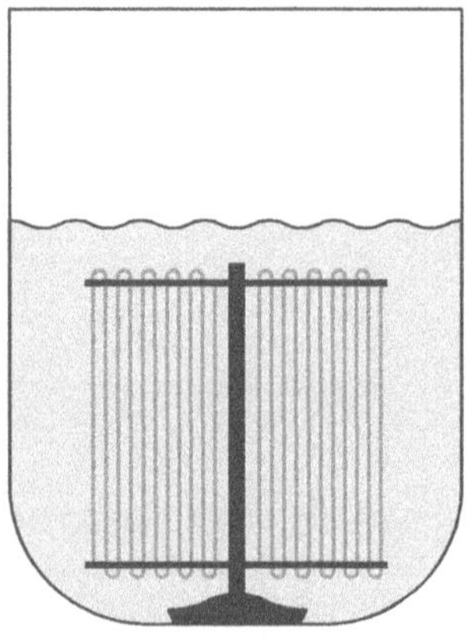

Figure 8.18. A stirred tank bioreactor with polymer tubing for oxygenation and mixing.

The membrane stirred tank was developed by Professor Jürgen Lehmann in the 1980s to overcome the problem of supplying oxygen in a high-serum medium (Figure 8.18).[6] The bioreactor had a rotating shaft with a number of extended upper and lower arms that rotated with it. Pieces of long microporous polypropylene tubing for supplying oxygen were wrapped around the rotating upper and lower arms. Air flowed through the tubing and exited above the surface of the medium without making direct contact with the medium. By adjusting the air pressure in the polypropylene tubing, the micropore could be expanded to allow the gas that is near bubble-bursting pressure to be in direct contact with the medium, thereby providing bubble-free aeration. The rotation of the tubing also provided gentle agitation to microcarriers or suspended cells. The design illustrates the challenge of supplying oxygen in a high protein content medium.

Membrane Bioreactors

The primary aim of all membrane reactors is to keep cells in a compartment that is isolated from mechanical stress caused by agitation, liquid pumping, or air sparging. The reactor, depending on its configuration, may also allow the product to accumulate in that compartment. All are plug-flow type reactors and rely on medium recirculation to supply oxygen to the cell, thus limiting the scalability of the reactor.

Hollow fiber bioreactor

A hollow fiber reactor consists of a bundle of microporous capillary fibers encased inside a cylindrical casing, thereby providing two different spatial compartments: the tube (fiber lumen) side and the shell (extracapillary) side. The culture media is pumped into the fiber lumen from one end, and exits from the fiber's other end (Figure 8.19).[7] As the medium passes through the fiber lumen, oxygen diffuses through the fiber wall into the shell side and supplies oxygen to the cells. Hollow fiber bioreactors are used primarily for the cultivation of suspension cells that grow in the extracapillary space on the shell side. The extracapillary space also has a medium flow loop that allows for additional medium supply, especially high molecular weight components of the medium, to the cell compartment and for harvesting the product. This basic configuration is somewhat similar to the hollow fiber cartridge used in kidney dialysis.

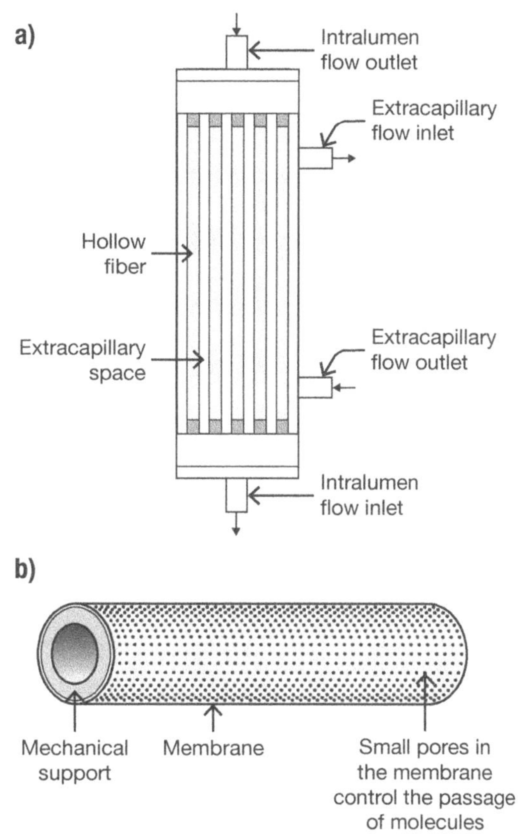

Figure 8.19. (a) A hollow fiber bioreactor and (b) a single fiber. Thousands of such fibers are encased in the reactor.

Each hollow fiber consists of a porous polymeric layer that provides mechanical support and is coated with a thin membrane that selectively allows molecules to pass through based on their size. In most cases, an ultrafiltration membrane is used. The molecular weight cut-off (MWCO) of the membrane differs according to the specific application, ranging from a few thousand to a hundred thousand daltons.

Supply of low-molecular-weight nutrients and oxygen to the cells and the removal of waste products occurs by diffusive transport across the membrane between the lumen and the shell spaces. The ultrafiltration membrane prevents free diffusion of secreted product molecules through the membrane, instead allowing them to accumulate in the extracapillary space to a high concentration. The medium recirculating in the lumen side is passed through an oxygenator before returning to the reactor. Although the use of hollow fiber bioreactors is now infrequent, they are still an easy and convenient tool for the production of small quantities of antibodies for research or diagnostic applications.

Multiple membrane plate bioreactor

The scaling up of a hollow fiber system is limited by the ability to extend the axial length of the fiber due to oxygen transfer limitations. The use of a very high flow rate to supply more oxygen is limited by the capacity of the pump and the mechanical strength of the membrane. Expanding the cartridge diameter to increase the capacity eventually causes an uneven flow distribution to the fibers. Furthermore, major resistance to oxygen transfer resides in the diffusion in the extracapillary, inter-fiber space (i.e., from the fiber wall to where the cells are). One more hurdle is in achieving an even distribution of cells upon their inoculation into the tortuous extracapillary space. In principle, one can use two different fibers to supply oxygen and medium separately but simultaneously, using one set of fibers for gas phase and another for medium. However, mixing different hollow fibers in a cartridge poses a major challenge in manufacturing. An alternative configuration can be made by using multiple flat membranes that allow for more even cell distribution in the cell chamber and better control of the oxygen diffusion distance to the cells. However, this seemingly versatile reactor also suffers from practical manufacturing complexities and is not used in large-scale operations. Overall, membrane reactors constitute only a very small fraction of all reactors used in cell culture. If one wishes to employ a membrane bioreactor for cell cultivation, the commercially available kidney dialysis hollow-fiber cartridges are most assessible.

Cell Support Systems that Shield Cells from Mechanical Stress

In addition to microcarriers developed for adherent cells, cell support systems were developed for suspension cells. The goal was to find a way to protect cells from possible damage caused by mechanical agitation and bubble aeration while retaining cells (along with the cell support mechanism) in the reactor during medium exchange. These systems invariably entailed encasing each cell in some form of bead. While these approaches dealt with the obstacles of the day, they also had limitations.

Many of the materials utilized prevented them from being used in stirred tanks unless the scale was small. Most methods generated particles of a fairly large size—in the range of millimeters—that required a large liquid pumping capacity to keep the particles in suspension. The cell mass was typically a very small fraction of the bead and an even smaller fraction of the reactor volume. Some of the systems later found applications in tissue engineering, for example, as a method of entrapping beta cells for insulin secretion. However, they often face the problem that only a small fraction of the total volume of implanted beads are active cells.

Agarose

Agarose entrapment of cells can be easily accomplished by adding cells in agarose suspension drop-wise into an immiscible oil phase, where it is allowed to gel. The beads formed usually have a large diameter of millimeters. To control the particle size, the cell suspension can be injected through a concentric double tube. The cell suspension flows into the inside tube and is ejected through an orifice into an air stream that flows in from the space between the tubes. The air jet stream carries the ejected cell suspension drop-wise into a chilled oil phase that allows the agarose to gel. In another method of cell entrapment, the cell-agarose suspension is allowed to drop onto the central region of a fast-rotating disk. The centrifugal spinning force causes the droplets to form and be dispensed on the outside of the disk. The cell-entrapped beads can then be packed in a column or suspended in a stirred tank for cultivation. Cells entrapped in agarose beads grow into clusters and produce the product.

Microencapsulation

In microencapsulation, cells are first entrapped in a polymeric matrix, such as alginate or chitosan, to form gelled spheres. Those cell-laden beads are then coated with a polymeric film that constrains the diffusion and release of the product molecules out of the microcapsule while allowing nutrients to diffuse through.[8] A polymer matrix commonly used for cell entrapment is calcium alginate. Cells are suspended in sodium alginate and added dropwise into a solution with a high calcium chloride concentration that causes the alginate to form a gel. The alginate beads are then coated with polylysine or other polymers to increase the mechanical and chemical stability of the beads (Figure 8.20). The alginate gel inside the beads coated with the polylysine membrane can be liquefied through treatment with a calcium chelator such as EDTA or citrate.

Large-scale application of microencapsulated cells is rare, since the beads are not mechanically strong enough to withstand the mechanical mixing encountered in scales beyond the pilot plant. However, the microcapsule provides a means of immunoisolation of transplanted cells or tissues, and could be suitable for some tissue engineering applications.

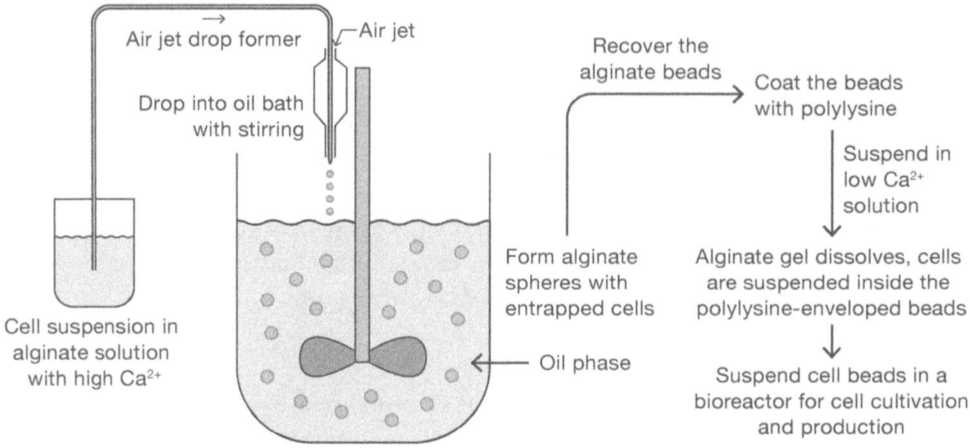

Figure 8.20. Preparation of a microencapsulation system for cell culture.

Concluding Remarks

This chapter highlighted the key differences between well-mixed and plug-flow bioreactors. Because of the low solubility of oxygen and the very limited oxygen-carrying capacity of media, virtually all industrial manufacturing of cell culture products uses stirred-tank bioreactors. When a stirred tank bioreactor is used for growing adherent cells, a supporting surface for cell attachment is provided through the use of microcarriers or macroporous microcarriers. Alternatively, adherent cells may be grown as aggregates in suspension. Microcarriers and a number of the cell culture bioreactors developed three decades ago were the product of intensive efforts to bring about cell culture manufacturing. Today, the time-tested stirred tank bioreactor for suspension cells remains the bioreactor of choice. The durability of the stirred tank bioreactor is due to its simplicity as well as advances in other aspects of cell culture, including cell adaptation and medium improvements. Even as disposable devices become more widely used, the basic type of disposable device still relies on extensive liquid mixing. With an increase in scale, even the disposables tend to take the form of the stirred tank. Some innovative bioreactors developed at the dawn of cell culture bioprocess may now be finding new applications in tissue engineering and cell therapy. As these new technologies advance and the demand for those specialized cells grows, we may find some lessons that can be learned from earlier bioreactor development efforts.

References

1. Hu, W.-S., *Engineering Principles in Biotechnology*. 2018: John Wiley & Sons.
2. Singh, V., *Disposable bioreactor for cell culture using wave-induced agitation.* Cytotechnology, 1999. **30**: p. 149–58.
3. Merten, O.W., *Advances in cell culture: anchorage dependence.* Phil. Trans. Soc. B, 2015. **370**: 20140040.
4. Telling, R.C. and R. Elsworth, *Submerged cultures of hamster kidney cells in a stainless steel vessel.* Biotechnol. Bioeng., 1965. 7: p. 417.
5. Nienow, A.W., *Reactor engineering in large scale animal cell culture.* Cytotechnology, 2006. **50**(1-3): p. 9–33.
6. Wagner, R. and J. Lehmann, *The growth and productivity of recombinant animal cells in a bubble-free aeration system.* Trends in Biotechnology, 1988. **6**: p. 101–04.
7. Piret, J.M. and C.L. Cooney, *Mammalian cell and protein distributions in ultrafiltration hollow fiber bioreactors.* Biotechnol. Bioeng., 1990. **36**: p. 902–10.
8. King, G.A., D.P. Faulkner, and M.F.A. Goosen, *Alginate-polylysine microcapsules of controlled membrane molecular weight cutoff for mammalian cell culture engineering.* Biotechnol. Progress, 1987. **3**: p. 231–40.

Nomenclature

α	Stoichiometric ratio
θ	Holding time
A	Area (area perpendicular to flow)
c	Concentration of tracer or reactant
F	Volumetric flow rate
L	Length of reactor
p	Product concentration
r	Reaction rate
t	Time
u	Linear velocity
V	Volume of reactor
z	Length dimension of the reactor

Contents

9
Fed-Batch Culture Processes

Introduction

THE CELLS USED IN CELL CULTURE PROCESSES are all derived from animals. Their genetic and epigenetic control circuits and their physiology have evolved to carry out the physiological functions of the tissue from which they were derived, not to sustain themselves and proliferate in *in vitro* culture. To gain insight into how to grow them well, it is instructive to first see how they are nurtured in their native environment. The human body is, in a way, the best bioreactor. More accurately, it is a reaction system consisting of many bioreactors. Figure 9.1a depicts a few organs in our body that perform complementary functions to sustain our body over a long period of time. The body operates as a batch system, with periodic input of nutrients and output of waste streams from excretions. An adult human body consists of about 3.7×10^{13} cells.[1] A modern bioreactor can reach $\sim 10^{10}$ cell/L at the end of a batch culture. In terms of quantity of cells, a human body is equivalent to an industrial bioreactor of ~400 L in size. In general, cells in a bioprocess reactor consume many times the daily recommended intake of both carbohydrates and amino acids for an adult. Of course, the comparison is simplistic since cells in the reactor are growing while most cells in an adult are not dividing and may require a smaller amount of nutrients. Nevertheless, our body is certainly more efficient in nutrient utilization. What are the major differences between growing cells in a culture and growing cells in a body?

The composition of the food we take in daily usually does not match the exact needs of our body. Furthermore, both the intake of nutrients and our bodies' needs for different activities change over time. Cell metabolism generates waste metabolites, some of which may become toxic if allowed to accumulate to high levels. Our bodies cope with such imbalances by using the liver as a metabolic factory to balance chemical compositions in the circulating blood and detoxify inhibitory metabolites by degrading them or converting them to less inhibitory compounds (Figure 9.1b). The liver also serves as a reservoir of glucose to maintain glucose levels in a range of ~4 to 7 mM in the blood. The kidney then purges the waste metabolite from the body and the lungs remove carbon dioxide from the bloodstream. These balancing actions by multiple organs keep the nutrient and metabolite levels in the bloodstream in check. With these organs keeping nutrient and metabolite levels in a relatively narrow range, cells evolved to thrive only in relatively invariant chemical environments, making them rather intolerant of large deviations from their normal chemical environments. This intolerance has narrowed the allowable range of nutrient and metabolite levels in bioreactors for optimal cell growth.

A drastic difference between a bioreactor and a human body is the absence of a metabolite-balancing mechanism. The accumulation of lactate, ammonium, and other metabolites of amino acid degradation in a bioreactor inhibits growth. In cell culture bioreactors, the nutrient is often added initially or fed periodically. Upon feeding, the nutrients accumulate to high levels and then decrease to low levels due to consumption. Bioreactor operation is largely based on the strategy of sustaining an

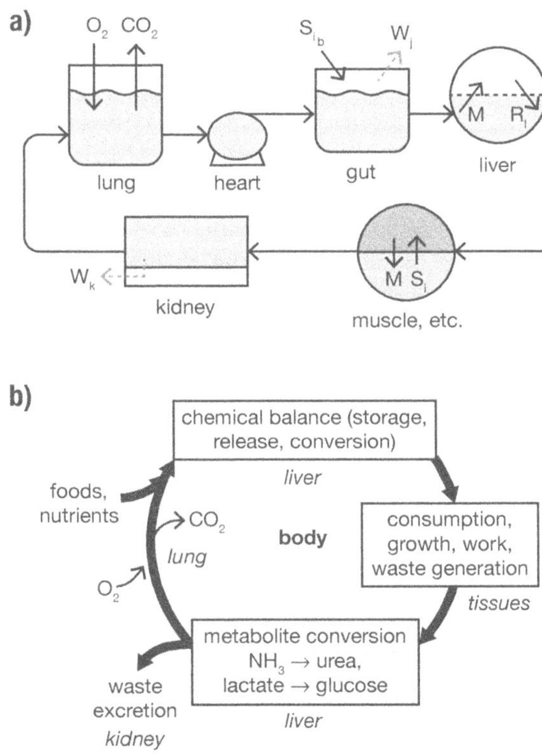

Figure 9.1. Organs as a bioreaction system. It consists of oxygen supply, carbon dioxide removal, a mixing provider, a chemical conversion factory, and a waste disposal mechanism. In the gut, nutrients (S) from food are absorbed into blood circulation, leaving some waste (W) for excretion. In the liver, nutrients and metabolites from different organs are processed to maintain a balanced state and some metabolites are chemically transformed (R) into waste. In muscle and other tissue, nutrients are consumed to generate energy for performing work and to grow more biomass, all the while producing metabolites. In the kidney, many metabolites (including R from liver metabolism) are excreted. Finally, CO_2 from various tissues is exhaled and fresh oxygen is supplied to the blood, where it recirculates back to different organs via blood pumping.

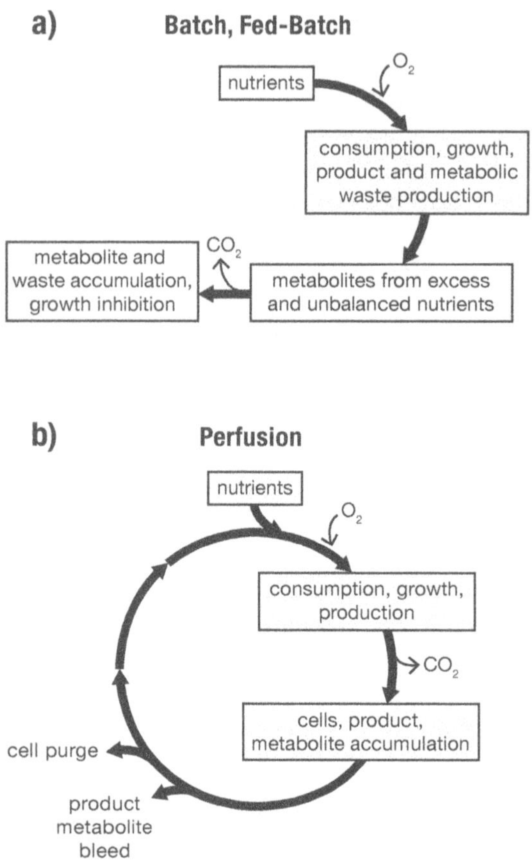

Figure 9.2. Cell culture processes lacking the chemical conversion and balancing capability of the organs' system. Balancing nutrient supply can minimize the accumulation of metabolites in batch/fed-batch cultures and alleviate the need for a high perfusion rate.

optimal chemical environment profile in order to maximize productivity and robustly deliver a high-quality product.

The initial level of nutrients in cell culture processes is typically higher than that seen in our bodily fluids. Even that level, however, is insufficient to grow cells to the density needed for industrial production. Hence, most industrial cell culture production processes are operated in a fed-batch mode. The main limiting factor for productivity in a fed-batch operation is the accumulation of metabolites and culmination of other growth inhibitory factors. A major cause of the metabolite accumulation is the unbalanced supply of nutrients (Figure 9.2a). A key factor in increasing the productivity of a fed-batch culture process is providing a balanced chemical environment that minimizes the production of metabolites and prolongs the time it takes for them to accumulate to inhibitory levels.

The operation of a bioreactor is generally classified as batch or continuous mode. In a continuous process, the feed is continuously being introduced to the reactor and the product stream is continuously being withdrawn. In a batch process, the medium, including all the nutrients (except oxygen), is

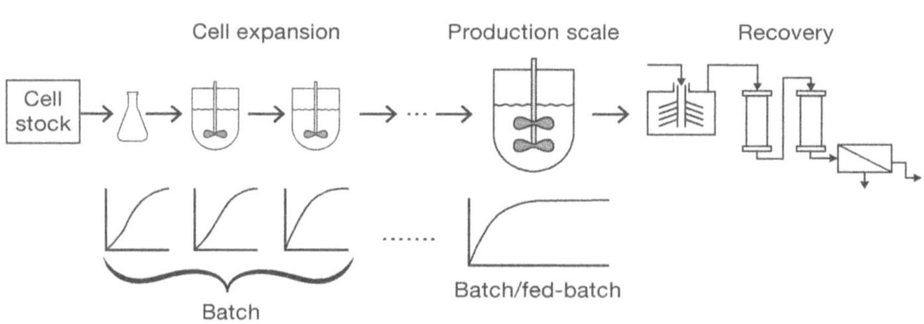

Figure 9.3. A typical cell culture process for the production of biologics combining batch and fed-batch cultures followed by downstream processing.

added at the beginning to support cell growth and production until the end of cultivation. A variant of batch process is the fed-batch process, in which additional nutrients are added during the cultivation period to sustain the growth and production longer than that of a batch culture. The vast majority of cell culture manufacturing processes employ batch culture for cell expansion and fed-batch culture in the production-scale reactor (Figure 9.3). The series of batch cultures for cell expansion is often referred to as a seed train. At the termination of a fed-batch reactor, the cell viability may be low and the product is recovered in downstream operations. Continuous culture is not commonly practiced in the cultivation of mammalian cells unless it is in conjunction with cell recycling. This process allows for continuous discharge of metabolites and product, thus alleviating the growth inhibition caused by metabolite accumulation and prolonging the culture (Figure 9.2b). The different types of continuous cultures will be discussed in the next chapter.

Batch Cultures

BATCH PROCESSES ARE SIMPLE and widely used. In fact, batch process culture is used much more frequently than fed-batch, though in smaller scales in seed trains. Batch culture is used for the production of viruses for vaccines and gene therapy applications as well as in cell production for immunotherapy and other therapies.

Kinetic description of batch culture

The growth behavior of a cell line in a batch culture can be described by setting up the material balance equations of the reactor on the cells, the key nutrients and metabolites, and possibly the osmolality, which all influence the growth rate (Panel 9.1). Note that material balance is performed on the quantity of the chemical species to be balanced, not the concentration of the species, as the volume of the system being balanced may change, thus rendering a balance on concentration invalid. In the case of a batch culture, the volume is constant (Eq. 9-1). We therefore eliminate the volume from the equations, thereby allowing balances on cells (x), substrate (i.e., nutrients; s), and the inhibitory metabolite (I) to be based on concentrations (Eqs. 9-2, 9-3, and 9-4). By incorporating a description of the dependence of the specific growth rate on the level of substrate and inhibitor (e.g., the Monod type of relationship (Eq. 9-2) and tying the production of the inhibitor to the consumption of the substrate by a stoichiometric coefficient, the kinetic model can be used to generate a growth curve for a batch culture (Figure 9.4). In most cases, the effect of inhibitory metabolite accumulation has a greater effect on cell growth than the reduction of substrate.

Panel 9.1. Material Balance for a Batch Bioreactor

$$\frac{dVx}{dt} = V\frac{dx}{dt} = V\mu x \quad \textbf{(Eq. 9-1)}$$

$$\frac{dx}{dt} = \mu x = \frac{\mu_{max}}{K_s + s} \cdot \frac{1}{1+\frac{I^2}{K_I}} x \quad \textbf{(Eq. 9-2)}$$

$$\frac{ds}{dt} = -q_s x \quad \textbf{(Eq. 9-3)}$$

$$\frac{dI}{dt} = \alpha_{I/s} q_s x \quad \textbf{(Eq. 9-4)}$$

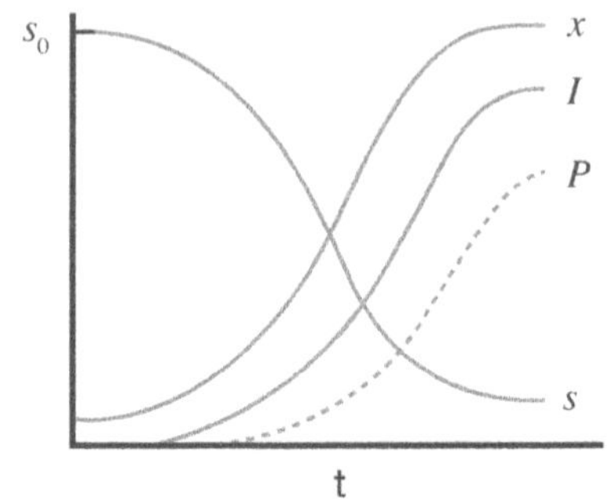

Figure 9.4. The kinetics of cell growth, substrate, product, and metabolites in a batch culture.

Fed-Batch Cultures

BATCH CELL CULTURES ARE LIMITED by the low cell and product concentrations achievable. To achieve a high cell and product concentration, the initial nutrient concentration in the medium must be high. Excessively high nutrient levels and the accompanying high osmolality causes growth inhibition; hence the use of fed-batch processes.

Types of Fed-Batch Cultures

Intermittent harvest

The simplest form of fed-batch culture involves intermittent harvest. At a late exponential growth stage of the culture, a portion of the cells and product are harvested, and the culture is replenished with fresh medium. This avoids metabolite inhibition of cell growth and replenishes nutrients for continued cell growth. This process is typically repeated several times before the culture is terminated (Figure 9.5). This simple strategy is commonplace for the production of viral vaccines produced by persistent infection, as it allows for an extended production period. It is also used in roller bottle processes with adherent cells. By intermittent harvesting, turn-around time involved in cleaning up and restarting the process is saved. Furthermore, after replenishing the medium

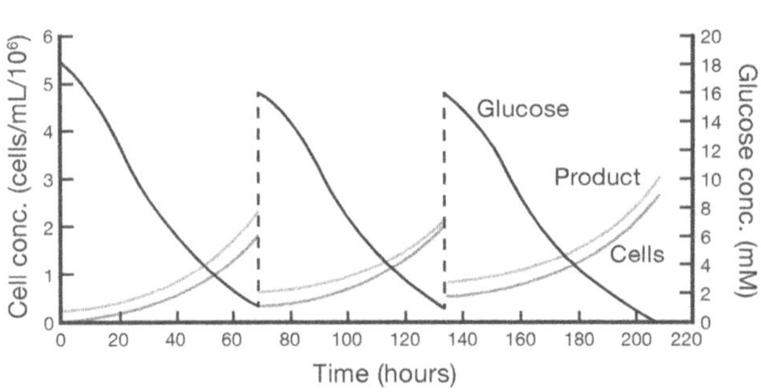

Figure 9.5. An intermittent fed-batch culture.

the culture reaches peak cell concentration faster than if the culture had been started all over. In some cases, this type of fed-batch culture is also used to supply inoculum to start the seed train for production for a period of time. Cell products, such as those to be used for cell therapy, cannot be subjected to highly stressed conditions such as in fed-batch culture and are therefore typically produced in batch cultures or intermittent fed-batch cultures.

Fed-batch culture with fortified feed

For production of recombinant proteins, a fed-batch culture is initiated by partially filling up the bioreactor with the startup medium (approximately 70% of the reactor capacity, or at a level sufficient to allow the impeller to be submerged) (Figure 9.6a). Nutrients, usually in a more concentrated solution (fortified feed) than in the startup medium, are added during the cultivation. The concentrated medium is added either continuously or intermittently. The additional nutrient supply prolongs the growth period and increases the cell concentration. It also extends the period of product synthesis and results in a substantially higher product titer. It is common practice to decrease the operating temperature after a period of rapid growth in fed-batch culture. The temperature shift reduces the growth and metabolic rates, extending the production period and increasing the product titer. During the late stage of a fed-batch culture, the osmolality in the medium is high due to the addition of fortified medium (which is commonly prepared with an additional alkali or acidic solution to facilitate the solution of solid nutrients) and a pH-neutralizing base. The osmotic stress also enhances the secretion of proteins.[2, 3] The practice of this type of fed-batch culture led to the enhancement of the productivity of cell culture processes that we saw in the decade around the turn of the century.

The increased cell concentration and prolonged process in a fed-batch culture amplifies the adverse effect of metabolite accumulation. In addition to the usual inhibitors such as lactate and ammonium, the degradation products of other compounds, for example aromatic and aliphatic amino acids, may also reach inhibitory concentrations. The high osmolality and accumulation of metabolites and reactive oxygen species all contribute to the eventual loss of viability and productivity of the culture. The resulting cell death and possibly cell lysis likely causes the release of cellular enzymes such as proteases and sialidase. The release of cellular enzymes has been attributed to the increased occurrence of structural and glycosylation variants at the end stage of fed-batch culture. Thus, most fed-batch processes do not operate until the viability becomes low.

As discussed in Chapter 3, at a stationary phase of culture cells may change from a high glucose flux/high lactate production state to a low

a)

b)

c)

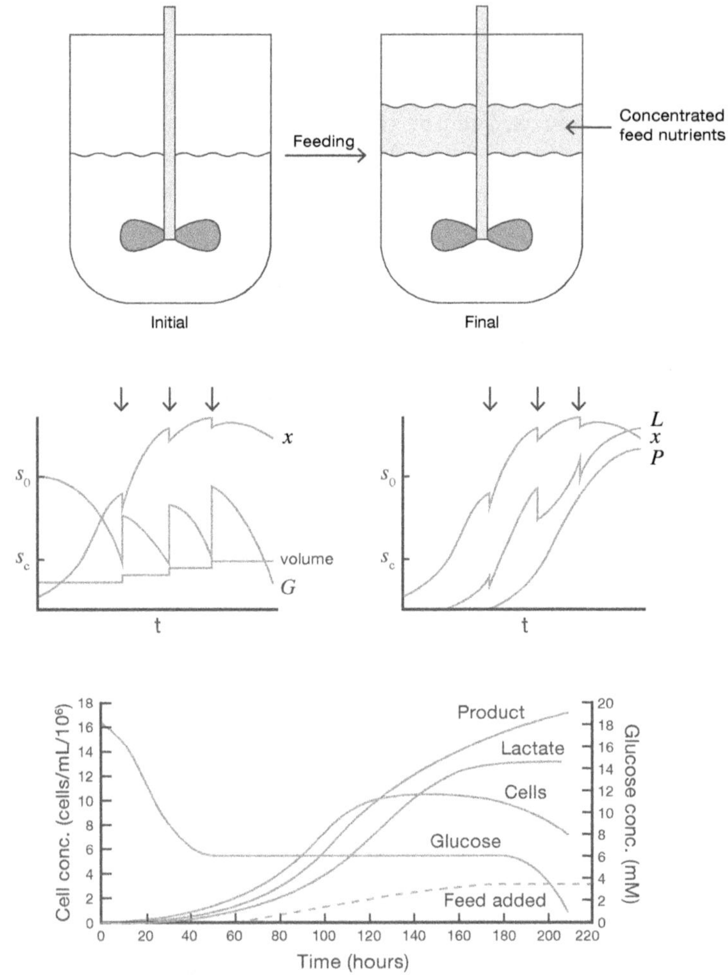

Figure 9.6. (a) Fed-batch culture with increasing volume during cell cultivation by (b) periodic feeding as indicated by ↓ and the step change in the concentrations, or (c) continuous feeding.

glucose flux/lactate consumption state. The metabolic shift is generally accompanied by a decreasing level of lactate, increasing pH, and a slowed increase of osmolality. In this period one also often observes a higher productivity.

Kinetic Description of Fed-Batch Culture

The most important design parameter of a fed-batch process is the feeding scheme, both in terms of the composition of the feed and the scheduling of feed addition. The main culprit of inhibitory metabolite accumulation arises from over or imbalanced supplies of nutrients. Balancing the supply of nutrients reduces metabolite accumulation and increases cell concentration and productivity. A kinetic model describing the state of the culture will greatly facilitate the formulation of a feeding scheme.

Panel 9.2. Material Balance for a Fed-Batch Bioreactor

$$\frac{dV}{dt} = F(t) \quad \textbf{(Eq. 9-5)}$$

$$\frac{dVx}{dt} = \mu xV \quad \textbf{(Eq. 9-6)}$$

$$\frac{dVs_i}{dt} = Fs_{f_i} - q_{s_i}xV \quad \textbf{(Eq. 9-7)}$$

$$\frac{dVI}{dt} = \alpha_{I/s_i}q_{s_i}xV \quad \textbf{(Eq. 9-8)}$$

In a fed-batch culture, the volume of the culture is changing over time (Eq. 9-5, Panel 9.2). The volume change is the result of feeding at a rate of F. Implicitly we assume that the density of the culture is constant, so the balance can be done on the culture's volume instead of its total mass. The differential equations for the material balance will include the culture volume, which considers the total amount in the reactor, instead of balancing on the concentration (Eqs. 9-6, 9-7, and 9-8). The biomass balance equation is the same as that for batch culture. For nutrients, we use subscript i to denote different medium components that are consumed by cells. If a component of the medium is not consumed, then the consumption term (the second term) on the right-hand side of Eq. 9-7 is zero. We assume a single feed is used to supply all nutrients. The production of metabolites (I) is related to a reference substrate (for example, glucose or glutamine) with stoichiometric ratio α between I and s_i (Eq. 9-8).

We consider two cases, one feeding continuously and the other intermittently. In the first case, one needs to determine the feeding rate as a function of time and feed concentration. In the latter, one determines the amount of feed to be added so that by the next feeding time the concentration can be maintained above a set level.

For the first case, we aim to maintain the substrate concentration constant. We thus set Eq. 9-7 equal to 0 and solve for the feeding rate. By the product rule (Eq. 9-8) and rearrangement, Eq. 9-7 becomes Eq. 9-11 (Panel 9.3). The feeding rate is dependent on the feed concentration s_{fi}. Using a constant feed concentration, the flow rate increases with increasing cell concentration as cells grow.

For the second case, the balance between two discrete time points, t_1 and t_2 (Eqs. 9-12 and 9-13, Panel 9.4) is used to determine the amount of nutrients that need to be fed. In the equation, we retain the symbol of feed as F, even though in practice it is added at only t_1. We then integrate from t_1 to t_2 and use M to denote the amount of feed added at t_1 (Eq. 9-14). By adding M amount, we aim to supply the substrate so that its concentration does not exceed or fall below a certain range. M is thus dependent on the amount of substrate consumed in the interval between feedings (Eq. 9-16). After the addition of the feed, the substrate concentration increases (Eq. 9-17). The higher the nutrient consumption in the

Panel 9.3. Continuous Feeding of a Fed-Batch Culture

From Eq. 9-7, to maintain constant s (i.e., $\frac{ds}{dt} = 0$)

$$V\frac{ds_i}{dt} + s_i\frac{dV}{dt} = 0 + s_iF = Fs_{f_i} - q_{s_i}xV \qquad \text{(Eq. 9-9)}$$

$$F\left(s_{f_i} - s_i\right) = q_{s_i}xV \qquad \text{(Eq. 9-10)}$$

$$F = \frac{q_{s_i}xV}{s_{f_i} - s_i} \qquad \text{(Eq. 9-11)}$$

Panel 9.4. Step (Intermittent) Feeding

$$\frac{\Delta(Vs_i)}{\Delta t} = Fs_{f_i} - q_{s_i}xV \qquad \text{(Eq. 9-12)}$$

$$\Delta Vs_i = Fs_{f_i}\Delta t - q_{s_i}xV\Delta t \qquad \text{(Eq. 9-13)}$$

Integrate between t_1 and t_2, feed is added in the amount of $F\Delta t$ at t_1. $F = 0$ at all other times. We use M to denote the total amount of feed added at t (this keeps the units correct in the balance equation).

$$M = F\Delta t \qquad \text{(Eq. 9-14)}$$

$$Vs_i\big|_{t_1} - Vs_i\big|_{t_2} = Ms_{f_i} - \int_{t_1}^{t_2} q_{s_i}xV\,dt \qquad \text{(Eq. 9-15)}$$

Can determine M to be added at t_1

$$M = \left(Vs_i\big|_{t_1} - Vs_i\big|_{t_2} + q_{s_i}xV\Delta t\right) / s_{f_i} \qquad \text{(Eq. 9-16)}$$

Eq. 9-16 is used to iterate M and Δt to find an operating combination.

After the addition of feed, the substrate concentration increases to s'_{i,t_1}

$$s'_{i,t_1} = \frac{s_{i,t_1}V_{t_1} + Fs_{f_i}\Delta t}{V_{t_1} + F\Delta t} \qquad \text{(Eq. 9-17)}$$

interval of feeding is (i.e., the higher the cell concentration or the longer the interval), the higher the concentration after the feeding will be. With a shorter interval (i.e., more frequent feeding), the amount of substrate cells consume is smaller, thus allowing a lower level of substrate concentration upon feed addition. A scenario of a fed-batch culture profile is shown in Figure 9.6.

Fed-Batch with Metabolic State Manipulation

With the realization that metabolite accumulation limits cell concentration and productivity in fed-batch culture came several strategies designed to alleviate the problem. The first approach induces a metabolic shift from a high glucose flux state to a low one during the rapid growth stage in order to eliminate lactate accumulation. This is accomplished by stoichiometric feeding of glucose, glutamine, and other amino acids using the strategy described in the preceding section. By keeping the concentration of glucose at a consistently low level, glycolysis is shifted to a low flux state characterized by the reduced production of lactate and ammonium (Figure 9.7).[4] The glucose concentration is in the vicinity of K_M for glucose transporter GLUT1 (see Chapter 3). The strategy thus requires continuous nutrient feeding to maintain glucose in a narrow range; too high will lead to a return to a high flux state, too low may cause glucose depletion and apoptosis. In another approach, the controlled feeding is only implemented when rapid cell growth is over. The feeding can be linked to pH control, as either lactate production or consumption can lead to control actions for pH fluctuation.[5]

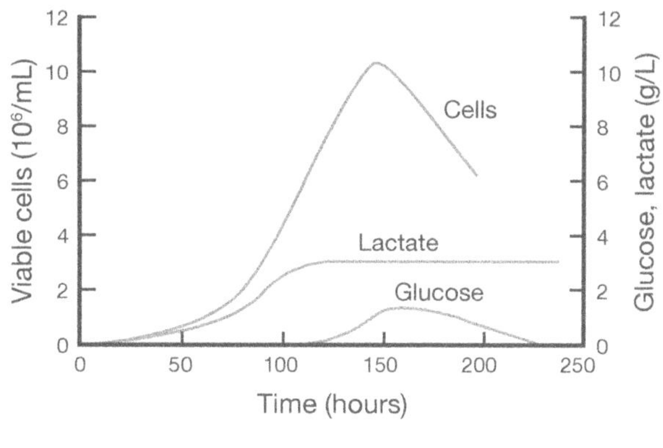

Figure 9.7. A fed-batch culture with continuous feeding to control nutrient levels and shift metabolism. Keeping glucose at low levels reduces glucose consumption and lactate formation. Reducing both glucose and glutamine levels reduces lactate and ammonium accumulation.

Stoichiometric Feeding

Critical to a successful fed-batch process is the formulation of feed media and a feeding strategy. Much of the inhibitory metabolite accumulation in fed-batch culture is due to an unnecessarily high level of nutrients caused by unbalanced supply. A sound feeding strategy is to provide a balanced nutrient medium in order to minimize the accumulation of inhibitory metabolites.

Under balanced growth conditions, cells consume various nutrients and produce different metabolites in some proportionality. These proportionalities are called stoichiometric ratios. The idea of stoichiometric feeding is that if all nutrients are consumed in relative proportions, one can feed all nutrients according to the stoichiometric ratios. If one supplies a balanced amount for one nutrient and keeps all the other nutrients at their correct stoichiometric ratios, then every nutrient will be supplied in a balanced amount and maintained within an optimal range. The stoichiometric ratio can thus be used as a basis for designing feed media.

Medium for Stoichiometric Feeding

A well-formulated feed media seeks to keep all nutrients at the same proportion as their consumption (i.e., at their stoichiometric ratios). The stoichiometric ratio is determined experimentally using the same cell line under relevant cultivation conditions. Typically, one medium component is chosen as a reference nutrient. The consumption rates of all other nutrients are related to the reference nutrient by its stoichiometric ratio. Common choices for reference nutrients are glucose, glutamine, oxygen, and lactate, as they are consumed or produced in larger quantities and are relatively easy to measure. As discussed in Chapter 5, the stoichiometric ratio is easily calculated from the cumulative consumption of nutrients and the production of metabolites. The cumulative consumption of a nutrient is plotted against that of the reference nutrient from the same culture. The slope on the plot gives its stoichiometric ratio.

Stoichiometric ratios are not necessarily constant. Cells may consume different nutrients in different proportions in different culture stages. The feed media composition may need to be adjusted according to known changes of stoichiometric ratios under different culture conditions or in different growth stages. The development of a stoichiometric feed medium takes a couple iterations. A medium with the first composition is tested and analyzed to derive a new set of stoichiometric ratios for a new feed medium, which is then tested in turn. It usually takes about three iterations for an optimal medium to emerge. In addition to the feeding medium, another variable is the optimal level a given nutrient ought to maintain in culture. Most essential amino acids need to be maintained above a critical level. Even some non-essential amino acids should be kept above certain levels because their deficiency may lead to amino acid misincorporation in the product or other physiological effects.

Once the composition is determined, it is ideal to make a highly concentrated feed medium to increase the amount of nutrients that can be delivered using a fixed volume of feed. However, this is limited by the solubility and kinetics of solution of the various components. Many components are slow to dissolve. A highly concentrated medium may also be prone to precipitation in storage. Typically, the concentration of nutrients in the feed medium is about 10 to 15 times that in basal medium.

Unconsumed Medium Components

Some medium components are not consumed or are consumed at such a minute rate that their concentration change in a typical culture is not detectable. Some examples include inorganic ions like sodium, calcium, and sulfate. These hardly-consumed medium components may not need to be added to the feed, but because the culture volume increases upon feeding they will become diluted. Consequently, feeding of these unconsumed components may be necessary if their dilution affects growth or product synthesis. When included in the feed, they are at basal medium levels so that they are not diluted by the expanding volume (Panel 9.5).

Panel 9.5. Feed Medium

Consumed Nutrients

- Present in the medium proportional to their stoichiometric consumption

Unconsumed

- Many components are consumed at very low levels, such that their concentration in the culture supernatant hardly changes
- They can be omitted from the feed or be present at basal medium levels to minimize dilution caused by culture volume expansion
- Feeding may cause an osmolality increase
- The "consumption" of hydrolysates, growth factors, etc., may not be easily quantified

Some ions, such as magnesium (complexed with ATP), phosphate (as free phosphate or in nucleotides), and potassium, are present at a much higher concentration intracellularly than in the medium. At high cell concentrations, the amount of these ions taken up by the cells may become significant. Therefore, it is often necessary to compensate for their consumption by supplying them in the feed medium.

In addition to basal medium components, protein hydrolysates, serum, insulin, transferrin, vitamins, and lipid additives are also used in the start-up medium. These additives may supply minute nutrients that are not identified and measured. In the absence of a reliable measurement of consumption rate for those additives, one has to rely on an order-of-magnitude estimate of the upper and lower limits of their consumption rate. The feeding rate of those additives can then be chosen to maintain their concentrations above a minimum threshold and below a maximum tolerable limit which is experimentally determined.

Another point of consideration in feed medium design is its effect on culture osmolality. In a basal medium, various salts and buffers contribute to the bulk of osmolality. The feed medium, in principle, should maintain osmolality balance in a culture. Immediately after feeding, the osmolality may increase, but, because the feed is stoichiometrically balanced, after the nutrients are consumed the osmolality should return to the original value unless additional salts are present in the feed. In practice, the feed typically has a high osmolality and high salt concentration in addition to the nutrients. Because some amino acids dissolve very slowly, an acid or base is often added during feed medium preparation to accelerate dissolution and then neutralized after solution. The added acid or base increases the salt content in the medium. Higher osmolality in the production stage has been shown to enhance productivity. However, overly high osmolality reduces cell viability. Keeping osmolality under an upper bound is thus part of the design consideration.

Delivery of Feed Medium

Intermittent Feeding

HOW THE FEED MEDIUM IS DELIVERED to a culture may affect the performance of a fed-batch process. In most manufacturing settings, the feed medium is added intermittently. The time point and amount of feeding is either prescribed in the operating protocol or based on the measurement of a variable (e.g., glucose concentration) that is to be kept within a certain range. Since the addition of feed increases osmolality, the first addition is performed after the initial rapid growth stage. Cells will then continue to grow at a slower rate. As shown in the balance equations, for the period between two consecutive feeding times, t_1 and t_2 (Eq. 9-12 to Eq. 9-15), the amount of nutrient delivered at t_1 (Ms_{if}) should be sufficient to supply the demand ($q_{si}xV\Delta t$) and sustain s_i above the set point level. At higher cell concentrations, more feed medium is added. That inevitably leads to a higher level of post-feed nutrients ($s'_{i,t1}$) which may cause excessively high levels of nutrients and magnify inhibitory metabolite accumulation. Alternatively, a smaller amount of feed can be added more frequently to keep the post-feed nutrient level within an acceptable range.

Continuous Feeding

Continuous feeding minimizes the medium composition's tendency to swing from high to low levels periodically as caused by intermittent feeding. It also offers the possibility of manipulating cell metabolism, for example to implement a more robust switch from a high flux glucose metabolism to lactate consumption in the late stage. As the concentration increases, the feeding rate has to increase to meet the increased demand. This requires frequent, ideally on-line, measurement of a process variable that can be used to adjust the feeding rate.

Feeding by Direct Measurement of Nutrient Consumption

Since nutrient feeding is based on stoichiometric balance, direct measurement of the concentration of the reference substrate is the most straightforward way to determine the amount of nutrient that has been consumed and how much feed should be added to balance the consumption. By measuring the reference nutrient concentration on-line, feed can be added in real time to maintain the nutrient at a constant level. Glucose and glutamine are the two nutrients most commonly measured and used as controls because their concentrations can be determined fairly rapidly in laboratories. Direct measurement of these compounds on-line requires using an auto-sampling device that is connected to a measurement device.

Proportional Feeding with Cell Concentration

A strategy of continuous feeding is to increase the flow rate proportional to cell concentration. Various laser turbidity probes can be used to determine the cell concentration in culture. Capacitance probes are increasingly being used because they measure total viable cell volume (i.e., viable cell concentrations and cell sizes) instead of the total cell concentration (both viable and dead cells) given by optical probes. Feeding with a simple proportionality to cell concentration assumes that all the other contributors to the cells' nutrient demands as shown in Eq. 9-11 are constant. This is an approximation that works well for a period of time. If the nutrient concentration does not need to be controlled in a narrow range, this method is easy to set up if an on-line cell concentration measurement is available.

Stoichiometric Feeding with Base Addition

A widely-used nutrient feeding strategy in microbial fermentation links nutrient feeding to acid or base addition for pH control. When the uptake of a nutrient or the production of a key metabolite is associated with the release or dissipation of H^+, a stoichiometric relationship can be established between them and the amount of acid or based added to neutralize pH can be used to determine the amount of nutrient consumed. In cell cultures, a large fraction of the glucose consumed is converted to lactic acid, and in the late stationary phase lactate may be consumed. To maintain pH, one mole of base or acid is added to compensate for the H^+ excreted or taken up with lactate through MCT transporters. From the amount of base/acid added, the rate of lactic acid production can be calculated. That in turn is used to feed the stoichiometrically balanced medium and maintain nutrient levels. In other words, the method uses the amount of base or acid added to maintain pH to determine the $q_{si}xV$ term in Eq. 9-11. It is simple and easy to implement. However, its response is sluggish because of the buffer capacity of the medium. The dissolved CO_2 level in the medium varies over a wide range throughout the culture time. Hence, the estimated lactate production is subject to errors. However, most manufacturing processes do not control nutrient levels in a narrow range, and the method is adequate even for reducing glucose flux.

Stoichiometric Feeding with Oxygen Uptake Rate (OUR)

The oxygen uptake rate (OUR) is a very sensitive method for assessing cellular metabolic activity that dictates the demand for nutrients and the feeding rate. A typical dissolved oxygen sensor can detect a change of 5% saturation with air at 1atm, or about 0.01 mmol/L. That is much more sensitive than typical laboratory measurement for glucose (about 0.3 mmol/L). In principle, pH measurement can detect very small differences of H^+ concentration, but the large buffer capacity in a biological fluid decreases its sensitivity for detection. OUR measurements, unlike pH, are not masked by buffers in the medium. OUR measurement is also relatively easy to implement on-line as discussed in Chapter 11.

With its high sensitivity, small changes of OUR can be confidently detected on-line and in real time, thereby providing an immediate indication of changes in metabolic rate. Its measurement is indicative of the overall metabolic activity, a combined effect of the specific rate and the cell concentration as opposed to the cell concentration measured using a capacitance probe. Through the established stoichiometric ratio, OUR measurement can be used to determine the feeding rate (Eq. 9-18, Panel 9.6). The metabolic rate change in cell culture processes is relatively slow. In most cases, intermittent determination of OUR in intervals of tens of minutes is sufficient to detect metabolic rate changes and adjust the feeding rate.

Panel 9.6. Stoichiometric Feeding by OUR

- OUR is a sensitive method for quantifying cells' metabolic load
  - Detect Δc ~0.02 mmol/L accurately (for glucose, Δc for detection ~ 0.3 mmol/L)

$$F = \frac{\alpha_{s/o}(\text{OUR})V}{s_{f_i} - s_i} = \frac{(\alpha_{s/o} q_{o_2} xV)}{s_{f_i} - s_i} \qquad \textbf{(Eq. 9-18)}$$

Product Quality in Fed-Batch Culture

The quality of the therapeutic protein produced in culture is the ultimate concern of biomanufacturing. Important quality attributes include the amino acid sequence, chemical modifications of the protein, and glycosylation. A number of intrinsic characteristics of fed-batch culture give its product a potentially wider range of heterogeneity compared to batch or continuous culture.

Quality Variation in Different Growth Stages

Seeking to enhance productivity, many fed-batch cultures employ culture conditions that incur high stress. The osmolality often increases from the 300 mOsm that is optimal for growth to over 400 mOsm in the final stage of production, as it increases product titer. Some fed-batch culture processes reduce the cultivation temperature once the rapid cell growth stage is over. A fed-batch culture thus has very distinct stages: the early stage, in which the chemical environment is optimal for growth; the middle stage, in which the environment is highly stressed; and the late stage, in which cells are non-proliferating due to the addition of a fortified medium and an accumulation of metabolites. During the rapid growth stage, cells are invariably in a high glucose flux and high lactate producing state. In many cases, a shift to low glucose flux and lactate consumption is seen in the late stage of fed-batch culture, as discussed in Chapter 3. Cells are at very different metabolic states in those stages and are also likely to have very different intracellular concentrations and many metabolic intermediates that may affect post-translational modification of proteins. Cells at these distinct stages with different chemical environments, temperatures, growth rates, and metabolic states, may produce products with different quality attributes.[6]

Overfeeding or undersupplying of nutrients affects product quality. Since the effect of errors is cumulative in a culture, the effects of over- or under-feeding is more profound in the late stage of culture. The depletion of a particular amino acid may cause amino acid misincorporation in the protein product. In place of the correct amino acid that is in short supply, the wrong amino acid is incorporated in the protein. Oversupply and consumption of aromatic and aliphatic amino acids increases the production of reactive oxygen species and other metabolites that may increase the number of protein structural variants.

Typically, the quality of the product produced is assessed by examining the quality attribute of the purified protein harvested at the end of the culture, which is the accumulation of all protein molecules produced from the beginning of the culture to the end. If we could isolate only the product produced during a given period of time in the culture, one would be able to assess how the quality of the product varies with the culture's stages. However, such information is rarely collected. If it becomes available, it will facilitate the development of a control strategy for product quality.

Quality and product residence time

Product molecules produced at different points of time in a fed-batch culture spend different durations of time in the reactor before being harvested, i.e., the product molecules have different "ages." As illustrated in Figure 9.8, product produced earlier spends more time in the culture, and the amount of product produced differs among the stages. Cells in different stages are at different metabolic states and may produce product of varying quality. In some cases, the culture environment causes the secreted product to undergo changes; product secreted at different times, therefore, will spend different durations in the culture and have a different probability of being modified. For example, sometimes the culture conditions involve a very high glucose concentration (15 g/L or 83 mM) (recall that the physiological range is 0.7–1.2 g/L) that increases the likelihood of glycation in the protein product. A prolonged culture under stressed conditions increases cell death and the release of cellular materials, including sialidase, which removes sialic acid from glycan and

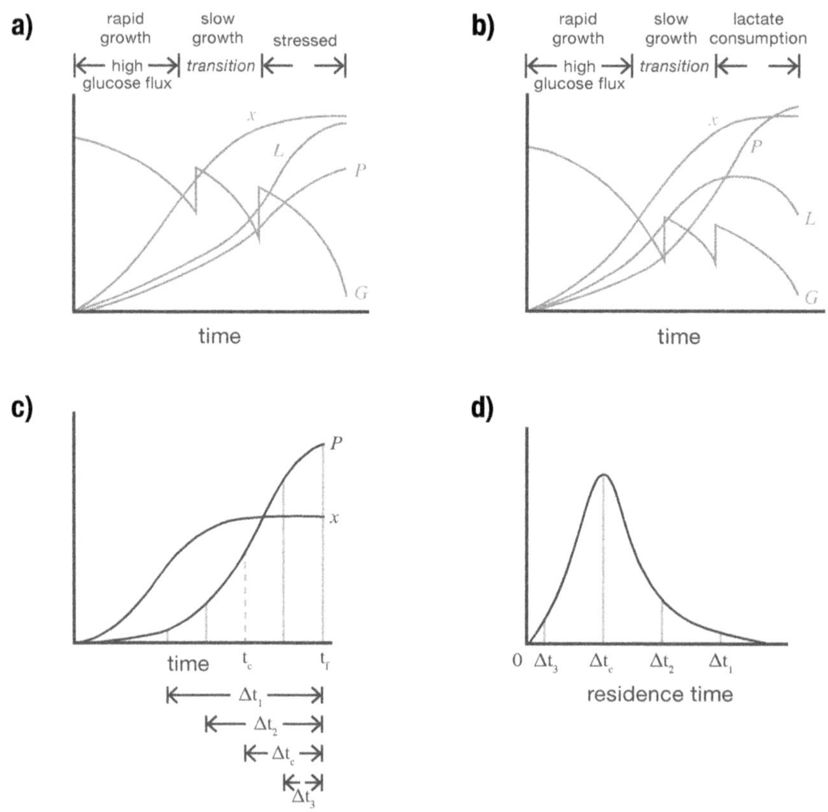

Figure 9.8. (a, b) Product proteins are produced at different stages of a culture with different physiological states. Different cultures may have different metabolic behavior; one (a) produces lactate until the end, while the other (b) switches to lactate consumption. The different physiological states may secrete proteins of different qualities. (c) The quality of the final protein product is the average of all the molecules produced at different times across the culture. t_c is the time point when half of the product molecules have been produced, and t_f is the end of culture. (d) Protein molecules secreted earlier have a longer residence time (or age) (Δt). Since they spend different amounts of time exposed to the culture fluid, they are subject to varying degrees of modification.

gives rise to asialilated glycans.[7] Various proteases and lytic enzymes released from dead cells may alter product protein quality. To minimize cell-death-related protein quality deterioration, some researchers maintain a long, slow growth phase and avoid long periods of low viability.

Quality and Culture Reproducibility

A fed-batch culture involves more process variables and is subject to wider variability than a batch culture. The presence of distinct culture stages and the disparate residence time among products, bring additional heterogeneity to the product quality. If the cells' physiological conditions change in different cultures, detectable alterations to the quality attributes of the product produced in different runs may be seen.[8] When such variations occur, the investigation of the cause may need to examine products produced at different stages of the culture to pinpoint the cause.

In some cases, different process runs under the same manufacturing conditions, with the same cell line, and at the same manufacturing plant can behave quite differently. For example, some runs switch from a high glucose flux state to a low flux and lactate consumption state, while others don't. These two types of culture would have very different metabolic states. The glycan produced under those two conditions appears to bear differences. Under different metabolic states, the pool size of nucleotide sugars may differ, thus causing glycoform inconsistency among different runs. A feeding strategy that can better guide cell metabolism to a consistent state will allow for a more robust production of a quality product.

Concluding Remarks

Fed-batch culture is the prevailing mode of cell culture in the final production of manufacturing recombinant proteins. It extends the cell growth and production periods, reaches high cell concentrations, and is operationally relatively simple. The design of the feed medium and the selection of the feeding strategy strongly influence the productivity of fed-batch culture. Compared to batch and continuous culture operations, cells in fed-batch cultures are subjected to wide variations in their chemical environment and are potentially physiologically different during different stages of their culture as well as in different runs. Enhanced process knowledge and better control of the chemical environment will enhance the consistency of product quality.

References

1. Bianconi, E., et al., *An estimation of the number of cells in the human body.* Ann. Hum. Biol., 2013. **40**(6): p. 463–71.

2. Oh, S.K.W., et al., *Substantial overproduction of antibodies by applying osmotic pressure and sodium butyrate.* Biotechnol. Bioeng., 1993. **42**: p. 601–10.

3. Lin, J., et al., *Enhanced monoclonal antibody production by gradual increase of osmotic pressure.* Cytotechnology, 1999. **29**: p. 27–33.

4. Zhou, W., J. Rehm, and W.-S. Hu, *High viable cell concentration fed-batch cultures of hybridoma cells through on-line nutrient feeding.* Biotechnol. Bioeng., 1995. **46**: p. 579–87.

5. Gagnon, M., et al., *High-end pH-controlled delivery of glucose effectively suppresses lactate accumulation in CHO fed-batch cultures.* Biotechnol. Bioeng., 2011. **108**(6): p. 1328–37.

6. Sumit, M., et al., *Dissecting N-glycosylation dynamics in Chinese hamster ovary cells fed-batch cultures using time course omics analyses.* iScience, 2019. **12:** p. 102–20.

7. Gramer, M.J. and C.F. Goochee, *Glycosidase activities in Chinese hamster ovary cell lysate and cell culture supernatant.* Biotechnology Progress, 1993. **9**: p. 366–73.

8. Le, H., et al., *Discerning key parameters influencing high productivity and quality through recognition of patterns in process data.* BMC Proceedings 2011, 2011. **5 (suppl 8)**: p. P91.

Nomenclature

α	Stoichiometric coefficient
μ	Specific growth rate
F	Feed medium volumetric rate
I	Inhibitor concentration
K_I	Half-saturation constant of growth inhibition
K_M	Half-saturation constant of substrate for reaction
K_s	Half-saturation constant of substrate for growth
M	Volume of feed medium
q	Specific rate
s	Substrate concentration
t	Time
V	Volume of reactor
x	Cell concentration

Contents

10

Continuous Cell Culture Processes

Introduction

In the past decade, there has been an increasing interest in the continuous manufacturing of pharmaceuticals. The initial focus, which was on chemical drugs, has extended to protein biologics in recent years. The drive was prompted by the success of continuous processes in the chemical industry. A large proportion of chemical reactors are operated in a continuous fashion: a feed stream continuously brings in the reactants while an effluent stream continuously draws the reactor contents out. A continuous reactor offers a number of attractive features. It uses the reactor more efficiently than a batch process by minimizing equipment turnover time and process start-up time. It provides a continuous product stream from the reactor at a product concentration comparable to a batch process, and it can be operated at a steady state over a long time to achieve a higher throughput than a batch process.

Many enzymatic biotransformation, waste treatment, and biodegradation processes have long been operated continuously. However, continuous culture is still not the dominant process type in microbial and cell culture bioprocessing.

The vast majority of biochemical processes that involve microbial or animal cell cultivation are batch processes for many reasons. Unlike the catalysts used in the chemical industry whose activities are stable over time, cells may mutate or change epigenetically. As a result, the makeup of the population, its productivity, and even the product quality changes over time. Additionally, the risk of microbial contamination makes long operations susceptible to failure. It is also notable that the current product capture and purification operations are all designed for batch mode. Even if the production bioreactor is operated in a continuous mode, the overall process is still cyclical and does not realize all of the advantages of a continuous process as seen in the chemical and other manufacturing industries.

A continuous culture is constrained by the maximum flow rate at which it can operate, as the continuous fluid flow must not wash out cells faster than cell growth can replenish them. On the other hand, cells in culture produce growth inhibitors and metabolites that must be continuously removed by media replenishment in order to avoid growth inhibition. For cell culture processes, at a flow rate that meets both constraints, the achievable cell and product concentrations are too low to be economically viable. To overcome this shortcoming, a cell recycling system is added to the continuous culture. By recovering cells from the effluent flow and returning them to the reactor, one can operate a continuous culture beyond the natural limitation of its dilution rate (the flow rate divided by the bioreactor volume). With a higher cell concentration in the reactor, the overall throughput of the reactor is also higher.

When cell culture was first adopted as the production vehicle for biopharmaceuticals, continuous operation was explored as an industrial process. Researchers realized that the growth of mammalian cells in batch culture was impeded by lactate and ammonium accumulation. Thus, a continuous process alleviates growth inhibition by removing metabolites through a continuous flow of medium. However, the high cost of continuously removing expensive medium with only a low concentration of product made continuous culture economically unaffordable.

To enable continuous operation, cells are retained in the reactor while the medium flow flushes out metabolites. This can be accomplished by either immobilizing cells on solid particles, which prevents them from being flushed out by the medium flow; or by separating cells from the effluent stream and recycling them back to the reactor. Thus, the general idea is continuous culture with cell retention. These processes are often called perfusion, which is reminiscent of the procedure of flowing fluid through an organ or tissue.

Table 10.1. Marketed Cell Culture Products Produced by Perfusion Processes

Product	*Company*
RecombinantTM, Antihemophilic factor (recombinant), (factor VIII)	Baxter
Kogenate-FS (factor VIII)	Bayer
Aldurazyme Naglazyme	BioMarin
ReoPro (IgG Fab Fragment) Remicade (IgG1) Simponi (IgG1)	Centocor (Janssen)
Xigris (Protein C)	Eli Lilly
Cerezyme Fabrazyme Myozyme/Lumizyme	Genzyme (Sanofi)
Gonal-F (follicle-stimulating hormone)	Serono (EMD)
Vpriv (velaglucerase alfa) Replagal (agalsidase alfa)	Shire (Takeda)
ReFacto (factor VIII)	Wyeth (Pfizer)

Panel 10.1. Continuous Cell Culture with Cell Retention

Use of Perfusion Culture
- For products that are labile or produced at very low levels
- Requires in-house expertise on handling continuous operation
- Recent advances in cell retention technology and the use of media with less complex components enabled its wider adoption

Potential Advantages of Perfusion Culture
- Possible steady-state operation
- More consistent product quality
- Lower metabolic levels and higher cell viability than at the conclusion of fed-batch culture

A number of biotherapeutic proteins are produced by perfusion processes (Table 10.1). Some of these products can be produced with a fed-batch process. The selection of the process mode is also dependent on the expertise available in-house and many other factors (Panel 10.1). If the product is labile and degraded or otherwise inactivated over time, a perfusion culture is certainly the process of choice. For products that accumulate only at very low concentrations, a perfusion process may also present a competitive advantage over a batch or fed-batch process by reducing the equipment turnover time. The advances in cell separation technology have greatly expanded the space of continuous culture in cell bioprocessing.[1] These advances have made a long-term perfusion culture achievable for many products. This, combined with the commercial availability of medium-production-scale disposable stirred tank bioreactors, has led to perfusion culture and hybrid perfused fed-batch culture becoming more widely adopted.

Analysis of Continuous Culture and Steady State

Simple Continuous Culture

We will use an idealized, continuous, well-mixed bioreactor to explain the basic concept of continuous culture and steady state, even though a simple continuous culture is rarely practiced. The stirred tank bioreactor receives a continuous feed stream and releases a continuous effluent stream consisting of medium and cells from the bioreactor (Figure 10.1a). Because of its "well-mixed" nature, the substrate from the feed stream is instantaneously mixed with the content of the bioreactor, and the effluent stream has the same composition as the content of the bioreactor. The flow rates of the feed and effluent streams are the same (i.e., $F_{in} = F_{out} = F$) to keep the reactor volume constant.

The material balance equations for cells and the rate-limiting substrate (Eqs. 10-1 and 10-5, Panel 10.2) reflect the balance of the amount of cell mass or substrate carried into and out of the reactor by the fluid flow. They also account for changes to cell or substrate concentrations caused by cell growth and nutrient consumption (ignoring cell death).

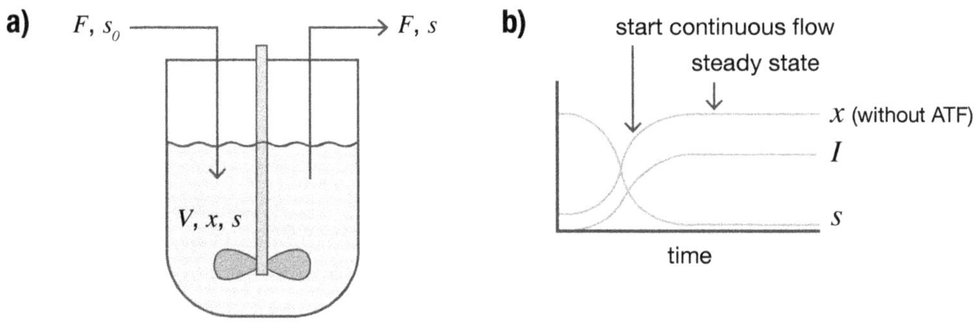

Figure 10.1. (a) A continuous culture bioreactor. (b) The starting up of a continuous culture to reach a steady state.

The flow rate into the reactor divided by the reactor volume is called the dilution rate (D, Eq. 10-2). Because the volume of the culture is constant, the material balance equation can be based on the concentration rather than on the total mass quantity in the culture (Eq. 10-3). The increase in cell mass due to growth is expressed as the multiplicative product of the specific growth rate (μ) and cell concentration (x). The consumption rate of the substrate is the cell concentration multiplied by the specific substrate consumption rate (q_s, Eq. 10-6).

Typically, a culture is started in batch mode. When the cell concentration approaches its maximum, continuous flow of the feed and effluent streams begins (Figure 10.1b). After some time, the cell concentration and substrate concentration become invariant. The culture is then regarded as having reached a steady state, as the most important culture variables (state variables) are no longer changing ($dx/dt = ds/dt = 0$). At a steady state, the specific growth rate and the dilution rate are equal (Eq. 10-4). This is a very important property of continuous culture. If a

Panel 10.2. Material Balance for a Continuous Culture

$$\frac{dxV}{dt} = V\frac{dx}{dt} = Fx_0 - Fx + \mu xV \qquad \textbf{(Eq. 10-1)}$$

$$D = \frac{F}{V}, \text{ cancel } V \qquad \textbf{(Eq. 10-2)}$$

$$\frac{dx}{dt} = \mu x - Dx \qquad \textbf{(Eq. 10-3)}$$

At steady state, $0 = \mu x - Dx$,

$$\therefore \mu = D \qquad \textbf{(Eq. 10-4)}$$

$$\frac{dsV}{dt} = Fs_0 - Fs - q_s xV \qquad \textbf{(Eq. 10-5)}$$

Divide by V and apply Eq. 10-2.

$$\frac{ds}{dt} = D(s_0 - s) - q_s x \qquad \textbf{(Eq. 10-6)}$$

At steady state,

$$D(s_0 - s) = q_s x \qquad \textbf{(Eq. 10-7)}$$

$$x = \frac{D(s_0 - s)}{q_s} \qquad \textbf{(Eq. 10-8)}$$

$$\frac{dI}{dt} = \alpha_{I/s} q_s x - DI \qquad \textbf{(Eq. 10-9)}$$

At steady state, $I = \dfrac{\alpha_{I/s} q_s x}{D}$ **(Eq. 10-10)** culture system is to be operated at a steady state, with cells comprising part of the effluent stream, then the cells must be growing and the specific growth rate must be equal to the dilution rate. Because of its simplicity, continuous culture has been an excellent tool for studying the kinetics of cell growth and metabolism.[2]

The cell concentration that can be achieved using a given feed substrate concentration at steady state can be calculated (Eqs. 10-7 and 10-8). In cell culture processes, inhibitor accumulation affects cell growth. Eq. 10-9 considers the case that the production of inhibitors is related to the consumption of substrate ($\alpha_{I/s}$), as evidenced in glucose consumption leading to lactate production. The concentration of the inhibitor (I) is determined by the cell concentration and the dilution rate (Eq. 10-10).

A continuous culture at steady state

To illustrate the steady state concept, we will use a simple Monod model to describe the relationship between the growth rate and the substrate concentration (Figure 10.2a, Eq. 10-11, Panel 10.3). We further assume that the specific substrate consumption rate is proportional to the growth rate (Eq. 10-12). $Y_{x/s}$ in Eq. 10-12 is the yield coefficient, i.e., the number of cells produced for each unit amount of substrate consumed. According to the Monod model, under a given substrate concentration (s) there is only one specific growth rate (μ). The material balance on cells and substrate gives a system of two equations that describes all steady states (Eqs. 10-13 and 10-14). With a given cell line, μ_m, $Y_{x/s}$, and

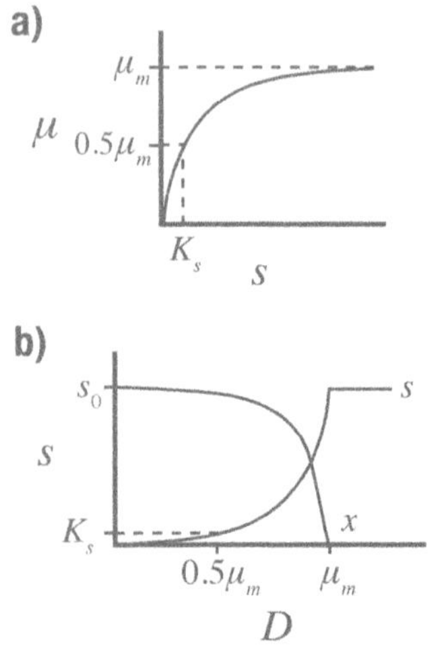

Figure 10.2. (a) The Monod model of growth kinetics. (b) The steady state cell and substrate concentration profiles of a continuous culture with Monod growth kinetics.

Panel 10.3. Steady State of a Continuous Culture with Monod Growth Kinetics

Assume $\mu = \dfrac{\mu_m s}{K_s + s}$ (Monod model) **(Eq. 10-11)**

and $q_s = \dfrac{\mu}{Y_{x/s}}$ **(Eq. 10-12)**

From Eq. 10-4:

$$\frac{\mu_m s}{K_s + s} = D \Rightarrow s = \frac{DK_s}{\mu_m - D} \quad \textbf{(Eq. 10-13)}$$

Plug Eq. 10-12 and Eq. 10-13 into Eq. 10-8:

$$x = Y_{x/s}\left(s_0 - \frac{DK_s}{\mu_m - D}\right) \quad \textbf{(Eq. 10-14)}$$

K_s are all fixed. We specify the feed substrate concentration, s_o, and the operating dilution rate, D. The system of equations then leaves only x and s as unknown.

Since the system has two linear equations (x and s do not multiply to themselves or to each other) and two unknowns, it has a unique answer. Once D and s_o are specified, the solution of x and s can be determined. In other words, if one uses the same cell dilution rate and feed concentration, the culture will always reach the same steady state (Figure 10.2b). With the assumptions of Monod growth kinetics and constant yield, the model predicts a relatively constant cell concentration over an increasing D until D approaches the maximum specific growth rate (and the critical dilution rate, D_c), whereupon it begins to decrease. It can be further shown that the steady state is stable if the system is subjected to small fluctuations in feed concentration or dilution rate, and the cell and substrate concentrations return to the same steady state after fluctuation subsides. Although the experimentally observed kinetic behavior differs somewhat from the model prediction,[2, 3] this model of cell growth in a continuous reactor still serves the purpose of exploring various process behaviors, especially regarding experiments that are difficult to perform.

Under a given set of operating conditions, a continuous culture may have more than one steady state or none at all. A simple system such as the one shown in Figure 10.2 has unique steady states. In some cases, non-linearity may arise from a more complex relationship between substrate concentration and growth rate, or if the conversion yield (i.e., metabolism) is dependent on the substrate level. In those cases, the system of equations may have more than one feasible (non-negative) solution under some conditions. For such a system, multiple steady states may be observed; in other words, in two systems with identical dilution rates and feed concentrations, the steady state reached may differ depending on the trajectory of the culture. This has been seen in many studies due to the complexity of metabolism in mammalian cells. For a system that has multiple steady states, one needs to control the culture condition to follow a specific trajectory to the target steady state.[4, 5]

Continuous Culture with Cell Recycle

The development of continuous cultures for the manufacturing of protein biologics was driven by the production of recombinant proteins that are labile and produced at low levels. The short product holding time that can be attained in a continuous culture is necessary for their production. However, a simple continuous culture is hindered by a low cell concentration and low productivity. In principle, increasing the substrate concentration in the feed, as shown in Eq. 10-8, can increase cell concentration. However, the correspondingly increased metabolite concentration (see Eq. 10-10) will lead to growth inhibition. To overcome this constraint, cell recycling is employed.

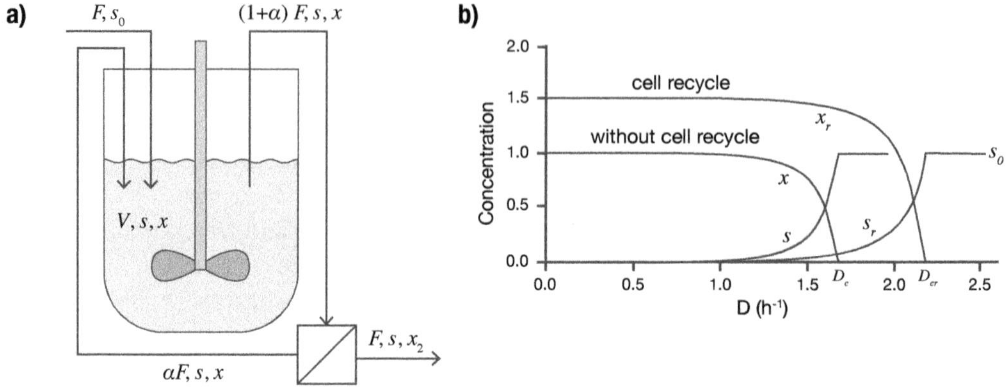

Figure 10.3. (a) A continuous culture with cell recycle. (b) Steady state cell and substrate concentrations at different dilution rates with and without cell recycle.

Panel 10.4. Continuous Culture with Cell Recycle

Balance on biomass for the reactor is

$$V\frac{dx}{dt} = \alpha Fcx - (1+\alpha)Fx + \mu xV \quad \textbf{(Eq. 10-15)}$$

Balance on the cell recycle system gives

$$(1+\alpha)Fx = \alpha Fcx + Fx_2 \quad \textbf{(Eq. 10-16)}$$

$$\frac{x_2}{x} = 1+\alpha-\alpha c \quad \textbf{(Eq. 10-17)}$$

The balance on the substrate is

$$V\frac{ds}{dt} = Fs_0 - \frac{\mu x}{Y_{x/s}}V - F(1+\alpha)s + \alpha Fs \quad \textbf{(Eq. 10-18)}$$

Defining $F/V = D$

$$\frac{\mu}{D} = 1+\alpha-\alpha c \quad \textbf{(Eq. 10-19)}$$

$c>1$, so $D>\mu$

At steady state,

$$0 = D(s_0 - s) - q_s x \quad \textbf{(Eq. 10-20)}$$

$$x = \frac{D(s_0 - s)}{q_s} \quad \textbf{(Eq. 10-21)}$$

A general system of continuous culture with cell recycle is shown in Figure 10.3a. The reactor volume is kept constant by having the same flow rate of fresh medium entering the reactor as there is culture content exiting it. The fresh medium stream is free of cells, and the effluent stream has the same cell and nutrient (substrate) concentrations as in the reactor. A cell separator divides the effluent stream into a concentrated cell stream (with a cell concentration of c_x and a flow rate of αF) that is recycled back into the reactor and a low cell density stream that is purged. Note that the effluent stream from the reactor has a flow rate of $(1+\alpha) F$, which is higher than the fresh feed. At steady state, the recycle stream has a flow rate of αF.

A material balance can be performed on the reactor and the cell recovery/recycling device for both cells and substrate in order to obtain the relationship among key process variables (Eqs. 10-15 to 10-21, Panel 10.4). The dilution rate is defined as the flow rate into the system (not the combined flow rate with the recycle stream into the reactor) divided by the reactor volume. Note that the relationship between biomass and substrate (Eqs. 10-20 and 10-21) is identical to that without cell recycle (Eqs. 10-7 and 10-8). However, with cell recycle, D and x are both higher than the corresponding specific growth rate and substrate concentration without cell recycle. By assuming Monod growth kinetics, one can plot out the steady state concentration profile as a function of the dilution rate (Figure 10.3b). With cell recycle, the dilution rate is higher than the specific growth (Eqs. 10-22 and 10-23, Panel 10.5). Furthermore, the cell concentration in the reactor is higher and the washout dilution rate is extended beyond the maximum growth rate (Eqs. 10-24 and 10-25, Panel 10.5). The enhanced performance of the system, or the extent of increase in the dilution rate from the specific growth rate, is described by the enhancement factor ϕ. Its value is affected by the recycling factor, α, and the cell concentration factor, c, as related by Eq. 10-22.

Panel 10.5. The Critical Dilution Rate of a Continuous Culture with Cell Recycle

Let $\phi = \dfrac{1}{1+\alpha-\alpha c}$, **(Eq. 10-22)**

$$\phi\mu = D \qquad \textbf{(Eq. 10-23)}$$

where ϕ = recycle enhancement factor

$$\phi\frac{\mu_m s_0}{K_s + s_0} = D_c \qquad \textbf{(Eq. 10-24)}$$

$$D_c \cong \mu_m \cdot \phi \qquad \textbf{(Eq. 10-25)}$$

$\therefore D_c > \mu_m$ for a cell recycle system, thus allowing a dilution rate higher than μ_m to be used.

Effects of a recycling factor

In a simple continuous culture, once the Monod model relationship between the specific growth rate and the substrate concentration is imposed (fixing K_s and μ_m), and the operating conditions of feed concentration (s_0) and dilution rate (D) are fixed, the steady state cell and substrate concentration for a given dilution rate is also fixed. This is because a system of two linear equations (Eqs. 10-13 and 10-14) gives only one unique set of s and x. In a system with cell recycle, the system of equations (Eqs. 10-19 and 10-21) has α and c as parameters in addition to x and s. The system will settle to a prescribed steady state when both α and c are specified. A perfusion culture inherently has more degrees of freedom than a simple continuous culture. To ensure that the same steady state is reached consistently, more operating parameters must be fixed.

The key factor that cell recycle introduces that increases productivity is the amount of biomass ($aFcx$ in Eq. 10-16) returned to the reactor. To return the same amount of biomass, one may use a less concentrated cell stream (lower cx) and a higher flow rate (larger αF) or a more concentrated cell stream with a lower flow rate. In Figure 10.4, each line represents a steady state that has the same dilution rate and cell concentration, which has been achieved by a different combination of recirculation flow rates (αF) and different degrees of concentration factor c. A low recycling rate (αF) can be employed by using a highly efficient cell separator with a concentrated recycle stream. Conversely, when using an inefficient cell separator that gives a low degree of cell concentration, a large recycling rate needs to be used. In other words, cells will need to be pumped out of the reactor and passed through the cell separator to be recycled more frequently.

In large-scale operations, the cell stream could potentially exist outside of the reactor for a prolonged time, so oxygen starvation is a concern. Consequently, the fluid stream out of the reactor is often chilled to reduce oxygen consumption. With a device that gives a low c value, cells will need to be subjected to more frequent environmental perturbations, such as going through the pump and being chilled. This factor should be considered when selecting a cell recovery method.

Perfusion Culture with an Independent Cell Purging Stream

Some cell recycling devices are sensitive to fluctuations of operating parameters, and return the recycling cell stream with somewhat varying cell concentration. The cell concentration factor, c, is also affected by the chemical and physical environment. As c changes, the recycling flow rate must also change in order to maintain a steady state. Increasingly more perfusion cultures employ an internal or external microfiltration membrane to withdraw a cell-free stream while using another purge stream to remove some cell mass to maintain a steady state. The cell purge rate is used as a means of controlling cell concentration at a steady value.

The material balance equations for a system (Figure 10.5) that also considers the accumulation of growth-inhibitory metabolites are shown in Panel 10.6 (Eqs. 10-26, 10-27, and 10-28). The steady state

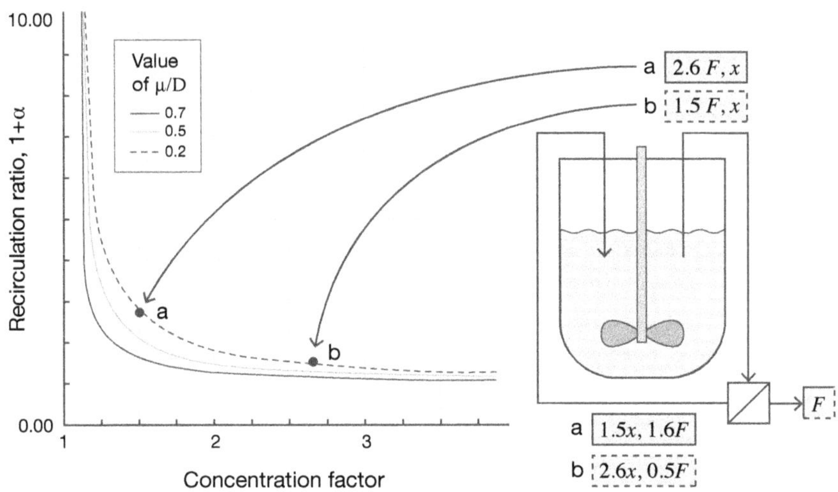

Figure 10.4. The relationship between the recirculation ratio and the cell concentration factor in a continuous culture with cell recycle. Two different recycler operating conditions (a and b) give the same cell concentration in the bioreactor at the same dilution rate.

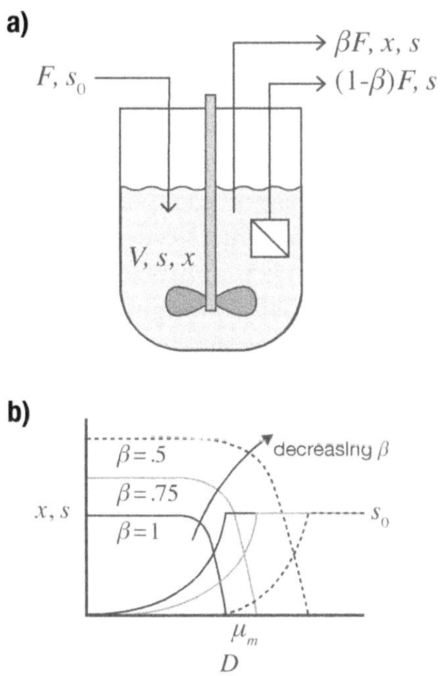

Figure 10.5. (a) A perfusion culture with a total cell retention device and cell purging. (b) Corresponding steady state profiles of cell and substrate concentrations.

Panel 10.6. Continuous Culture with Cell Retention and a Purge Stream

$$\frac{dx}{dt} = \mu x - \beta D x \qquad \textbf{(Eq. 10-26)}$$

$$0 < \beta \le 1$$

$$\frac{ds}{dt} = D(s_0 - s) - q_s x \qquad \textbf{(Eq. 10-27)}$$

$$\frac{dI}{dt} = \alpha_{I/s} q_s x - DI \qquad \textbf{(Eq. 10-28)}$$

At steady state

$$\mu = \beta D \qquad \textbf{(Eq. 10-29)}$$

$$x = \frac{D(s_0 - s)}{q_s} \qquad \textbf{(Eq. 10-30)}$$

$$I = \frac{\alpha_{I/s} q_s x}{D} \qquad \textbf{(Eq. 10-31)}$$

concentrations of cells and substrate for Monod growth kinetics are shown in Figure 10.5b. The dilution rate is higher than the growth rate, as shown in Eq. 10-28; D is always larger than μ since $\beta < 1$. The system does not have a steady state if there is no purge stream (i.e., $\beta = 0$ in Eq. 10-26) while the growth rate is not zero. Cells in suspension generally do not remain in a static, non-growing state over a long period. Thus, a purge stream that withdraws cells continuously or periodically is necessary. Cell retention increases the cell concentration in the reactor, as seen in the previous case of using an external cell recycling device. Such a complete cell retention device (for example, microfiltration, discussed later in this chapter) gives the system only one more degree of freedom in its operation compared to a simple continuous culture. For a system where the relationship between the growth rate and substrate concentration behaves like Monod kinetics, as long as the purge rate (βF), feed substrate concentration, and dilution rate are fixed, the steady state is also determined. We also noted that the balance equation between cell concentration and substrate concentration (Eq. 10-30) remains unchanged from the other cases discussed (Eqs. 10-8 and 10-21).

In a stirred tank bioreactor with cell retention or cell recycle, the inhibitor concentration can be kept at a low level to allow for a high feed concentration, which can subsequently support a high cell concentration (Eq. 10-31). By increasing only the medium components that are consumed by cells and balancing the feed concentration and the consumption rate, excess accumulation of unconsumed substrates and an increase in osmolality can be avoided. Such a stoichiometric balancing strategy is applied to both steady states and the transient stage of a ramping-up state designed to bring the cell concentration up to the target steady state level. It should be noted that overfeeding a fortified medium may also lead to an inhibitory accumulation of substrates, especially during the start-up transient period. Prior to the implementation of a perfusion culture, one may use an empirically derived relationship between the growth rate, lactate, and osmolality to simulate the kinetic behavior of the culture and optimize the feeding strategy.[6]

Perfused Fed-Batch Culture

When employing a perfusion culture, one aims for a long period (up to months) of steady state operation. In order to sustain a culture for a long period, cells are nourished to grow at a moderate rate. Another mode of operation is the employment of perfusion only to sustain a fed-batch culture, thereby enhancing its productivity by reaching a very high cell concentration and then extending the period that the reactor is at that concentration. This method is thus a combination of the fed-batch culture discussed in Chapter 9 and the perfusion culture described previously in this chapter. The stoichiometric principle of balancing the feed with the cells' demand still applies. Overall, a perfusion or perfused fed-batch culture is performed in a small-scale reactor, often a disposable one. In contrast, fed-batch cultures can be operated on a scale of tens of thousands of liters.

The balance equations on cells, product, substrate, and metabolites are shown in Panel 10.7. A perfused fed-batch culture may undergo a number of stages: initial start-up, intermittent feeding, perfused growth rate, perfused stationary, and decline (Figure 10.6). The balance equations are the same for all of the stages, but the values of the parameters for those equations can vary widely. For example, in the biomass equation the discharge term is removed and a cell death term is added, since cell death is prominent at the last stage (Eq. 10-32, Figure 10.6). A balance equation is also added for the product (Eq. 10-33), since the vast majority of the product is most likely produced in the perfusion stages. The specific product formation rate is strongly affected by the chemical environment. Note that in this case, the system is not operated at a steady state, because it does not have cell purging; the viability and productivity are therefore likely decreasing. In some cases, the cell concentration is so high that even at the full capacity of the cell recycle device a continued accumulation of metabolites is unavoidable. The optimization effort for a perfused fed-batch culture may thus focus on identifying the chemical environment that gives the highest productivity and the longest production period.

Panel 10.7. Material Balance on a Perfused Fed-Batch System

$$\frac{dx}{dt} = \mu x - \mu_d x \qquad \textbf{(Eq. 10-32)}$$

$$\frac{dP}{dt} = q_p x - DP \qquad \textbf{(Eq. 10-33)}$$

$$\frac{ds}{dt} = D(s_0 - s) - q_s x \qquad \textbf{(Eq. 10-27)}$$

$$\frac{dI}{dt} = \alpha_{I/s} q_s x - DI \qquad \textbf{(Eq. 10-34)}$$

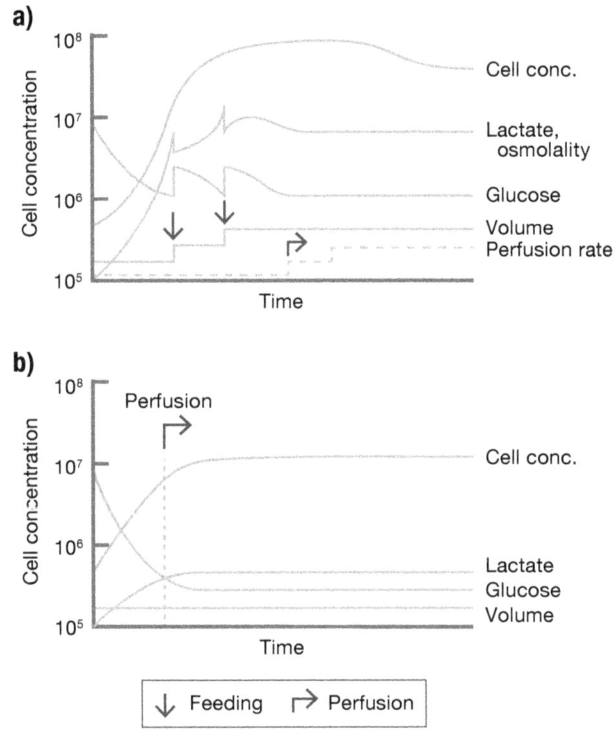

Figure 10.6. Kinetic behavior of (a) a hybrid perfused fed-batch culture and (b) a perfused culture.

Methods of Cell Retention

A CELL RETENTION DEVICE FOR PERFUSION needs to meet more stringent requirements than those for cell separation in downstream processing. It must maintain the aseptic condition for an extended period of time and be mechanically reliable or easily replaceable in a manufacturing setting. These devices give different ranges of the concentration factor (c), thus requiring different recycling factors (α) to achieve the same degree of recycle enhancement (ϕ) (Eq. 10-22). The device that gives a low concentration factor will have to circulate the culture fluid more often, i.e., require a larger flow rate out of the reactor ($(1+\alpha)F$).

Cell Separation Using Settling Velocity

Sedimentation

The simplest cell separator is perhaps the conical settler (Panel 10.8).[7] The cell stream from the bioreactor enters the settler in a region with a large cross-sectional area designed to reduce the flow velocity. In this region, the fluid flow separates into an upward and a downward stream. The flow rate of the upward stream is adjusted so that its vertical velocity is smaller than the cells' settling velocity (Figure 10.7). Hence, cells are separated from the upward stream and carried by the downward flow to the recycle stream that is returned to the reactor.

Panel 10.8. Conical Settler

- Selective removal of dead cells
- Low separation efficiency
  - Cell settling velocity ~2–10 cm/h
  - Good for large cells, aggregates, microcarriers
- Long residence time outside the bioreactor

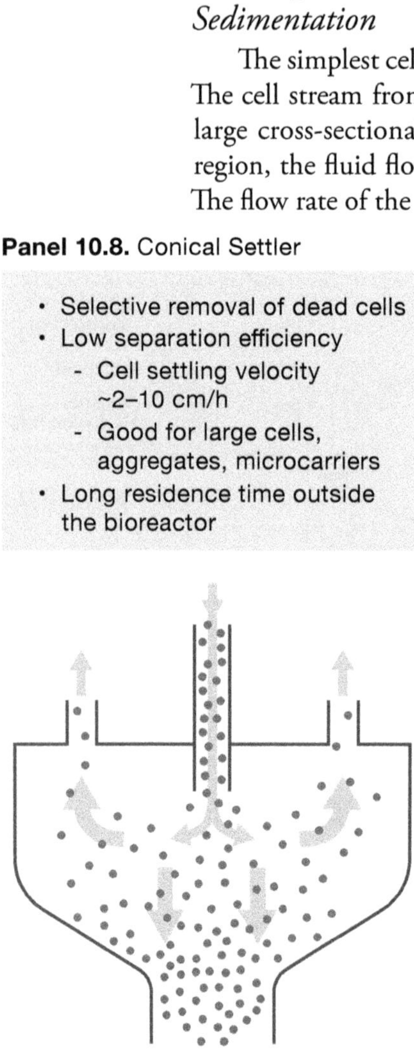

Figure 10.7. A settling cone for cell recycle.

It is important to note that the separation behavior is sensitive to the size of the settler and the scale of the reactor. As the reactor scale increases, the recycling flow rate also needs to increase proportionally. However, the cross-sectional area of the settler increases only with 2/3 the power of the settler volume. The efficiency of cell separation thus decreases as the scale increases. The method is particularly useful in separating large particles. Some industrial processes employ microcarriers with particle diameters ranging from 0.2 mm to 2 mm, which are separated from the medium stream via sedimentation. In some cases, cells are grown as large clumps, or aggregates, of 1–2 mm in size, thus making sedimentation readily applicable.

Inclined settling

In a simple settling tank, the direction of fluid flow and cell settling are along the same axis (both vertical). A sufficiently long transient zone is necessary to separate the cell and cell-free

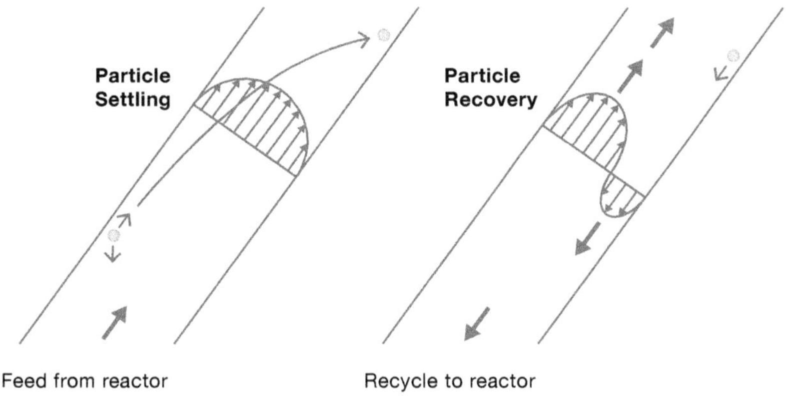

Figure 10.8. Cell separation in an inclined settler for cell recycle.

Panel 10.9. Inclined Settling

- While a particle is moving upwards with the flow, it also settles towards the bottom plate
- It is "collected" upon hitting the bottom
- Eventually, the particle-rich zone has a higher fluid density and begins to move downward
- The particle-rich stream is recycled to the bioreactor

streams. To enhance the separation efficiency, the settler is often inclined so that the fluid flow direction and the particle settling direction are at an angle (Figure 10.8, Panel 10.9).[8] The feed cell stream enters at the bottom and moves upward. Inside the settler, cells begin to "settle" down vertically due to gravity. If a cell particle hits the surface of the lower plate, it is "collected," since the fluid velocity on the surface is zero. The effluent stream exiting the settler from the top carries fewer cells than when it enters. Eventually, the cells that settled on the bottom plate form a layer of fluid that has a higher density than the stream above. This heavy stream then moves downward, carrying the cells along with it. At a steady state, there are three streams in the system: the feed stream, the effluent stream (carrying unsettled cells), and the concentrated cell stream exiting at the bottom.

In industrial design, multiple inclined plates are used in a single settler. In such designs, the feed stream and the returning cell stream are partitioned in different zones to avoid intersecting with each other. In some cases, mechanical vibration is applied to the plates to prevent settled cells from sticking to the surface and being lysed.

The residence time in the settler has to be at least as long as the particle settling time. With a high cell concentration in the stream, oxygen starvation is a major concern, as it may induce apoptosis and cell lysis. Therefore, the stream passing through the settler is often chilled to reduce the cells' metabolic rates.

Acoustic cell separator

To enhance settling, an acoustic resonance device was developed (Figure 10.9).[9] This device uses acoustic energy to enhance cell agglomeration.

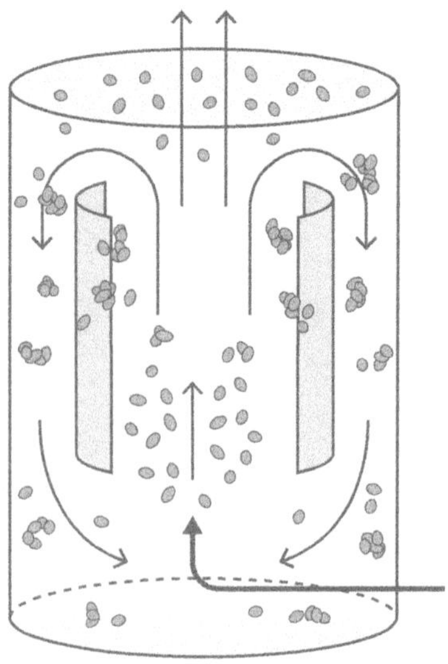

Figure 10.9. An acoustic cell agglomeration device for cell recycle.

Panel 10.10. Centrifugation

Centrifugation Technology
- Excellent separation efficiency
- High perfusion capacity
- Little clogging
- Easy scale-up
- Vulnerable to mechanical failure during long-term continuous operation

Different Designs
- Disk type
  - Continuous cell recycle back to fermenter
- Disposable separation unit

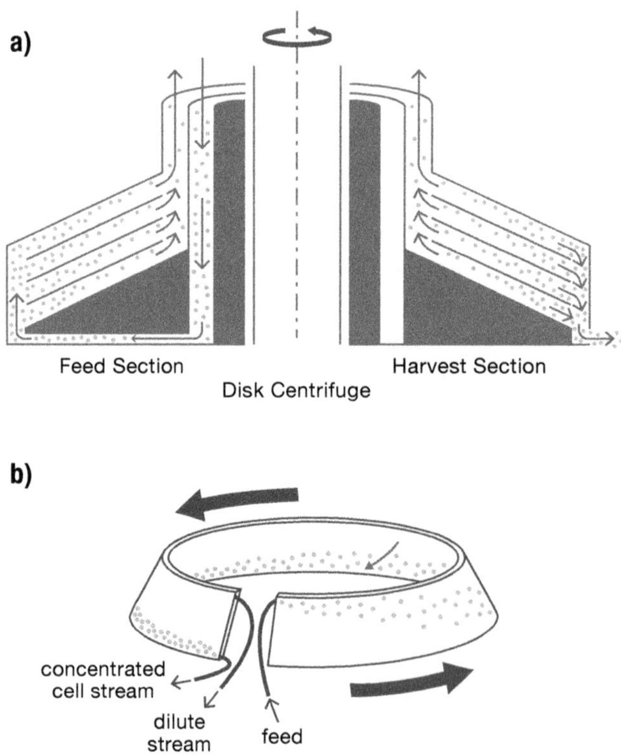

Figure 10.10. (a) A steam-sterilizable disk-type centrifuge. (b) A disposable bag for a Centritech centrifuge for continuous cell separation.

Cells are induced to agglomerate as they pass through the acoustic chamber, giving rise to a faster settling velocity. With an increased settling velocity, sedimentation is easily accomplished without resorting to a large settling device with a long settling time.

Centrifugation

Centrifugation is a standard unit operation in many downstream recovery processes (Panel 10.10). However, most centrifuges are not designed for long-term continuous and aseptic operations. In the early stages of perfusion culture development, in-line centrifuges were used for intermittent cell recycling. They were also used to remove serum-containing spent medium and replenish it with fresh medium for production. As a result, a number of steam-sterilizable centrifuges were subsequently developed. These centrifuges and the disposable bag-based centrifuges are all capable of processing up to hundreds of liters of medium a day and are used in perfusion culture (Figure 10.10). The disk-type centrifuge is analogous to a multiple parallel plate settler, except that the parallel plates are rotating and generating a centrifugal field for cell settling. The disposable bag system employs three tubes: a feed tube, an outflow tube for the heavy (cell-rich) stream, and an outflow tube for the light stream.[10] The unique design of an inverted question mark allows the three tubes to rotate along with the centrifuge without becoming twisted.

Centrifugal-Force-Enhanced Cell Sieves

Spin filter: Centrifugal filter and rotating cage

A spin filter device is a cylindrical rotating cage with a high porosity filter with relatively large openings (~20–100 μm) installed on the inner wall. The whole device is submerged in the culture fluid (Figure 10.11). A pump draws medium from inside the cage as the purge stream. The culture fluid passes through the filter and enters the cage, though some cells are excluded and remain outside. Thus, the cell concentration inside the cage is lower than that in the bulk culture fluid which achieves the overall retention of cells in the reactor. The cage rotates along the center shaft of the impeller agitator at a low speed. The

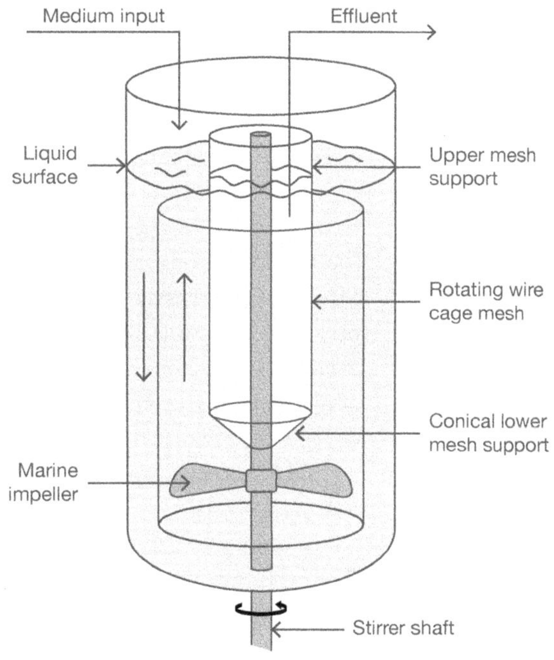

Figure 10.11. A spin filter bioreactor. The culture fluid for effluent flow is withdrawn from inside the wire cage.

centrifugal field is typically insufficient to push cells away along the outside wall of the cage. It is plausible that the liquid layer around the cage has a lower cell concentration than in the bulk. As a result, the fluid drawn across the filter has a lower cell concentration than in the bulk. The system has been employed in moderate scales of up to hundreds of liters.

The rotating cage is difficult to scale up, as its operating mechanism is not well understood. Later modifications of the spin filter increased its rotation rate up to hundreds of rpm, thus allowing it to operate like a centrifugal filter. The centrifugal force pushes the cells away from the surface of the filter, thus drawing liquid through at a lower cell concentration than that in the bulk. In some variations, the centrifugal filter is installed outside of the reactor and used as an external cell retention device.

Membrane Separation

Microfiltration and alternating tangential filtration

Microfiltration uses membranes of different configurations, including parallel plates and hollow fiber devices, which have a pore size of around 2–4 μm. Microfiltration was among the first techniques used for cell retention. Its widespread use was impeded by membrane fouling, which is especially severe when a complex medium with a high concentration of proteins is used in the culture or when cell viability is low. With the increased use of low protein medium in the last decade, the problem of protein fouling has lessened, but the clogging by debris from dead cells remains problematic. Most microfiltration devices used for perfusion culture are of the tangential flow type; as the fluid passes through the fiber, a small portion permeates through the fiber and the cell-free permeate is collected in the extracapillary space and discharged. Meanwhile, the cell-containing retentate stream is returned to the reactor.

Microfiltration devices are configured for tangential flow, meaning that the feed stream flows in a direction parallel to the membrane while the filtrate flows across it. The more recent use of a pulsatile flow system, often called alternating tangential filtration (ATF), uses a diaphragm pump to periodically reverse the flow direction. This generates a transmembrane pressure difference which pushes the medium filtrate out while retaining cells in the lumen side of the device, which is returned to the reactor (Figure 10.12).

ATF has become commonly used in perfusion and hybrid perfused fed-batch culture in the past few years.[1] Detailed analysis of its operating parameters has not yet emerged. For example, the optimal range of the Reynolds number for luminal flow and transmembrane pressure drop, as well as its effect on membrane fouling, have not been reported. The transmembrane pressure drop drives the filtrate flux. At a high cell concentration, a high transmembrane flux may carry cells toward the membrane pores and cause damage. A low transmembrane flux will require a higher recirculation rate (i.e., the number of times the culture fluid

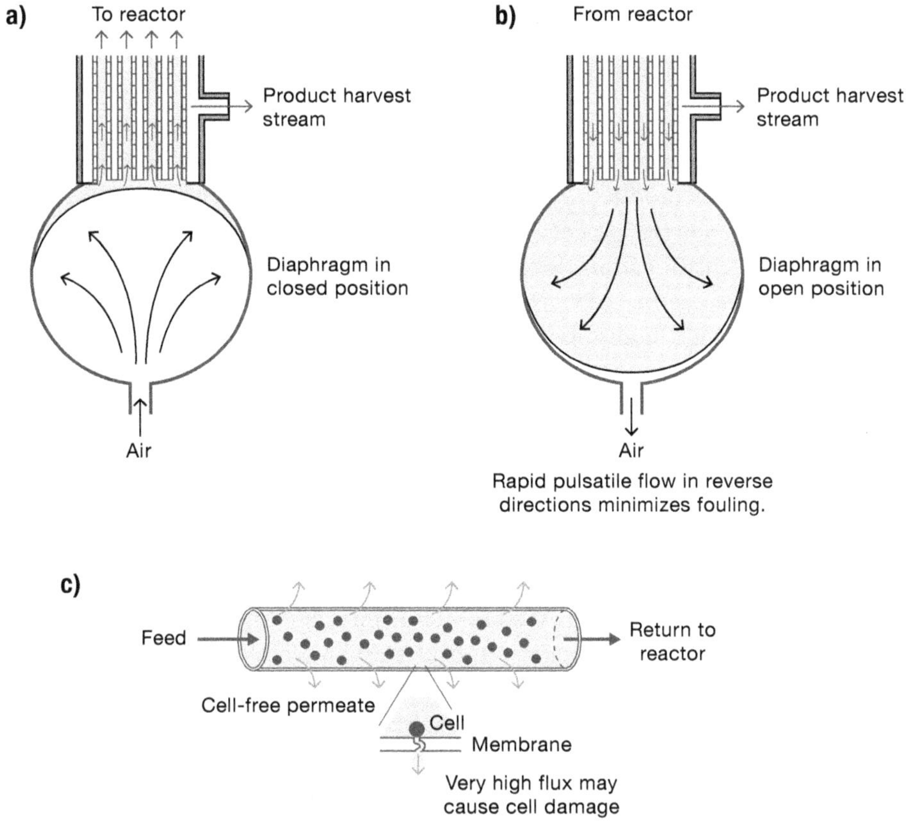

Figure 10.12. (a, b) A tangential flow hollow fiber filter device in its pumping cycle. (c) A depiction of transmembrane flux.

passes through the device each day) to achieve the same perfusion rate; in turn, this may result in more frequent exposures to high shear stress regions near the entrance to the membrane. Given that the Reynolds number in blood flow ranges from a very small number in the capillaries to a few thousand in the highly turbulent regions of the arteries, neither the flow in the fiber lumen nor the entry into the reservoir area are likely to cause severe cell damage. Nevertheless, careful analysis should be undertaken in developing processes using a membrane cell retention device.

Both membrane- and settling-based cell retention methods face challenges in scaling up. The capacity of a settling device is dependent on the surface area perpendicular to the direction of settling for cell collection, while the capacity of a membrane separator is dictated by the membrane area available for the permeate to pass through. As the volumetric flow rate of perfusion increases, the area required for cell separation increases proportionally. Increased transmembrane flux can somewhat increase the capacity, at the expense of increased membrane fouling and potential cell damage. In some cases, multiple units of ATF are used to meet the demand of removing metabolites.

Continuous Culture & Continuous Process

WITH THE USE OF A MEMBRANE SYSTEM for cell retention, the effluent stream is cell-free. This allows for the isolation or capture of a continuous product. As illustrated in Chapter 1, a typical cell culture product recovery process involves cell removal using membrane separation or centrifugation. Product isolation entails adsorption chromatography to reduce the process volume, followed by product purification by ion exchange, hydrophobic interaction chromatography or other chromatography, virus inactivation/filtration, and final polishing (for an overview, see Chapter 13 in *Engineering Principles in Biotechnology* (reference [11])). Since the effluent stream from the bioreactor is cell-free, it eliminates the need for cell removal. By operating at a very high cell density, the resulting product concentration may reach a level that does not require further concentration using ultrafiltration before feeding into an adsorption column such as a protein A chromatography. This enables almost direct feeding from the bioreactor into the downstream process, barring a medium particulate removal step and perhaps a buffer exchanging diafiltration step.

There have also been attempts to make the product capture adsorption step continuous. This is usually done by using a split-column strategy.[1] As an illustration, we consider a column that is operated with a cycle of three stages: feed-stream loading, washing, and elution (Figure 10.13). The three stages have an equal time interval of θ. Now we split the column into three sections. We can use the same flow rate of feed, washing, and elution stream as before, with the first column being loaded with feed stream while the second and third columns are being washed and eluted. Since the capacity of each column is only one-third of the original volume, after a period of $1/3\theta$ the first column will be fully loaded, the second washed, and the third eluted. The feed stream is then switched to load the third column, while the first column is washed and the second eluted. In the original single-column operation, a complete cycling time is 3θ, but the feed flow is only on for a period of θ. In the split-column fashion, the feed stream is being continuously fed into one of the three columns at the same flow rate as before. This method bears much similarity to moving-bed chromatography. The split-column method also has an additional advantage because it avoids unadsorbed product loss due to column breakthrough. In the adsorption process, it is inevitable to have a breakthrough, because at the end of the adsorption the flow pattern in the front is not like a piston flow, but rather a gradient decrease. To avoid product loss, feeding is stopped as soon as the unadsorbed product begins to elute at the end of the column. This reduces the column

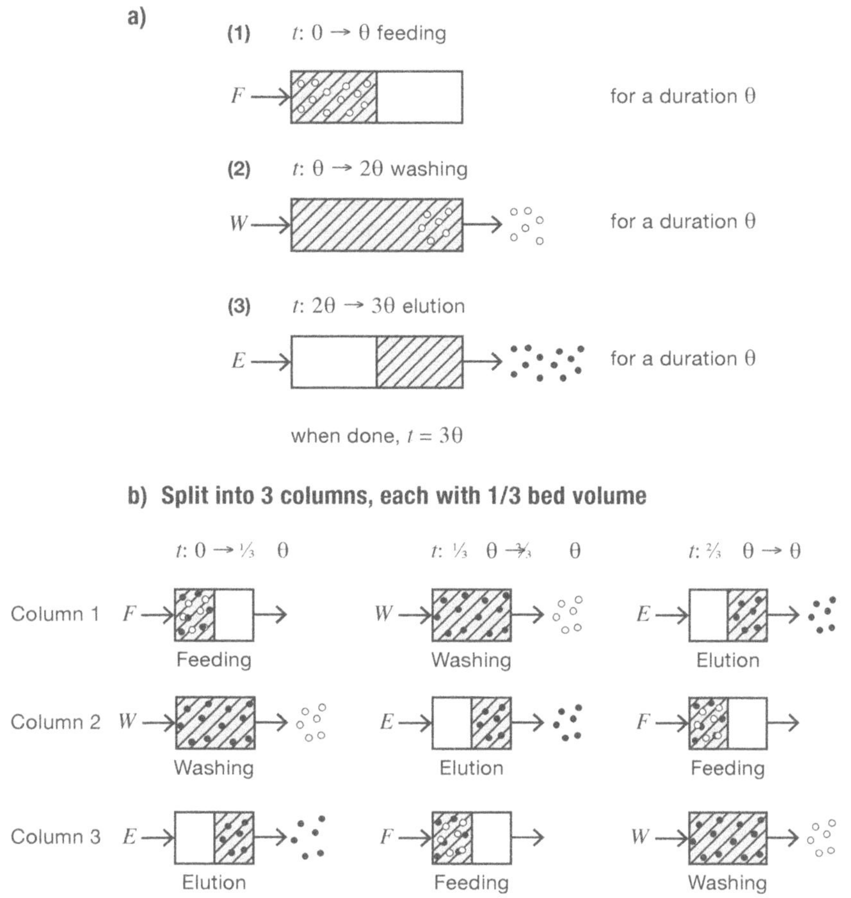

Figure 10.13. Simulated continuous adsorption chromatography. A hypothetical three-stage split column is shown cycling through the feeding (adsorption), washing, and elution stages.

utilization efficiency. Alternatively, one can use the entire column capacity by letting the product elute until the column is fully loaded. But in that case, the unadsorbed product is lost or has to be collected and loaded again in the next adsorption process—an option that is not available if each run is a separate product lot, since different lots of material cannot be mixed. Using the split-column method, the unadsorbed product is directed to the second column and is not lost.

The control of column switching may not be as complicated as in moving-bed chromatography, but it still requires extensive optimization. The product capture step reduces much of the process volume. There are efforts to make the downstream unit operations even more continuous, though their use of smaller process volumes may mean that a conversion to a continuous process would not yield much benefit.

Concluding Remarks

CELL GROWTH IN A BIOREACTOR is limited by the accumulation of many inhibitory metabolites in the culture. Unlike our bodies, which have multiple organs working to convert the metabolites, balance the nutrients, and excrete the remaining metabolites, cells in culture lack these capabilities. Fed-batch culture has been the prevailing mode of cell culture processing in the production of recombinant proteins. Continuous culture, traditionally used mostly for the production of labile products, is increasingly being adopted to prolong cultivation periods and increase productivity. The increased adoption of perfusion has been facilitated by advances in cell retention technology and the development of less complex media. This has also led to an increased adoption of perfused fed-batch culture.

A very important design parameter for both fed-batch and continuous culture is the feeding strategy. Foremost, the feed medium, usually with a fortified nutrient composition, should be stoichiometrically balanced to minimize the accumulation of excess nutrients and inhibitory metabolites, and to reduce the increase of osmolality. It is likely that perfusion will be increasingly adopted for moderate-scale processes, especially those employing disposable bioreactors, whereas very large-scale fixed-tank processes will continue to operate in fed-batch mode. It is important to remember that a system will robustly reach a steady state only when the operation variables are sufficiently specified such that the system of equations describing the culture has solution(s) at a steady state. For a perfusion culture, the number of operation variables is greater than that for a simple continuous culture, and one must specify more operation variables in order to make the process robust. Given the intrinsic advantages of continuous operation and the advances in cell retention technology, we may begin to see a more widespread application of perfusion culture in the coming years. It has considerable potential to increase the capacity of high-throughput processes, reduce reactor sizes, and possibly minimize product quality fluctuations throughout steady state operations.

References

1. Warikoo, V., et al., *Integrate continuous production of recombinant therapeutic proteins.* Biotechnol. Bioeng., 2012. **109**: p. 3018–29.
2. Miller, W.M., H.W. Blanch, and C.R. Wilke, *A kinetic analysis of hybridoma growth and metabolism in batch and continuous suspension culture: Effect of nutrient concentration, dilution rate, and pH.* Biotechnol. Bioeng., 1988. **32**: p. 947–65.

3. Frame, K.K. and W.-S. Hu, *Kinetic study of hybridoma cell growth in continuous culture: I. A model for non-producing cells.* Biotechnol. Bioeng., 1991. **37**: p. 55–64.

4. Gambhir, A., et al., *Analysis of cellular metabolism of hybridoma cells at distinct physiological states.* J. Biosci. Bioeng., 2003. **95**(4): p. 317–27.

5. Yongky, A., et al., *Mechanism for multiplicity of steady states with distinct cell concentration in continuous culture of mammalian cells.* Biotechnology & Bioengineering, 2015. **112**(7): p. 1437–45.

6. Gambhir, A., et al., *Analysis of the use of fortified medium in a continuous culture of mammalian cells.* Cytotechnology, 1999. **31**: p. 243–54.

7. Kitano, et al., *Production of human monoclonal antibodies by heterohybridomas.* Appl. Microbiol. Biotechnol., 1986. **24**: p. 282–86.

8. Searles, J.A., P. Todd, and D. Kompala, *Viable cell recycle with an inclined settler in the perfusion culture suspended recombinant Chinese hamster ovary cells.* Biotechnol. Progress, 1994. **10**: p. 198–206.

9. Trampler, F., et al., *Acoustic cell filter for high density perfusion culture of hybridoma cells.* Bio/Technology, 1994. **12**: p. 281–84.

10. Johnson, M., et al., *Use of the Centritech Lab Centrifuge for perfusion culture of hybridoma cells in protein-free medium.* Biotechnol. Progress, 1996. **12**: p. 855–64.

11. Hu, W.-S., *Engineering principles in biotechnology*. 2018: Wiley.

Nomenclature

α	Stoichiometric ratio
β	Cell purge fraction
μ	Specific growth rate
μ_d	Specific death rate
ϕ	Enhancement factor in cell recycle
c	Concentration of tracer or reactant
D	Dilution rate
F	Volumetric flow rate
I	Inhibitor concentration
K	Half-saturation constant
P	Product concentration
q	Specific rate
s	Substrate concentration
t	Time
V	Volume of reactor
x	Cell concentration
Y	Yield coefficient

Contents

11

Oxygen Transfer in Cell Culture Bioreactors

Introduction

OXYGEN IS THE MAIN ENERGY DRIVER for all mammals. It drives the oxidation of organic nutrients from food and extracts energy from them for cellular functions and growth. The requirement of oxygen is universal in cells derived from mammals. Blood circulation brings oxygen to cells in different tissues to sustain their viability and functions. The solubility of oxygen in water is very low. In Chapter 8 we briefly discussed the tendency of oxygen to deplete rapidly in a plug flow reactor due to its low solubility. The artery is essentially a plug flow system. Our body overcomes the limitations of plug flow by using red blood cells to increase the oxygen carrying capacity of blood. Even then, blood must be frequently recirculated back to the lungs to be re-oxygenated. Ultimately, in a system containing living cells that consume oxygen, there must be a source of oxygen in a gas phase to continuously replenish oxygen in the liquid phase in which cells reside. For cell culture process, gas bubbles in the culture medium constitute the gas phase.

When an air bubble is placed in a cell suspension, oxygen will diffuse from the bubble into the liquid (Figure 11.1). To cross the gas–liquid boundary, oxygen crosses a relatively stagnant film of gas on the gas side and another liquid film on the liquid side. The convection inside the bubble makes the oxygen concentration in the bulk (outside the gas film) relatively uniform. Similarly, within the bulk liquid and outside the liquid film the concentration of oxygen can be considered uniform. However, across the film the oxygen concentration decreases, and oxygen diffuses from the gas side to the liquid side. Once in the liquid phase, oxygen travels to reach cells. If cells are growing in a stagnant medium, an appreciable concentration difference between the bubble's surface and the cell may be seen, and oxygen molecules may diffuse from the bubble to the cell. In a stirred tank, the bulk oxygen level can be considered to be uniform. At the cell's surface, oxygen diffuses through the cell membrane and is mostly consumed in the mitochondria. However, in the case that cells form aggregates, the diffusion of oxygen through the aggregates may be slow, causing oxygen starvation in the interior of the aggregate. For the vast majority of industrial cell culture, the rate-limiting step of oxygen transfer in the bioreactor is in the diffusion from the gas to the liquid. In this chapter, we will focus on oxygen transfer across the gas phase and liquid phase interface.

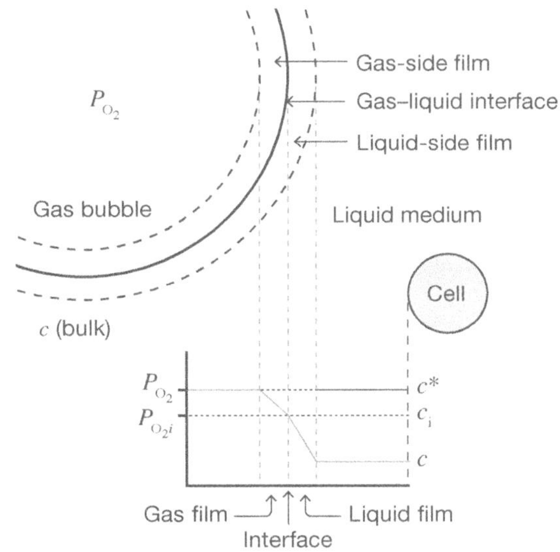

Figure 11.1. Transfer of oxygen across an air bubble in culture medium.

Oxygen Concentration across the Gas–Liquid Interface

Equilibrium and Oxygen Transfer

IF A GAS PHASE AND A LIQUID PHASE are brought in contact and a component in the gas phase is soluble in the liquid phase, given enough time the component gas will reach equilibrium between the two phases. Before equilibrium is reached, molecules of the soluble species are exchanged between the liquid phase and the gas phase; once equilibrium is reached, the liquid phase becomes "saturated" with the component gas and there is no longer any net transfer in either direction.

As discussed above in the interfacial transfer, oxygen is hypothesized to cross a gas film and a liquid film (Figure 11.1). In between the two films is a boundary where oxygen is in equilibrium between gas and liquid. The oxygen level on the gas-side surface of the gas film (P_{O2}) is higher than that at the interface (P_i), so oxygen diffuses from the bulk gas phase towards the interface. On the liquid side of the film, the oxygen concentration is lower than that in the interface, and oxygen diffuses toward the bulk liquid. However, the thickness of the film and the concentration at the interface are difficult to measure. In the subsequent discussion, we will treat the interfacial boundary (i.e., the gas film and the liquid film) as one homogeneous interface.

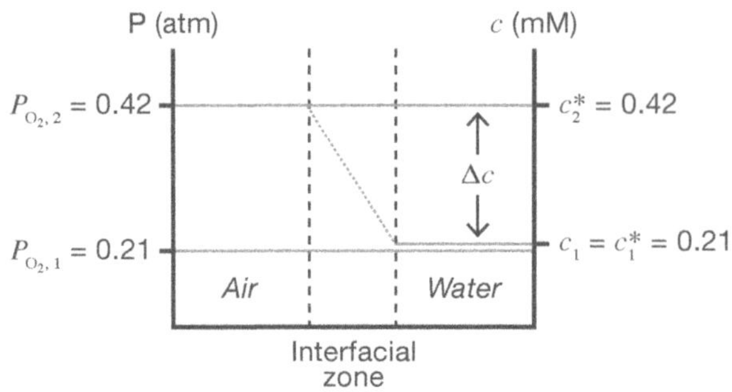

Figure 11.2. Oxygen concentration profile across the air–liquid interface.

Consider the case that a cup of water is brought in contact with air consisting of 79% N_2 and 21% O_2 at 1 atm initially. A concentration profile of oxygen across its air–water interface is depicted in Figure 11.2. Let us assume that the volume of the air is very large, so that losing a small quantity of N_2 and O_2 from their diffusion into water will not change the composition of the air. Given a long enough time, the water will become saturated with oxygen, or the oxygen level in the air and water will reach an equilibrium. Let us denote the concentration of oxygen in the bulk liquid (i.e., the liquid at some distance away from the interface) as c, and that of the saturation as c^*. In the bulk gas phase, the oxygen partial pressure ($P_{O2,1}$) is 0.21 atm. After the two phases reach an equilibrium, c_1 is the same as c_1^*. The nitrogen solubility in water is very low, about 20 mg/L. At such a low level, the dissolution of nitrogen into water does not interfere with the transfer of oxygen. If the oxygen content in the gas phase is increased to 42%, then the partial pressure of oxygen becomes 0.42 atm ($P_{O2,2}$) and the oxygen in the liquid phase is no longer saturated. An oxygen concentration difference across the interface will arise and is sketched qualitatively. Oxygen will then begin to diffuse from the gas into the liquid phase and the dissolved oxygen concentration in the water will gradually increase until the water reaches a new equilibrium (c_2^*) with the new air containing 42% oxygen. Conversely, if the oxygen level in the gas phase is decreased to 10%, then oxygen will escape from the liquid into the gas phase until a new equilibrium is reached. In the above discussion, the composition of the gas phase is changed and the total pressure remains at 1 atm. One can achieve the same change of oxygen partial pressure by changing the total pressure while keeping the gas phase composition static.

Description of Oxygen Concentration and Its Solubility

Oxygen level in a gas phase is usually described as a mole fraction (y_{O2}) or as partial pressure (P_{O2}), which is the mole fraction of oxygen multiplied by total pressure. If the ambient pressure is 1 atm (or 760 mm Hg) and the oxygen mole fraction is 0.21 ($y_{O2} = 0.21$), then the partial pressure of oxygen in the air is 0.21 atm (or 159.6 mm Hg). The oxygen concentration in a liquid phase (or dissolved oxygen concentration) is often expressed as mmol/L (mM). Since 0.21 mmol/L in water is the equilibrium value with an oxygen partial pressure of 159.6 mm Hg, some describe the concentration in water using the equivalent of gas phase and express it as 159.6 mm Hg, especially in the healthcare profession (Panel 11.1). For example, water which is at 50% of saturation with air at 1 atm and 37 °C has a dissolved oxygen concentration of 0.105 mmol/L, or 79.8 mm Hg. The difference between concentration in the liquid phase, *c*, and the saturation concentration, c^*, is the driving force for oxygen transfer. The driving force can also be described in terms of liquid phase concentration (i.e., mM), or gas phase concentration (mm Hg, atm).

For sparingly soluble species, the solubility of the gas solute in the liquid phase is described by Henry's law (Panel 11.2). Henry's law states that, at equilibrium, the concentration of a gas in a liquid solvent is proportional to the partial pressure of that gas (Eq. 11-1, Panel 11.2 and example in Panel 11.3). The concentration in each phase can be described in different ways. Commonly used units are partial pressure and the molar or mass concentration (mmol/L, mg/L). The value of Henry's law constant is, thus, dependent on the units used to describe the concentration.

Henry's law describes oxygen concentration in an ideal solution. Despite all of its components, cell culture medium is close to an ideal solution. The saturation concentration calculated using an aqueous solution applies to cell culture media. However, the gas phase in a cell culture bioreactor also contains carbon dioxide that is intentionally added to provide the pH buffer (see Chapter 7), thus reducing the oxygen level from the 21% in ambient air. Additionally, in a bioreactor the interior pressure

Panel 11.1. Dissolved Oxygen Concentration in Water at 1 Atm of Air

- Solubility in water at 37 °C with air of 1 atm, 21% O_2
  - 0.21 mmol / L
  - 6.7 mg / L
- Since air in equilibrium contains 159.6 mm Hg or 0.21 atm, sometimes solubility is expressed also in mm Hg, especially for blood oxygen level

Panel 11.2. Henry's Law

$$x_A = \frac{P_A}{H} \qquad \textbf{(Eq. 11-1)}$$

- Oxygen solubility is not affected by other dissolved species in the medium. Its solubility is virtually identical to PBS and water.

Panel 11.3. Henry's Law Example

- At 37 °C, in 1 atm air (P_{O2} = 0.21 atm), O_2 concentration in H_2O is:

$$c^* = \frac{0.21\,\text{atm}}{\left[5.18 \times 10^4 \frac{\text{atm} \cdot \text{mole H}_2\text{O}}{\text{mole O}_2}\right]} \cdot 55.5 \frac{\text{mole H}_2\text{O}}{\text{L H}_2\text{O}}$$

$$= 0.21 \frac{\text{mmol O}_2}{\text{L H}_2\text{O}}$$

- In the bracket is Henry's law constant for water at 37 °C
- In ambient air (oxygen partial pressure = 0.21 atm), c^* ~0.21 mmole / L in H_2O at 37 °C = 160 mmHg = 6.7 mg / L (6.7 ppm)

is usually greater than 1 atm. These effects on oxygen solubility need to be accounted for.

The solubility of carbon dioxide can also be calculated using Henry's law. However, in an aqueous solution carbon dioxide associates with water molecules and then dissociates into HCO_3^- and H^+. The value calculated using Henry's law is the concentration of $CO_{2(aq)}$. In addition to $CO_{2(aq)}$, water also contains HCO_3^- and CO_3^{2-}, and the amount varies with pH. The total content of CO_2 and related species (HCO_3^- and CO_3^{2-}) is affected by the pH of the solution.

The height of a large microbial bioreactor (~300 m^3 in volume) can reach tens of meters. Large cell culture bioreactors, about 20 m^3, reach a few meters vertically. At the bottom of the bioreactor, the hydrostatic pressure incurred by the liquid mass is substantially higher than that at the top. At the bottom of the bioreactor, the saturation concentration (c^*) of oxygen and carbon dioxide is higher. The hydrostatic pressure enhances the driving force for oxygen transfer but diminishes that for carbon dioxide removal.

Panel 11.4. Effects of Hydrostatic Pressure

- 10 m liquid height gives a 1 atm hydrostatic pressure at the bottom of the tank
- Up to almost 9 atm has no adverse effects on hybridoma cell growth and specific glucose consumption

Cells circulating in the bioreactor will travel to different regions and periodically reach an area where the hydrostatic pressure is high (Panel 11.4). Experimental data has shown that cells grow normally at up to nine atmospheric pressure and exhibit little discernible metabolic differences compared to normal conditions. It thus appears that in the range of hydrostatic pressures practiced in bioprocess, cell growth is not visibly affected.

Oxygen Transfer at the Gas–Liquid Interface

THE BOUNDARY BETWEEN A GAS PHASE and a liquid phase can be modeled as an interface separating a gas film and a liquid film that connect to the bulk liquid phase and gas phase as shown in Figure 11.1. The distance between the liquid phase concentration and equilibrium is c^*-c. The larger the deviation is from equilibrium, the faster the transfer. The magnitude of the driving force for oxygen transfer across the interface can thus be described as c^*-c.

In addition to the concentration difference, the rate of oxygen transfer is affected by the area of the interface that is available for oxygen to diffuse across (Panel 11.5). Just like the evaporation of water, the same amount of water evaporates faster when it is spread out on a large plate than when it is contained in a tall cup with a small top surface. The interfacial area (a) is typically expressed as interfacial area per volume of liquid. It therefore has a unit of inverse of length (like cm^{-1}).

The mass transfer coefficient is a descriptor that quantifies the mass transfer rate at the interface. It reflects the combined effects of the liquid film and the gas film at the interface and is often referred to as the overall mass transfer coefficient. It is affected by the physical and chemical features of the molecule that is crossing the interface and the liquid. The overall mass transfer coefficient (K_L) has the same units as velocity (like cm per second). Under vigorous mixing, the resistance for mass transfer is lower and the value of K_L is higher. Again, consider the evaporation of water placed in a plate, where stirring the liquid increases the mass transfer coefficient and therefore speeds up evaporation.

Panel 11.5. Oxygen Transfer Rate (OTR)

Three Factors Affecting OTR

- Mass transfer coefficient (K_L)
- Specific transfer area (i.e., the interfacial area)
- Driving force (i.e., the gradient across the interface)

$$\mathrm{OTR} = K_L a \cdot (c^* - c) = K_L a \Delta c \qquad \textbf{(Eq. 11-2)}$$

$$[\mathrm{mmol/L \cdot hr}] = [\mathrm{cm/hr}] \cdot [1/\mathrm{cm}] \cdot [\mathrm{mmol/L}]$$

- To transfer oxygen or CO_2 between the gas and liquid phases through the interface, the two phases must not be in equilibrium

Note that the “L” in K_L stands for liquid, as opposed to “G” for gas. The resistance of oxygen transfer across the interface is mostly contributed by the liquid phase. Hence, L is used in the notation by convention.

The transfer rate of oxygen (OTR) or other gas species with a low solubility is described as multiplicative product of three factors: the driving force, the mass transfer coefficient, and the interfacial area (Eq. 11-2, Panel 11.5). All three are leveraged in a bioreactor to modulate the supply rate of oxygen (Panel 11.6). The driving force can be increased by augmenting the oxygen content in the air or by increasing the head pressure in the reactor. In a bioreactor where the dissolved oxygen level is controlled at a set point, the set point can be set at a lower level (e.g., at 30% of saturation as opposed to 60% of saturation) to increase (c^*-c). The interfacial area can be increased by supplying air as small bubbles, and K_L can be enhanced by increasing the turbulence at the gas–liquid interface.

Panel 11.6. Means to Enhance Oxygen Transfer

- Increase K_L (make the interface more “turbulent”)
- Increase a (use smaller bubbles for sparging, or use silicone tubing)
- Increase Δc (use oxygen-enriched air or maintain dissolved oxygen (c) at a low/optimal level)

Methods of Supplying Oxygen

Surface aeration

Due to its low solubility, oxygen must be supplied to the culture medium continuously from a gas phase. In a T-flask or small spinner flask, the open space above the culture medium supplies oxygen for cell growth.

Panel 11.7. Methods of Supplying Oxygen

- Surface aeration
  - Used in laboratory culture wares
  - Effectiveness diminishes as the culture volume increases
  - Contribution to oxygen transfer is significant in small reactors, but insignificant in large reactors
- Highly oxygen-permeable membrane tubing
  - Provides bubble-free aeration
  - Only used in small reactors with a low oxygen demand
- Sparging
  - Gas holdup in the reactor provides interfacial areas for oxygen transfer
  - Less effective in very small-scale laboratory reactors due to the short hold time of air bubbles in the liquid phase
  - The magnitude of the interfacial area can be estimated roughly by the following process:
    - Estimate the gas holdup, V_g (difference of total reactor volume before and after aeration starts)
    - Estimate the bubble size, d_p; surface area per bubble volume is $6/d_p$.

$$a = \left(\frac{V_g}{V}\right)\left(\frac{6}{d_p}\right) \quad \text{(Eq. 11-3)}$$

The top surface area relative to the total culture volume deceases as culture volume increases. Surface aeration alone is insufficient to support cell growth at a high cell concentration even in a laboratory bioreactor. In most process-scale reactors, the contribution of surface aeration to overall mass transfer is insignificant and the direct sparging of air into the culture medium is the most effective means of supplying oxygen (Panel 11.7).

Indirect contact with gas phase

Oxygen transfer by air sparging in small bioreactors is inefficient due to the short air bubble holding time in the liquid. Hence, a high aeration rate in terms of volume of air per volume of liquid per minute (vvm) is necessary to supply oxygen. A high air flow rate may lead to excessive foaming or even cause cell damage. Excessive sparging is also avoided in microcarrier cultures. Microcarriers covered with cells on their external surface have a propensity to stick to air bubbles, possibly due to the somewhat hydrophobic nature of the cells' surfaces. As the bubbles rise to the liquid surface, microcarriers sticking to the bubble are carried along. The fluid shear caused by bubble rising can cause cell damage.

To avoid direct air sparging, oxygen can be supplied through silicone or Teflon tubings. Air is passed through the tubing, which is submerged in the medium in the reactor. Oxygen diffuses through the tubing membrane and dissolves into the medium. The method is effective in the small scale, but the long length of tubing required makes it impractical beyond small laboratory reactors.

Sparging

In an aerated stirred tank reactor, air is typically introduced through a sparger located near the bottom of the reactor and released as swarms of bubbles into the culture medium. As the bubbles rise, the rotating impeller further breaks them up and increases the total bubble surface area, thus enhancing oxygen transfer.

The efficiency of oxygen delivery, or the amount of oxygen that is transferred into the liquid phase per unit volume of air supplied, is affected by the total amount of interfacial area provided by the amount of gas entrapped in the liquid phase. It takes a finite amount of time for air bubbles to rise from the bottom to the top of the culture broth. In a laboratory-scale cell culture bioreactor, where air bubbles rise quickly to exit from the liquid's surface, the holding time in the culture broth is short, providing only a small amount of interfacial area. Conversely, as the scale increases, the bubble holding time and the efficiency of oxygen utilization per volume of air supplied also increases. The total reactor content thus includes both the liquid and the air bubbles that are retained in the liquid. As the aeration rate increases, the liquid level rises because more air is trapped, and the interfacial area increases.

Bubble size, interfacial area, and the mass transfer coefficient

The air bubbles traveling through the bioreactor are the main contributor to the interfacial area for oxygen transfer. As the height of the bioreactor increases, the bubble traveling time increases. Given the same gas provision rate (typically denoted by volume gas/volume culture broth/min, or vvm), the gas-hold up also increases. In a microbial fermenter, a gas holdup of 25% (meaning every liter of medium contains 250 mL of gas bubbles) is not unusual, although for cell culture bioreactors the gas-hold is typically only in the range of a few percent.

The interfacial area can be estimated from the gas holdup volume divided by the specific surface area of the bubble (Panel 11.7). The specific surface area of the bubble is inversely proportional to bubble diameter ($6/d_p$). Smaller bubbles give larger surface area for a given volume of gas. Small bubbles also rise slower in a stagnant liquid. Small bioreactors tend to employ microspargers that generate very fine bubbles.

Under an idealized condition of a single bubble being slowly released from a submerged orifice in a stagnant fluid, the bubble leaves the orifice when the buoyant force just overcomes the surface tension exerted on the bubble. The force balance between buoyancy and surface tension indicates a modest decrease of bubble diameter, d_p, with a smaller orifice diameter, d_o (Panel 11.8). Using orifice size to control bubbles is thus not very effective. As the scale increases and multi-orifice spargers and higher flow rates are used, inertial effects of local liquid velocity, medium properties, and the momentum of gas flow become dominant, making the initial bubble size

Panel 11.8. Effect of Orifice Diameter on Bubble Size

- A bubble bursting into a stagnant liquid from an orifice:

$$d_p = \left(\frac{6 d_o \sigma}{\Delta \rho} \right)^{1/3}$$

- Orifice diameter has a weak effect on bubble diameter
- In a stirred tank, the fluid dynamics have a stronger effect on bubble diameter

almost independent of the orifice size. Bubbles agglomerate as they rise to the liquid surface and coalesce into larger bubbles. They are also broken up by the impeller. Thus, in large-scale reactors, using fine bubbles to increase the interfacial area for oxygen transfer is not effective.

Bubble size also affects bubble rising velocity and the mass transfer coefficient. In principle, larger bubbles have a higher terminal rising velocity and a higher K_L if bubbles are considered to be rigid spheres. However, as bubble diameter increases, the resistance from the liquid increases and the bubble tends to deform and become flatter on the top, thus decreasing its rise velocity. Experimentally, it has been observed that both rising velocity and K_L only increase with bubble diameter when the bubble size is small (<3 mm). Beyond that, they are relatively invariant. Hence the effect of bubble size on K_L is relatively moderate.

Oxygen Consumption and CO_2 Production

Oxygen, like other nutrients, needs to be maintained at a certain level in the medium to allow cells to grow and function normally. For most mammalian cells, an oxygen level of ~30% saturation with ambient air, or ~0.06 mmol/L, is sufficient for optimal growth (Panel 11.9). The oxygen demand of cells is described as the specific oxygen consumption rate (q_{O2}). Its magnitude is affected by metabolic activities, the growth status (proliferating or quiescent), the temperature of cultivation, cell type and size, etc. Most cells have a q_{O2} in the order of 1–5 x 10^{-10} mmol/cell-h (Panel 11.9). The cell concentration that can be reached in a reactor before oxygen supply becomes limiting can be calculated from the oxygen transfer capacity and specific oxygen consumption rate (Panel 11.10).

Panel 11.9. Optimal Dissolved Oxygen Level and Specific Oxygen Consumption Rate

Optimal Oxygen Concentration

- Most cells grow well at a dissolved oxygen level of 30% saturation with ambient air (1 atm)

Oxygen Demand by Cells

- On the average, $q_{O2} \approx$ 1.0 x 10^{-10} mmole/cell-h

Most oxygen taken up by cells is used to oxidize carbon sources. The most important carbon sources for mammalian cells in culture are glucose and glutamine. The oxidation reaction of both compounds yields carbon dioxide at a stoichiometric ratio (CO_2:O_2) of 1.0 and 1.1, respectively. The mole ratio of CO_2 produced to that of O_2 consumed is called the respiratory quotient (RQ). Since the carbon source consumed in the largest quantity by cells is glucose and the next is glutamine at about 1/5 of that of glucose, the RQ of cells in culture is very close to 1.0 (Panel 11.11). The CO_2 produced by cells needs to be removed from the culture to avoid excessive accumulation, which would cause a drop in pH or an inhibition of growth. A key role of aeration in cell culture is thus the removal of carbon dioxide, in addition to the supplying of oxygen. The removal of carbon dioxide generated by cell metabolism requires the presence of a gas phase just like the supply of oxygen, except that the transfer through the gas–liquid interface is in the opposite direction (Panel 11.12).

Panel 11.10. Oxygen Transfer Rate and Achievable Cell Concentration

- K_L is about 3–6 cm/h for surface aeration in a cell culture vessel

What cell concentrations can be reached in a 400 mL spinner flask with a diameter of 8 cm?

Estimate K_L = 5 cm/h.

Surface area is:

$$A = (4\text{cm})^2 \cdot \pi = 50.2\text{cm}^2$$

$$A = 50.2\ \text{cm}^2 / 500\ \text{cm}^3 = 0.1\ \text{cm}^{-1}$$

Assume D.O. is almost zero at maximum cell concentration (maximum driving force)

$$\Delta c = 0.18\ \text{mmol/L}$$

$$q_{O_2} = 1.0 \text{ x } 10^{-10}\ \text{mmol/cell} \cdot \text{h}$$

$$q_{O_2} \cdot x = \text{OTR} = \frac{5\ \text{cm}}{\text{h}} \cdot \frac{0.1}{\text{cm}} \cdot \frac{0.18\ \text{mmol}}{\text{L}} = \frac{0.09\ \text{mmol}}{\text{L} \cdot \text{h}}$$

$$x = \frac{0.09\ \text{mmol}}{\text{L} \cdot \text{h}} \div \frac{1 \text{ x } 10^{-10}\ \text{mmol}}{\text{cell} \cdot \text{h}} = \frac{9 \text{ x } 10^{8}}{\text{L}} = \frac{9 \text{ x } 10^{5}\text{cell}}{\text{mL}}$$

Panel 11.11. CO_2 Production by Cells

- Respiratory quotient of cells is about 1.0, i.e., 1 mole CO_2 is produced for every mole of O_2 consumed, or $q_{CO2} \approx 1.0 \times 10^{-10}$ mmol / cell-h
- A high CO_2 concentration is inhibitory
- The inhibitory level usually starts at 15% to 20% CO_2

Panel 11.12. Stripping of CO_2 by Aeration

- Aeration supplies oxygen as well as stripping CO_2 from culture medium
- CO_2 is required for cell growth (e.g., in fatty acid synthesis) in addition to providing a pH buffer *in vivo*
- Cells can grow well even in sodium-bicarbonate-free medium, especially in a stationary culture, because CO_2 produced by cells accumulates in the medium to condition their growth
- When cell concentration is low in a sodium-bicarbonate-free medium, excessive aeration may strip off CO_2 to inhibit cell growth, unless some CO_2 (~0.5–1%) is in the gas

The accumulation of carbon dioxide to a high level causes growth inhibition. A medium containing 44 mM of $NaHCO_3$ as a pH buffer, like Dulbecco's modified Eagle's medium (DMEM) is typically used in conjunction with air containing 10% CO_2 to maintain a neutral pH. Thus, 10% CO_2 is within the optimal range for cell growth. The level of carbon dioxide that begins to exert growth-inhibitory effect is usually higher than 0.15 atm of P_{CO2} (15% CO_2). Carbon dioxide is also required for growth because of its participation as a substrate in a number of biosynthetic reactions, notably fatty acid synthesis. If carbon dioxide is completely stripped from the medium, cell growth may be inhibited.

Balancing Oxygen Transfer and Consumption

The oxygen content in cell culture fluid is the balance between oxygen supply and cell consumption. The rate of change in the dissolved oxygen level is the difference between the rate of transfer into the culture broth from the gas phase and the rate of oxygen uptake by the cells (Eq. 11-4, Panel 11.13). The oxygen uptake rate (OUR) is the specific oxygen consumption rate multiplied by cell concentration (Eq. 11-5), and the oxygen transfer rate (OTR) is shown in Eq. 11-2. Both OUR and OTR are expressed on a per culture volume basis. The total of oxygen consumption and oxygen transfer in the reactor is thus those terms multiplied by the culture volume (Eq. 11-5). It is convenient to express the balance of oxygen on a per unit culture volume basis (Eq. 11-6).

Panel 11.13. Balance of Oxygen in a Reactor

$$V\frac{dc}{dt} = \text{OTR}\cdot V - \text{OUR}\cdot V \qquad \textbf{(Eq. 11-4)}$$

$$V\frac{dc}{dt} = K_L aV(c^* - c_2) - q_{O_2} xV \qquad \textbf{(Eq. 11-5)}$$

$$\frac{dc}{dt} = K_L a(c^* - c) - q_{O_2} x \qquad \textbf{(Eq. 11-6)}$$

In a cell culture bioreactor, the dissolved oxygen level changes relatively slowly once an appropriate air flow rate and agitation condition are set. As cells grow, the dissolved oxygen level decreases over time, but the rate of change is rather slow because the cell doubling time is long. In contrast, if the air supply is shut off, the dissolved oxygen level will decrease rapidly due to consumption by cells. Similarly, if cells are rendered incapable of oxygen consumption suddenly, the dissolved oxygen level will rise quickly once the culture is deprived of oxygen consumption but continues to receive the supply. At a given moment, the dissolved oxygen level in a culture can be considered to be at a quasi-steady state because the time constant of OUR and OTR is in minutes, while that for change in the reactor is in hours. In general, OUR and OTR in a cell culture bioreactor can be considered equal in magnitude (Eq. 11-7, Panel 11.14). With the steady state assumption, OTR and OUR are equal, as can be seen in Eq. 11-4.

Panel 11.14. Oxygen Transfer and Uptake in a Stirred Tank Bioreactor

- A reactor can be considered to be in a quasi-steady state due to the slow change of the oxygen level
- OTR ≈ OUR

$$K_L a(c^* - c) \approx q_{O_2} x \qquad \textbf{(Eq. 11-7)}$$

Measurement of $K_L a$

The capacity of oxygen transfer in a bioreactor is often the limiting factor, or at least one of the limiting factors, that dictates the maximum cell concentration or productivity achievable in a bioreactor. Knowing the oxygen transfer capacity of a bioreactor is important in process design and scale translation. The magnitude of K_L and a is highly dependent on the properties of the liquid and gas, the flow conditions, power input, etc. Except under idealized conditions, such as a single bubble rising in stagnant fluid, it is very difficult to theoretically predict the value of K_L and a. Experimentally, K_L and a are extremely difficult to measure independently. As a result, mass K_L and a are lumped together as "K_La" in experimental measurements and reported as a single value. K_La is sometimes called the volumetric mass transfer. The K_La of a reactor varies under different operating conditions, such as different aeration rate, media composition (e.g., protein content, surfactant concentration), and agitation rate. Thus, ideally the oxygen transfer capacity should be measured under the cultivation conditions for process evaluation. However, few process engineers have the opportunity to perform K_La measurements under manufacturing conditions.

The K_La of a reactor is usually measured in the absence of cells using water, culture medium, or a buffer solution. First, the liquid is aerated to reach an initial dissolved oxygen level. Then the gas phase is switched to nitrogen under the same aeration and other operating conditions to strip oxygen off the liquid. The time profile of decreasing dissolved oxygen level is recorded and plotted in a semilogarithmic graph of $\ln(c^*-c)$ or $\ln(c)$ vs. time (Panel 11.15). The slope of the plot gives $-K_La$ (Figure 11.3). One can also perform the experiment by first stripping oxygen from the medium and then starting the measurement at a low level of c. By sparging air into the liquid, the dissolved oxygen level begins to rise. The time profile of the dissolved oxygen level is then plotted on a semilogarithmic graph. The slope in the semilogarithmic plot is then K_La. Multiple measurements of K_La under different operating conditions can be carried out to develop correlations between K_La and process conditions.

Panel 11.15. Measurements of K_La

- In the absence of cells

$$\frac{dc}{dt} = K_La(c^*-c) \qquad \textbf{(Eq. 11-8)}$$

Initial condition: $t = 0$, $c = c_O$

- After integration: $\ln\frac{c}{c_O} = -K_Lat$
- Plot $\ln(c^*-c)$ vs. t, the slope is $-K_La$

It should be noted that the measured K_La is the combined contribution of bubble aeration and surface aeration from the top surface of the liquid phase (Panel 11.16) where the gas composition may be the same or different from that in the reactor. In small-scale reactors, the contribution of the top surface is significant.

Measurement of Specific Oxygen Consumption Rate and OUR

OUR and the specific oxygen consumption rate can be easily measured off-line using culture broth sampled from the bioreactor (Figure 11.4). By depriving the cell suspension of any gas phase to eliminate any oxygen supply, the dissolved oxygen level will decrease over time

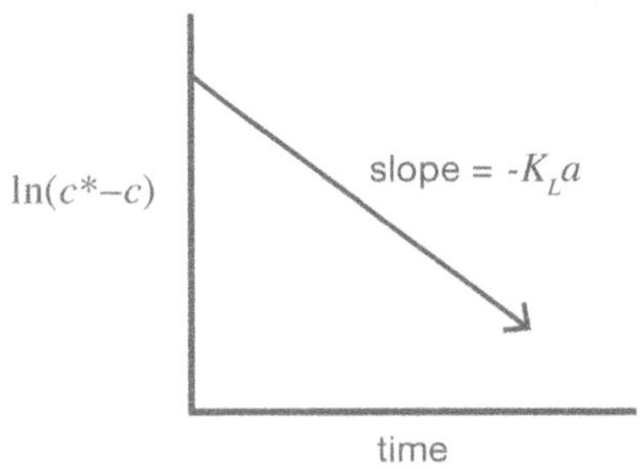

Figure 11.3. Measurement of K_La by degassing, or stripping of oxygen from the liquid phase.

Panel 11.16. OTR Contribution from Gas Sparging and Surface Aeration

$$\mathrm{OTR} = K_{L_s}a_s\left(c_s^* - c\right) + K_La(c^* - c)$$

$$= K_{L_s}\frac{A}{V}\left(c_s^* - c\right) + K_La\Delta c \qquad \textbf{(Eq. 11-9)}$$

$$a_s = \frac{A}{V} \qquad \textbf{(Eq. 11-10)}$$

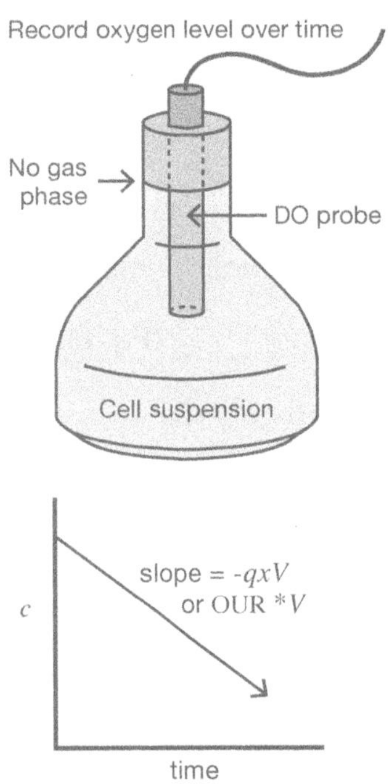

Figure 11.4. Measurement of oxygen uptake rate off-line.

due to consumption by cells. This can be done by filling up a container with a cell suspension without leaving room for a gas phase. The change in the dissolved oxygen level will be relatively linear over time. The slope of the linearly decreasing dissolved oxygen level gives the oxygen uptake rate (OUR) of the culture (Figure 11.4). The specific oxygen consumption rate can be obtained by dividing OUR by the cell concentration in the cell suspension.

OUR measurement can be practiced in real time using in-line dissolved oxygen measurement. A commonly used method is referred to as dynamic OUR measurement because it relies on changing gas composition or allowing aeration to be temporarily turned off to allow the dissolved oxygen level to change. It can be easily practiced in large-scale manufacturing bioreactors, as it requires no additional instrumentation. In a large reactor, the surface area of the liquid top is relatively small compared to the interfacial area generated by gas sparging. Thus, if the gas sparging is turned off, the oxygen transfer rate into the culture fluid is negligible and the dissolved oxygen concentration will decrease linearly with time. The slope of the decreasing dissolved oxygen level gives the OUR (Figure 11.5). When the method is used in a small bioreactor, the contribution from surface aeration cannot be neglected. In that case, $K_{Ls}a_s$ of surface aeration needs to be determined separately, and the transfer of oxygen into the liquid from surface aeration must be subtracted from the measured slope. Such an on-line measurement of OUR is a very sensitive method of determining changes in total oxygen consumption in a bioreactor, far more sensitive than measurements of the concentration changes of nutrients or metabolites. It is even more sensitive than detecting pH changes caused by proton release from lactic acid production. The pH buffering effect of the medium renders the pH less sensitive. OUR has been shown to detect the rapid change in cell vitality during virus production, faster than detecting the cell viability change caused by virus replication.[1–2] Its high sensitivity also allows for continuous feeding of glucose to control its concentration at rather low levels, thus eliciting metabolic shift and guiding continuous culture to steady states with low flux metabolic states.[3]

When the method is used in a small bioreactor, the contribution from the surface aeration cannot be neglected. In that case, $K_{Ls}a_s$ of the surface aeration needs to be determined separately and the transfer of oxygen into the liquid from surface aeration must be subtracted from the measured slope.

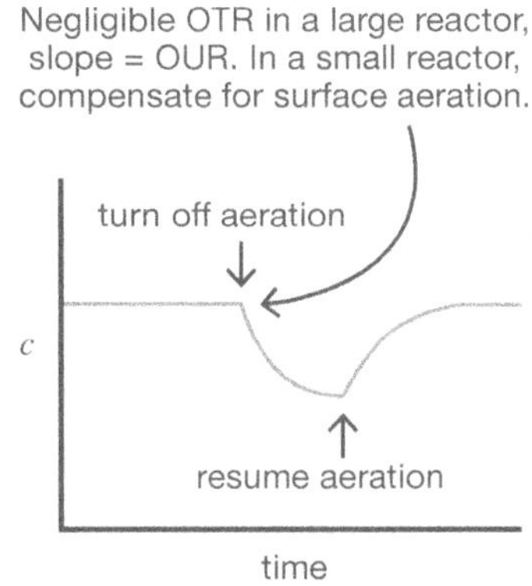

Figure 11.5. Dynamic in-line measurement of oxygen uptake rate in a bioreactor.

Damage to Cells by Gas Sparging

In an aerated mixing vessel, the many different types of force exerted on cells may cause cellular structural reorganization, deformation, or biochemical alterations that lead to cell damage or death. Cells may be squeezed by forces normal to their surface from opposite sides (Figure 11.6). They may impinge on moving mechanical objects, or be subjected to a fluid shear field with fluid moving faster on one side of the cell than the other (Figure 11.7). In a short time scale, the stress may cause deformation in cell shape. For example, squeezing cells from opposite sides changes their relatively spherical shape to an ovoid. At long time scales, cells may reorganize their cytoskeleton and redistribute the lipid bilayer membrane. For example, exposing endothelial cells to fluid flow may induce them to align in the flow direction. The effects of mechanical force on cell behavior can thus be physical or biological. If the rate of

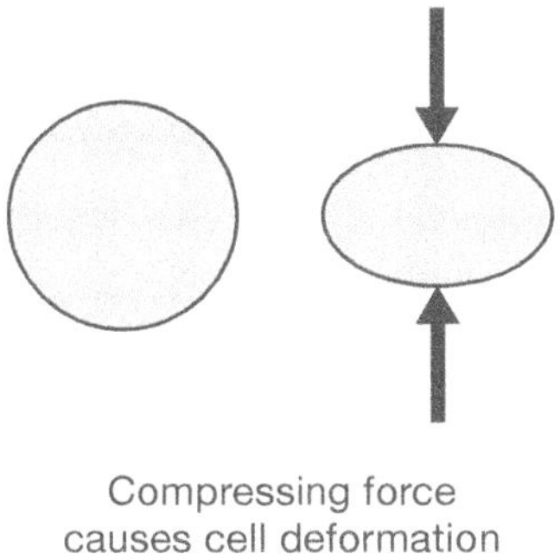

Figure 11.6. Deformation of a cell by a compressing force.

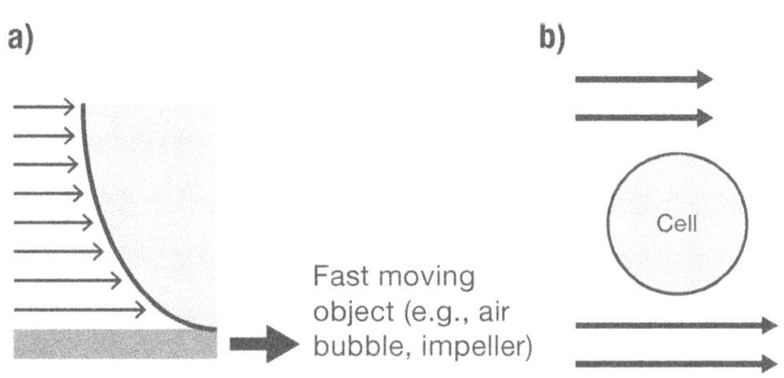

Figure 11.7. A difference in the liquid velocity on opposite sides of a cell causes shear stress. (a) The liquid velocity profile near the surface of a moving plate. (b) The velocity gradient around a cell in a high shear zone.

deformation is fast and the extent is extensive, the cytoskeletal network and the membrane integrity can be compromised, leading to stress response or even cell death. In bioreactors, the concern over mechanical forces has been mostly on the physical effects, especially cell damage. In most studies, the damage was measured in cell death. One should keep in mind that preceding cell death, cell damage may be incurred on other physiological events, such as metabolism, gene expression, or productivity. However, the damage in those respects has been less well documented.

Vigorous sparging of air bubbles through the culture causes cell damage. The mechanism most likely to cause cell damage is a fluid shear field caused by bubbles. Several possible events may dissipate a high-intensity burst of mechanical energy during the bubbles' time traversing the suspended cells in the bioreactor (Figure 11.8).[4] As the bubble escapes from the orifice, the volume originally occupied by the orifice-attached bubble suddenly becomes void, causing the surrounding, cell-containing fluid to rush in to fill the space. A large velocity gradient and high shear field that may cause cell damage inevitably occurs. As the bubble rises, a velocity gradient develops around the bubble: very high in proximity to the bubble's surface, but decreasing rapidly with distance. Similarly, when bubbles coalesce or break up, a liquid velocity gradient develops. Bubbles bursting from the liquid surface cause the volume originally occupied by gas to be rapidly replaced by liquid, thereby creating a strong shear field around the region of shear bursting.

Under typical operating conditions, the event that generates the greatest shear stress is the bursting of gas bubbles from the liquid surface (Panel 11.17). As the bubbles burst off the liquid surface, the high liquid velocity at the air–liquid interface decreases to the velocity in the bulk

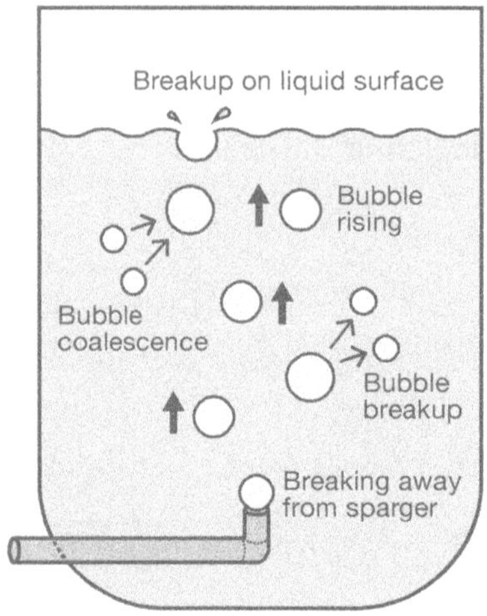

Figure 11.8. Various points during bubble rising in the reactor that may impart damaging shear force on cells.

Panel 11.17. Damage to Cells by Gas Sparging

- It is generally thought that bubble breakup at the liquid surface dissipates the most energy and is most damaging to cells
- Cells and cell-laden micro-carriers adhere to rising gas bubbles
- Energy dissipation in gas sparging causes cell damage
- Protective agents are added to the medium via the sparger

liquid over a very small distance. The momentum of the liquid movement caused by the bursting bubbles causes the liquid to rise above the liquid level surface (Figure 11.9). As a result, the liquid moving downward to fill the empty space left by the bubble reverses its direction to rise upward. In the upward-moving liquid cone, the velocity is highest in the center and lowest on the surface of the rising cone. When the rising liquid reaches its maximum height, it quickly reverses direction and drops down, causing the velocity gradient to change direction. The

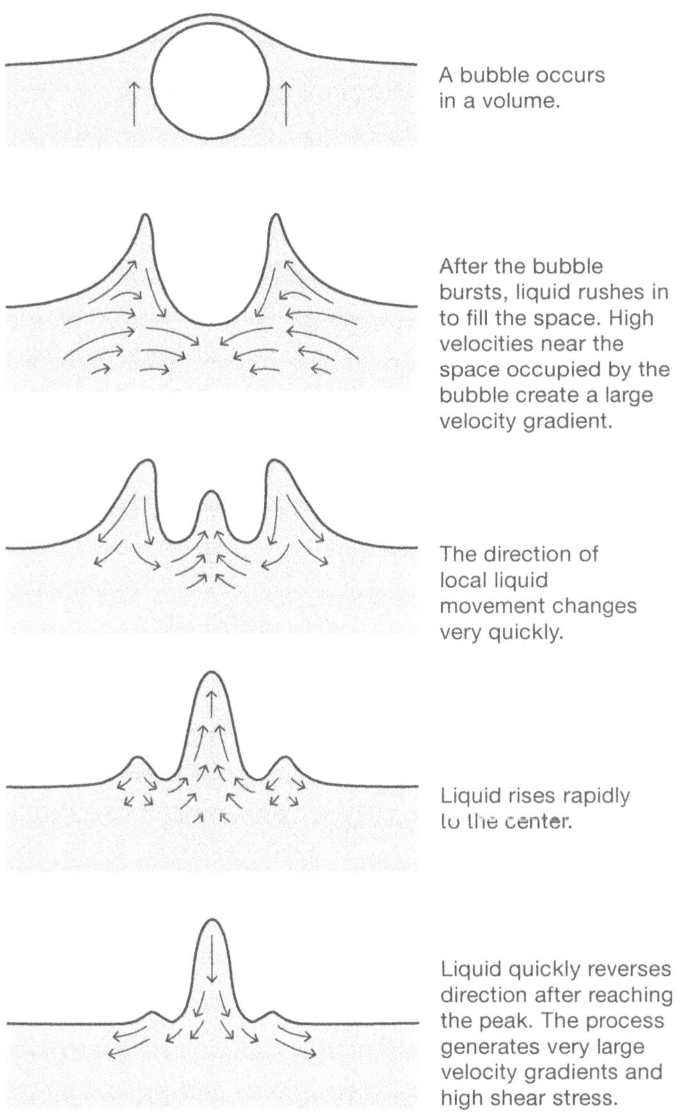

Figure 11.9. Conceptual depiction of fluid movement surrounding the point at which a bubble emerges from the liquid surface.

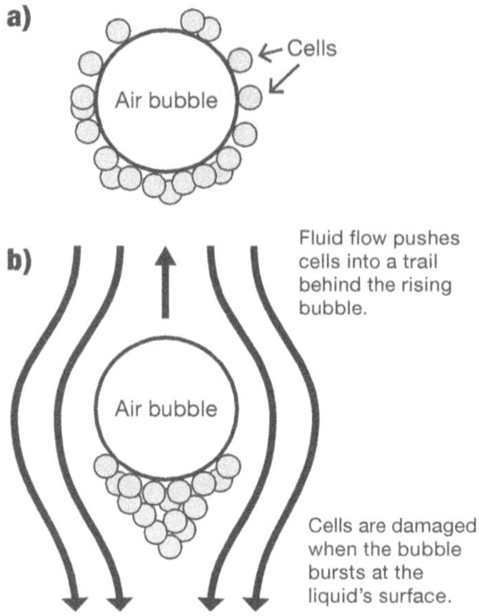

Figure 11.10. Accumulation of cells to a bubble's surface due to hydrophobic interactions and their accumulation to the tail end while the bubble rises.

energy released is very high and its damaging effects on cells entrapped in the bubble-bursting zone can be tremendous.

A bubble submerged in a cell suspension is often surrounded by a layer of cells under microscopic observation, possibly due to hydrophobic interactions between the cells and the bubble's surface (Figure 11.10). Once the bubble begins to rise, cells can be seen trailing the bubble in its tail, or the wake. As the bubble bursts from the liquid surface, cells can be severely damaged, if not outright killed. The block copolymer Pluronic F-68, which is commonly used in culture medium in bioprocessing, reduces the number of cells adhering to rising bubbles through its surface tension modulation effect.[5–6] This reduction of cell attachment to bubbles likely contributes to the "protective" effect of Pluronic F-68 (Poloxamer 188 (P188)) on cell damage from sparging.

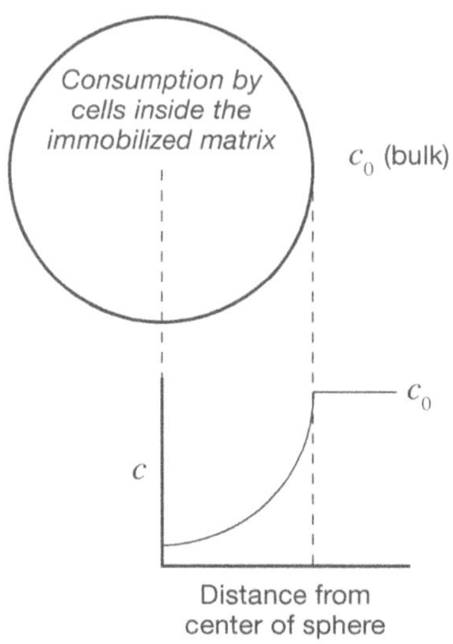

Figure 11.11. Concentration gradient of oxygen in a cell aggregate or an immobilized cell matrix particle.

Oxygen Transfer in Aggregates and Membrane Bioreactors

Cell Aggregates

Many cell lines and differentiated cells grow as aggregates or organized three-dimensional structures. For example, embryonic stem cells can grow as embryoid bodies and liver cells can grow as spheroids. In some cases, adherent cells are grown inside macroporous microcarriers. In these cases, oxygen must be supplied from a gas phase to the culture medium, travel to the aggregate, and then diffuse from the surface of the particle to the cells in the interior. As oxygen diffuses into the interior, it is consumed by cells. The concentration of oxygen thus decreases toward the interior from the particle surface (Figure 11.11). The steepness of the decrease is dependent on how fast the cells consume oxygen, and on the geometry and size of the particle. Further into the interior, oxygen may become depleted or become too low to support optimal growth. Oxygen deprivation causes cell death. The use of viability stain shows that dead cells often accumulate in the interior of large cell aggregates. To minimize oxygen limitations, one might reduce the size of particles or maintain a higher oxygen concentration in the culture medium.

Plug-Flow-Like Membrane Bioreactors

Some bioreactors for cell cultivation are comprised of multiple compartments separated by nutrient- and oxygen-permeable membranes. The most commonly seen of such devices is the hollow fiber bioreactor (Figure 11.12, also Figure 8.19). Cells reside in the cell chamber as adherent cells or as suspension cells so that they are not carried out of the reactor by fluid flow in the medium chamber. A medium stream flows through the medium chamber (the intraluminal space of the fiber) to allow nutrients and oxygen to diffuse across the membrane and reach cells in the extracapillary space (i.e., the shell side). In the lumen of the fiber, medium flows through in a plug-flow fashion. The oxygen is supplied by the medium entering from one end and exiting at the other end of the reactor. After exiting the bioreactor, the medium is re-oxygenated in an external oxygenator, usually also a hollow fiber membrane device with oxygen-enriched air in the shell side, to replenish oxygen in the medium before it is recirculated back to the bioreactor (Figure 11.13). The amount of oxygen carried in by fluid flow must be sufficient to supply for cell consumption (Panel 11.18).

It can be easily shown that, at a cell concentration of 2 x 10^{11}/L, a very high circulation rate is needed to supply enough oxygen. Once started up, a hollow fiber bioreactor can be continuously operated over months to produce a product. It is a convenient device for producing antibodies for diagnostic use.

Figure 11.12. Photographs of a hollow fiber bioreactor. Large-diameter fibers are shown for illustration.

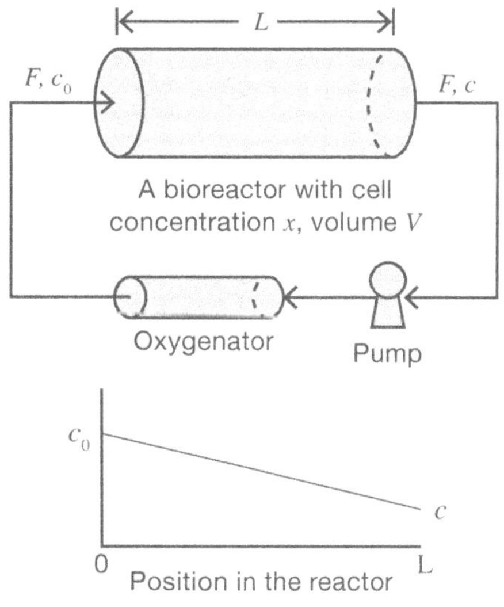

Figure 11.13. Medium recirculation and oxygen concentration gradient in the axial direction of fluid flow in a plug flow reactor.

Panel 11.18. Oxygen Supply by Medium Recirculation

Balance of oxygen concentration:

$$V\frac{dc}{dt} = F\left(c_{\text{in}} - c\right) - q_{O_2}xV \qquad \textbf{(Eq. 11-11)}$$

Example: The medium flow rate required to supply oxygen in a hollow fiber bioreactor with 0.2 L of cell chamber at 2 x 10^{11} cell/L is:

$$q_{O_2} \cdot x \cdot V = \frac{5 \text{ x } 10^{-10} \text{ mmol}}{\text{cell} \cdot \text{h}} \cdot \frac{2 \text{ x } 10^{11} \text{ cell}}{\text{L}} \cdot 0.2 \text{ L}$$

$$= 20 \frac{\text{mmol}}{\text{L} \cdot \text{h}}$$

Assume $c_{\text{in}} = 0.4 \text{ mmol/L}$ (saturated by air)

$$F = \frac{20}{0.4} = 50 \frac{\text{L}}{\text{h}}$$

$$F/V = 250 \text{ h}^{-1}$$

The recirculation rate is 250 times per hour.

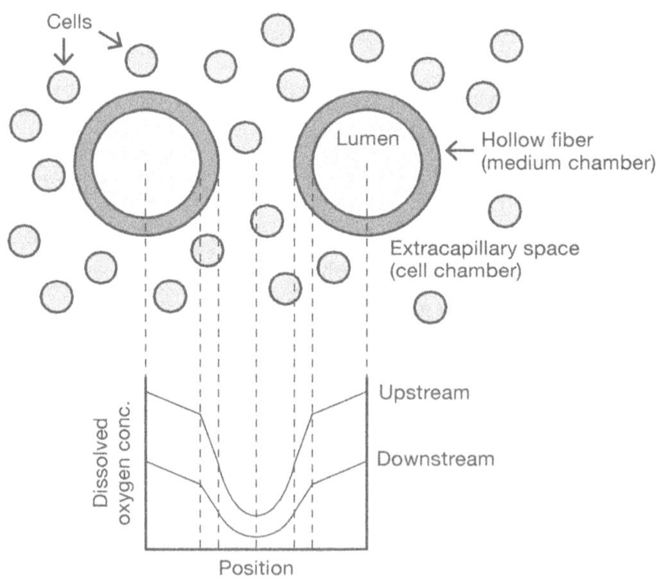

Figure 11.14. Oxygen gradient in the radial direction of a fiber in a hollow fiber bioreactor.

In a hollow fiber bioreactor, the oxygen concentration gradient exists in both axial and radial directions. It was realized very early on, when various reactors were being explored for cell culture bioprocessing, that oxygen transfer limitations would restrict the widespread application of hollow fiber bioreactors. Hence, oxygen transfer in hollow fiber bioreactors was extensively studied.[7] As the medium flows downstream in the lumen, oxygen is diffused through the membrane into the cell compartment. Oxygen concentration therefore decreases along the axial direction of the medium flow (Figure 11.13). In the radial direction, oxygen first diffuses from the bulk liquid in the fiber lumen to the wall of the fiber, and then through the fiber wall into the cell chamber. In the cell chamber, oxygen travels through cell clusters where it is consumed by cells before reaching the cells distal to the fiber. At a high cell concentration, the drop in oxygen concentration is steep in both the axial and radial directions (Panel 11.19). Figure 11.14 is a qualitative description of oxygen concentration in a single fiber, in the direction of medium flow. Under some conditions, an oxygen concentration difference can be seen between the center point of the fiber lumen and the inner wall. The fiber wall often presents a large resistance, causing oxygen concentration to decrease across significantly. In the cell chamber, oxygen decreases rapidly due to a high consumption rate. While the dissolved oxygen concentration in a stirred tank bioreactor is often maintained at 30% of the saturation, that in the exit stream of a hollow fiber bioreactor is typically kept at a much higher level to avoid cells in the downstream suffering from oxygen starvation. To improve oxygen transfer, one may increase the flow rate of medium recirculation or reverse the flow direction periodically to alternate the upstream and downstream ends. Even with those practices, the longitudinal length of a hollow fiber bioreactor is limited due to oxygen transfer limitations. A hollow fiber bioreactor with two sets of hollow fibers, each supplying medium and oxygen respectively, can alleviate the problem of oxygen transfer limitation. The demand for such bioreactors comes during scale-up.

Panel 11.19. Oxygen Transfer in a Hollow Fiber Bioreactor

- Oxygen transfer limitation occurs in both axial and radial directions
- Periodic reversal of the direction of flow helps

Concluding Remarks

THE SOLUBILITY OF OXYGEN IS VERY LOW in cell culture medium. It must be supplied continuously from a gas phase to sustain cellular demand. Since the respiratory quotient for cells in culture is close to 1.0, almost equal molar of carbon dioxide to that of oxygen consumption should be removed from the culture as cells grow. The most direct and effective method of supplying oxygen and removing carbon dioxide is by direct sparging of air into the culture broth in the bioreactor. The transfer rate of oxygen through the liquid–gas interface is affected by the mass transfer coefficient, the interfacial area, and the driving force, or the extent of deviation from equilibrium. All three factors can be manipulated to enhance oxygen transfer.

In some cases, such as when a medium containing a high concentration of serum is used, direct sparging is to be avoided because of excessive foaming or other adverse effects. Oxygen can then be supplied from an air stream that is separated from the medium by means of tubing or a membrane with high oxygen permeability, such as silicone or Teflon membranes. In a hollow fiber bioreactor, a recirculating medium passes through an oxygenator before re-entering the reactor to supply oxygen and other nutrients. The oxygenator is often a membrane device such as a hollow fiber oxygenator. The culture medium that exits from the bioreactor is re-oxygenated in the oxygenator and then recirculated back into the bioreactor.

References

1. Galvez, J., et al., *Optimization of HEK-293S cell cultures for the production of adenoviral vectors in bioreactors using on-line OUR measurements.* J. Biotechnol., 2012. **157**(1): p. 214–22.
2. Kussow, C.M., et al., *Monitoring of mammalian cell growth and virus production process using on-line oxygen uptake rate measurement.* Enzyme & Microbial Technology, 1995. **17**: p. 779–83.
3. Europa, A.F., et al., *Multiple steady states with distinct cellular metabolism in continuous culture of mammalian cells.* Biotechnology Bioengineering, 2000. **67**: p. 25–34.
4. Papoutsakis, E.T., *Fluid-mechanical damage of animal cells in bioreactor.* Trends in Biotechnology, 1991. **9**: p. 427–37.
5. Ma, N., et al., *Quantitative studies of cell-bubble interactions and cell damage at different Pluronic F-68 and cell concentrations.* Biotechnol. Progress, 2004. **20**: p. 1183–91.
6. Michael, J.D., et al., *Analysis of cell-to-bubble attachment in sparged bioreactors in the presence of cell-protecting additives.* Biotechnology Bioengineering, 1995. **47**: p. 407–19.
7. Piret, J.M. and C.L. Cooney, *Model of oxygen transport limitations in hollow fiber bioreactors.* Biotechnology Bioengineering, 1991. **37**: p. 80–92.

Nomenclature

$\Delta\rho$	Difference in density between liquid and gas
σ	Surface tension
A	Surface area of cross-section of the reactor
a	Interfacial area per culture volume
a_s	Interfacial area per culture volume for surface aeration
c	Oxygen concentration
Δc	Oxygen concentration gradient (i.e., the driving force)
c^*	Solubility of oxygen
c_s^*	Saturation oxygen concentration with the gas composition in the headspace of the reactor
d_O	Orifice diameter
d_p	Bubble diameter
F	Medium flow rate
H	Henry's law constant
K_L	Mass transfer coefficient
K_{Ls}	Mass transfer coefficient for surface aeration
P	Partial pressure
q	Specific consumption rate
t	Time
V	Culture volume
V_g	Gas holdup
x	Mole solution
y_{O2}	Mole fraction of oxygen in the gas phase

Contents

12

Scaling Up and Scaling Down for Cell Culture Bioreactors

Introduction

WHEN DESIGNING A CELL CULTURE PROCESS for a new product, one uses a laboratory-scale model to predict the process performance in manufacturing reactors. In some cases, this may involve designing new reactors. More often, it involves taking the new process to an existing manufacturing plant or to a reactor that is designed for generic purposes. The staggeringly high cost of constructing a new manufacturing plant, the trend to decentralize the manufacturing facility and divide it among multiple sites, the increasing use of platform process with disposable reactors, and continued improvement of productivity (thus freeing up manufacturing capacity for new products) all contribute to the increased importance of taking a new process to a pre-existing manufacturing reactor. Understanding the principle of reactor scale translation and developing the ability to evaluate key factors that are sensitive to scale in laboratory equipment is important for cell bioprocess technologists (Panel 12.1).

Panel 12.1. Scale Translation

Objective

- Predict process performance in large-scale reactors
- Specify operating conditions in small-scale equipment that simulates conditions at the large scale

Major Effects of Scale

- Oxygen (gas) transfer
- Heat transfer
- Hydrodynamic effects
- CO_2 removal
- May impact the chemical environment and physiological variables

When the scale of a reactor changes by its length dimension, many associated physical parameters (e.g., volume and total surface area) also change. Changes in physical parameters may influence the chemical environment or have a direct impact on cells' behavior. Chemical changes may cause biological alterations. As will be discussed below, an increase in the scale of a stirred tank reactor is accompanied by a reduction in the agitation rate. As a consequence, liquid pumping, shear stress, mixing time, etc., are all affected. Some changes to the physical parameters have a direct effect on cells. For example, the changes in the mechanical agitation rate may influence the extent of potential mechanical damage to cells. Fundamentally, the various surface areas related to the reactor (e.g., the heat transfer surface area in the cooling water jacket or the cross-sectional area of the reactor) all decrease with increasing scale on a per reactor volume basis. Changes to the ratio of surface area to volume influence the kinetics of temperature shift that are often practiced at the end of the rapid cell growth stage of production. The physical constraints of decreasing the cross-sectional area to reactor volume ratio leads to a less-than-proportional increase in the aeration rate relative to the scale increase. This affects the gas–liquid mass transfer, which in turn causes changes to the accumulation of CO_2, the pH and the addition of a base for controlling the pH, and the consequential osmolality of the culture fluid. Hence, changes to the physical parameters when scaling up inevitably lead to changes in the chemical environment that may affect cell behavior.

In this chapter, we will discuss the scaling up of a bioreactor based on geometrical similarity. We will keep constant the ratio of the geometrical lengths in the reactor when comparing different scales. In other words, when scaling up all different lengths will scale proportionally. For example, in scaling 10 liters to 10,000 liters, the reactor height, reactor and impeller diameter, and relative size of the internal parts (e.g., impeller, flow diverter) will all be increased by a factor of 10. By scaling all length dimensions of the reactor proportionally, all surface areas will increase with the length dimension to the second power, while the volume will increase to the third power. As a result, the scale-related surface area per unit volume of equipment will decrease. In microbial fermentation, the decrease in the surface area to volume ratio creates an impediment for the removal of heat generated by the metabolism and mechanical agitation. In mammalian cell processing, the generation of metabolic heat is less of a concern. However, the process may still be sensitive to other variables related to scale change.

As the scale of equipment changes, the physical and mechanical parameters may not remain constant. As will be discussed, it is not possible to keep all key operating parameters constant between different scales. Instead of striving to keep scale-related parameters constant, one ought to define the operating range of scale-sensitive physical and mechanical parameters, so that the cellular physiological state and productivity can be maintained within an acceptable range.

Mechanical Agitation

THE STIRRED TANK BIOREACTOR is the standard vessel for growing mammalian cells in industrial processes. Figure 12.1 shows an example of a typical industrial cell culture bioreactor. One or more impellers are mounted on a shaft that is driven by a motor mounted at the top of the tank. The impeller diameter is usually around 35–45% of the tank diameter and includes three or four blades. On the inner tank wall, four or six baffle plates are mounted to break the circulation flow pattern around the tank and generate better mixing of the fluid. The sparger is typically mounted some distance from the bottom impeller. Mechanical agitation provides the lifting force that keeps cells in suspension (Panel 12.2). This agitation sustains the fluid flow that mixes the chemical components and creates a more homogeneous chemical environment. In many cases, it also creates a flow pattern that helps reduce the size of gas bubbles and increase their retention time in the culture fluid to enhance oxygen transfer. For cells grown as aggregates, agitation also helps reduce the formation of oversized particles. In many fermentations of mycelial mold or actinomycete, extensive agitation is used to overcome the high viscosity of culture fluid and to reduce the mycelial pellet size to enhance oxygen transfer.

Among all the reasons for providing mechanical agitation, supplying oxygen is the most important. In microbial fermentation, oxygen demand is high (often exceeding 120 mmol /L-h). Mechanical agitation

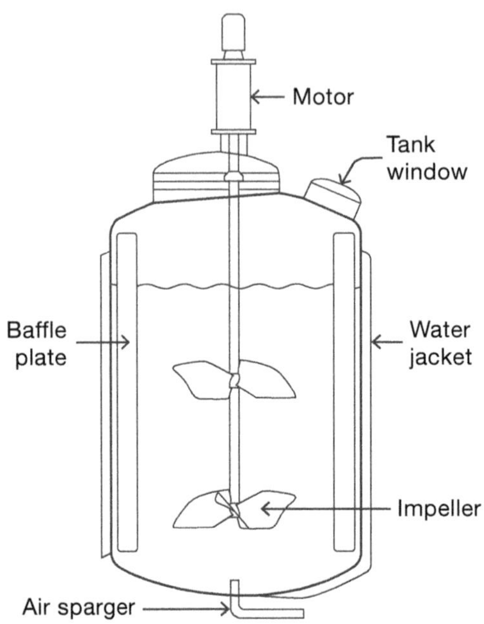

Figure 12.1. Schematic of a typical stirred tank bioreactor for cell culture process.

Panel 12.2. Purposes of Agitation

- Gas–liquid mass transfer
  - The higher shear field near the impeller tip produces small bubbles, thereby increasing the gas–liquid interfacial area
- Suspension of solids (e.g., microcarriers, cell aggregates) or dispersion of liquid
- Liquid–liquid, liquid–solid mass transfer (e.g., quick mixing of pH-neutralizing base)
- Minimization of pellets or aggregates
  - Pellets are cell aggregates of mycelial microorganisms (streptomyces, molds)
- Mixing, especially for viscous fluid (e.g., xanthan gum)

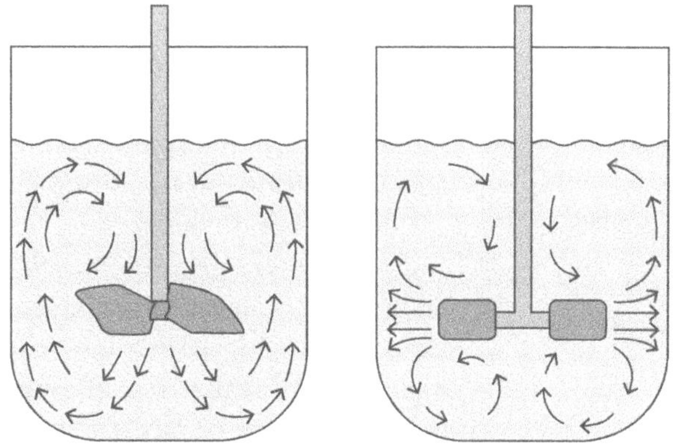

Figure 12.2. Two different kinds of fluid flow patterns in a stirred tank reactor: axial and radial flow.

breaks up air bubbles to achieve the level of oxygen transfer capacity needed for a high process productivity. In cell culture processes, the oxygen demand is at least one order of magnitude lower than that in microbial fermentation. Even then, the most important role of mechanical agitation is supplying oxygen.

Mechanisms of Agitation

Impellers can generally be classified into two types according to the flow pattern they generate: axial flow and radial flow (Figure 12.2). An axial flow impeller primarily generates upward or downward flows due to its pumping action. With a radial flow impeller, the liquid moves primarily outward, toward the walls of the vessel. In cell culture processing, impellers that generate axial flow are used because the shear fields generated by axial flow patterns are lower than those generated by radial flow patterns (Figure 12.3a and b).

Rushton disk turbines, which are often used with multiple installations in large reactors, are the predominant type used in microbial fermentation (Figure 12.3c). Rushton turbines are usually used with an air sparger placed directly underneath it. Gas bubbles from the sparger rise, hit the disk, and are directed outward. The blades, which rotate at a fast speed, then break the bubbles up.[1] Very high-energy dissipation in the area immediately surrounding the blades causes a high degree of turbulence in the fluid, further contributing to bubble break-up. The agitation also creates a high shear zone surrounding the impeller that can potentially damage cells. Despite this, it has been shown that mammalian cells grown in suspension can proliferate well with a Rushton turbine-type impeller rotating at a moderate agitation rate. Many cells are perhaps more tolerant to fluid flow stress than commonly thought.

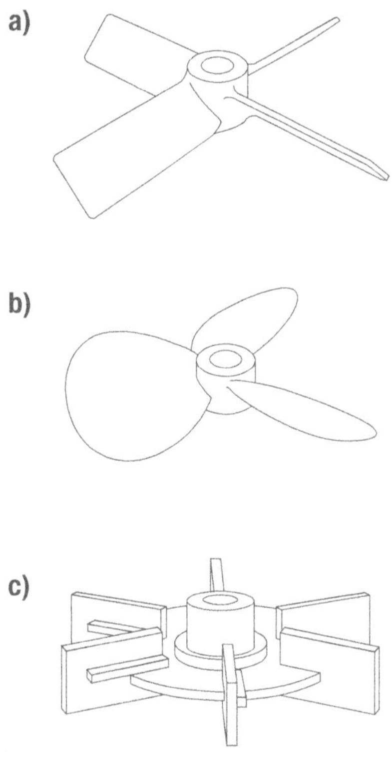

Figure 12.3. Three types of impellers commonly seen in stirred tank reactors: (a) pitched-blade, (b) propeller, and (c) Rushton turbine.

Propeller or pitch blade-type impellers focus on moving the liquid to create lift for mixing and suspending solids. Various designs of axial flow impellers are used in cell culture. As in microbial fermentation, in large bioreactors two to three axial flow impellers are often installed to provide adequate lifting force and to retain air bubbles for oxygen transfer.

The agitation needed to suspend microcarriers for cell growth is somewhat more delicate than for suspension cells. Microcarriers are generally more than ten times larger in diameter than suspension cells. They have a much higher settling velocity (>100 times) and are subjected to higher fluid shear stress. The impeller of a stirred tank for microcarrier culture tends to be larger relative to the tank diameter, but rotates at a slower rate. While the "propeller three blades" is used extensively in microbial fermentation to enhance oxygen transfer, the "axial flow three blades" provides less shear stress and a more uniform velocity in the entire discharged area. Some have attempted to use more specialized impellers, such as elephant ear or "sail" impellers. The latter, which rotates at a very slow rate of only a few rotations per minute, has been used at a pilot plant scale for cell aggregates and microcarriers.[2]

Power Consumption for Mixing Tanks

Power Consumption of Impeller-Agitated Bioreactors

In designing equipment and processes involving fluid flow, heat transfer, and mass transfer, engineers often employ a relationship established using dimensionless parameters. These dimensionless parameters are combinations of descriptors of physical properties such as force, time, and mass. The combination of those physical parameters cancels out the unit of time, mass, and length. Such parameters are therefore called dimensionless numbers. Fundamentally, these dimensionless numbers represent the ratio of a physical quantity in a system, for example the ratio of inertial forces to viscous forces exerted on an object. An important dimensionless quantity that is used widely to describe the dynamics of fluid flow in pipes is the Reynolds number. The number takes the ratio of inertial force ($(\rho D^3)(v/t)$) to viscous force ($\mu v D$). It was used to establish a unified relationship of the dimensionless energy loss of fluid flow in different viscosities, through different scales and flow conditions. The dimensionless number has since been called the Reynolds number (Re) (Panel 12.3). The quantity that describes the energy loss of fluid flow in a pipe is called the friction factor (f). The relationship between the Reynolds number and the friction factor has distinct regions: in the laminar flow (low Re) region, f decreases linearly with increasing Re; after a critical region, f becomes less dependent on Re; and in the turbulent (high Re) region, it becomes relatively constant. In the turbulent flow region,

Panel 12.3. Dimensionless Quantities Used to Describe Energy Loss in Fluid Flow in Pipes

Reynolds Number

$$\mathrm{Re} = \frac{\text{inertial force}}{\text{viscous force}}$$

$$= \frac{(\rho D^3 v)/t}{\mu v D} [=] \frac{(\mathrm{kg \cdot m^{-3}})(\mathrm{m^3})(\mathrm{m \cdot s^{-1}}) \cdot \mathrm{s^{-1}}}{(\mathrm{kg \cdot m^{-1} \cdot s^{-1}})(\mathrm{m \cdot s^{-1}})(\mathrm{m})} \rightarrow \text{dimensionless}$$

Fanning Friction Factor

$$f = \frac{\text{sheer stress}}{\text{kinetic energy density}} = \frac{\text{force/area}}{\text{kinetic energy/volume}}$$

$$= \frac{(\mathrm{kg \cdot m \cdot s^{-2}}) \cdot \mathrm{m^{-2}}}{(\mathrm{kg \cdot m^{-3}})(\mathrm{m \cdot s^{-1}})^2} = \frac{\mathrm{kg \cdot m^{-1} \cdot s^{-2}}}{\mathrm{kg \cdot m^{-1} \cdot s^{-2}}} \rightarrow \text{dimensionless}$$

pipes of different surface roughness give somewhat different relationships. These are typically presented as multiple lines representing the different surface roughnesses in high Re regions in a f vs. Re plot. Correlations between the Reynolds number (Re) and various dimensionless numbers (in addition to the friction factor) were developed and used in the design of equipment and processes involving fluid flow, mass transfer, heat transfer, etc.

During the development of design principles for stirred tank reactors, the concept of a fundamental relationship among dimensionless numbers was also adopted. A correlation was established between power consumption and the Reynolds number in a stirred tank reactor. In this case, the Reynolds number is denoted as Re_I (Impeller Reynolds number) to indicate that the characteristic length is the diameter of the impeller (Panel 12.4). The dimensionless number for power consumption by impeller is the power number, N_p.

The plot of N_p vs. Re_I compiled from experimental data resembles that of the friction factor vs. Re. At

Panel 12.4. The Impeller Reynolds Number and Power Number

Impeller Reynolds Number (Re_I): $\dfrac{N D_I^2 \rho}{\mu}$

Power Number (N_p): $\dfrac{P_o}{N^3 D_I^5 \rho}$

- With an aqueous medium, a bioreactor is always operated in the turbulent region
- In turbulent regions, N_p is constant, independent of Re_I:

$$\frac{P_o}{N^3 D_I^5 \rho} = K \qquad \textbf{(Eq. 12-1)}$$

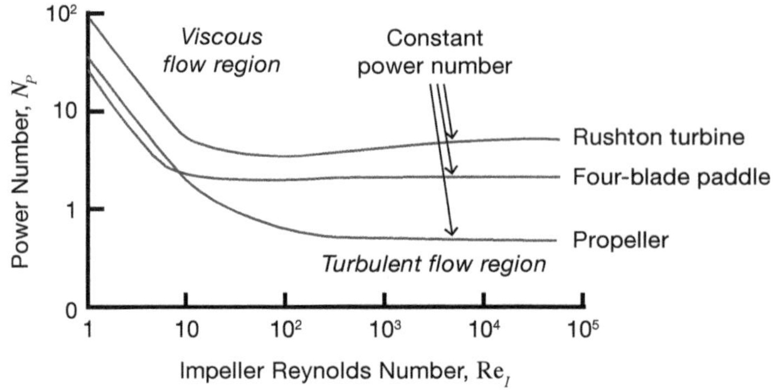

Figure 12.4. Relationship between impeller power number and impeller Reynolds number for different types of impellers.

low Re, N_p decreases linearly until the Reynolds number is ~2000, where a short break is seen, and then continues at a relatively constant value in the high Re region (Figure 12.4). The first region is the laminar (or viscous) flow region, and the constant value region is the turbulent flow region. In the turbulent region, different types of impellers give different correlations. They all exhibit a similar behavior of having a relatively constant N_p over a wide range of Re_I.

In a stirred tank, laminar flow is only seen when the viscosity of the fluid is very high, such as when glycerol is used. The flow in a cell culture bioreactor is always in the turbulent regime. Since NP is constant in the turbulent region for a given type of impeller, we consider it to be constant in the scaling up of cell culture bioreactors. From the definition of N_p, one can see that the impeller power (P) divided by $N_3 D_I^5$ (N is the agitation rate, D_I is the impeller diameter) is constant.

Other Scale-Sensitive Variables for Stirred Tank Reactors

In addition to power consumption, a number of other properties of a stirred tank may profoundly affect cell growth and productivity. For example, the liquid velocity and its associated shear rate and shear stress influence the size of cell aggregates and exert mechanical stress on the cells. The volumetric flow rate (or pumping rate) affects the time that the content of the reactor is mixed, called the mixing time (Figure 12.5). For our discussion, we will choose a representative quantity in the reactor and call them the characteristic liquid velocity, pumping rate, and mixing time, and use them to discuss the effects of scaling up.

The maximum liquid velocity that occurs at the tip of the impeller can be taken as the characteristic liquid velocity in a stirred tank. It can be represented by the multiplicative product of the rotation rate of the impeller and its diameter, ND_I (Eq. 12-2, Figure 12.5). (Note: we will

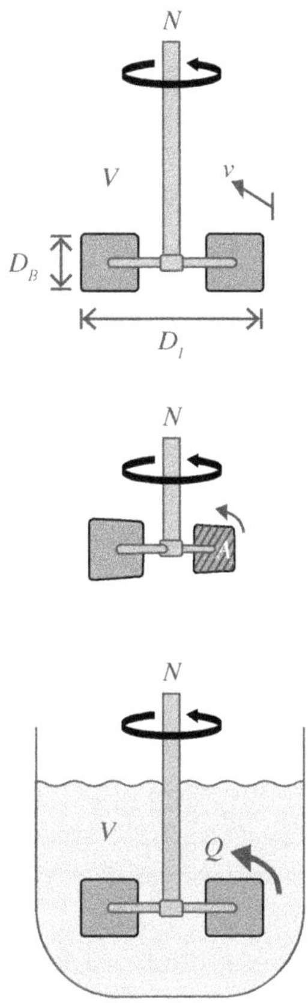

Maximum velocity

$$v = ND_I \qquad \textbf{(Eq. 12-2)}$$

Pumping

$$Q = \text{pumping area} \cdot \text{pumping velocity} = A \cdot v$$

$$= D_B^2 \cdot ND_I \propto ND_I^3 \qquad \textbf{(Eq. 12-3)}$$

Mixing time

$$\theta = \frac{V}{Q} \propto \frac{D_I^3}{ND_I^3} = \frac{1}{N} \qquad \textbf{(Eq. 12-4)}$$

Figure 12.5. The characteristic liquid velocity (impeller tip velocity), characteristic pumping rate, and characteristic mixing time in a stirred tank bioreactor.

ignore π in our discussion of the perimeter, area of circle, etc. The constant value π is cancelled out when comparing different scales.) ND_I is also the impeller tip speed.

The amount of fluid that the impeller moves with its rotational motion is directly dependent on its rotating velocity and the area of the impeller blades. Since we are considering scale translation under the conditions of geometrical similarity, we may use any length dimension of the reactor for comparison of reactors of different scales. We will use the length of the impeller, instead of the length of the impeller blade, to represent the length scale for our discussion. The pumping, then, is the projected area (D_I^2) of the impeller multiplied by the velocity of its rotation (ND_I), which gives ND_I^3 (Eq. 12-3, Figure 12.5).

How quickly a solute can be mixed after it is added to the fluid in a stirred tank is dependent on the volume of the fluid in the tank (D_T^3) and the amount of liquid the impeller can pump (ND_I^3). Since we are considering the case of geometrical similarity, the ratio of the length dimensions can be represented by D_I. The changes in mixing time in scaling up can be estimated by dividing the reactor volume by the pumping. The characteristic mixing time in a stirred tank is the inverse of its rotation speed ($1/N$) (Eq. 12-4, Figure 12.5).

As the scale changes, the characteristic fluid velocity (v), liquid volumetric flow rate (Q), and mixing time (θ) also change. By comparing changes in these characteristic quantities across different scales, we can gain much insight into the effects scaling up has on many process-related physical effects.

Effects of Scale on Physical Behavior in Bioreactors

We will use the correlations between the power number and the impeller Reynolds number and the three characteristic quantities described above to explore the effects of a changing scale. We will assume that the equipment between the different scales will remain geometrically similar (Figure 12.6). The effects of different reactor sizes can then be compared using characteristic length D (the tank diameter). If the tank diameter increases by 10-fold, all the other reactor parts (tank height, impeller diameter, etc.) will increase by the same proportion of 10-fold. Hence, in the subsequent discussion we will use only one length symbol, D, without specifying whether it is tank diameter or impeller diameter. As noted above, the constant π will be dropped from our equations.

When scaling up different processes, one ought to keep the most important variable(s) constant or within an acceptable range. Common

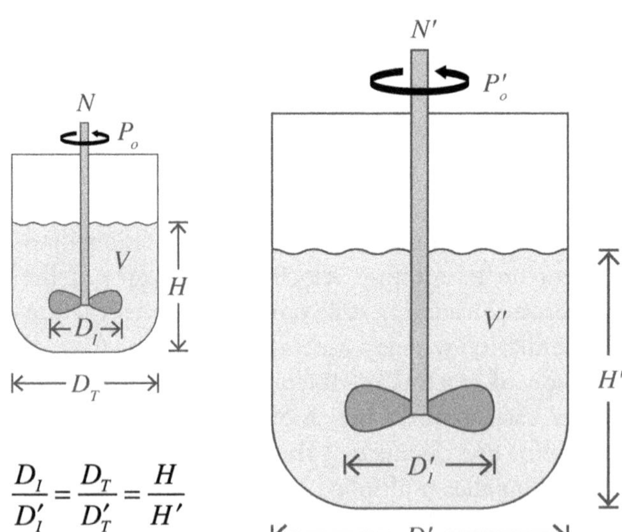

Figure 12.6. Notation of an impeller-based mixing reactor and geometrical similarity in scaling up. H: liquid height, V: liquid volume, N: impeller rotation rate, P_O: agitation power, D_I: impeller diameter, D_T: tank diameter.

design objectives for scaling up are: (1) keep K_La constant, so that mass transfer can be maintained; (2) keep impeller tip speed constant, thus maintaining a critical value of high shear velocity to avoid mechanical damage to cells, or to break up agglomerating pellets of mycelial cells in fungal fermentation; (3) keep power input per volume constant, usually for less power-intensive processes such as crystallization and blending; and (4) keep mixing time constant (Panel 12.5).

Panel 12.5. Scale Translation Approaches

- Constant K_La
- Constant impeller tip speed (ND_I)
- Constant power per unit volume (P_o/V)
- Constant mixing time ($1/N$)

Consider the case of scaling up by keeping the power input per reactor volume constant. Recall that the power number in a turbulent region is constant (Eq. 12-1, Panel 12.4), so the power input (P) is proportional to N^3D^5 (Eq. 12-5, Panel 12.6). The reactor volume is described by πHD^2. Because of geometrical similarity, we represent H by D and ignore the constant π. The reactor volume (V) is thus represented by D^3 (Eq. 12-6,

Panel 12.6. Scaling Up Geometrically Similarly by Keeping Power Per Unit Volume Constant

Constant parameters, such as π, can be dropped in the discussion of the effect of scale (D_I). One can use a characteristic length D to represent the length dimension.

$$\frac{P_o}{N^3D^5\rho}$$ is constant in a turbulent region.

The density of water, ρ, is constant. Thus,

$$P_o = KN^3D^5 \quad \textbf{(Eq. 12-5)}$$

The volume of the reactor can be expressed as the characteristic length raised to the third power,

$$V = \pi HD^2 = cD^3 \quad \textbf{(Eq. 12-6)}$$

The power per unit volume is described as

$$\frac{P_o}{V} = \frac{KN^3D^5}{cD^3} = K'N^3D^2 \quad \textbf{(Eq. 12-7)}$$

This leads to the conclusion that, when power per unit volume is kept constant, $N^3D_I^2$ is also constant.

$$N^3D^2 = \text{constant} \quad \textbf{(Eq. 12-8)}$$

$$N = \frac{\text{constant}}{D^{2/3}}$$

Panel 12.7. Effect on Agitation Rate

Comparing scale 1 and scale 2:

$$N_1^3 D_1^2 = N_2^3 D_2^2 \quad \text{(Eq. 12-9)}$$

$$\frac{N_1}{N_2} = \left(\frac{D_2}{D_1}\right)^{2/3} \quad \text{(Eq. 12-10)}$$

The agitation rate N decreases with increasing scale. When the diameter increases eight times, the agitation rate is reduced to 1/4 in the larger scale.

Panel 12.8. Effect on Impeller Tip Speed

Tip speed is described by N multiplied by D, from Eq. 12-6. By keeping P/V constant,

$$N_1^3 D_1^2 = N_2^3 D_2^2$$

$$\frac{N_1^3 D_1^3}{D_1} = \frac{N_2^3 D_2^3}{D_2}$$

$$\frac{N_1^3 D_1^3}{N_2^3 D_2^3} = \frac{D_1}{D_2}$$

$$\frac{N_1 D_1}{N_2 D_2} = \left(\frac{D_1}{D_2}\right)^{1/3} \quad \text{(Eq. 12-11)}$$

Tip speed increases with increasing scale, but only at 1/3 power of the length of scale.

Panel 12.6). By keeping P_o/V constant, N^3D^2 is also constant across different scales (Eq. 12-7). As D increases, the rotation speed must decrease by $1/D^{2/3}$ (Eq. 12-8). The relationship can be written even more explicitly by comparing N across different scales (Eq. 12-9, Panel 12.7). By keeping power per volume constant in scaling up, it is inevitable that larger reactors will need to be operated at lower rotation speeds.

By similar algebraic manipulation, one can also see that scaling up by keeping power input per reactor volume (P/N^3D^2) constant will lead to changes in impeller tip speed (ND) (Panel 12.8), total pumping (ND^3) (Panel 12.9), and mixing time (Panel 12.10). The impeller tip speed increases with the length scale factor to 1/3 power (Eq. 12-11). The liquid pumping also increases with the length scale factor to 7/3 power (Eq. 12-12, Panel 12.9). However, pumping per volume will decrease as the scale increases (Eqs. 12-13 and 12-14, Panel 12.9). Since mixing time can be represented by reactor volume divided by total pumping, mixing time increases with scale (Eq. 12-15, Panel 12.10).

Table 12.1 explores the effects of scaling up a geometrically similar bioreactor by a factor of 10 using three different criteria: constant power per volume (P/D^3), constant agitation rate (N), constant pumping rate, and constant tip speed. The calculation is based on the power number remaining constant when scaling up. By keeping the power input per volume constant, the agitation rate is reduced to $10^{-2/3}$ (≈ 0.21), but the impeller tip speed is increased by 2.15-fold. The total pumping is increased, but pumping per volume and mixing time increase. If the impeller tip speed is kept constant when scaling up, the decrease in the agitation rate will become much steeper with respect to the scale. Instead of decreasing with $D^{-2/3}$ as in constant power per volume, it decreases proportionally to the scale. Consequently, the power input and the liquid pumping per volume are

Panel 12.9. Effect on Liquid Pumping

The capacity of liquid pumping can be described by the impeller tip speed, ND_I, and by the area that it moves against the liquid, D_{I2}.

At constant power per volume,

$$N_1^3 D_1^2 = N_2^3 D_2^2$$

Multiply both the numerator and the denominator by the diameter to the seventh power.

$$\frac{N_1^3 D_1^9}{D_1^7} = \frac{N_2^3 D_2^9}{D_2^7} \quad \textbf{(Eq. 12-12)}$$

$$\frac{N_1 D_1^3}{N_2 D_2^3} = \left(\frac{D_1}{D_2}\right)^{7/3}$$

Liquid pumping capacity increases with scale. By dividing both sides by the characteristic length raised to the third power, we can obtain the pumping capacity on a per volume basis.

$$\frac{ND^3}{D^3} = \text{pumping per volume} = \frac{Q_P}{V} \quad \textbf{(Eq. 12-13)}$$

$$\frac{Q_{P_1}}{V_1} \Big/ \frac{Q_{P_2}}{V_2} = \left(\frac{D_1}{D_2}\right)^{-2/3} = \left(\frac{D_2}{D_1}\right)^{2/3} \quad \textbf{(Eq. 12-14)}$$

The pumping capacity per volume decreases with increasing scale.

much reduced. This is accompanied by an equal magnitude of increase in mixing time. When the agitation rate is kept constant, the pumping rate and mixing time will also be constant. However, a tremendous increase in power consumption would be needed to keep the agitation rate constant, and as such cannot be implemented in practice.

This example illustrates a very important point in the discussion of scaling up: all physical properties cannot be kept constant at different scales. If one property is chosen to be kept constant, many other properties must change.

Panel 12.10. Effect on Mixing Time

Pumping per volume decreases with scale. Thus, mixing time increases when scale increases.

From Eq. 12-4 and Eq. 12-10:

$$\frac{\theta_1}{\theta_2} = \frac{N_2}{N_1} = \left(\frac{D_2}{D_1}\right)^{2/3} \quad \textbf{(Eq. 12-15)}$$

Mixing Time

THE PURPOSE OF AGITATION is to provide a homogenous chemical environment. Spatial heterogeneity of nutrients may subject cells to an oscillating environment as they traverse the reactor, leading to inferior growth or production performance. Since nutrients that are present from the beginning of the culture eventually become homogeneous in a stirred tank, mixing is potentially a concern only for the components that are added continuously or intermittently during the cultivation. In most reactors, the feeding stream is added in a fixed position(s). It takes a finite amount of time for the added materials to become uniformly distributed throughout the reactor. When mixing is inadequate, cells may encounter regions of excessively high concentrations of added solute(s), potentially causing physiological changes. When areas with especially low concentration are encountered, cells may experience nutrient starvation.

Nutrient Starvation Time

When cells are grown in a closed, nutrient-rich environment without any external supply, oxygen is always the first nutrient to be depleted. Because of its low solubility in medium, oxygen concentration in the medium is orders of magnitude lower than glucose. The ratio of the molar specific consumption rate of oxygen to glucose ranges from about 1.0 (when most glucose is converted to lactate) to close to 6.0 (when most glucose is converted to CO_2). With a cell concentration commonly seen

Table 12.1. Effects of Scaling Up Based on Different Criteria. N_p (P/N^3D^5) is constant.

		Small scale: D=1, V=1	*Large scale:* D=10, V=1000		
			Constant power per volume	Constant agitation rate	Constant impeller tip speed
Power	P	1	1000	10^5	100
Power per volume	P/V	1	1	100	0.1
Rotation rate	N	1	0.21	1	0.1
Tip speed	ND	1	2.15	10	1
Pumping	ND^3	1	215	1000	100
Pumping per volume	Q_p/V	1	0.215	1	0.1
Mixing time	θ	1	4.76	1	10

Table 12.2. Comparison of Oxygen and Glucose Saturation Times in a Typical Culture (for 10^{10} cells/L)

	Oxygen	*Glucose*
Concentration	0.1 mM (50% saturation with air space)	1 g / L (5.6 mM)
Specific consumption rate	1×10^{-10} mmol / cell-h	$0.15–1.0 \times 10^{-10}$ mmol / cell-h
Volumetric consumption	1 mmol / L-h	0.15–1 mmol / L-h
Time to depletion	0.1 h (6 min)	12 h

in industrial processes and no external supply, oxygen is depleted within minutes, whereas the depletion time for glucose is orders of magnitude longer (Table 12.2). To avoid adverse physiological consequences or even apoptosis, nutrients must be replenished before they are reduced to below their critical levels.

Mixing Time

The mixing time of a reactor can be quantified using a tracer solute (Panel 12.11). First, a solution of dye is injected into the reactor and allowed to be carried by the fluid and dispersed. A sensor placed in a fixed position in the reactor then records the change of dye concentration over time (Figure 12.7). The concentration will fluctuate, initially over a wide range, and then gradually ebb to a steady value. The time needed for the concentration to reach within a bounded range of steady value is considered the "mixing time." For example, if two ranges are +/-10% and +/-5% of the final steady value, then the dye concentration begins to enter the range at 90% and 95% mixing time, respectively.

Panel 12.11. Mixing Time Measurement

Measurement

- At $t = 0$, add tracer to the reactor and measure the fluctuation of tracer concentration over time
- Measure terminal mixing time, defined as the point when an arbitrarily chosen uniformity (e.g., 90% or 95%) is reached

Let us denote the final, well-mixed steady dye concentration as c_s, the concentration at any time point as c, and the concentration deviation from its final steady value as Δc. The semilogarithmic plot of the time profile of $\ln(\Delta c)$ vs. t can often be approximated by a constant slope, suggesting that the kinetic behavior is first order. The slope, which is a kinetic constant of mixing, can be taken as the pumping rate (Q_p) divided by the volume of the reactor. The slope is a rate constant and its inverse is a characteristic time (Eq. 12-16, Figure 12.7). After one characteristic time, the tracer concentration will be $0.37\Delta c$ from the final c_s, and after three characteristic times it will be $0.02\Delta c$ from c_s. In other words, the content of the reactor can be considered well-mixed after three characteristic mixing times.

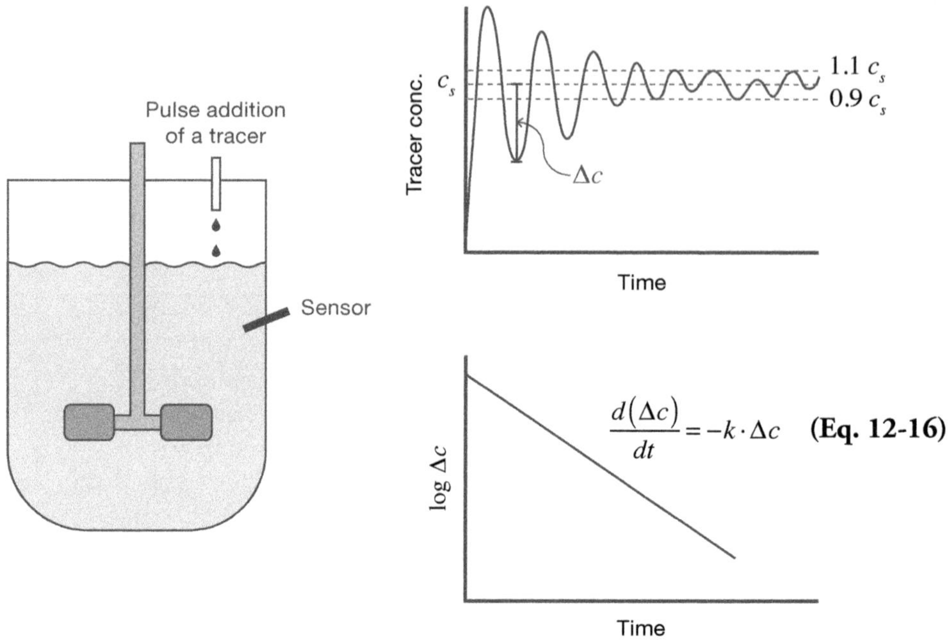

Figure 12.7. Measurement of averaged mixing time in a stirred tank.

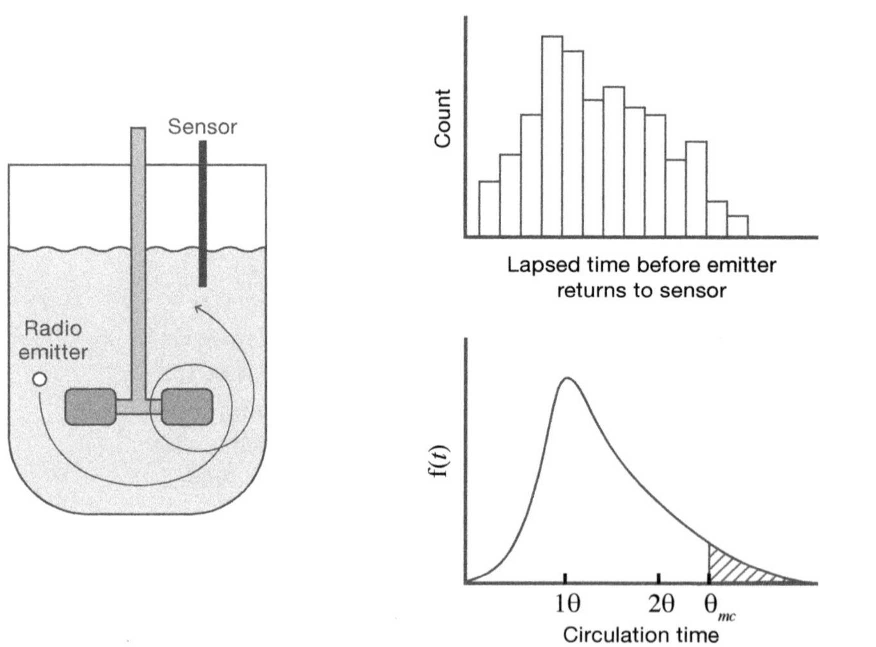

Figure 12.8. Circulation time distribution in a stirred tank and its measurement.

In scale-up, Q_p/V decreases as the scale increases, meaning that the time to reach complete mixing lengthens. To avoid nutrient deprivation, one may keep the set point nutrient concentration (c_s) higher so that the level will not drop too low before the next nutrient addition time. One may also feed more frequently. For very large reactors or if the nutrient level cannot be raised (possibly because of toxicity), multiple nutrient addition points in the reactor are sometimes implemented.

Mixing Time Distribution

Panel 12.12. Medium Circulation Time Distribution

- The circulation in a tank is not uniform. If the average circulation time is 6 minutes, many fluid elements will have a circulation time shorter than 6 minutes, and others will have longer times.
- The average circulation time should be shorter than the critical circulation time

The mixing time described above is an averaged value measured at a fixed position in the reactor. Consider the case that a nutrient solution is added at a fixed position in a reactor. A fluid element with a certain volume and cell concentration passes by the position and carries an amount of the added nutrient as it traverses the reactor via the fluid flow created by the mixing mechanism. At some point, the fluid element will return to the feeding position, acquire more nutrient, move away to circulate around the reactor again, and finally returns to the feeding position again. The time elapsed between the two consecutive returns to the feeding position, or the circulation time/medium re-enrichment time, is not uniform. Instead, it is distributed over a range; sometimes the time is short, other times it is long (Panel 12.12). As discussed above, the amount of nutrient that the fluid acquires at the nutrient feeding zone must be sufficient to sustain the metabolic needs of the cells before it returns to the feeding zone. In other words, the nutrient re-enrichment time must be shorter than the nutrient depletion time. The nutrient depletion time is dependent on cell concentration and the consumption rate. If the nutrient re-enrichment time is longer than the nutrient depletion time, a time period of nutrient depletion will appear during the circulation of the fluid element.

To measure the distribution of the circulation time (defined as the time elapsed between two consecutive feedings), imagine that a signal-emitting ball with an identical density to the fluid is being carried freely by fluid motion (Figure 12.8, Panel 12.13). A sensor at the position where the nutrient is added to the reactor picks up the signal when the ball is close by and records the time interval between consecutive detections of the signal. The recorded time interval is not uniform, and instead distributes over a range because sometimes the ball returns to the sensor shortly after it moves away, while other times it roams around the reactor for a while before returning to the sensor position. The

Panel 12.13. Mixing Time Distribution Measurement

- Add a radio emitter to the reactor. A sensor picks up the signal when the emitter, in the course of circulating around the reactor, passes by.
- Measure the circulation time for each encounter and plot the frequency distribution of the circulation time
- Determine the mean and median circulation time and the standard deviation σ
- The distribution of the circulation time can be plotted as a population density function. The portion of the distribution with a circulation time lying between t and $t + \Delta t$ is the area under the curve between t and $t + \Delta t$.

distribution function thus describes the frequency of the circulation time or the nutrient feeding interval.

In general, the distribution follows a logarithmic normal distribution. The mean or median of the distribution function is a descriptor of the mixing characteristics of the reactor, but it does not present the entire picture of mixing. Two reactors that have the same median or mean nutrient feeding interval may still have a very different frequency distribution. Cells that circulate with the fluid element receive nutrients only when the fluid returns to that feeding position. If the circulation time is longer than the nutrient depletion time, then the nutrient level will fall below the critical value. If the duration of nutrient depletion exceeds a critical time period, apoptosis or other metabolic anomalies may occur. A wide distribution of the circulation time can be a concern. Even low occurrences of long circulation times may cause nutrient starvation once cell concentration becomes high. When very high cell concentrations are used—as is often the case during the search for ever-higher target productivities—the oxygen depletion time can be only a few seconds long. The effect of circulation time distribution should therefore be taken into consideration when scaling up.

Scaling Up and Mechanical Forces on Cells

CELLS IN A STIRRED BIOREACTOR may encounter shear stress caused by fluid flow, or be subjected to direct impact by moving mechanical parts such as the impeller. Since cells are neutrally buoyant particles, they tend to follow the motion of relatively large eddies. In general, direct impact of cells on the impeller is therefore minimal, since cells are usually carried away by fluid before direct mechanical impact. Damage to cells due to mechanical agitation in a stirred tank is more likely to arise from shear stress caused by large fluid velocity gradients, or from the transference of kinetic energy from swirling pockets of fluid (called eddies) in a turbulent flow than from mechanical impacts (Figure 12.9).

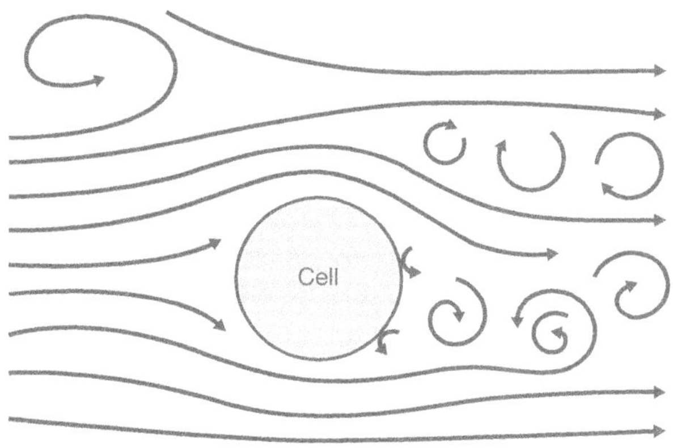

Figure 12.9. Fluid flow and eddies surrounding a cell. Eddy size increases with scale if the power per unit volume decreases with scale.

The maximum fluid velocity in a stirred tank occurs at the tip of the impeller (Panel 12.14). As the scale increases, the impeller tip speed increases. The shear rate (i.e., the steepness of the fluid velocity change over a unit distance), increases with the agitation rate. Because of this, shear damage was a considerable concern when cell culture was undergoing the transition from a research platform to a major biomanufacturing technology in the 1980s and early 1990s. Nevertheless, large stirred tanks of up to tens of cubic meters in volume have been commonly used for cell culture-based manufacturing since the mid-twentieth century. Agitation conditions well below those causing significant cell damage can be identified and employed in industrial processes for the vast majority of cell lines of industrial interest. In general, the conditions of mechanical agitation that are commonly employed in cell culture processes are much less energy-intensive and impart much less mechanical stress than those in microbial fermentation. It is advantageous to employ a higher agitation rate when possible, as it provides more efficient oxygen transfer and carbon dioxide removal.

Panel 12.14. Effects of Mechanical Agitation on Cells

- The most velocity and energy released in a stirred tank occurs at the impeller tip
- An increasing agitation rate increases the impeller tip speed and the shear rate
- Most suspension cells are sensitive to mechanical stress. Satisfactory, non-cell damaging conditions are identified and used in large-scale cell culture.
- More intensive agitation may be advantageous in enhancing oxygen and CO_2 transfer in high density cell culture process
- Most studies employ cell viability and growth as criterion for evaluating cell damage. Evaluating the non-lethal effects on cells' physiology is also important.

It has long been reported that cells are not as fragile as they were thought to be at the dawn of cell culture bioprocessing. Even Rushton turbines were shown to support normal growth in laboratory bioreactors, and elephant ear impellers with high agitation rates have been successfully used in small reactors. Many of the early studies on the effects of mechanical agitation focused on cell death—an irreparable event—as the outcome variable. Before cell death is caused, however, mechanical stress may elicit changes in the cell metabolism, stress response, and other physiological processes. Many studies have been undertaken on the physiological differences among endothelial cells that have been exposed to different flow conditions. Cells grown under different hydrodynamic conditions show gene expression changes in some functional classes, including the stress response. But, in general, our knowledge of non-lethal cellular changes in industrial cell lines under different mechanical agitation conditions is still relatively scarce.

Cells grown on microcarriers are more susceptible to damage caused by forces exerted by fluid flow. With their much larger particle diameters, microcarriers have a higher velocity relative to the fluid compared to suspension cells. By using the energy dissipation rate to compare the agitation conditions that cause cell damage, a study reported that the energy dissipation rate that causes cell damage in microcarrier cultures is at least two orders of magnitude lower than that for suspension cells.[3] It should be noted that different cell lines have different sensitivities to mechanical agitation. Even for suspension cell lines, the maximal shear-stress tolerance range spans over two orders of magnitude, suggesting that the sensitivity to fluidic dynamic damage is highly dependent on the cell line, reactor, and many other factors. Nevertheless, the low tolerance of microcarrier cultures to high mechanical energy dissipation is generally reflected in their higher sensitivity to mechanical agitation when scaling up.

A mechanistic understanding of cell damage caused by fluid flow in a stirred tank is still lacking, despite attempts to link Kolmogorov micro-eddies to damage on cells grown on microcarriers. Further complicating the matter is the nature of cell damage. Cells in a reactor roam around different locales that have vastly different energy dissipation rates. It has been reported that a vast amount of energy is often released in a very small region. The energy released in the bulk can differ from the high energy release zone by several orders of magnitude. Cells may be damaged by a constant exposure to moderate energy dissipation or by a short exposure to a very high level of mechanical stress. Analyses based on the average energy dissipation consideration may not reflect all possible causes that can lead to cell damage in scale-up. The potential of cell damage or other biological changes caused by hydrodynamics that affect productivity in scaling up is still a concern.

Oxygen and Carbon Dioxide Transfer

THE DESIGN OBJECTIVE IN SCALING UP a process is to produce product in quantities proportional to the scale. To meet that goal, all nutrients are supplied proportionally to the scale. For liquid nutrients, this is easily met. However, for oxygen supply and CO_2 removal, scaling up presents a challenge due to the constraints of physical factors.

Scale-Up and Oxygen Transfer

Oxygen balance on liquid phase

In Chapter 11, we established that the oxygen transfer rate (OTR) into the culture medium is dependent on the overall mass transfer coefficient (K_La) and the driving force (c^*-c). The dissolved oxygen content in the reactor is informed by the balance between the transfer rate through the gas–liquid interface and the oxygen uptake rate (OUR) (Eqs. 12-17 and 12-18, Panel 12.15). OUR is the specific oxygen consumption rate multiplied by the cell concentration. In a reactor, the oxygen content (i.e., the oxygen partial pressure) in the incoming air is different from that in the air exiting the reactor. Which c^*, then, is the one to use for the driving force calculation? For small reactors, we can assume that culture content in the reactor is well-mixed. In cases like these, the concentration at the outlet is the same as the concentration in the reactor. The c^* for calculating the magnitude of the driving force should thus be that at the air outlet (Eq. 12-19, Panel 12.15). For a large reactor, it is assumed that

Panel 12.15. Liquid Phase Oxygen Balance

The dynamics of the dissolved oxygen concentration are described by the balance between OTR and OUR at a quasi-steady state.

$$V\frac{dc}{dt} = K_La\left(c^*-c\right)\cdot V - \text{OUR}\cdot V = \text{OTR}\cdot V - \text{OUR}\cdot V \qquad \textbf{(Eq. 12-17)}$$

At a steady state, OTR = OUR

$$K_La\left(c^*-c\right) = \text{OUR} = q_{O_2}x \qquad \textbf{(Eq. 12-18)}$$

For small-scale bioreactors, one can assume both the liquid phase and the gas phase are well-mixed. The gas phase in the reactor is thus the same as that in the exit gas stream, and c^* is related to the oxygen concentration at the exhaust gas by Henry's law constant:

$$c^* = \frac{P_{O_2,\text{out}}}{H_{O_2}} = \frac{Py_{O_2,\text{out}}}{H_{O_2}} \qquad \textbf{(Eq. 12-19)}$$

For large-scale bioreactors, the inlet and outlet oxygen concentrations may be very different. One uses the logarithmic mean driving force described below:

$$K_La\left(c^*-c\right)_{LM} = K_La\frac{\left(c^*_{\text{in}}-c\right)-\left(c^*_{\text{out}}-c\right)}{\ln\dfrac{c^*_{\text{in}}-c}{c^*_{\text{out}}-c}} \approx K_La\left(\frac{c^*_{\text{in}}+c^*_{\text{out}}}{2}-c\right) \qquad \textbf{(Eq. 12-20)}$$

the gas phase behaves like in a tubular reactor (plug flow), and a logarithmic mean of the driving force is used (Eq. 12-20, Panel 12.15). In most cell culture processes, though, the difference between the logarithmic mean or a simple average is rather small.

As the reactor volume increases, the magnitude of OTR ought to be maintained in order to supply sufficient oxygen for growth. The increased liquid height in a large-scale reactor increases the hydrostatic pressure, and thus also the value of c^*. Recall that 10 m of water height gives an additional 1 atm of hydrostatic pressure. The effect of hydrostatic pressure on c^* should be accounted for in the calculation of OTR, although we do not express it explicitly in the equation.

Oxygen balance on gas phase

In an aerated bioreactor, air bubbles travel through the culture fluid, dispersing oxygen as they go (Figure 12.10). When air exits the reactor, the amount of oxygen it carries is less than that at the gas inlet at the bottom of the reactor. While oxygen is transferred into the liquid phase, water vapor and CO_2 produced by cells in the medium are stripped out of the culture broth, potentially causing the volume flow rate at the outlet to differ from that at the inlet. However, on a dry air basis, the molar (or volumetric) flow rate at the inlet and outlet can be considered to be the same because the RQ (respiratory quotient, or CO_2 produced/oxygen consumed) is very close to 1.0. The oxygen transferred into the liquid is simply the difference between the rate of oxygen entering the reactor at the inlet and exiting at the outlet (Eq. 12-21, Panel 12.16). Assuming ideal gas behavior ($PV = nRT$), the molar flow rate of the oxygen at the inlet and the outlet is the total air flow rate (PQ/RT) multiplied by the molar fractions of oxygen at the inlet and the outlet ($y_{O2,\,in}$ and $y_{O2,\,out}$), respectively. The difference between the inlet and the outlet is the amount retained in the reactor. It is also the amount that is transferred into the liquid. Therefore, the OTRs determined from the gas and liquid phase transfers are equal (Eq. 12-22, Panel 12.17).

In a cell culture process, the operating conditions are such that the dissolved oxygen level does not change rapidly because of the balance of OTR and OUR. The dissolved oxygen level can be taken as being at a quasi-steady state (i.e., the change in dissolved oxygen (dc/dt) is ~0). The OUR can be approximated from the OTR calculated from the gas phase balance or liquid phase balance.

Aeration Rate and Superficial Velocity

Consider again scaling up a reactor by geometrical similarity. All length dimensions (diameter, height, etc.) increase by the same proportion factor. The reactor volume, V, thus increases proportionally with the tank diameter to the third power, while the cross-sectional area of the

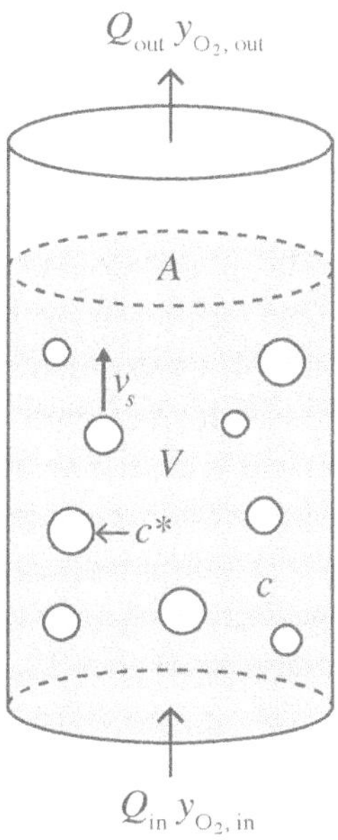

Figure 12.10. Gas sparging for aeration in a stirred tank bioreactor. Oxygen content in the gas at the inlet and outlet is different.

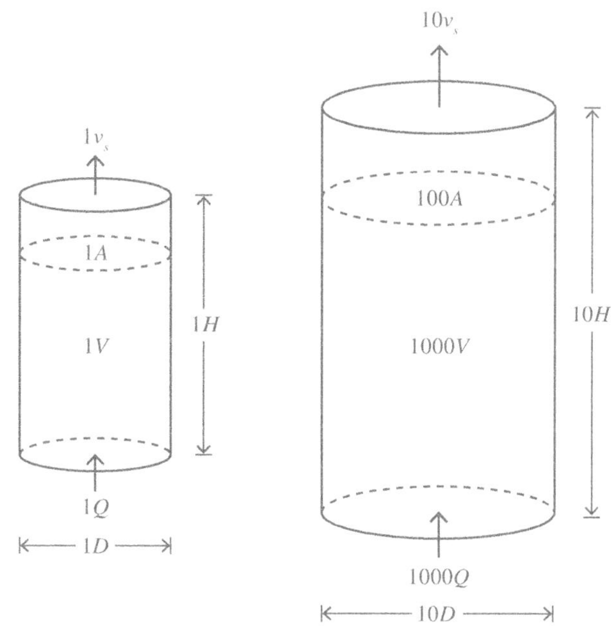

Figure 12.11. Depiction of scaling up 1000 times with geometrical similarity. Gas flow is increased 1000 times to meet the demand of 1000 times the culture volume.

Panel 12.16. Gas Phase Oxygen Balance

On the gas side, the oxygen transferred from the gas side to the liquid side is reflected in the difference of oxygen concentration between the gas inlet and outlet.

$$\text{OTR} \cdot V = \left(Q_{\text{in}} N_{O_2,\text{in}} - Q_{\text{out}} N_{O_2,\text{out}}\right) \quad \textbf{(Eq. 12-21)}$$

$$= \frac{P}{RT}\left(Q_{\text{in}} y_{O_2,\text{in}} - Q_{\text{out}} y_{O_2,\text{out}}\right)$$

Panel 12.17. Steady State Gas Phase and Liquid Phase Balance

On the liquid side, OTR and OUR can be assumed to be at a steady state since the rate of change of dissolved oxygen is small. In cell culture process, the air flow rate can be considered to be the same at the inlet and outlet. Overall, the relationship is described as:

$$\text{OUR} = K_L a\left(c^* - c\right) = \frac{PQ}{RT}\left(y_{O_2,\text{in}} - y_{O_2,\text{out}}\right) \quad \textbf{(Eq. 12-22)}$$

Note: P/RT converts volume flow rate Q to a molar flow rate using the ideal gas law.

Panel 12.18. Gas Superficial Velocity in Scale-Up

Superficial gas velocity = gas flow rate/reactor cross-sectional area

$$v_s = \frac{Q}{A} = \frac{Q}{\pi D_t^2} \quad \textbf{(Eq. 12-23)}$$

The reactor volume increases with the length scale raised to the third power, while the cross-sectional area increases only by the second power.

$$\frac{V_1}{V_2} = \frac{D_1^3}{D_2^3} \quad \textbf{(Eq. 12-24)}$$

$$\frac{A_1}{A_2} = \frac{D_1^2}{D_2^2} \quad \textbf{(Eq. 12-25)}$$

When scaling up, one may choose to increase the air flow rate proportionally to the increasing reactor volume:

$$\frac{Q_1}{Q_2} = \frac{V_1}{V_2}$$

One can see that

$$\frac{v_{s_1}}{v_{s_2}} = \frac{Q_1/A_1}{Q_2/A_2} = \frac{V_1/A_1}{V_2/A_2} = \frac{D_1}{D_2} \quad \textbf{(Eq. 12-26)}$$

Superficial gas velocity will increase linearly with increasing scale.

tank, A, increases with the diameter to the second power (see Figure 12.11 for an example of scaling the reactor volume by 1000-fold). To supply a sufficient amount of oxygen, we provide an airflow rate (Q) in a proportional fashion to the reactor volume ratio. The velocity of the air blowing through the liquid surface in the reactor (i.e., the air flow rate divided by the cross-sectional area, referred to as the superficial velocity) increases with the same factor as the tank diameter (Panel 12.18). For a case of scaling up 1000 times in volume (i.e., 10 times in diameter), the superficial air velocity will be 10 times higher in the large scale.

A very high superficial velocity can cause flooding of the impeller, a situation in which the moving impellers encounter more air rather than liquid. Additionally, as superficial velocity increases, foaming becomes a more serious concern. To avoid these problems, airflow per reactor volume generally decreases with increasing scale.

Given the same OUR, if a lower air flow rate per reactor volume is used in scaling up, then the exiting gas will have a lower oxygen content. This, in turn, will reduce OTR through a lower c^*_{out} and a lower driving force, as shown in Eqs. 12-19 and 12-20. The decreased driving force must be compensated for by increasing K_La in order to maintain the same oxygen supply rate, or by enriching the air with oxygen (Panel 12.19, Table 12.3). The decreased air flow rate per reactor volume (typically expressed as vvm, or volume air per volume liquid per minute) is somewhat compensated for by the increased holding time and hydrostatic pressure in the large-scale reactor. The greater height of the large-scale reactor increases the travel time for air bubbles reaching the top of the reactor, thus increasing the gas hold-up and the interfacial area for transfer with a given gas flow rate.

Panel 12.19. Effect of Air Flow Rate in Scaling Up on Oxygen Transfer Driving Force

Considering gas phase balance in the scale up from scale 1 to scale 2:

$$\frac{Q_1}{V_1}\left(y_{O_2,\,in,\,1}-y_{O_2,\,out,\,1}\right)=\frac{Q_2}{V_2}\left(y_{O_2,\,in,\,2}-y_{O_2,\,out,\,2}\right)$$

$y_{O_2,\,in}$ (oxygen concentration in the inlet air) is the same in both cases.
If Q_2/V_2 is smaller, $y_{O_2,\,out,\,2}$ and c_2^* will be smaller to keep $OUR_1 = OUR_2$.

Considering the liquid phase, OTR needs to be maintained at the same level in the two scales:

$$(K_L a)_1\left(c_1^*-c_1\right)=(K_L a)_2\left(c_2^*-c_2\right)$$

This can be accomplished by:

- With a smaller c_2^*, $(K_L a)_2$ must increase to keep $c_2 = c_1$.
- If $(K_L a)_2$ is to be kept the same as $(K_L a)_1$, the concentration driving force of oxygen must be increased (i.e., $(c_2^*-c_2) > (c_1^*-c_1)$) by either
  - Allowing c_2 to decrease
  - Increasing c_2^* by using enriched oxygen

Table 12.3. Effects of Scale on Oxygen Transfer

	Reference scale	*Constant air flow*	*Constant superficial velocity*
Scale (volume)	1	1,000	1,000
Cross-sectional area	1	100	100
Air flow rate	1	1,000	100
Superficial air velocity	1	10	1
O_2 consumption	1	1,000	1,000
CO_2 production	1	1,000	1,000
$Q(y_{in}-y_{out})$	1	1	$(y_{in}-y_{out})$ ~10 times larger (y_{out} is very low)
Comments		May reach flooding, excessive foaming	Need to increase $K_L a$ or power input; overall driving force for O_2 transfer is low

Effects of Scaling Up on CO_2 Removal

The respiratory quotient for cells using glucose and glutamine as their main sources of energy is very close to 1.0. Every mole of oxygen consumed by the cell generates one mole of CO_2. A highly active culture with a high cell concentration may produce nearly 100 mmol/L of CO_2

per day at the peak of its metabolic activities. In comparison, a cell culture medium initially has about 20–40 mM of sodium bicarbonate. The CO_2 produced by cells will therefore exceed the amount that is included in the medium initially.

Many cells, such as hepatocytes, are rather tolerant to CO_2, but others are more sensitive. The growth of most cells may begin to be affected once the CO_2 concentration exceeds 15% (~65 mM HCO_3^-). The CO_2 produced needs to be stripped off from the culture medium to prevent it from accumulating to high levels that can affect growth or productivity.

The rate of CO_2 stripping is dependent on the concentration difference of CO_2 in the liquid and gas phases. The direction of the driving force for CO_2 is in the opposite direction than that of oxygen transfer (Figure 12.12, Panel 12.20). A higher concentration of CO_2 in the gas phase reduces the driving force of CO_2 and thus the stripping efficiency. When scaling up, the airflow rate per reactor volume often decreases somewhat with increasing scale. As a consequence, less air per culture volume is used to strip CO_2 off the medium. The CO_2 level in the gas stream exiting the reactor will thus be higher in the large-scale reactor, causing the CO_2 in the liquid phase to also be higher.

To compensate for the reduced O_2 driving force when scaling up, one can use O_2-enriched air. However, such a measure cannot compensate for the diminished efficiency of CO_2 stripping. Therefore, CO_2 accumulation patterns in large-scale bioreactors can differ significantly from small-scale reactors.

The overall mass transfer coefficient (K_La) for CO_2 is about 15%–30% lower than that for oxygen (depending on the temperature and solution composition) due to its larger molecular size. The difference,

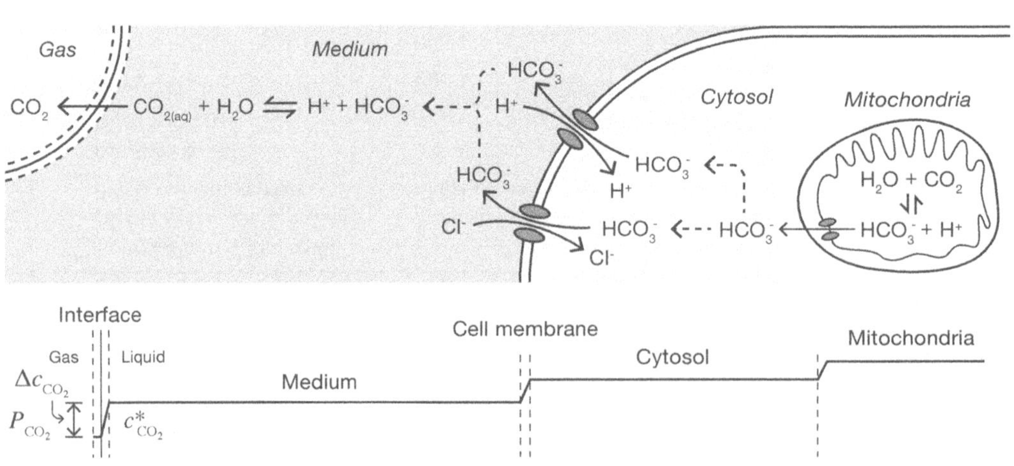

Figure 12.12. Schematic of the removal of carbon dioxide produced by cells.

Panel 12.20. CO_2 Removal from Culture Medium

- The mass transfer coefficients for oxygen and carbon dioxide are about the same
- R.Q. (moles of CO_2 produced / moles O_2 consumed) for mammalian cells is very close to 1.0. Thus, the oxygen uptake rate (OUR) and carbon dioxide evolution rate (CER) are about equal.
- The inhibitory level of CO_2 is around 15–20% (110–150 mm Hg)

Gas phase CO_2 balance:

$$\text{CER} = \frac{PQ}{RT}\left(y_{\text{CO}_2,\,\text{out}} - y_{\text{CO}_2,\,\text{in}}\right) = \frac{QH_{\text{CO}_2}}{RT}\left(c^*_{\text{CO}_2,\,\text{out}} - c^*_{\text{CO}_2,\,\text{in}}\right) \quad \textbf{(Eq. 12-27)}$$

Liquid side balance:

$$\text{CER} = K_{L,\,\text{CO}_2}a\left(c_{\text{CO}_2} - c^*_{\text{CO}_2,\,\text{out}}\right) \quad \textbf{(Eq. 12-28)}$$

- A high level of CO_2 in the exit air can be caused by a high metabolic rate / high cell concentration, or by a low air flow rate. It diminishes the driving force for CO_2 stripping and leads to the accumulation of CO_2/HCO_3^- in the medium.
- The driving force for CO_2 removal decreases with increasing scale

approximately the square root of their molar weight ratio, is significant but not huge.

Unlike O_2, the solubility of CO_2 in an aqueous solution is very high. At the gas bubble interface, O_2 in the liquid phase can be assumed to be in equilibrium with the gas phase, so (c^*-c) is a good estimate of the driving force. The same assumption is not valid for CO_2. CO_2 in the medium exists as CO_2, HCO_3^-, and CO_3^{2-}, with HCO_3^- being the dominant form at a neutral pH. At the interface, CO_2 crosses the film and is transferred out of solution, but HCO_3^- does not diffuse into the gas phase. HCO_3^- must dissociate to CO_2 before being transferred to the gas phase. The kinetics of the dissociation of HCO_3^- to $CO_{2(aq)}$ is slow. Because of this, the actual driving force is smaller than it would be were the kinetics faster. An enzyme carbonic anhydrase is expressed in the mitochondria and in many tissues to facilitate the hydration of CO_2. A similar mechanism facilitating the kinetics of $HCO_3^-/CO_{2(aq)}$ conversion is not available in a reactor.

As the scale increases, the hydrostatic pressure exerted by the liquid height increases the solubility of CO_2. All factors combined, the stripping of CO_2 from the culture medium becomes less efficient as the scale increases.

Variation in Chemical Environment in Scale Translation

As discussed earlier in this chapter, when scaling up many physical parameters cannot be kept constant. Changes in some of those parameters, such as maximal fluid velocity in the reactor and the frequency and nature of bubbles bursting through the liquid surface, can directly impart physical stress on cells and cause metabolic or other physiological changes. In other cases, a change in scale necessitates the alteration of operating conditions that in turn elicit changes in the culture's chemical environment and then in cellular physiology. As a notable example, the efficiency of CO_2 removal and the time profile of CO_2 concentration in culture changes with the scale of the reactor. CO_2 concentration is not controlled explicitly, but is affected by the aeration rate that is used to control the dissolved oxygen level. Over time, the CO_2 concentration in a culture distributes over a wider range than dissolved oxygen does.

CO_2 is a major contributor to the pH buffer in culture and the intracellular environment. The vast majority of CO_2 production in the cell takes place in the TCA cycle in the mitochondria. The CO_2 produced by cells is exported from the mitochondria and excreted from the cell. The excretion of CO_2 is necessary to maintain a physiological range of intracellular pH.

Carbon dioxide is transported through the plasma membranes as CO_2 and HCO_3^-. CO_2 can diffuse through cellular membranes, while HCO_3^- transport is mediated by transporters. A symporter co-transports HCO_3^- and H^+, while another antiporter co-transports Cl^- and HCO_3^- in the opposite direction. In some tissues, Na^+/HCO_3^- symporters play a major role in intracellular pH regulation.

When scaling up, the CO_2 level in culture is likely to increase in the late culture stage. This has an effect on the intracellular CO_2/HCO_3^- level and possibly the intracellular pH. As CO_2 builds up in the culture medium, extracellular HCO_3^- levels increase, thus reducing the concentration difference between the intracellular and extracellular levels of HCO_3^- (or the driving force of HCO_3^- excretion). To compensate, a higher Cl^- gradient is needed to achieve the necessary driving force for excreting HCO_3^- out of the cell. However, experimental evaluation of the effects of scale on cellular levels of CO_2 and intracellular pH is still lacking.

The increased accumulation of CO_2 in the medium also increases the need for a base addition to neutralize the pH and the associated change in osmolality. This further changes the chemical environment and possibly influences cell physiology.

Process Scale-Up and Scale-Down Models

THE DISCUSSION ABOVE FOCUSES on the reactor aspects of scale translation. Besides the reactor, many operations and pieces of auxiliary equipment are also affected by the process scale. For example, the fluid flow pattern and time duration involved in liquid transfer, the kinetics of the dissolution of media components, and the variability of raw materials all differ with the process scale. In cell recycle, the amount of time needed for cells to settle in the cell separator, and the frequency of passing through the device per day (thus, the frequency of cells' subjugation to shear stress incurred by pumping and temperature fluctuation), are likely to vary with the scale of operation. With the increasing practice of perfusion process, the impact of cell retention devices in large-scale operations should be examined. The throughput of the cell retention device should be kept in proportion to the process volume. In principle, this is accomplished by keeping the cell settling surface area in a settling device, or the transmembrane flow area in a membrane filtration device, proportional to the culture volume. This constraint can be met in small-scale reactors, but is difficult to implement at larger scales. In a high-throughput perfusion operation, multiple units of hollow fiber cell retention devices are sometimes needed in order to provide a sufficient surface for filtration. Another approach is to devise a more balanced nutrient feed that minimizes the accumulation of inhibitory metabolites and reduces the volumetric flow rate in large-scale operations.

A rational approach to scale translation is to establish a scale-down model that predicts the process kinetics and performance in large-scale operations or that reproduces large-scale process conditions (Panel 12.21). To establish a scale-down model, it is critical to identify the major physical and chemical factors that affect process performance. The physical factors that may affect cell growth, productivity, metabolism, and overall process performance include: the mechanical stress of agitation, the mechanical stress of aeration, the heat transfer feature (temperature gradient, speed of temperature change), and the mixing time. The chemical parameters include: the CO_2 level, the osmolality, the bulk salt profile (e.g., sodium), and the antifoam concentration. It is not possible to keep all physical parameters constant. A better understanding of the sensitivity or insensitivity of the cell and process to these parameters will allow for the selection of certain pivotal variables to be examined in a scale-down model. For example, one can vary the value of a pivotal variable over a given range in both small- and large-scale reactors while allowing less important variables to vary over a wider range. For chemical parameters, one can reproduce in the scale-down model a profile that was seen at a larger scale by the manipulation of the medium or through dynamic feeding.

Panel 12.21. Establishing a Scale-Down Model of a Production Process

- Identify the sensitivity/insensitivity of a cell line to major physical (e.g., impeller tip speed, energy input) and chemical (e.g., CO_2, lactate, hydroxybutyrate) factors in terms of the effects on growth, productivity, and metabolism, in both the early and late growth stages
- Evaluate the effects of major physical and chemical factors on quality attributes if the mode of action of the product calls for its evaluation
- For creating a scale-down model of an established production process:
  - Obtain the time profile and its range of variability of the physical and chemical factors with a high sensitivity
  - Reproduce the time profile of those factors, individually or in combination, in a scale-down model bioreactor while keeping the insensitive factors at conditions typically used in small scales
  - Evaluate the similarity of the growth and metabolic characteristics between the large-scale process and the scale-down model
  - Once a scale-down model has been established, it can be used to explore new process conditions
- For predicting the performance of a new process in a large-scale bioreactor:
  - Use the time profiles of the aeration rate, gas composition, dissolved oxygen and CO_2 levels, base addition, and osmolality in the large reactor to estimate the mass transfer characteristics of the large reactor
  - Obtain the metabolic profile of the cell line in a small bioreactor, including its metabolic rate of glucose, lactate, etc., oxygen uptake, and CO_2 production
  - Compute the time profile of the base addition and osmolality, CO_2, and key metabolites in a large-scale process
  - Perform experiments in the scale-down model to reproduce the computed time profile of the chemical environment
  - Perform experiments in the scale-down model to reproduce the profile of the physical factors with a high sensitivity
  - Use the data to define the range of operation conditions in the large-scale reactor

In the long term, a rational approach to scale translation will be to combine *in silico* process simulation with experimentation in a scale-down model (Figure 12.13). The process model will likely be a multiscale model that integrates intracellular metabolic metabolism with an empirical cell growth model and a reactor operation model. The metabolic model predicts how cells change their chemical environment through the consumption of nutrients and the production of metabolites. It also describes how cell metabolism adjusts to changing chemical environments. It is thus the heart of the model. The growth model charts out the trajectory of the process. The reactor model describes how external controls of oxygen, CO_2, base additions, and nutrient levels affect the chemical environment. The integrated multiscale model can combine experimentally determined metabolic, growth, and production behavior with the knowledge of large-scale reactor data on mass transfer characteristics to prescribe the operating conditions in a scale-down model that would optimize large-scale operations, or to predict their outcome.

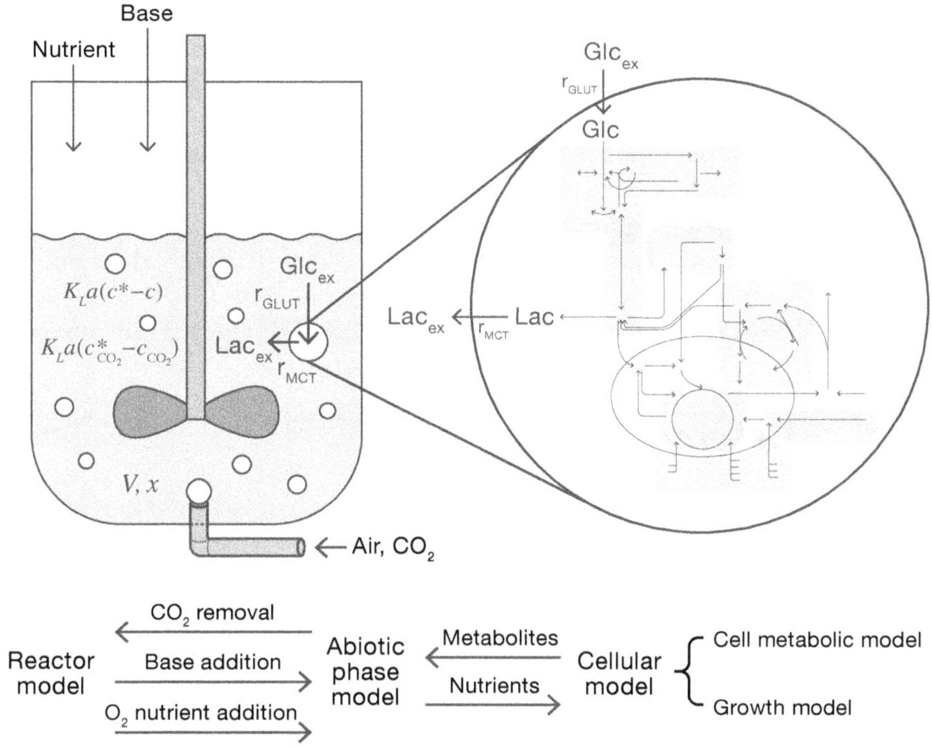

Figure 12.13. Depiction of a mathematical model for a cell culture process.

Concluding Remarks

In scale translation, the relationship of geometry-related physical parameters cannot all be kept constant. One must identify the relevant parameters that affect their cell's growth, physiology, and productivity, and focus on keeping those parameters in the acceptable range. In most cases, one chooses not to scale-up with complete geometrical similarity. Most large-scale reactors have a larger height to diameter ratio than smaller-scale ones. Nevertheless, the physical constraints on scaling up are the same regardless of whether one scales up geometrically similarly or not. In scaling up, the gas flow rate is likely to change in its proportion to the reactor volume. This causes the mass transfer characteristics to be different for different scales. While the dissolved oxygen can be controlled at the same level, the CO_2 concentration profiles are likely to differ for reactors of different scales. Differences in CO_2 concentration in the reactor cause pH control actions (including base and CO_2 addition and stripping) to vary at different scales. Differences in pH control actions may further change the chemical environment of the culture. Given that the physical and chemical parameters related to scaling up cannot easily be manipulated or controlled, one may resort to adapting cells empirically so that they are less sensitive to those parameters. An understanding of scale-sensitive parameters and a sound knowledge of how to estimate the range of those critical parameters will greatly facilitate the scale translation of cell culture processes.

In process development involving scale translation, one should aim to reproduce or to predict the conditions of physical constraints, as well as the resulting chemical environment. It may not be possible to replicate all physical and chemical parameters across drastically different scales. Ultimately, one should identify and aim to control critical physical parameters, which will minimize variations in the chemical environment.

References

1. Nienow, A.W., *Reactor engineering in large scale animal cell culture.* Cytotechnology, 2006. **50**(1–3): p. 9–33.
2. Tolbert, W.R. and J. Feder, *The large-scale cultivation of mammalian cells.* Scientific American, 1983. **248**(1): p. 36–43.
3. Hu, W., C. Berdugo, and J.J. Chalmers, *The potential of hydrodynamic damage to animal cells of industrial relevance: current understanding.* Cytotechnology, 2011. **63**: p. 445–60.

Nomenclature

θ	Mixing time
μ	Fluid viscosity
ρ	Density
a	Interfacial area for oxygen transfer
A	Cross-sectional area of the tank
c	Concentration
c^*	Solubility of oxygen
c_s	Steady state concentration
CER	Carbon dioxide evolution rate
D	Characteristic length
D_B	Blade diameter
D_I	Impeller diameter
D_T	Tank diameter
f	Friction factor
H	Height of culture liquid
H_{O_2}	Henry's law constant for oxygen in water

k	First order rate constant
K	Proportionality constant
K_L	Overall mass transfer coefficient
N	Impeller rotation rate
N_{O_2}	Molecular concentration of oxygen in the gas phase
N_p	Power number
OUR	Oxygen uptake rate
P	Impeller power
P_o	Ungassed impeller power
Q	Volumetric gas flow rate
Q_p	Liquid pumping
R	Ideal gas constant
Re	Reynolds number
T	Temperature
t	Time
v	Tip velocity
v_s	Superficial gas velocity
V	Culture volume
y_{O2}	Fraction of oxygen in the gas

Contents

Appendix A

Cell Culture Process Data Analysis

Overview of Process Data

Modern bioprocesses, from design and development to manufacturing, are all data-driven. Data guides strategy development and decision-making. Cell culture data analysis does not start during process development. Even before cell line development, the knowledge database on the cells' gene-trait relationship and the product's sequence-quality interconnection guides early stage product development. Using a well-developed knowledge base, a process can be designed based on predicted cell performance derived from the cell's genotype even before extensive cell performance data becomes available. Enhancement of productivity and product quality is aided by historical data of the process performance. Even after commercial production begins, the process continues to evolve, enhancing product quality consistency and productivity (Figure A.1). Data from cell lines, process development, and manufacturing of the same and different products are aggregated and analyzed, and the knowledge is compiled and disseminated for further product development.

Cell Line

Data management and analysis should aim to interconnect different segments of the product life cycle. The data on the integration locus/loci of a transgene can be revealing in regards to its genomic structural stability, transcriptional activity, and epigenetic accessibility (Figure A.1). The transcript profiles of genes pertaining to metabolism, protein processing, and post-translational modification can offer information on product quality robustness. Currently, cell line selection is based on a cell's

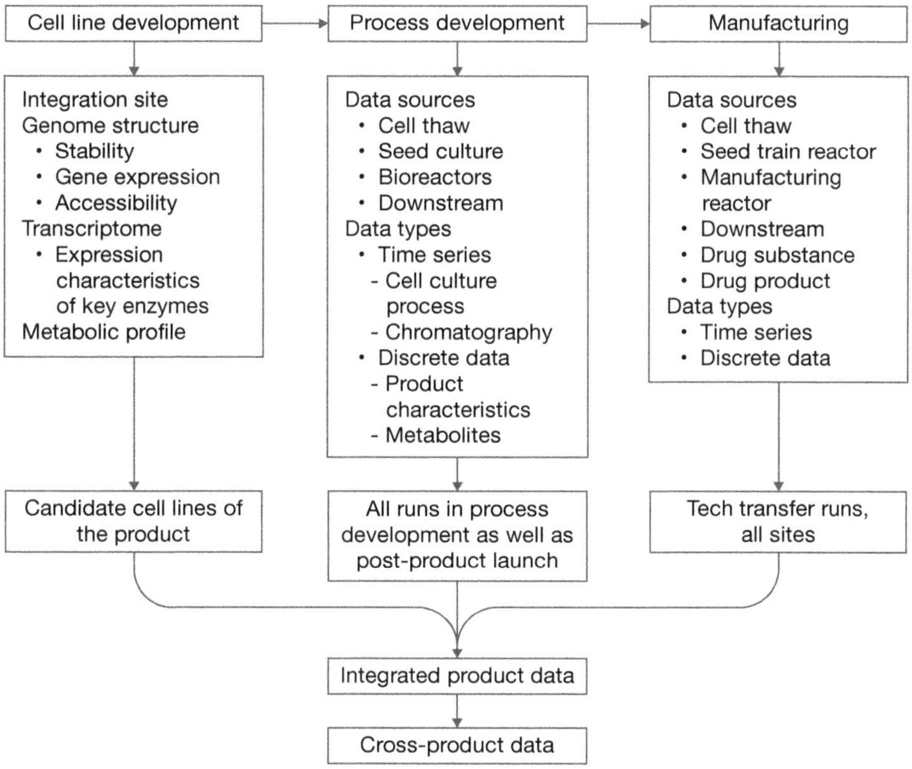

Figure A.1. The role of data analysis in different stages of process development.

growth rate and productivity. Increasingly, the genomic, transcriptomic, and metabolic characteristics of candidate cell lines are also factoring into cell line selection. The success of such an approach hinges on a well-developed knowledge base, and a well-developed knowledge base hinges on data of other cell lines and their process. Even though the focus of this appendix is on the analysis of process data, the importance of integration and analysis of data on different process segments and different types cannot be over-emphasized.

Process

Modern process development and production facilities are electronically monitored and strictly controlled. From reactors in a seed train to downstream unit operations, all process data, including material inputs and outputs, in-line measurements, and many control actions, are recorded. Off-line data from samples taken during process, concentrations of cells, chemical species and products, and even quality attributes, are logged. The first step of data analysis is the acquisition and organization of on-line and off-line measurement data pertaining to process and product (Figure A.2). The data inevitably contains a mixture of time series with profiles of "continuous" and "discrete" types. The data is then

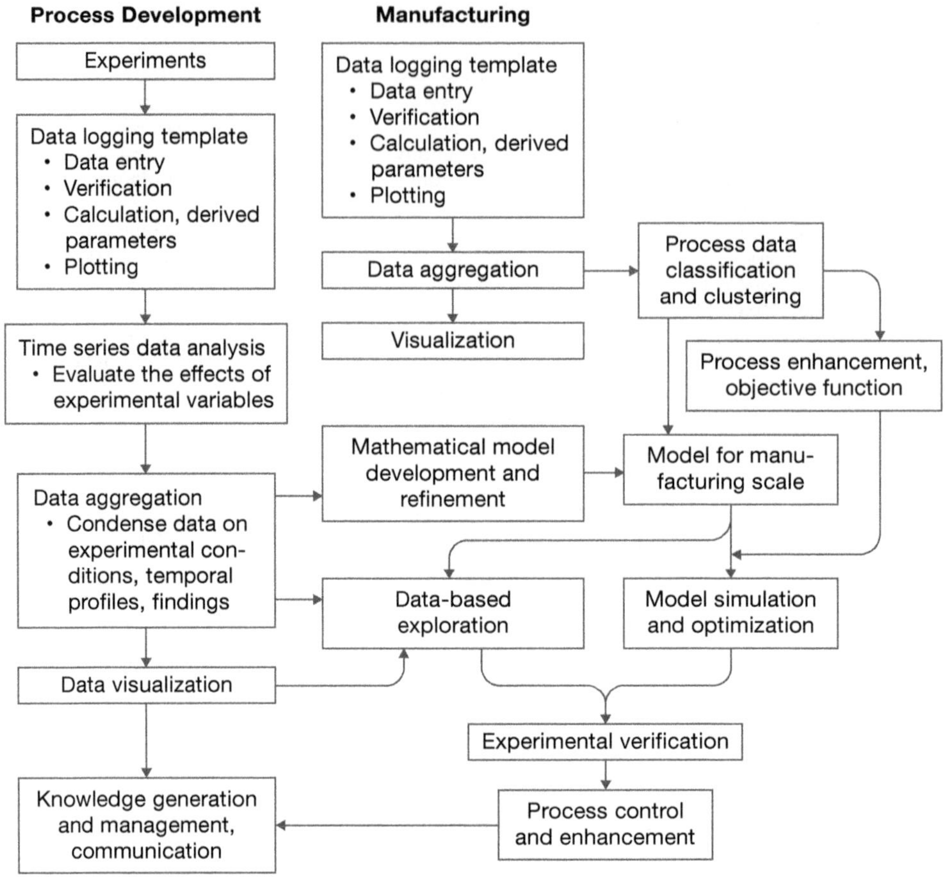

Figure A.2. Outline of data management during process development and manufacturing.

processed to obtain various specific rates and stoichiometric values which are indicative of the cells' physiological state. From this, the process outcome can be analyzed and related to cellular physiology, and key observations can be recorded.

Process development data and manufacturing data are different in nature (Figure A.2). Process development focuses on investigating the effects of different process variables. Replicate cultures for each condition studied are performed. Data from different conditions are treated as longitudinal (time series) data, and the effects of the study variables on growth and production kinetics and other characteristics are statistically compared. Manufacturing strives for consistent process. At first glance, manufacturing runs appear to be replica cultures generated with the same operating protocols. However, variability inevitably arises due to factors such as raw material lot changes, fluctuations in various process parameters, and equipment differences on different sites. Standard operating protocols also allow for some adjustments of process variables.

Thus, analysis focuses on identifying the underlying factors that influence the process outcome and on devising a control strategy.

Over time, the manufacturing process will evolve to attain better productivity and more robust product quality. In many cases, manufacturing occurs at multiple sites. The aggregation of these process development and manufacturing data provides a wealth of information. Mining these data to recognize hidden patterns of behavior can lead to further process enhancement. The data pipeline thus starts with the acquisition of various process data, continues to data analysis, and progresses to the aggregation of the process development and manufacturing data for process data mining (Panel A.1). The knowledge and insight obtained from the analysis of process data can be integrated into a predictive mathematical model. Two important utilities of data analysis are its integration with process monitoring and control and its ability to serve as a core of process analytical technology (PAT) for enhancing bioprocess robustness.

Panel A.1. Cell Culture Data Pipeline

- Acquisition of in-line and off-line measurement data
- Data processing
- Data analysis and visualization
- Data aggregation
- Process pattern identification

Data Processing Pipeline

Standardized Templates for Data Logging and Processing

A TYPICAL DATASET acquired in a cell culture process includes the concentrations of cells, glucose, and other nutrients, lactate and other metabolites, the pH, and the dissolved oxygen level at different time points. In many cases, additional data such as the concentrations of glutamine and other amino acids, the oxygen uptake rate, and the osmolarity are available. Data obtained from different sources, such as different analytical instruments, may report the same variable in different units or formats. A standardized template will homogenize the nomenclature of variables, units of quantities, data formats (Panel A.2). These measurements are plotted automatically in the template as the data are entered, allowing for a visual inspection to detect potential outliers possibly caused by faulty measurements or data-entry error. An algorithm can be built into the data-logging spreadsheet to alert the user of possible incorrect values. Many of these data processing steps are often repeated over and over in experiments, in both manufacturing and research settings. Setting a standardized data format and uniform data processing protocol is critical for increasing efficiency and minimizing frustration.

Panel A.2. Standardized Templates in Data Processing

- Automate calculations and regressions
- Generate standardized plots
- Ensure consistency in data logging and minimize data entry errors
- Facilitate data aggregation and platform migration

Panel A.3. Process Variables

Measured variables

- Discrete measurements
  - Concentrations of cells, glucose, glutamine, other sugars and amino acids, lactate, ammonium, and carbon dioxide
  - Osmolality
- In-line continuous measurements
  - pH, DO, temperature
  - OUR (may also be discrete)

Derived (calculated) variables

- Cumulative consumption (e.g., glucose, lactate, amino acids)
- Specific rates (growth, glucose consumption, lactate production, glutamine and amino acid consumption)

a)

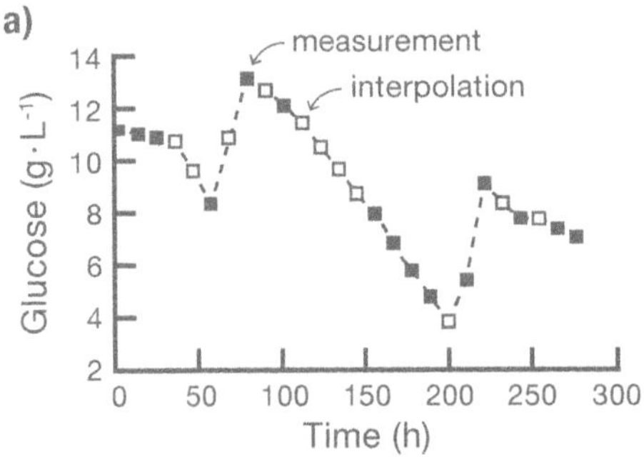

b)

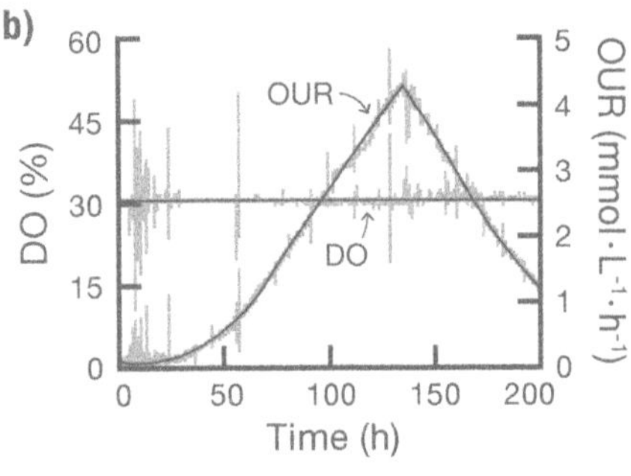

Figure A.3. Example of data processes. (a) Fill in missing or spare data points by interpolation. (b) Use a moving average window to take discrete time points for data recording.

Cell Culture Data Processing

The data-logging template should be designed to perform automatic calculations on process variables. This ensures that all calculations are conducted in the same manner and the data models employed are identical. The dynamics of derived process variables are then also automatically plotted for visualization.

Data collected from a process may be discrete or continuous in nature (Panel A.3). Off-line measurement of samples, such as the concentrations of cells and glucose, are discrete time-point values. Some control action data, such as base addition for controlling the pH, are also discrete in nature. Continuous data are mostly from in-line measurements and calculations, such as the dissolved oxygen level and on-line calculated oxygen uptake rate (OUR). The standardized template is normally constructed for recording discrete data suitable for measurement. The in-line data acquired continuously is noisy and very large. A moving window average can be performed and only discrete points recorded for further analysis (Figure A.3). In different experimental runs, samples may be taken at different time intervals and frequency. Such time series data can be interpolated to render it more amenable to further analysis.

Calculation of derived variables

Cellular metabolism profoundly influences productivity and quality attributes. Glucose and amino acid metabolism are the main cause of the production of metabolites and CO_2, which in turn affects oxygen demand, base addition for pH control, and osmolality change. It is important to evaluate the consumption or production of glucose, amino acids, and key metabolites when assessing the performance of a cell culture process. To that end, the material inputs and outputs of the reactor should be an integral part of data entry, and the inputs and outputs of cells should be a key derived physiological variable.

Thus, the composition of the initial medium (and, in fed-batch and perfusion cultures, the composition of the feed medium and the volume of feed and withdrawal) should be recorded and used in the calculation of physiological variables.

Cumulative nutrient consumption and metabolite production

One of the first calculations of the process measurement data is the cumulative nutrient consumption, metabolite accumulation, and cell biomass production. For batch culture, these quantities can be readily obtained by calculating the difference between the current concentration and the initial concentration. For fed-batch and perfusion cultures, the volume of the culture may vary over time, and therefore the amounts of nutrients consumed and metabolites produced should take into account the periodic medium addition and withdrawal and the resulting culture volume change. For fed-batch culture, the cumulative nutrient consumption is calculated time-segment-wise on consecutive time points of volume change (Figure A.4, Panel A.4). Whenever a step change of volume occurs, the concentrations before and after the volume change are needed to calculate the material input and volume dilution. In principle, the concentration should be measured before and after the medium change. If the measurement is not available, then the material input (i.e., the amount of substrate added by feeding) should be available for the calculation of the cumulative consumption. A drawback of presenting the cumulative data in total quantity is that a larger reactor volume will also give a larger value. It is thus common practice to divide the cumulative data by the final reactor volume and express them as normalized concentrations. In the example shown in Figure A.4, the plot still has units of concentration.

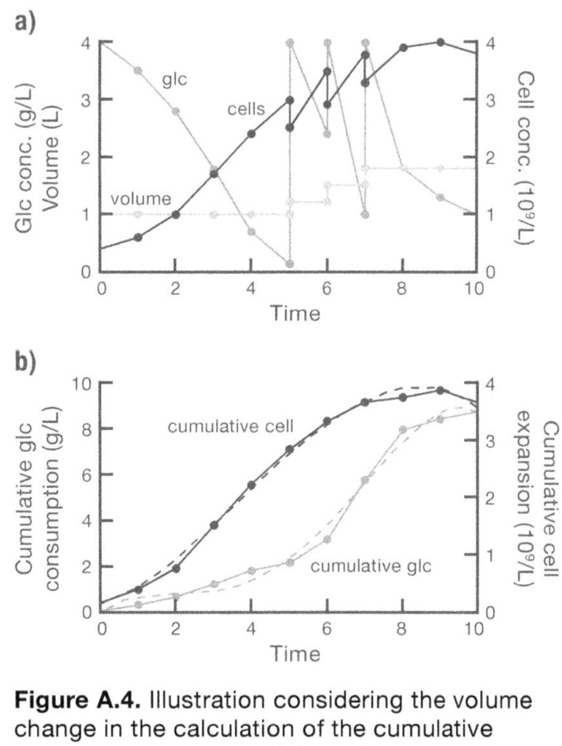

Figure A.4. Illustration considering the volume change in the calculation of the cumulative nutrient consumption of the substrate.

For continuous cultures at a steady state, nutrient consumption is easily calculated by finding the difference between feed concentration and residual nutrient concentration. However, most perfusion cultures are not operated at a true steady state. Often, the exit flow rate is adjusted according to the cell concentration measurement using a capacitance probe, and therefore varies over time. For example, with an alternating

Panel A.4. Derived Variables

- Cumulative nutrient consumption, metabolite production / product formation
  - Material balance is based on the total amount in the reactor when the volume changes over time
  - Fed-batch cultures have step changes in volume when feed medium is added. Concentrations before and after medium addition should be measured or calculated for material balance.
  - Converting the concentration curve into a cumulative consumption curve allows for regression of the curve
- Stoichiometric ratio
  - The ratio of the cumulative consumption or specific rate of a pair of nutrients or metabolites
  - Obtained by plotting the cumulative consumption of two nutrients
  - The stoichiometric ratio may change over time
  - Provides a guide for designing feed medium composition
- Specific rate
  - The slope of the cumulative consumption divided by the cumulative cell concentration at that time point

Panel A.5. Cumulative Cell Growth and Substrate Consumption

Fed-batch:

$$X_{t_j} = \frac{\int_0^t x \cdot V\,dt}{V_{t_f}} = X_{t_{j-1}} + \left(x_{t_j} \cdot V_{t_j} - x_{t_{j-1}} \cdot V_{t_{j-1}}\right)\frac{1}{V_f}$$

$$S_{i,t_j} = \frac{\int_0^t q_{i,t} \cdot x \cdot V}{V_{t_f}} = S_{t_{j-1}} + \left[s_{t_{j-1}} \cdot V_{t_{j-1}} - s_{t_j} \cdot V_{t_j} + M_j\, s_f\right]\frac{1}{V_f}$$

Perfusion:

$$X_{t_j} = \frac{\int_0^t x \cdot V\,dt}{V_{t_f}} = X_{t_{j-1}} + \left[x_{t_j} \cdot V_{t_j} - x_{t_{j-1}} \cdot V_{t_{j-1}} + \frac{F\left(x_{t_j} + x_{t_{j-1}}\right) \cdot \left(t_j - t_{j-1}\right)}{2}\right]\frac{1}{V_f}$$

$$S_{i,t_j} = \frac{\int_0^t q_{i,t} \cdot x \cdot V}{V_{t_f}} = S_{t_{j-1}} + \left[s_{t_{j-1}} \cdot V_{t_{j-1}} - s_{t_j} \cdot V_{t_j} + F\left(s_f - \frac{S_{t_j} + S_{t_{j-1}}}{2}\right)\left(t_j - t_{j-1}\right)\right]\frac{1}{V_f}$$

tangential flow (ATF) device for cell retention, the flow rate of the cell-free stream varies over time as the membrane resistance increases. In some cases, the cell purge stream is operated periodically rather than continuously. As in fed-batch culture, the material balance of cells, nutrients, and metabolites should be performed based on total quantities, not concentrations.

The time profile of cumulative data is monotonic, except for the nutrient or metabolite that is consumed at one time and produced at another (such as lactate and alanine). Regression can then be performed on the cumulative profile over time (Panel A.4, Panel A.5). The regression can be built in the standardized spreadsheet template. Often, a third-order polynomial fits most cumulative data, but inspection is necessary to ensure a good fit.

Stoichiometric ratios

The cumulative data of a pair of nutrients or metabolites can then be plotted against each other; the slope will give the stoichiometric ratio between them (Figure A.5). The stoichiometric ratios that are most physiologically relevant are lactate to glucose (ΔL/ΔG), oxygen to glucose, glutamine to glucose, ammonium to glutamine, and ammonium to total amino acids. Stoichiometric ratios may vary with culture time as cell metabolism changes in different growth stages. The molar stoichiometric ratio of ΔL/ΔG is an indicator of a high or low flux metabolism. Oxygen to glucose in the range of 1–2 mol/mol indicates a glycolytic high flux state, and increases to 5–6 at a low flux state. The glutamine to glucose ratio reflects the contribution of anaplerosis to energy metabolism. Stoichiometric ratios serve as the basis of rational medium design, especially for feed medium for fed-batch and continuous cultures.

Specific rates

From the regression curve of the cumulative data, specific rates can be calculated (Panel A.4, Panel A.6). The slope of the time profile of the cumulative concentration (recall that the cumulative data is divided by the final culture volume to give a normalized concentration) at a given time point is

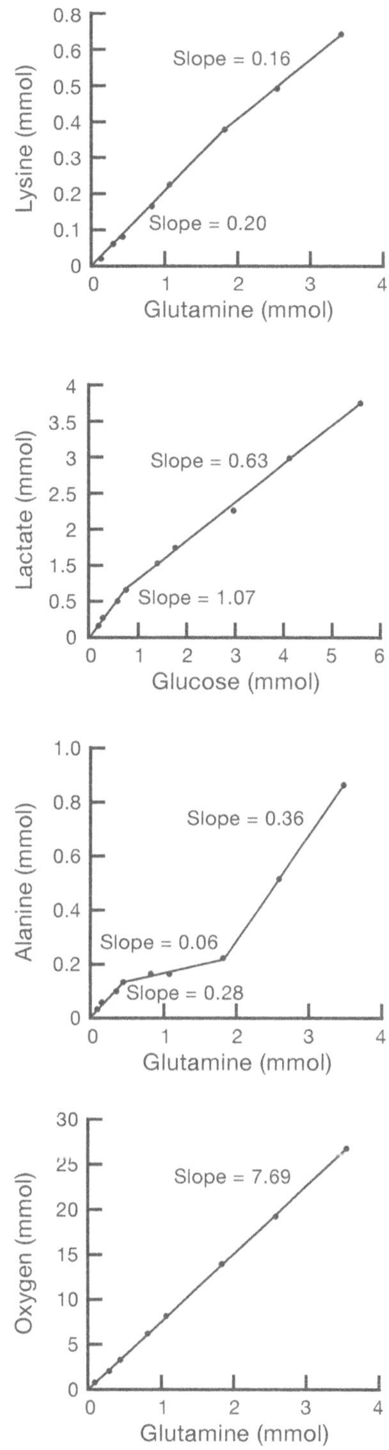

Figure A.5. Example of stoichiometric ratios from a fed-batch culture.

Panel A.6. Specific Rates

Two-point specific nutrient consumption rate calculation for fed-batch culture:

$$q_s = \frac{1}{x \cdot V} \cdot \frac{dS}{dt} \approx \frac{1}{\frac{x_2 \cdot V_2 + x_1 \cdot V_1}{2}} \cdot \frac{S_2 - S_1}{t_2 - t_1}$$

divided by the cumulative cell concentration to give the specific rate.

Specific rates can also be calculated from consecutive time points. The specific curve rate obtained this way is cumulative data prone to large fluctuations due to measurement errors.

In an industrial setting, some calculate the specific product formation rate by taking the final product concentration (g/L) and dividing it by the cumulative curve of cell concentration (with units of cell number-time per volume (cell-h/L)) or integral cell concentration. The value is a gross, time-averaged productivity. The value obtained is different from the specific rate determined as described above and lacks time dynamics.

Data Plotting and Further Analysis

Cell culture processes often extend over a long duration, from 5–15 days to a couple of months for perfusion cultures. Using the data template, in-process analysis takes place immediately after entry of measurement data. The measurement data, cumulative curves, and specific rates are all automatically plotted as the culture progresses. Such in-process data analysis allows for timely identification of potential outliers and faulty conditions so that corrective measures can be taken if necessary. After the process, data from cultures under different experimental conditions are compared. Automated data plotting greatly facilitates this analysis.

In data analysis, the key process characteristics are cell and product concentration profiles. Stoichiometric ratios and specific rates are physiologically relevant indicators. In addition to the specific rates of material exchange between the cells and their environment, it is sometimes useful to understand how those nutrients are distributed in internal pathways by performing metabolic flux analysis (MFA).

Metabolic Flux Analysis

MFA IS A TOOL that uses measured nutrient consumption and metabolite production rates to estimate fluxes in the cellular reaction network. It provides a bird's-eye view on where (and through which pathways) the nutrients have traveled. Glucose, glutamine, and other amino acids account for an overwhelming majority of the total carbon consumed by cells. The remaining organic nutrients, including vitamins, nucleosides, and lipid precursors, typically contribute to only a few percentages of carbon. The major outputs of cell growth are cell biomass, protein product, lactate, ammonium, and CO_2 (and H_2O). Under different growth conditions, cells may consume glucose and glutamine at different proportions, distribute them through different pathways, and

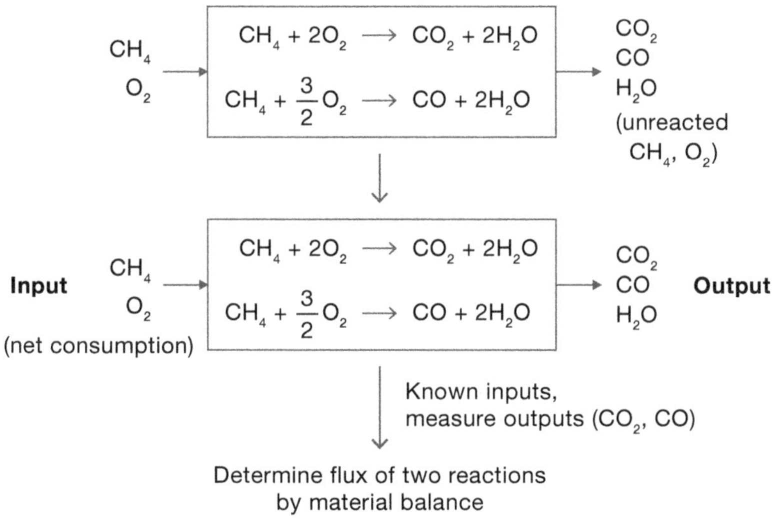

Figure A.6. Example of how to determine the flux of a reaction through a material balance of the combustion of methane.

produce different metabolites. Some types of flux distribution may result in better productivity or even product quality than others. Knowing how different combinations of nutrient utilization lead to different flux distributions may give insight on how to redistribute metabolic fluxes to enhance productivity.

An MFA problem is similar to a material balance on a furnace that takes in oxygen and methane for combustion. The outlet consists of unreacted methane and oxygen and the combustion products CO, CO_2, and H_2O. Two reactions are involved in the combustion of methane to CO and H_2O, and to CO_2 and H_2O. For each reaction, the extent (or the flux) is unknown. By measuring the flow rate and concentration of different species in the inlet and outlet gases, one can solve the material balance equations and determine the flux of each reaction (Figure A.6).

Biochemical Reactions and Material Balance

The first step in MFA is to list all of the reactions to be considered and set up material balance equations for all the intracellular components involved in the reaction system (Figure A.7). The balance on each chemical component involved in the intracellular reaction network is the sum of its intake (i.e., specific uptake rate; positive value), secretion (i.e., production rate; negative value), synthesis from upstream reaction(s) (positive value), and consumption by downstream reaction(s) (negative value) (Figure A.8). For compounds that are not taken up or excreted to the surrounding environment, the balance is only the synthesis and consumption by the metabolic reaction. Since cells are growing and expanding in volume, all the components are also being diluted. However, in general, the time constant of reactions is much shorter than the doubling

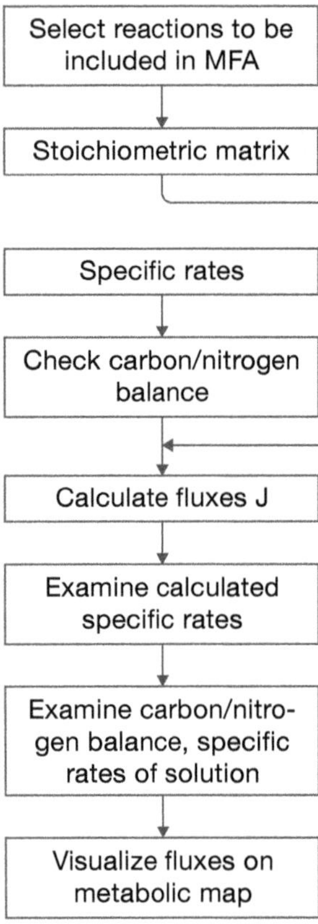

Figure A.7. Steps in MFA.

time of cells, rendering the dilution effect caused by cell growth negligible.

Overall, the glycolysis pathway has the highest flux for cultured cells. Only a few dozen reactions have a flux greater than 5% of that of glycolysis. The reactions with a larger flux, including those in glycolysis, the pentose phosphate pathway, the TCA cycle, amino acid degradation, and the biosynthesis of building blocks for biomass and product formation, are the focus of most MFA on cells in culture.

In MFA, a series of reactions that has no branching reaction in between is given a flux designation. For example, in Figure A.8, although there are three reaction steps between 3PG and Pyr, only one flux is assigned because all three reactions (3PG $\rightarrow$2PG, 2PG $\rightarrow$PEP, PEP $\rightarrow$Pyr) have the same flux. The reaction rate of each component compound in the reaction step is the flux multiplied by its stoichiometric coefficient. In the example shown in Figure A.8, the flux of CO_2 is J_8, while that of NADPH is $2J_8$.

Compartmentalization

Energy metabolism takes place in both the cytoplasm and the mitochondria. Pyruvate, glutamine, and components of the malate-aspartate shuttle cross the two

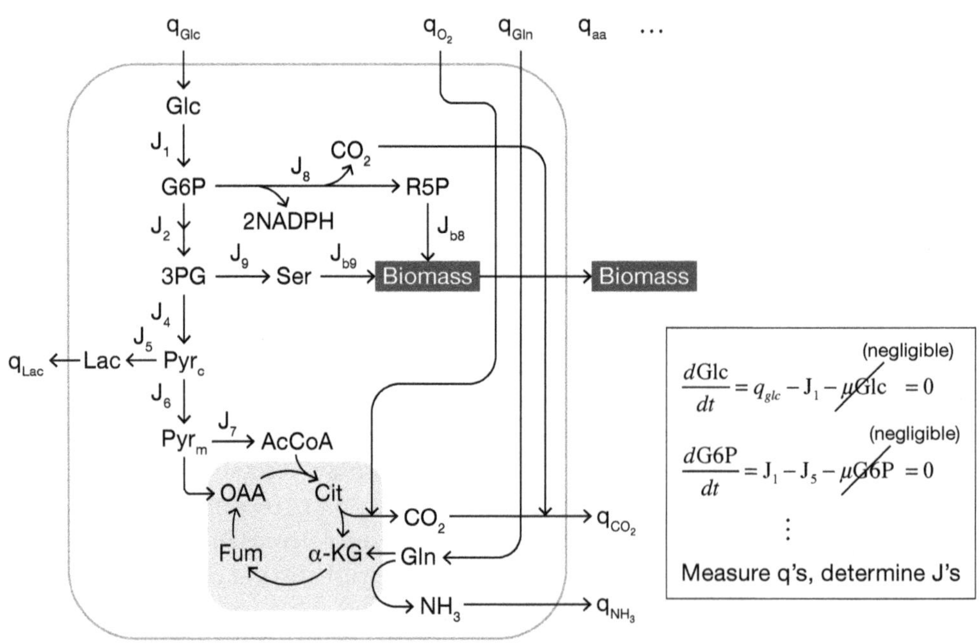

Figure A.8. Example of a flux map.

compartments at high fluxes. The malate-aspartate shuttle that transports the reducing equivalent of NADH into the mitochondria poses additional constraints on flux distribution. In MFA, those compounds present in both compartments are treated as two species. For example, Pyr_c and Pyr_m denote pyruvate in the cytosol and in the mitochondria, respectively (Figure A.8).

Material balance

Considering the inevitability of errors in the measurement, a complete accounting of carbon and nitrogen in the input and output is difficult (Panel A.7). This is compounded by the difficulty in measuring the production of CO_2, a major metabolic product. If oxygen consumption data is available, one can assume that RQ is 1.0 and set CO_2 production to be equal to oxygen consumption. A divergence in the carbon balance is often observed when analyzing experimental data. This may be attributed to errors in measurement and in estimating biomass production. It is also likely that some excreted metabolites are unaccounted for. The difficulty in closing material balance equations contributes to the uncertainty of the result of MFA.

Some cell culture media contain complex components, such as serum or plant hydrolysate. These components provide peptides, lipids, and fatty acids for growth. The extent of their consumption is hardly measured, making it even more difficult to close the material balance.

Panel A.7. Material Balance on a Metabolic Network

- MFA deals with the reactions of major material flows
- The flux of a very large number of reactions is very small
- CO_2 is an unmeasured major metabolite, but is necessary for balancing carbons
- Assume RQ = 1 and use OUR to estimate CO_2 formation
- The mitochondria and the cytoplasm are treated as separate compartments

Biomass and Product Equations

In cell culture bioprocess, cell growth and product synthesis account for a significant fraction of the total material flux. Their synthesis reactions should be considered in MFA (Panel A.8). The product synthesis reaction is easily formulated from its amino acid composition. Normally, cell concentration is measured by cell number. For MFA, it needs to be converted to cell mass. In principle, a large number of reactions contribute to the synthesis of cellular components that make up cell biomass. To simplify MFA, all biosynthetic reactions contributing to the formation of cells' mass are lumped into one equation.

Cell size and composition vary with cell line and culture conditions. Ideally, biomass composition should be determined for each cell line. However, this is rarely done. Most MFA relies on literature values. Based on an elemental analysis of the C, N, O, and H of a mouse hybridoma cell line, a general compositional formula is given as: $CH_{1.975}N_{0.2605}O_{0.489}$. The general range of cellular composition of lipids, proteins, nucleic acids, and polysaccharides is available in many general cell biology textbooks.

Panel A.8. Biomass Equation

- Biomass composition (e.g., proteins, lipids, nucleic acid) is difficult to measure, and data is unavailable for most cells
- The biomass equation is difficult to construct; most studies rely on literature
- When applying the literature values, cell size should be taken into account
- Under fast-growing and lactate-production conditions, biomass formation constitutes only a small fraction of the total carbon
- Under a low-flux glycolytic state, biomass can be a major fraction of the total material consumed. The biomass equation affects the flux estimation.

A simplified cellular reaction system may consider only the metabolism of glucose and amino acids, while lumping lipid and nucleotide synthesis into a biomass formation equation. The resulting reaction network consists of about fifty fluxes involving a similar number of compounds. One may also include reactions of the synthesis of nucleotides, lipids, and other compounds contributing to biomass, and construct a biomass synthesis equation that takes those building blocks to form biomass. Overall, the available data on cell biomass is very limited. Composition and cell size are expected to vary significantly among cell lines, and when biomass synthesis accounts for a large fraction of the total material intake, the error in MFA can be large.

Under most culture conditions, the vast majority of carbon-containing nutrients taken up by cells are converted to lactate and CO_2. Only a small portion are incorporated into biomass. Hence, an error in the estimation of biomass may not affect the major fluxes in MFA. However, this may not be the case under a low-flux metabolic state, where the amount of carbon channeled into biomass becomes a larger portion of the overall material flow.

Quasi-steady state

Material balance on cellular metabolites gives a set of differential equations. In general, a change in a cell's metabolic state will take at least tens of minutes. For a short duration of time, the concentration of all species in the cell can be considered to be invariant. When assuming the system is at a quasi-steady state, the material balance equations become a system of linear algebraic equations.

Stoichiometric Matrix, Flux Vector, and Solution

The balance equation for each intracellular component describes its production or consumption rate in all the biochemical reactions in the network, and the rate of import (nutrient consumption) or export (metabolite production). The equations can be organized in a table such that each row is the balance of a component in the network and each column represents a reaction. On the right-hand side of the equation is a column of net material exchanges with the environment for each compound (i.e., the specific rate). If a compound is not consumed or excreted, the balance is zero.

To solve a large set of equations such as in MFA, one generally uses matrix operations. A stoichiometric matrix is set up. An example of a

		1	2	3	4	…	42	43	44	45				
Stoichiometric Coefficient Matrix		GLC-PYR	PYR-LAC	PYR-ACCOA	OAA-CIT	…	PYRc-PYR	CIT-αKG	THR-GLY	TRY-Alanine				
1	Antibody	-	-	-	-		-	-	-	-		J_1		1 q_p
2	Acetyl-CoA	-	-	1	-1		-	-	1	-		J_2		2 0
3	α-Ketoglutarate	-	-	-	-		-	1	-	-		J_3		3 0
4	Alanine	-	-	-	-		-	-	-	1		J_4		4 q_4
⋮											×	⋮	=	⋮
45	$Glutamine_m$	-	-	-	-		-	-	-	-		J_{29}		45 0
46	Tryptophan	-	-	-	-		-	-	-	-1		J_{30}		46 q_{46}
47	$NADH_m$	-	-	1	-		-	1	1	-		J_{31}		47 0
48	NAD^+	-2	1	-1	-		-	-1	-1	-		J_{32}		48 0

Figure A.9. Example of a stoichiometric matrix, a flux vector, and an external vector. See Appendix B for the complete dataset.

reaction system with 48 compounds and 32 reaction steps is shown in Figure A.9 and Appendix B. Each row of the (48 × 32) matrix represents the balance of a chemical species i (i = 1, I). Each column represents a reaction step j (j = 1, J). The coefficient in each column is thus the stoichiometric coefficient of the chemical species of reaction j. The stoichiometric coefficient has a negative sign for reactants and a positive sign for reaction products. The stoichiometric matrix is multiplied by a flux vector of J elements to give a vector of I elements on the right-hand side. For chemical species that have no exchange with the extracellular environment (such as glucose 6-phosphate and fructose 6-phosphate), the element in the vector is 0. For those that are taken up from the medium or excreted, the value of the element is their specific consumption or production rate.

Solution and Analysis

Unless the system includes only a small section of the metabolic network, it is unlikely that the system of equations will have a unique solution (i.e., the number of unknown fluxes is the same as the number of independent equations) (Panel A.9). A system of algebraic equations may be overspecified (having more independent equations than unknowns) or underspecified (having more unknowns than independent equations). In the flux analysis of cellular energy metabolism, the resulting system is invariably underspecified, as there are more reactions whose fluxes are to be determined than chemical species on which material balance is to be performed. Hence, many different solutions are possible in MFA. In

Panel A.9. System of Equations for MFA

- The balance of intracellular metabolites involved in the reaction network can be presented in a stoichiometric matrix, flux vector, and external rate vector
- The system of equations is typically underdetermined (there are more unknowns than the number of measurements and equations) unless only a small section of reactions is being analyzed
- A solution is obtained by a pseudoinverse / least square regression approach
- The solution is a best fit of the flux and external rate (specific rate). The specific rate of the solution should be compared to the measurement to ensure no gross deviation.

general, a pseudoinverse approach is taken to find a best-fit solution using software such as Mathematica and MATLAB®. It is important to remember that the results of MFA are influenced by the reactions considered and the solution method used. The best-fit solution will also give a new estimated value for the specific rate. The values given by the solution should not deviate hugely from the measured specific rate. MFA results in the most probable flux distribution based on a given set of specific rates and reaction networks. If a more definitive determination of fluxes is necessary, isotope-labeled glucose, glutamine, or other metabolites can be used to determine the intracellular distribution of some key nodes.

Linking Process Data to a Visualization of Flux Distribution

Once an algorithm for MFA is established, routine application of MFA becomes easy. A pipeline that performs C and N material balances on transfer data, data from the standardized process spreadsheet, and the specific rates of cell growth, nutrient consumption, and metabolite production can be constructed. The data can be further linked to a metabolic map for visualization. Presentation of the flux data on a pathway map makes it easy to compare metabolic changes over time or under different conditions. An example of flux distribution from MFA in two cultures with different glucose consumption rates and metabolic states is shown in Figure A.10. By plotting fluxes onto a metabolic map, it is easily seen that a high glucose consumption rate is accompanied by high lactate production and high glutamine consumption, but a similar TCA cycle flux.

Time Series Data Analysis

Automated cell culture systems are standard industrial facilities for multi-parameter experimentation. Replicate cultures are performed to evaluate the effects of test variables (Panel A.10). The outcome variables of high interest may be discrete values, such as product titer and product quality, or time course profiles, such as cell or nutrient concentration. Statistical comparison of these time series data to appraise the effects of the test variables is called for.

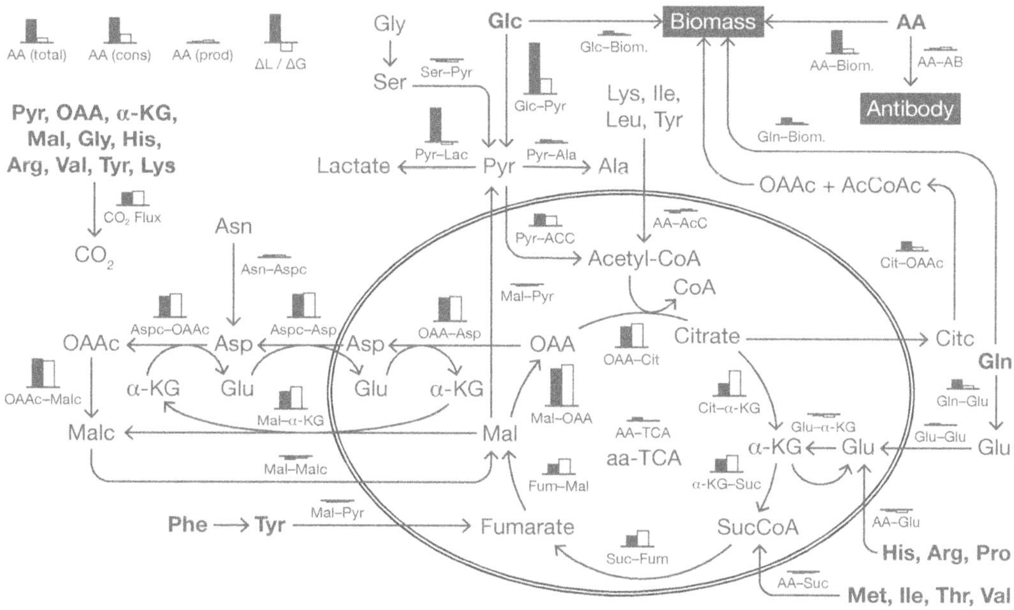

Figure A.10. Metabolic flux distribution in a cell with high or low glycolytic flux using a model considering glucose and amino acids as inputs (bold).

Panel A.10. Time Series (Longitudinal) Data Analysis

- Cell culture process data are longitudinal
- Multiplex culture systems enable replica cultures to be used for statistical evaluation of the effects of particular variables
- Step changes are removed by using cumulative data
- F-test is used to evaluate time-regressed data

The time course profile sometimes includes a step change. For example, the nutrient concentration profile encounters a step increase upon nutrient addition. As discussed earlier, for such time profiles the cumulative data are used to remove the step changes. In some cases, different runs may have different sampling time points and intervals (Figure A.3). Interpolation of data is performed to align the time points. The discussion below focuses on data which has already received such treatments.

Typical longitudinal data (as time series data are referred to) encountered in cell culture process comes from replicate runs of a control and a treatment case. The null hypothesis is thus that the treated case and the control case are not different. One approach is to take a polynomial of degree P to fit all time series data, including the control and treatment. An F-test is then performed on the two sets of data by computing the sum of the squared residuals (SSR) considering the degrees of freedom. The P-value is then obtained from the F-ratio to determine whether the null hypothesis should be rejected (thus ascertaining that the treatment is significantly different from the control).[1]

One may also take an ANOVA (analysis of variance) approach to analyzing time series data by presenting the response variable (e.g., cell concentration, glucose concentration) as the combined contributions from

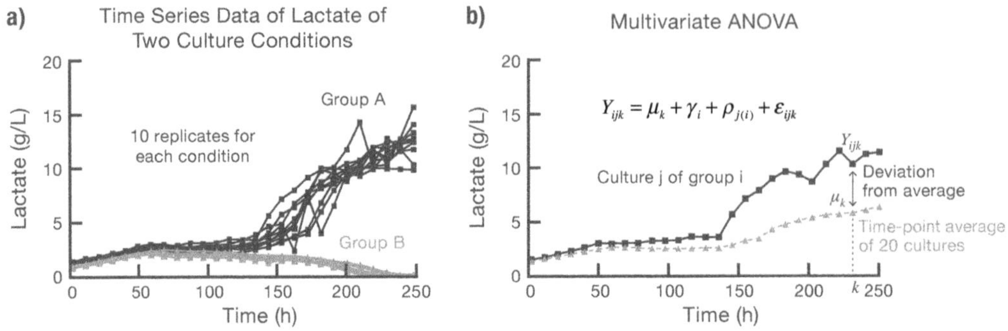

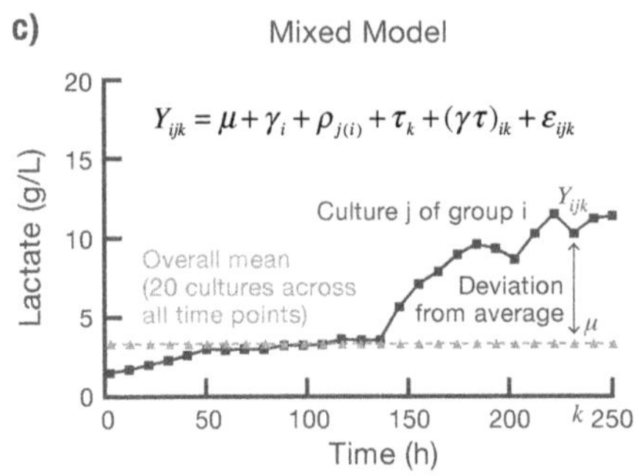

d)

Lactate profiles of two groups of culture

Source	*Nparm*	*DFNum*	*DFDen*	*F Ratio*	*Lactate Prob > F*
Time	26	26	468.0	162.3342	<0.0001*
Group	1	1	18.0	788.0066	<0.0001*
Group*Time	26	26	468.0	285.0375	<0.0001*

Y_{ijk}	Lactate at time k	μ_k	Average of 20 runs at time k
γ_i	Group effect i	μ	Average of 20 runs across all time points
$\rho_{j(i)}$	Random effect of replicate culture j	τ_k	Time effect
ε_{ijk}	Experimental errors	$(\gamma\tau)_{ik}$	Group by time effect

Figure A.11. An example of time series data analysis. (a) Experimental data. (b) Analysis using multivariate ANOVA. (c) Analysis using a mixed model. (d) Statistics from using a mixed model.

the mean, treatment condition, random effect, and experimental errors. In one approach, the mean of all data at the same time point is used (μ_k), as shown in the lactate profiles of two experimental groups of ten cultures each (Figure A.11a and b). The mean thus varies with the time point of the culture. In another approach (called mixed model), a population average (μ) of all data from all time points is taken, but additionally a term for time effect and another for condition and time are included (Figure A.11c). A model based on the combined data from the control and treatment is constructed alongside another with only the data from the treatment. F-statistics is then performed to test the null hypothesis that the two models are not significantly different. Many statistical packages, such as JMP and R, can be used to analyze the data. In these cases, the output from JMP, which uses a mixed model, gives significance to the time, group, and group-time effects (Figure A.11d).

In a cell culture experiment, multiple response variables, including the concentrations of cells, glucose, and lactate, are measured. The F-test calls out significance only on a response variable. The effects of treatment may be significant on one or more of the response variables but not on others. The overall significance of the effect of treatment is dependent on the weight of each response variable on the process or product to be evaluated. Understanding these results still requires insight on cell physiology, the process, and the product.

Aggregation and Analysis of Process Data

Data Aggregation

In the course of process development, culture performance data on different culture conditions accumulates. In a manufacturing setting, data from a large number of runs with similar conditions accumulates over time. These runs may involve process modifications over the lifetime of the product, or involve data from different manufacturing sites. By analyzing a large number of runs, hidden data patterns may be correlated to process performance, allowing for the generation of hypotheses for process enhancement. In some cases, additional data on metabolomes, transcriptomes, and product quality attributes may be available. Analyzing the aggregated data from diverse conditions gives a holistic view of the process and facilitates continued process improvement (Panel A.11).

Bioprocess data analysis thus starts with individual studies, continues to comparisons of multiple experiments, and ends in an integrated analysis of a large set of data from diverse sources, including a meta-analysis of all historically available runs of the process. In a manufacturing setting, data accumulates over time. While process development data are often handled and plotted in standard spreadsheet, manufacturing data are often archived in a different format. Furthermore, while spreadsheets are versatile for simple data calculation and plotting, the visualization and analysis of a large dataset is usually carried out with specialized or custom software. Unfortunately, different software have different data structures and formats. Homogenization of data formats between different software may be difficult, as many software are developed for a wide range of industrial sectors and not specialized for the biotech or pharmaceutical industries. In the past few years, calls for exploitation of big data have been growing louder. As one embarks on such an exploitation, the data pipeline—including standardized models for data management, protocols for data migration, visualization, analysis, and mining

Panel A.11. Data Aggregation

- Aggregation of bioprocess development data generates an enhanced view of a wide range of variables studied in the past
- Aggregation of manufacturing data allows for mining of hidden patterns of variables affecting product and process
- Integration of non-routine data (metabolome, transcriptome, etc.) further enhances process insight
- Data analysis and visualization software facilitates large data analysis
- Data format conversion is important in data migration and analysis

platforms—is an integral part of data management efforts. Community efforts to standardize data management, such as those for clinical research data (www.cdisc.org), will greatly enhance our capacity to leverage the wealth of information in process data.

Data Visualization and Exploratory Analysis

The scope of process data analysis is large, spanning data from a single experiment to aggregated data of all historical experimentations and manufacturing runs. For single runs and compilations of a few runs, the analysis can be exploratory and heuristic (Panel A.11). This might be nothing more than comparing graphs of process kinetics (e.g., cell and product concentrations, specific rates over time) under different treatments, or manufacturing runs in a campaign. As the dataset grows larger, especially for an established manufacturing process, a systematic approach becomes necessary. At a more advanced level, aggregated data can be used to model the process and develop a process control strategy.

At all levels of data analysis, efficient visualization of data is critical. Even the most basic data-logging template should automatically plot kinetic curves and stoichiometric and specific rate profiles. How specific rates and stoichiometric relationships change among different manufacturing runs over time may yield information that is not revealed by visual inspection of the growth and nutrient curves alone. Such exploratory analysis largely relies on one's physiological understanding of cell growth and metabolic and production behavior. As the data accumulates, the datasets of individual runs are compiled, organized, and migrated into a visualization and analysis software, such as Spotfire. Such specialized software offers a means of quickly filtering runs, conditions, or the range of variables on display. They also allow for the quick alteration of the dimensions of a plot, simple calculation of data, and cluster analysis.

Supervised or unsupervised clustering classifies different runs into groups of similar characteristics. One may string the time profiles of the concentrations of cells, glucose, lactate, etc., one after another, thus presenting multiple time series data of different parameters as one long time series. The linked series from different runs can then be clustered using a variety of clustering algorithms or subjected to principal

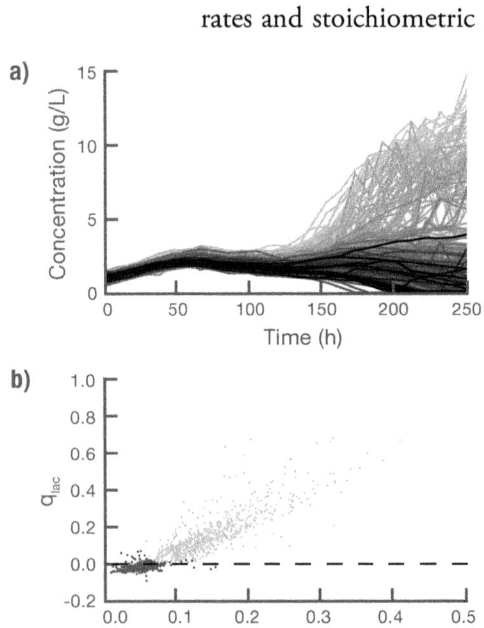

Figure A.12. (a) Lactate concentration profile from multiple runs of cell culture. The shading indicates the final product concentration: black, high titer; gray, low titer. (b) The relationship between the specific lactate consumption rate and the specific glucose consumption rate in the late stage of culture. Black, high titer; gray, low titer.

component analysis (PCA). Again, a visualization tool is useful for inspecting the results and generating insight. Hypotheses are then generated which prompt experimental testing for process enhancement. For example, when lactate concentration profiles from over 200 runs are colored by the final product titer, one can see that high-titer runs mostly consume lactate in the late stage of culture, while the low-titer runs still produce lactate (Figure A.12a).[2] If one plots the top 10 and bottom 10 runs with the highest and lowest product titer, the profiles resemble the two groups in Figure 3.1. By plotting the specific rates of glucose consumption and lactate production at different time points, it can be seen that lactate consumption (negative values) occurs only when glucose consumption is also low.[2] The data further show that, when lactate is being consumed, glucose consumption does not stop, but cell metabolism is at a low-flux state. With a visualization tool, the data analysis process is made easy and fun.

Model-Based Analysis

In chemical manufacturing, industry process control usually relies on a process model. The input into a reactor is adjusted by a controller. A sensor measures the output variable from the reactor and transmits it to the controller. The controller then checks the deviation of the output from the set point and computes the control action, i.e., the amount of adjustment of the input necessary to keep the reactor output at the target level. The computation requires a process model that relates the input to the output; this model is key to process control. Process control in biologics manufacturing centers around controlling the physical and chemical parameters that can be directly measured by sensors, such as the pH and dissolved oxygen level. Few aim to control physiological variables such as the glycolytic flux state. Well-characterized process models for controlling physiological variables are scarce. This can be attributed in part to the lack of a large dataset for model development and evaluation of the values of model parameters.

In the past few years, the potential of exploiting "big data" has led to increased emphasis on data management infrastructure. Tools for process data analysis are becoming more available. These developments, together with the desire to better control product quality through process consistency, is likely to drive model-based process design forward.

In a cell culture process, the growth rate and metabolism are arguably the two most physiologically important factors affecting productivity and product quality. Glucose metabolism influences the consumption of other major nutrients and the production of lactate, CO_2, and amino acids. These metabolites, in turn, affect the pH and osmolality and drive the transition of culture stages. Models describing the specific growth rate and glycolytic flux are thus the foundation of any process model (Figure A.13). Although a mechanistic metabolism model is attainable,[3]

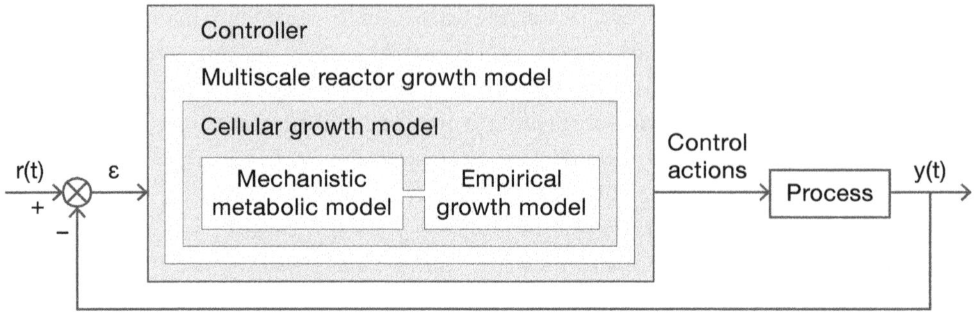

Figure A.13. Integration of a cellular metabolic and growth model with a multiscale reactor model for process control.

computationally a reduced model is more accessible for process control applications. The growth kinetic model is likely to be an empirical model that simply describes the growth rate as a function of major environmental parameters, such as lactate, ammonium, and osmolality.[4] With large aggregates of manufacturing data available, the model parameters can be identified. The metabolism model and growth model can then be combined into a cellular growth model which considers both metabolism and growth. This combined model can be further integrated with the reactor model for use in predicting process outcomes. The resulting model is a multiscale model which deals with growth and metabolism in the scale of minutes as well as that of the reactor in days.[3]

The process of model development is iterative. First, early data is used to set the framework of the model and fit the parameter values. The model is then refined as more data becomes available. Next, a working model is used to analyze the process and aid in data mining, hypothesis generation, and testing. Over time, a model with predictive value can be used to evaluate different process scenarios. Although process modeling for biomanufacturing is still not widely practiced, a well-organized process data aggregate can greatly facilitate model development and application.

Concluding Remarks

Analysis of process data is key to understanding and improving productivity and enhancing product quality consistency. A pipeline for data processing should encompass data logging, calculating derived variables, and plotting various data. Ideally, the data would be organized such that it is ready for integration with algorithms for MFA, data aggregation, and advanced analysis. With the increasing capabilities of multiplexing experimentation, time series data analysis for statistical testing of the effects of culture conditions on process performance is gaining importance. Key to data analysis is an efficient means of data visualization.

A robust data pipeline will facilitate the use of multiple sources of data for process model development for process control applications. A good practice of data processing and analysis is therefore fundamental to process analytical technology.

References

1. Storey, J.D., et al., *Significance analysis of time course microarray experiments.* Proc. Natl. Acad. Sci. U.S.A., 2005. **102**(36): p. 12837–42.
2. Le, H., et al., *Multivariate analysis of cell culture bioprocess data-lactate consumption as process indicator.* J. Biotechnol., 2012. **162**(2–3): p. 210–23.
3. Mulukutla, B.C., et al., *Multiplicity of steady states in glycolysis and shift of metabolic state in cultured mammalian cells.* PLOS One, 2015. **10**(3).
4. Gambhir, A., et al., *Analysis of the use of fortified medium in continuous culture of mammalian cells.* Cytotechnology, 1999. **31**: p. 243–54.

Nomenclature

μ	Mean of a series of values
F	Feed medium flow rate
j	Denotes a time point
J	Flux
M_j	Volume of feed medium added between time j and j–1
q	Specific nutrient consumption rate or specific metabolite production rate
s	Concentration of a nutrient
s_f	Nutrient concentration in the feed
S	Cumulative normalized volumetric nutrient consumption
t	Time
V	Total culture volume
x	Cell concentration
X	Cumulative normalized volumetric cell growth

Appendix B

Stoichiometric Coefficient Matrix		*1* GLC-PYR	*2* PYR-LAC	*3* PYR-ACCOA	*4* OAA-CIT	*5* αKG-SUCCOA	*6* SUCCOA-FUM	*7* FUM-MAL	*8* MAL-OAA	*9* GLN-GLU	*10* GLU-αKG	*11* MAL-PYR	*12* ALA-PYR	*13* SER-PYR	*14* GLY-SER	*15* CYS-PYR	*16* OAA-ASP	*17* ASN-ASPc	*18* HIS-GLU	*19* ARG-GLU	*20* PRO-GLU	*21* ILE-Acetyl/SUCCOA
1	Antibody	-	-	-	-	-	-	-	-	-	-	-	-	-	-	-	-	-	-	-	-	-
2	Acetyl-CoA	-	-	1	-1	-	-	-	-	-	-	-	-	-	-	-1	-	-	-	-	-	1
3	α-Ketoglutarate	-	-	-	-	-1	-	-	-	-	1	-	-	-	-	-	1	-	-	-1	-	-1
4	Alanine	-	-	-	-	-	-	-	-	-	-	-	1	-	-	-	-	-	-	-	-	-
5	Arginine	-	-	-	-	-	-	-	-	-	-	-	-	-	-	-	-	-	-	-1	-	-
6	Asparagine	-	-	-	-	-	-	-	-	-	-	-	-	-	-	-	-	-1	-	-	-	-
7	Aspartate	-	-	-	-	-	-	-	-	-	-	-	-	-	-	-	1	-	-	-	-	-
8	Biomass	-	-	-	-	-	-	-	-	-	-	-	-	-	-	-	-	-	-	-	-	-
9	CO_2	-	-	1	-	1	-	-	-	-	-	1	-	-	1	-	-	-	-	2	-	1
10	Cysteine	-	-	-	-	-	-	-	-	-	-	-	-	-	-	-1	-	-	-	-	-	-
11	Fumarate	-	-	-	-	-	1	-1	-	-	-	-	-	-	-	-	-	-	-	-	-	-
12	Glucose	-1	-	-	-	-	-	-	-	-	-	-	-	-	-	-	-	-	-	-	-	-
13	$Glutamine_c$	-	-	-	-	-	-	-	-	-1	-	-	-	-	-	-	-	-	-	-	-	-
14	$Glutamate_m$	-	-	-	-	-	-	-	-	-	-1	-	-	-	-	1	-1	-	1	2	1	1
15	Glycine	-	-	-	-	-	-	-	-	-	-	-	-	-	-2	-	-	-	-	-	-	-
16	Histidine	-	-	-	-	-	-	-	-	-	-	-	-	-	-	-	-	-	-1	-	-	-
17	Isoleucine	-	-	-	-	-	-	-	-	-	-	-	-	-	-	-	-	-	-	-	-	-1
18	Lactate	-	1	-	-	-	-	-	-	-	-	-	-	-	-	-	-	-	-	-	-	-
19	Leucine	-	-	-	-	-	-	-	-	-	-	-	-	-	-	-	-	-	-	-	-	-
20	Lysine	-	-	-	-	-	-	-	-	-	-	-	-	-	-	-	-	-	-	-	-	-
21	Malate	-	-	-	-	-	-	1	-1	-	-	-1	-	-	-	-	-	-	-	-	-	-
22	Methionine	-	-	-	-	-	-	-	-	-	-	-	-	-	-	-	-	-	-	-	-	-
23	NH_3	-	-	-	-	-	-	-	-	1	1	-	-	1	1	-	-	1	2	2	2	-
24	Oxaloacetate	-	-	-	-1	-	-	-	1	-	-	-	-	-	-	-	-1	-	-	-	-	-
25	Phenylalanine	-	-	-	-	-	-	-	-	-	-	-	-	-	-	-	-	-	-	-	-	-
26	Proline	-	-	-	-	-	-	-	-	-	-	-	-	-	-	-	-	-	-	-	-1	-
27	$Pyruvate_c$	2	-1	-	-	-	-	-	-	-	-	-	-1	1	-	1	-	-	-	-	-	-
28	Serine	-	-	-	-	-	-	-	-	-	-	-	-	-1	1	-	-	-	-	-	-	-
29	Succinyl-CoA	-	-	-	-	1	-1	-	-	-	-	-	-	-	-	-	-	-	-	-	-	1

⋮

22	*23*	*24*	*25*	*26*	*27*	*28*	*29*	*30*	*31*	*32*	*33*	*34*	*35*	*36*	*37*	*38*	*39*	*40*	*41*	*42*	*43*	*44*	*45*
VAL-SUCCOA	MET-SUCCOA	THR-Acetyl	PHE-TYR	TYR-FUM	LYS-ACCOA	LEU-ACCOA	Biomass Synthesis	Antibody Synthesis	OAAc-MALc	CIT-CITc	CITc-OAAc	ASPc-OAAc	NADH-NAD$^+$	ASP-ASPc	MALc-MAL	GlUc-GLU	$FADH_2$-FAD	MALc-PYR	PYR-OAA	PYRc-PYR	CIT-αKG	THR-GLY	TRY-Alanine
-	-	-	-	-	-	-	0.000	1.000	-	-	-	-	-	-	-	-	-	-	-	-	-	-	-
-	-	1	-	-	1	1	0.000	0.000	-	-	-	-	-	-	-	-	-	-	-	-	-	1	-
-1	-	-	-	-1	-2	-1	0.000	0.000	-	-	-	-	-	-	-1	-	-	-	-	-	1	-	-
-	-	-	-	-	-	-	-0.013	-0.011	-	-	-	-	-	-	-	-	-	-	-	-	-	-	1
-	-	-	-	-	-	-	-0.007	-0.005	-	-	-	-	-	-	-	-	-	-	-	-	-	-	-
-	-	-	-	-	-	-	0.000	-0.007	-	-	-	-	-	-	-	-	-	-	-	-	-	-	-
-	-	-	-	-	-	-	0.000	0.000	-	-	-	-	-	-1	-	-	-	-	-	-	-	-	-
-	-	-	-	-	-	-	1.000	0.000	-	-	-	-	-	-	-	-	-	-	-	-	-	-	-
2	1	-	-	1	2	1	0.000	0.000	-	-	-	-	-	-	-	-	-	1	-1	-	1	-	-
-	1	-	-	-	-	-	0.000	-0.005	-	-	-	-	-	-	-	-	-	-	-	-	-	-	-
-	-	-	-	1	-	-	0.000	0.000	-	-	-	-	-	-	-	-	-	-	-	-	-	-	-
-	-	-	-	-	-	-	-0.021	0.000	-	-	-	-	-	-	-	-	-	-	-	-	-	-	-
-	-	-	-	-	-	-	-0.038	-0.010	-	-	-	-	-	-	-	-	-	-	-	-	-	-	-
1	-	-	-	1	2	1	0.000	0.000	-	-	-	-	-	1	-	1	-	-	-	-	-	-	-
-	-	1	-	-	-	-	-0.016	-0.014	-	-	-	-	-	-	-	-	-	-	-	-	-	1	-
-	-	-	-	-	-	-	-0.003	-0.003	-	-	-	-	-	-	-	-	-	-	-	-	-	-	-
-	-	-	-	-	-	-	-0.008	-0.005	-	-	-	-	-	-	-	-	-	-	-	-	-	-	-
-	-	-	-	-	-	-	0.000	0.000	-	-	-	-	-	-	-	-	-	-	-	-	-	-	-
-	-	-	-	-	-	-1	-0.013	-0.014	-	-	-	-	-	-	-	-	-	-	-	-	-	-	-
-	-	-	-	-	-1	-	-0.010	-0.014	-	-	-	-	-	-	-	-	-	-	-	-	-	-	-
-	-	-	-	-	-	-	0.000	0.000	-	1	-	-	-	-	1	-	-	-	-	-	-	-	-
-	-1	-	-	-	-	-	-0.003	-0.003	-	-	-	-	-	-	-	-	-	-	-	-	-	-	-
-	-	-	-	-	-	-	0.000	0.000	-	-	-	-	-	-	-	-	-	-	-	-	-	-	-
-	-	-	-	-	-	-	0.000	0.000	-	-	-	-	-	-	-	-	-	-	1	-	-	-	-
-	-	-	-1	-	-	-	-0.005	-0.007	-	-	-	-	-	-	-	-	-	-	-	-	-	-	-
-	-	-	-	-	-	-	-0.008	-0.015	-	-	-	-	-	-	-	-	-	-	-	-	-	-	-
-	-	-	-	-	-	-	0.000	0.000	-	-	-	-	-	-	-	-	-	1	-	-1	-	-	-
-	-1	-	-	-	-	-	-0.010	-0.027	-	-	-	-	-	-	-	-	-	-	-	-	-	-	-
1	-	-	-	-	-	-	0.000	0.000	-	-	-	-	-	-	-	-	-	-	-	-	-	-	-

⋮

Stoichiometric Coefficient Matrix		*1*	*2*	*3*	*4*	*5*	*6*	*7*	*8*	*9*	*10*	*11*	*12*	*13*	*14*	*15*	*16*	*17*	*18*	*19*	*20*	*21*
		GLC-PYR	PYR-LAC	PYR-ACCOA	OAA-CIT	αKG-SUCCOA	SUCCOA-FUM	FUM-MAL	MAL-OAA	GLN-GLU	GLU-αKG	MAL-PYR	ALA-PYR	SER-PYR	GLY-SER	CYS-PYR	OAA-ASP	ASN-ASPc	HIS-GLU	ARG-GLU	PRO-GLU	ILE-Acetyl/SUCCOA
	(cont.)																					
30	Threonine	-	-	-	-	-	-	-	-	-	-	-	-	-	-	-	-	-	-	-	-	-
31	Tyrosine	-	-	-	-	-	-	-	-	-	-	-	-	-	-	-	-	-	-	-	-	-
32	Valine	-	-	-	-	-	-	-	-	-	-	-	-	-	-	-	-	-	-	-	-	-
33	$Malate_c$	-	-	-	-	-	-	-	-	-	-	-	-	-	-	-	-	-	-	-	-	-
34	Citrate	-	-	-	1	-	-	-	-	-	-	-	-	-	-	-	-	-	-	-	-	-
35	$Citrate_c$	-	-	-	-	-	-	-	-	-	-	-	-	-	-	-	-	-	-	-	-	-
36	$Oxaloacetate_c$	-	-	-	-	-	-	-	-	-	-	-	-	-	-	-	-	-	-	-	-	-
37	O_2	-	-	-	-	-	-	-	-	-	-	-	-	-	-	-2	-	-	-	-	-	-
38	$Acetyl\text{-}CoA_c$	-	-	-	-	-	-	-	-	-	-	-	-	-	-	-	-	-	-	-	-	-
39	$\alpha\text{-}Ketoglutarate_c$	-	-	-	-	-	-	-	-	-	-	-	1	-	-	-	-	-	-	-	-	-
40	GLU_c	-	-	-	-	-	-	-	-	1	-	-	-1	-	-	-	-	-	-	-	-	-
41	ASP_c	-	-	-	-	-	-	-	-	-	-	-	-	-	-	-	-	1	-	-	-	-
42	FADH	-	-	-	-	-	1	-	-	-	-	-	-	-	-	-	-	-	-	-	-	1
43	$Pyruvate_m$	-	-	-1	-	-	-	-	-	-	-	1	-	-	-	-	-	-	-	-	-	-
44	FAD	-	-	-	-	-	-1	-	-	-	-	-	-	-	-	-	-	-	-	-	-	-1
45	$Glutamine_m$	-	-	-	-	-	-	-	-	-	-	-	-	-	-	-	-	-	-	-	-	-
46	Tryptophan	-	-	-	-	-	-	-	-	-	-	-	-	-	-	-	-	-	-	-	-	-
47	$NADH_{Mm}$	-	-	1	-	1	-	-	1	-	-	1	-	-	1	-	-	-	-1	1	-	1
48	NAD	-2	1	-1	-	-1	-	-	-1	-	-	-1	-	-	-	-	-	-	1	-1	-	-1

22	*23*	*24*	*25*	*26*	*27*	*28*	*29*	*30*	*31*	*32*	*33*	*34*	*35*	*36*	*37*	*38*	*39*	*40*	*41*	*42*	*43*	*44*	*45*
VAL-SUCCOA	MET-SUCCOA	THR-Acetyl	PHE-TYR	TYR-FUM	LYS-ACCOA	LEU-ACCOA	Biomass Synthesis	Antibody Synthesis	OAAc-MALc	CIT-CITc	CITc-OAAc	ASPc-OAAc	NADH-NAD$^+$	ASP-ASPc	MALc-MAL	GlUc-GLU	FADH$_2$-FAD	MALc-PYR	PYR-OAA	PYRc-PYR	CIT-αKG	THR-GLY	TRY-Alanine
-	-	-1	-	-	-	-	-0.008	-0.016	-	-	-	-	-	-	-	-	-	-	-	-	-	-1	-
-	-	-	1	-1	-	-	-0.004	-0.008	-	-	-	-	-	-	-	-	-	-	-	-	-	-	-
-1	-	-	-	-	-	-	-0.010	-0.019	-	-	-	-	-	-	-	-	-	-	-	-	-	-	-
-	-	-	-	-	-	-	0.000	0.000	1	-1	-	-	-	-	-1	-	-	-1	-	-	-	-	-
-	-	-	-	-	-	-	0.000	0.000	-	-1	-	-	-	-	-	-	-	-	-	-	-1	-	-
-	-	-	-	-	-	-	0.000	0.000	-	1	-1	-	-	-	-	-	-	-	-	-	-	-	-
-	-	-	-	-	-	-	0.000	0.000	-1	-	1	1	-	-	-	-	-	-	-	-	-	-	-
-	-	-	-1	-2	-	-	0.000	0.000	-	-	-	-	-1	-	-	-	-1	-	-	-	-	-	-
-	-	-	-	-	-	-	0.000	0.000	-	-	1	-	-	-	-	-	-	-	-	-	-	-	-
-	-	-	-	-	-	-	0.000	0.000	-	-	-	-1	-	-	1	-	-	-	-	-	-	-	-
-	-	-	-	-	-	-	-0.001	-0.011	-	-	-	1	-	-1	-	-1	-	-	-	-	-	-	-
-	-	-	-	-	-	-	-0.026	-0.008	-	-	-	-1	-	1	-	-	-	-	-	-	-	-	-
1	-	-	-	-	1	1	0.000	0.000	-	-	-	-	-	-	-	-	-2	-	-	-	-	-	-
-	-	-	-	-	-	-	0.000	0.000	-	-	-	-	-	-	-	-	-	-	-1	1	-	-	-
-1	-	-	-	-	-1	-1	0.000	0.000	-	-	-	-	-	-	-	-	-2	-	-	-	-	-	-
-	-	-	-	-	-	-	0.000	0.000	-	-	-	-	-	-	-	-	-	-	-	-	-	-	-
-	-	-	-	-	-	-	0.000	0.000	-	-	-	-	-	-	-	-	-	-	-	-	-	-	-1
1	1	1	-1	-	5	2	0.000	0.000	-	-	-	-	-2	-	-	-	-	-	-	-	1	1	-
-1	-1	-1	1	-	-5	-2	0.000	0.000	1	-	-	-	2	-	-	-	-	-	-	-	-1	-1	-

$$
\times
\begin{bmatrix}
J_1 \\ J_2 \\ J_3 \\ J_4 \\ J_5 \\ J_6 \\ J_7 \\ J_8 \\ J_9 \\ J_{10} \\ J_{11} \\ J_{12} \\ J_{13} \\ J_{14} \\ J_{15} \\ J_{16} \\ J_{17} \\ J_{18} \\ J_{19} \\ J_{20} \\ J_{21} \\ J_{22} \\ J_{23} \\ J_{24} \\ J_{25} \\ J_{26} \\ J_{27} \\ J_{28} \\ J_{29} \\ J_{30} \\ J_{31} \\ J_{32}
\end{bmatrix}
=
\begin{bmatrix}
1 & q_p \\
2 & 0 \\
3 & 0 \\
4 & q_4 \\
5 & q_5 \\
6 & q_6 \\
7 & q_7 \\
8 & \mu x \\
9 & q_{CO_2} \\
10 & q_{10} \\
11 & 0 \\
12 & q_{glc} \\
13 & q_{13} \\
14 & q_{14} \\
15 & q_{15} \\
16 & q_{16} \\
17 & q_{17} \\
18 & q_{lac} \\
19 & q_{19} \\
20 & q_{20} \\
21 & 0 \\
22 & q_{22} \\
23 & q_{NH_3} \\
24 & 0 \\
25 & q_{25} \\
26 & q_{26} \\
27 & 0 \\
28 & q_{28} \\
29 & 0 \\
30 & q_{30} \\
31 & q_{31} \\
32 & q_{32} \\
33 & 0 \\
34 & 0 \\
35 & 0 \\
36 & 0 \\
37 & q_{O2} \\
38 & 0 \\
\vdots & \\
39 & 0 \\
40 & q_{40} \\
41 & q_{41} \\
42 & 0 \\
43 & 0 \\
44 & 0 \\
45 & 0 \\
46 & q_{46} \\
47 & 0 \\
48 & 0
\end{bmatrix}
$$

Index

B

D

E

F

H

I

N

O

P

Q

R

S

T

U

V

W

X

Y

Z

For Product Safety Concerns and Information please contact our EU
representative GPSR@taylorandfrancis.com
Taylor & Francis Verlag GmbH, Kaufingerstraße 24, 80331 München, Germany

www.ingramcontent.com/pod-product-compliance
Ingram Content Group UK Ltd.
Pitfield, Milton Keynes, MK11 3LW, UK
UKHW050057200926
475702UK00019B/2879
9781498762854